This book is to be returned on or before
the last date stamped below.

Determination of Organic Substances in Water

Volume 1

T. R. CROMPTON

A Wiley–Interscience Publication

JOHN WILEY & SONS

Chichester · New York · Brisbane · Toronto · Singapore

Copyright © 1985 by John Wiley & Sons Ltd.

All rights reserved.

No part of this book may be reproduced by any means, nor
transmitted, nor translated into a machine language
without the written permission of the publisher.

Library of Congress Cataloging in Publication Data:

Crompton, T. R. (Thomas Roy)
  Determination of organic substances in water.
  'A Wiley-Interscience publication.'
  1. Water chemistry.   2. Organic compounds—Analysis.
I. Title
GB855.C76   1984   628.1'61   84-7443
ISBN 0 471 90468 6 (v.1)

British Library Cataloguing in Publication Data:

Crompton, T. R.
  Determination of organic substances in water.
  Vol. 1
  1. Organic water pollutants—Analysis
  2. Water—Analysis
  I. Title
  628.1'68   TD427.07

  ISBN 0 471 90468 6

Phototypeset by Dobbie Typesetting Service, Plymouth, Devon
Printed by Page Bros. (Norwich) Ltd. Norwich

Contents

Preface vii

1. Hydrocarbons 1
   Oils and petroleums 1
   Aromatics 116
   Polyaromatics 129
   Greases 200
   References 205

2. Detergents 218
   Anionic detergents 218
   Cationic detergents 247
   Non-ionic detergents 249
   References 277

3. Pesticides and PCBs 280
   Organochlorine insecticides 280
   Carbamate insecticides 361
   Organophosphorus insecticides 371
   PCB and PCB–organochlorine insecticide mixtures 416
   References 483

4. Herbicides 495
   Introduction 495
   Triazine type herbicides 498
   Substituted urea type herbicides 503
   Phenoxyacetic acid type herbicides 514
   Miscellaneous herbicides 529
   Mixture of herbicides and organochlorine insecticides 542
   References 545

Index 549

Preface

The presence of concentrations of organic substances in water is a matter of increasing concern to the water industry, environmentalists and the general public alike, from the point of view of possible health hazards presented to both human and animal life represented by domesticated and wild animals, bird and fish life. This awareness hinges on three facts: the increasing interest of the scientist and the public alike in matters environmental, an increased usage of organic materials in commerce coupled with the much wider variety of organic substances used nowadays, and finally, the availability of analytical methods sensitive enough to determine very low concentrations of these substances, the presence of which we were formerly unaware.

It has been estimated that river waters can contain up to two thousand different organic substances over a wide concentration range, many of which survive processing in the water works and occur in potable water with possible health implications. The Food and Drug Administration in America, amongst others, is systematically working its way through screening tests on the substances so far identified in water, but this is a process that will take many years to complete.

As well as organics occurring in water as a direct result of industrial activity there are those which arise more indirectly, such as haloforms produced in the chlorination stage of water treatment process, organio-metallic compounds produced by conversion of inorganic salts by biological activity in rivers and nitrosamine formation by conversion of inorganic nitrites. There are also, of course, naturally occurring organic substances in water.

The purpose of this work is to draw together and systematize the body of information available throughout the world on the occurrence and determination of organics in all types of water and effluents. A particular feature is the presentation of detailed procedures in the case of many of the more important procedures, so that reference to a very scattered literature can in many cases be avoided. Complete coverage is given of all the major instrumental techniques now available.

The contents are presented in as logical a fashion as possible, starting in Volume 1 Chapter 1 with a discussion of hydrocarbons and polyaromatic hydrocarbons. Chapter 2 deals with the various types of surface active agents whilst Chapters 3 and 4 respectively discuss the numerous types of organochlorine and organophosphorus insecticides and herbicides and polychlorinated biphenyls now in use in agriculture and which are finding their way into the water courses. Volume 2 deals with organometallic compounds, types of compounds classified under elements including carbon, oxygen, halogen, nitrogen, sulphur and phosphorus and concludes with chapters on ozonization products and natural pigments.

Particular groups of substances which are causing concern by their presence in the environment are discussed in detail, e.g. polyaromatic hydrocarbons, chlorine and phosphorus insecticides, herbicides, polychlorinated biphenyls, haloforms and organometallic compounds.

As well as discussing the analysis of river, surface and underground waters and potable water, various sections include discussion, where relevant, of ocean and beach waters, sewage and trade effluents and muds and sediments. In certain instances, the analysis of fish, crustaceae and plant life for organic pollutants is also discussed. Such measurements are very useful as these will reflect the general level of pollution that has occurred over a period of time, as opposed to spot measurements obtained by analysis of water samples.

Examination for organic substances combines all the exciting features of analytical chemistry. First, the analysis must be successful, and in cases such as spillages must be completed quickly. Often the nature of the substances to be investigated is unknown; they might occur at exceedingly low concentrations and might indeed be a complex mixture. To be successful in such an area requires analytical skills of a high order and the availability of sophisticated instrumentation.

The work has been written with the interests of the following groups of people in mind: management and scientists in all aspects of the water industry, river management, fishery industries, sewage effluent treatment and disposal, land drainage and water supply; also management and scientists in all branches of industry which produce aqueous effluents. It will also be of interest to agricultural chemists, agriculturalists concerned with the ways in which organic chemicals used in crop or soil treatment permeate the ecosystem; also to the biologists and scientists involved in fish, insect and plant life, and to the medical profession, toxicologists and public health workers and public analysts. Other groups or workers to whom the work will be of interest include oceanographers, environmentalists and, not least, members of the public who are concerned with the protection of our environment.

Finally, it is hoped that the work will act as a spur to students of all subjects mentioned and assist them in the challenge that awaits them in ensuring that the pollution of the environment is controlled so as to ensure that by the turn of the century we are left with a worthwhile environment to protect.

T. R. CROMPTON

Chapter 1

Hydrocarbons

OILS AND PETROLEUMS

In many areas oil has become the most frequently encountered water pollutant, and oil pollution incidents are becoming more numerous. This reflects the expanding and widespread consumption of petroleum products, a consumption that will continue to increase in the foreseeable future. Oil pollution has harmful effects on aquatic life and lowers the aesthetic appearance of an inland water. Occasionally it necessitates the closure of waterworks intakes. Existing legislation concerning oil pollution cannot be effective unless there are adequate analytical means of detecting oil pollutants, identifying them with regard to their sources and determining their concentration. Many analytical techniques have been proposed, but the references are scattered throughout the literature, and require collation and assessment in relation to inland water pollution.

Although marine oil pollution has received much attention in recent years, this has not been the case with oil pollution of inland waters, and for some time now there has been need for an assessment of the analytical and related problems in this field. This chapter discusses these problems. Particular attention is given to the identification of the polluting oil.

Wherever oil is produced, stored, transported by vehicle or pipeline, or consumed, there exists a potential source of oil pollution, either directly by surface drains and surface run-off, or indirectly by seepage into the ground. Unlike some pollutants, oil pollution is generally unpredictable as to location and time, and usually exists as a surface phenomenon. Heavy pollution is obviously unwelcome, but even thin ephemeral films representing only small amounts of oil may cause complaints and require investigation if continually present on surface water.

The effects of oils on water are manifold, and generally only the acute effects are understood. Long-term chronic effects are slightly comprehended, if at all. Oil, directly or indirectly, seriously lowers the aesthetic appearance of inland waters and interferes with their recreational use. Since oil can readily form a visible thin film of about 7.6×10^{-5} mm thickness, a small quantity of oil may be important and produce a heavily polluted appearance.

It is often claimed that oil affects the transfer of oxygen into water, but the significance of the effect in the case of thin films is not fully understood. Thick viscous layers do affect the transfer.

Photosynthetic activity can be affected by increased reflection and possibly by the absorption of light by the oil. On rivers, where a continuous, prolonged oil slick is improbable, the effect would probably not be significant.

Biodegradation of oil in inland waters depletes the oxygen content, but unless frequent oil pollutions occur in a river, such as in heavily industrialized areas, or in non-flowing waters, the depletion would probably do little damage.

Most oil products are not considered particularly toxic. Petrol, white spirit and similar volatile products are generally regarded as the most toxic. Heavy surface oil pollution and oil which is emulsified, dissolved, or associated with suspended solids in the body of the water may injure or kill aquatic animal life. Plant life, both aquatic and riparian, can be destroyed by heavy pollution. Prolonged pollution will denude most plant and animal life by a combination of the above mentioned effects. Long-term toxicity effects are not really understood although carcinogenicity and general health considerations have been studied in some detail.

Recent information suggests that sedimented oils in a lake or river can act as carriers of toxic non-polar organic chemicals, e.g. DDT, owing to the high solvent power of sedimented oils for this type of compound. The significance of this effect in rivers is unknown.

Oil contamination is most undesirable in waterworks treatment systems. High concentrations of oil can impair the filters, while even trace quantities can produce taste and odour problems.

Practically the total range of petroleum products is encountered as pollutants on inland water. Crude oils are very seldom found inland, but petrol, paraffin, gas oils, heating fuels, lubricating oils, transformer oils and cutting fluids have given problems. In most areas, heating fuel, due to its widespread industrial and domestic utilization, and diesel fuel are the most commonly occurring oil pollutants, Petrol, although used in greater quantity than most other petroleum products, does not often pollute inland waters. Probably this is due to its high volatility on water surfaces, the strict regulations concerning its storage, and the general public awareness of its dangers. Its relatively high water solubility may be a lesser factor. Lubricating oils give pollution problems especially in highly industrialized areas. Since lubricating oils are seldom stored or used in large quantities, pollutions tend to be of a smaller nature, but are often responsible for intermittent surface films on inland waters, which are regarded as insignificant. However, the increase in concern over the aesthetic appearance of our inland waters may soon render them significant. Lubricating oils are common pollutants at sewage works. Heavy fuel oils are occasionally met as pollutants, and owing to their high viscosity can cause extensive soiling of banks and riparian structures as well as being extremely difficult and troublesome to remove.

Natural oils, such as essential oils, vegetable, animal, and fish oils, are only infrequently found as significant pollutants. However, when they do occur they can pose extremely difficult problems.

Characterization/identification

The identification procedure for oils in water samples, whether river or ocean can be divided into three stages:

(1) isolation of the hydrocarbon components from the pollutant sample.
(2) identification of the sample in terms of the petroleum product, for example, crude oil, petroleum, gas oil etc.
(3) identification of the specific source of pollution, such as an individual tanker, tank truck, factory or domestic fuel tank etc.

Stage (2), a general classification of the oil, is often satisfactorily achieved by gas chromatographic techniques possibly coupled with mass spectrometry or infrared spectroscopy applied to a sample of the oil pollutant. Stage (3), the true identification, invariably requires samples from potential sources for comparion with the pollutant. This is often attempted again using gas chromatography, by comparison of the resulting chromatograms, but in a less satisfactory and confident manner. Generally, when the comparisons of chromatograms are reasonably similar, the perpetrator of the pollution accepts liability in the face of accumulated circumstantial and scientific evidence, and introduces the recommended remedial measures.

Existing gas chromatographic techniques can, in the majority of cases, classify petrol, paraffin, light fuel oils, intermediate fuel oils and, with less ease, lubricating, transformer and cutting oils. Higher boiling products with little volatility are not amenable to conventional gas chromatographic techniques and recourse has to be taken to other techniques such as the use of capillary columns or non-g.l.c. techniques.

Techniques other than gas chromatography or, more commonly, combinations of techniques have been used to characterize oil spills, as discussed below.

Nearly every known analytical technique has been used or suggested for oil pollutant identification, but, certainly, no one has emerged of such superiority that all the others can be considered as redundant. Therefore, earlier attempts of oil characterization have been performed by a multimethod approach; the particular combination of analytical techniques depends on the facilities and the experience existing in a laboratory and the expenditure which is justified to identify any unknown source.

Representative examples of these overall approaches are reported in Table 1. They include analytical determinations such as the infrared spectra, asphaltene and paraffin contents, etc., that provide a general classification of the pollutants (crude oils, fuel oils, oil sludges, etc.) and others, such as the Ni/V ratio, sulphur content, chromatographic profiles, etc., that permit, by comparison with reference samples, their precise identification.

Table 1 Overall approaches for identification of oil pollutants

IP Method (1974)	Sp. gr.
	Asphaltenes
	S, Ni, and V contents
	T.l.c./u.v.
	G.l.c.
DGMK Method (1973)	I.r. spectroscopy
	S, Ni, and V(Ca, Ba, and Zn)
	Column chromatography
	G.l.c.
US Coast Guard (1974)	I.r. spectroscopy
	Fluorescence spectroscopy
	T.l.c./u.v.
	G.l.c.

Reprinted with permission from Rasmussen[57]. Copyright (1976) American Chemical Society.

Table 2 Multiparametric methods for fingerprinting oil pollutants

Analytical methods	No. of parameters	References
Trace elements	3	Brunnock et al.[1]
	22	Duewer et al.[2]
I.r. spectroscopy	3	Kawahara and Ballinger[3]
	18	Lynch and Brown[4]
	23	Mattson et al.[5]
Gas chromatography	3	Erhardt and Blumer[6]
	19	Clark and Jurs[7]
	36	Rasmussen[8]

Reprinted with permission from Rasmussen.[57] Copyright (1976) American Chemical Society.

However, another approach involving only one analytical technique, but increasing the number of parameters considered, has been emphasized recently as is exemplified in Table 2 for trace analysis, infrared spectroscopy and gas chromatography. In these cases a multiparametric profile is used for identification, instead of a combination of different analytical determinations and pattern recognition techniques have, often, been applied to improve the diagnostic performance.

The main requirements that must fulfil these fingerprinting parameters besides their specificity, is that they must remain unaltered during the sea weathering processes affecting the pollutant, namely by evaporation, solution, photo-oxidation, and biodegradation. In consequence, both conditions, specificity and stability, need to be investigated in order to evaluate the reliability and the usefulness of any proposed method.

General weathering effects occasionally affect identification, and chemists are conscious of the possibility that weathering action on the polluting oil may prevent identification or lead to an incorrect inference. In most cases, since an oil would be unlikely to remain on an inland surface water for any length of time prior to analysis, these effects are probably limited to evaporation of the more volatile components, and to a lesser extent, of the more water-soluble components. With practice most analysts can allow for small evaporation effects during the visual inspection of the chromatograms. Little is known of the significance or nature of solution of components on identification. With material of much less volatility than petrol the effect would probably be small.

Quantitative analysis of oils polluting the surface of a river is seldom attempted, as the information is not particularly useful. The pollution is obvious, and the value obtained meaningless, because it is impossible quantitatively to determine oil films or slicks, and relate the determination to the total volume of water, the total film area, and the film thickness.

Quantitative results for oil are, however, required on effluents entering rivers. The method often used is a gravimetric procedure[9] which determines 'non-volatile ether-extractable matter'. The method is non-specific for petroleum oils, gives erroneous results when soluble oils (cutting oils, etc.) are present, and is inaccurate below 5 mg oil/litre range. Many analysts are dissatisfied with this procedure. Thin-layer chromatography and infrared analysis offer specificity and accuracy in the low concentration range and will be discussed later.

The problems associated with the characterization of oils and the determination of oils are discussed in detail below under separate headings.

In the case of heavy oil pollution, a sufficient sample of the neat oil is readily taken from the surface. Smaller quantities of oil, for example, thin films and emulsions, can be liquid-liquid extracted with carbon tetrachloride,[10,11] chloroform,[12] pentane,[11-13] ether,[14,15] nitrobenzene,[16,17] methanol/benzene (for sediments),[18,19] hexane,[20] toluene/xylene, toluene,[21] and iso-octane.[13] Solvents are chosen in order that no significant interference occurs to the components of the pollutant, and that loss of volatile components during evaporation of the solvent, if required, is minimal. For very volatile hydrocarbons in aqueous samples, head-space analysis and degassing techniques are recommended.[22-24] Goma and Durand used an ultrasonically prepared emulsion for their gas chromatographic work, thus avoiding quantitative liquid-liquid extractions.[25]

Most authors recommend collection and storage of samples in glass containers. Plastic containers may introduce organic contaminants which may be significant,[14] especially with trace amounts of oil. Metal containers may affect subsequent determinations of nickel and vanadium in crude in residual petroleum oils.[21] Pollutants on the surface can be collected by skimming with a suitable container. Kawahara[12] suggests a glass, wide-mouthed filter-funnel fitted with Teflon tubing and a two-way stopcock. The surface oil is ladled into this and is repeatedly separated from the lower aqueous phase. Alternatively, a paint-free dustpan with a stopcock attached to the handle, or even a mop with a fitted wringer, have been employed with success.

Samples can be preserved to a limited extent by the addition of solid carbon dioxide to expel air, by refrigeration,[12] or by the addition of mercuric chloride which inhibits bacterial activity.[24]

Invariably, dual packed columns have been employed, and one of the earliest articles devoted to the identification of petroleum products is that of Lively,[13] who used dual 4 ft × ¼ in. columns packed with 20% SE-30 as the liquid phase on a Chromosorb solid support. Most subsequent workers in this field have employed the same or a similar liquid phase, usually at a lower loading of 4–5% and somewhat different column dimensions, 4–12 ft × ⅛ – ¼ in.[10-12,16,20,21,26-31] Chromosorb, or occasionally a similar solid support[29] was usually the solid support utilized after first being acid washed and treated with hexamethyl disilazane (HMDS) or dimethyl dichlorosilane (DMCS). Liquid phases of similar properties that have been employed are 5%[21] and 10%[31] OV-1, 20% SE-52,[27] 5% E301,[28] and 2.2%,[32] 10%,[33] and 20%[30] Apiezon L. These liquid phases are essentially non-polar substances, but more polar phases, 5%[17] and 10%[16] polyethylene glycol 1500, have been used for investigations of the water-soluble components in diesel and gas oil[17] and in petrol.[16]

In almost all recent publications, flame ionization detectors and temperature programming were employed, the latter especially in the identification of less volatile oils. For routine investigations, isothermal conditions have been adopted in order to allow quicker analysis, but at the sacrifice of some fine detail in the chromatogram.[30]

Columns containing relatively non-polar liquid phases, separate homologous series of hydrocarbons virtually in order of boiling point. In practice, this results in the *n*-alkanes of petroleum products being separated in order of increasing number of carbon atoms in the molecule. Branched-chain, saturated and aromatic hydrocarbons follow a similar pattern although possibly less precisely. With products such as petrol, paraffin, diesel oil, gas oils, TVO, and intermediate fuel oils, *n*-alkanes produce a predominant series of well separated peaks on the chromatogram, indicative of the boiling range of the sample. The *n*-alkanes on the chromatogram usually reside on a mound of poorly resolved branched and cyclic paraffins, aromatic hydrocarbons, and heterocyclic compounds. In the case of lubricating, cutting and transformer oils, which are generally lacking in *n*-alkanes, only fairly featureless mounds are obtained, Pollutants are classified as to type principally by examination of the carbon number range and proportion of the *n*-alkanes on the chromatogram. Specific *n*-alkane peaks are usually identified by comparison with standard solutions of some known compounds and a 1% w/v solution has been recommended.[11] Identification of the source of pollution is achieved by comparison of pollutant and suspect oil chromatograms, usually by matching minor component peaks. It has been stated that while *n*-paraffins tend to be indicative of the processing of a product, the minor components are more specific and can indicate the original crude oil source. With gas chromatography small evaporation effects can be allowed for by expecting irregularities in the volatile component peaks.

Detailed salient points involved in classifying and identifying oils have been compiled.[11,21,27,28,34,35] Most distillate products e.g. petrol, paraffin, fuel oils, diesel oil, TVO etc., and even very similar products such as turbo-jet fuel and kerosene,[27] can be classified in the absence of excessive weathering effects. Various chromatograms have been published: kerosene, turbo-jet fuel, steam-cracked naphtha;[27] lubricating oil/gas oil/weathered paraffin;[35] white spirit, turpentine substitute, paraffin, 30 second fuel oil;[30] standard gasoline;[13] petrol;[16] diesel fuel and gas oil,[17] various gasolines, diesel fuels, and aviation fuels.[33]

Lubricating, cutting, transformer oils etc., heavy fuel oils, asphaltic and bituminous materials are difficult or impossible to classify with any certainty by gas chromatography.

Characterization of hydrocarbons in river water

A polar liquid phase was found more suitable for studying the major components of petrol,[16] gas oil, and diesel oil,[17] forming true solutions in water. With such a phase, saturated hydrocarbons tended to elute before aromatic hydrocarbons which were found to be the principal components in true solution, and therefore their investigation was facilitated, in the case of gas oil and diesel oil, forming true solutions in water. With such a phase, saturated hydrocarbons tended to elute before aromatic hydrocarbons which were found to be the principal components in true solution, and therefore their investigation was facilitated. In the case of gas oil and diesel oil, no saturated hydrocarbons could be detected in solution. These authors reached the important conclusion that the determination of the origin of oil components in true aqueous solution could be more difficult because of selective solution of certain components. This effect was likely to apply to the lower distillates, which tended to be relatively more water soluble, rather than to the non-volatile petroleum products. Distinction between petrol and gas oil or diesel oil seemed possible, but appeared difficult between similar products such as gas oil and diesel oil.

Capillary columns provide greater resolution and therefore more detail for comparison between a polluting oil and a suspect sample. The enormous separation power available has been demonstrated in their application to petroleum analysis. Sanders and Maynard,[37] using a 200 ft × 0.01 in. column containing squalane, separated approximately 240 hydrocarbon components of gasoline, of which 180 were identified. Gouw[38] describes a versatile 10 m × 0.01 in. capillary column coated with CV-101, and its application to the separation of hydrocarbon mixtures in the C_4–C_{58} n-alkane range. Columns of 500 ft × 0.01 in. coated with l-octadecene and operated at 30 °C have been recommended for identification of crude oils.[12] Hydrocarbons of a selected boiling range, e.g. C_7–C_8 saturates, were collected by prefractionation in a short packed column, trapped, and finally examined on the capillary column. Alkylbenzenes up to C_{10} were examined in a similar manner on a 800 ft × 0.01 in. column coated with more polar polyethylene glycol and operated at 60 °C. It is claimed that gasoline and other volatile products can be analysed and identified. However, the technique in no way allows for evaporation of volatile components, which would have a serious effect. Columns of the stated length would be impractical for general routine use.

Cole[39] found that in the identification of kerosene and aviation fuels, the usual packed columns provided sufficient detail only up to C_{13} *n*-alkane, a range readily altered by evaporation effects. A suitable capillary column was developed which revealed extra detail of minor components above C_{13} *n*-alkane, and consisted of a 45 m × 0.25 mm stainless steel column coated with OV-101. This gave satisfactory resolution and stability in the 50–310 °C temperature range. Naturally abundant saturated terpenoid substances, e.g. farnesane (C_{15}), pristane (C_{19}) and phytane (C_{20}) were determined in fractionated crude oils and used for identification purposes. The use of a similar column should be applicable to general identification of the commonly occurring oil pollutants of inland waters up to the C_{25}–C_{30} *n*-alkane range. This seems a logical and valuable extension of the present gas chromatographic techniques. Unfortunately the application to analysis of lubricating oils, etc., does not seem promising. The cost and stability of such a capillary column seems satisfactory for routine uses.

An interesting gas chromatographic technique of identifying petroleum products, including lubricating oils, is that of Dewitt Johnson and Fuller.[31] A gas chromatographic column effluent was split in order that it could be simultaneously sensed by a double-headed flame photometric detector, specific for both sulphur and phosphorus compounds, and by dual-flame ionization detectors for carbon compounds in general. Since most petroleum products are claimed to contain sulphur and phosphorus compounds, three chromatographic traces were obtained for the products examined. In the cases of lubricating oils, which normally give poorly resolved peaks on chromatograms with flame ionization detection, sulphur and phosphorus detection gave considerably more detail for identifying purposes. The apparatus is, however, complex and expensive for routine application.

Lysyj and Newton[40] have described a multicomponent pattern recognition and differentiation method for the analysis of oils in natural waters. The method is based on that described earlier by Lysyj[41] which depends on the thermal fragmentation of organic molecules followed by gas chromatography. Dried algae and outboard-motor oil were analysed and a specific pattern or numerical 'fingerprint' was obtained for each by pyrographic means. The algal pattern comprised three specific peaks and seven peaks common to those of the oil, whereas the oil pattern comprised two specific peaks and seven peaks common to those of the algae.

Kawahara[42] has discussed the characterization and identification of spilled residual fuel oils on surface water using gas chromatography and infrared spectrophotometry. The oily material was collected by surface skimming and extraction with dichloromethane, and the extract was evaporated. Preliminary distinction between samples was made by dissolving portions of the residue in hexane or chloroform. If the residue was soluble in chloroform but not in hexane it was assumed to be a crude oil, a grease, a heavy residual fuel oil or an asphalt; if it was soluble in both solvents it was assumed to be a light or heavy naphtha, kerosene, gas oil, white oil, diesel oil, jet fuel, cutting oil, motor oil or cutter stock. The residue was also examined by infrared spectrophotometry; wavenumber values of use for identification purposes are tabulated. Classified, volatile petroleum products such as naphtha, gasoline, jet fuel, kerosene, various fuel oils, crude oil, petroleum jelly and some lubricating oils, were identified by gas chromatography of the residue on, e.g., a column of 5% of of DC-200 on Gas-Chrom Q temperature programmed from 110 to 224 °C at 10 ° per minute, or a capillary column coated with 0.5 μm of DC-200 and temperature programmed from 80 to 170 °C at 3° per minute, both operated with helium as carrier gas and flame ionization detection, or a column of 5% of QF-1 and 3% of DC-200 on Gas Chrom Q operated at 180 °C with ^{63}Ni electron-capture detection. For heavier products, the pentafluorobenzyl derivatives were used for gas chromatography.

Jeltes and den Tonkelaar[43] investigated problems of oil pollution, the nature of the contaminants and the chemical methods of oil pollution, the nature of the contaminants and the chemical methods used for their detection. In particular, the use of gas chromatography to obtain 'fingerprint' chromatograms of oil pollutants in water, and of infrared spectrophotometry to determine the oil contents of soils and sediments, is discussed.

Ahmadijian and Brown[44] investigated the feasibility of remote detection of water pollutants and oil slicks by laser-excited Raman spectroscopy. They showed that, by use of a system of lenses and mirrors, laser-excited Raman spectra can be recorded for samples containing oil at a distance of 21 ft from the instrument.

McMullen et al.[45] discussed the principle and operation of a system to detect and measure the pollution of water surfaces by oil. The apparatus, which is described in detail with the aid of photographs and drawings, utilizes buoyant sampling heads which float on the water surface to record surface tension changes caused by the presence of either soluble or insoluble monolayers. A battery-operated central control unit and pen recorder are attached at the end of a 60 ft floating cable, and a trigger circuit can be used to activate an alarm when surface tension changes are detected. Field trials, using oil as the polluting agent, on various canals are reported, and it is concluded that the apparatus could be used for the detection of detergent and crude sewage pollutants.

Jeltes et al.[46] have applied capillary gas chromatography to the analysis of hydrocarbons in water and soil. The advantages in capillary columns over packed columns in obtaining practically useful information on environmental pollution by hydrocarbons was demonstrated by these workers. Improved separation on capillary columns gives fine-structured chromatographic fingerprints useful for source recognition. These workers pointed out that the ratios of *n*-alkanes and isoprenoids are important in studies of biodegradation of oil pollution. The compounds were not separated on packed columns.

Cole[47] has investigated the use of gas chromatography in the identification of slop oils resulting from oil refinery leaks. Oils leaking into the waste-water system were examined by gas chromatography on a column (6 ft × 0.25 in. o.d.) of 20% of SE-52 on silanized Chromosorb W (60–80 mesh) temperature programmed from 50 to 300 °C at 15° per minute, with nitrogen (60 ml min^{-1}) as carrier gas and flame ionization detection; and by infrared spectrophotometry. Reference is then made to a library of the results of similar tests on samples of all the refinery process streams. Only one ml of oil sample is required, and identification normally takes less than 1 hour.

Vos *et al.*[48] have carried out a detailed study of the analysis of oil contaminated ground water to ascertain the rate of filtration of oil components, and the effects of their biodegradation, under conditions very close to those in a natural aquifer. Large scale lysimeter experiments are reported in a sand dune area where the ground-water level could be adjusted with an external overflow device. Details are given of hydrocarbon concentrations determined by adsorption onto Amberlite XAD-4 resins, and investigations using gas chromatography, mass spectroscopy, high resolution gas chromatography, infrared spectroscopy and ultraviolet spectroscopy.

Garria and Muth[49] have used gas chromatography to characterize crude, semi-refined, and refined oils.

Characterization of oils in the marine environment

In the marine environment gas chromatography has been employed to identify petroleum products.[12,21,28,34,35] Here the pollutants are crude oil, marine residual fuel oil and crude oil sludge consisting of a concentrated suspension of high-melting-point paraffin wax in crude oil. Although weathering of marine oil pollutants can be such that the oil is rendered unrecognizable, the time required to achieve this was found to be so long as to be insignificant in regard to pollutant identification.[35] Ramsdale and Wilkinson[28] employed a specially constructed injection device which permitted direct introduction of samples contaminated with water, seaweed, sand, or sediments. Pollutants were generally readily identified.

Occasionally the mound of unresolved components on the chromatogram supporting the superimposed *n*-alkane peaks, confuse the true *n*-alkane profile. This has been overcome by separating off the *n*-alkanes using molecular sieves, prior to g.c.[34] However, separation of *n*-alkanes in this way, or by urea complex formation,[36] is reported as being more applicable to distillates rather than residual materials, and also the separation is not entirely specific.[35] In the case of marine pollutants it has been found advantageous to chromatograph a distilled residue, b.p. > 343 °C[34], or fraction, b.p. 245–370 °C,[20] which avoids problems caused by evaporation of lower ends by weathering.

Chromatograms of marine pollutants have been published; crude oils, 200 second fuel oil, 2000 second fuel oil, pollution samples;[34] sludge wax, crude oil sludge, Bahrein fuel oil, 3000 second diesel fuel, pollution samples;[28] marine diesel fuel, sludges, crude oil, residual fuel oil.[20]

Freegarde et al.[50] have discussed the identification, determination, and ultimate fate of oil spilt at sea.

Erhardt and Blumer[51] have developed a method for the identification of the source of marine oil spills, by gas chromatographic analysis and results for eight different crude oils are given. Distinguishing compositional features are still recognizable after weathering for more than 8 months. The method was used for the tentative source identification of samples of beach tar.

Zafirion et al.[52] have shown that commercial oil spill emulsifiers can interfere with the gas chromatographic detection of the source of oil spills.

Boylan and Tripp[53] determined hydrocarbons in sea water extracts of crude oil and crude oil fractions. Samples of polluted seawater and the aqueous phases of simulated samples (prepared by agitation of oil-kerosene mixtures and unpolluted seawater to various degrees) were extracted with pentane. Each extract was subjected to gas chromatography on a column (8 ft × 0.06 in.) packed with 0.2% of Apiezon L on glass beads (80-100 mesh) and temperature programmed from 60 to 220 °C at 4° per minute. The components were identified by means of ultraviolet and mass spectra. Polar aromatic compounds in the samples were extracted with methanol-dichloromethane (1:3).

Investigations on pelagic tar in the North West Atlantic have been carried out by Bulten et al.[54] using gas chromatography. Their report collects together the results of various preliminary investigations. It is in the Sargasso Sea, where the highest concentrations (2-40 mg m^{-2}) occur, and on beaches of isolated islands, such as Bermuda. These workers discuss the occurrence, structure, possible sources, and possible fate of tar lumps found on the surface of the ocean.

Zafirion and Oliver[55] have developed a method for characterizing environmental hydrocarbons using gas chromatography. Solutions of samples containing oil were separated on an open-tubular column (50 ft × 0.02 in.) coated with OV-101 and temperature programmed form 75 to 275 °C at 6° per minute; helium (50 ml min^{-1}) was used as carrier gas and detection was by flame ionization. To prevent contamination of the columns from sample residues, the sample was injected into a glass-lined injector assembly, operated at 175 °C, from which gases passed into a splitter before entering the column. Analysis of an oil on three columns gave signal-intensity ratios similar enough for direct comparison or for comparison with a standard. The method was adequate for correlating artificially weathered oils with sources and for differentiating most of 30 oils found in a sea port.

The identification of hydrocarbon oil spills poses a difficult analytical problem. Not only are these samples extremely complex but, in many cases, there are only slight differences in their compositions. The action of the environment is another complicating factor that changes compositions by the loss of light ends, the action of microbes, the action of sand and sea water, and contamination by other organic matter.

Hertz et al.[56] have discussed the methodology for the quantitative and qualitative assessment of oil spills. They describe an integrated chromatographic technique for studies of oil spills. Dynamic head-space sampling and gas chromatography and coupled-column liquid chromatography are used to quantify petroleum-containing samples, and the individual components in these samples are identified by gas chromatography and mass spectrometry.

Rasmussen[57] has described gas chromatographic methods for the identification of hydrocarbon oil spills. The spill samples are analysed on a 100 ft Dexsil-300 support coated open tube (SCOT) column to obtain maximum resolution yet retain a high upper temperature limit. The chromatograms are mathematically treated to give 'g.c. patterns' that are a characteristic of the oil, yet are essentially unaffected by moderate weathering. Both liquid and solid oil spill samples were analysed with very little sample preparation in less than 2 h.

Rasmussen[57] describes a gas chromatographic analysis and a method of data interpretation that he has successfully used to identify crude oil and bunker fuel spills. Samples were analysed using a Dexsil-300 support coated open tube (SCOT) column and a flame ionization detector. The high resolution chromatogram was mathematically treated to give 'g.c. patterns' that were a characteristic of the oil and were relatively unaffected by moderate weathering. He compiled the 'g.c. patterns' of 20 crude oils. Rasmussen[57] uses metal and sulphur determinations and infrared spectroscopy to complement the capillary gas chromatographic technique.

Rasmussen[57] used a Perkin-Elmer model 3920 and a Varian model 1400 gas chromatograph, both with flame ionization detectors with helium as carrier gases. A 250 ft × 0.01 in. i.d. Dexsil-300 wall coated open tube (WCOT) column was used initially but was found to have too short a life at 400 °C. The 100 ft SCOT column proved to be much more stable and most analyses were performed using this column. The oven was programmed from 40 to 400 °C at 4° min^{-1} to obtain high resolution. The preferred sampling procedure was to inject 0.1 μl of the undiluted oil sample. Solids were introduced with a solid injection probe. No inlet splitter was used and peak areas were measured with an IBM 1800 computer. The gas chromatograms of most oil samples examined had similar basic features. All were dominated by the n-paraffins, with as many as 13 resolved, but unidentified, smaller peaks appearing between the n-paraffin peaks of adjacent carbon numbers. Each oil had the same basic peaks, but their relative size within bands of one carbon number varied significantly with crude source. To aid in the comparison of chromatograms, the peaks were labelled as shown in Figure 1 and the relative areas of the smaller peaks were expressed as a percentage of the preceding n-paraffin peak. In other words, n-$C_{19}H_{40}$ was assigned the peak number 19.0 and the 12 peaks eluting between n-$C_{19}H_{40}$ and n-$C_{20}H_{42}$ were assigned the peak numbers 19.1–19.12. The areas of the peaks 19.1–19.12 were then expressed as a percentage of peak 19.0 to form the C_{19} segment of the 'g.c. pattern' of the oil. Generally, the C_{14}–C_{35} portion of the chromatogram was treated in this manner.

The partial chromatograms in Figure 1 are typical of analyses performed on

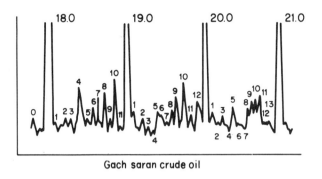

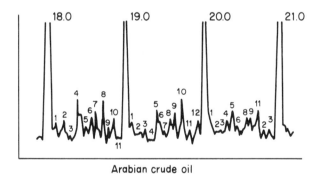

Figure 1 Partial chromatograms of Gach Saran, Cabinda, and Arabian Crudes. Reprinted with permission from Rasmussen.[57] Copyright (1976) American Chemical Society

the Dexsil-300 SCOT column and demonstrate the peak numbering system. Differences between the three crude oils are evident from the chromatograms and their 'g.c. patterns' in Table 3. The repeatability of 'g.c. patterns' was determined to be 5% at the 95% confidence level.

The preferred approach to an oil spill identification programme was to analyse the oil spill sample and all potential sources followed by the comparison of their

Table 3 Partial g.c. patterns of Gach Saran, Cabinda, and Arabian crude oils

	Partial g.c. patterns		
Peak number	Gach Saran crude	Cabinda crude	Arabian crude
18.0	100	100	100
18.1	2	1	5
18.2	2	3	6
18.3	3	6	2
18.4	9	7	13
18.5	3	3	4
18.6	6	3	7
18.7	7	8	8
18.8	9	6	11
18.9	3	4	2
18.10	12	24	6
18.11	1	0	1
19.0	100	100	100
19.1	5	2	6
19.2	3	4	3
19.3	2	2	4
19.4	1	2	2
19.5	6	4	10
19.6	5	2	6
19.7	3	4	3
19.8	5	3	8
19.9	9	8	9
19.10	12	12	13
19.11	4	7	3
19.12	8	10	6
20.0	100	100	100
20.1	5	5	8
20.2	2	2	2
20.3	4	5	4
20.4	2	3	7
20.5	7	5	11
20.6	2	1	2
20.7	2	2	2
20.8	6	4	8
20.9	8	8	8
20.10	8	13	6
20.11	9	6	11
20.12	3	5	3
20.13	3	6	4

Reprinted with permission from Rasmussen.[57] Copyright (1976) American Chemical Society.

gas chromatographic patterns. When it was not possible to analyse all potential sources, the spill sample was compared to a library of gas chromatographic patterns of known crude oils and products.

The effect of weathering on the applicability of this identification procedure was investigated. A Cabinda crude oil was artificially weathered in agitated salt water at 20 °C for up to 30 days and aliquots were analysed at regular intervals. The weathered crudes showed the expected loss of light ends; however, the gas chromatographic patterns above C_{14} were essentially unchanged and each sample was recognizable as Cabinda. The stability of the gas chromatographic patterns was attributed to the fact each pattern was derived from the ratio of components of similar boiling points. In this way, the effect of weathering was minimized.

Rasmussen[57] quotes an example of a bunker oil spill. The spill sample and samples of three potential sources were analysed. Their gas chromatographic patterns were calculated and a comparison confirmed that the oil spill was bunker fuel from a nearby power plant (source A). The chromatograms are shown in Figures 2–5 and segments of the g.c. patterns are compared in Table 4. Key areas of comparison are marked with an asterisk.

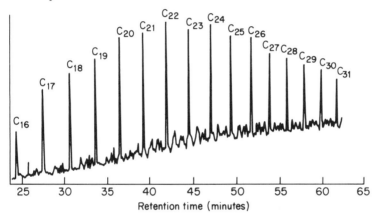

Figure 2 Oil spill sample C_{16}–C_{31}: segment of chromatogram. Reprinted with permission from Rasmussen.[57] Copyright (1976) American Chemical Society

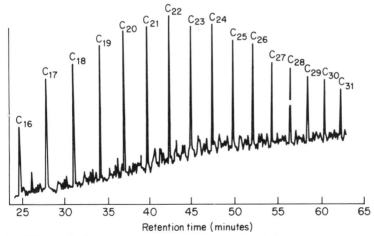

Figure 3 Oil sample from source A: C_{16}–C_{31} segment of chromatogram. Reprinted with permission from Rasmussen.[57] Copyright (1976) American Chemical Society

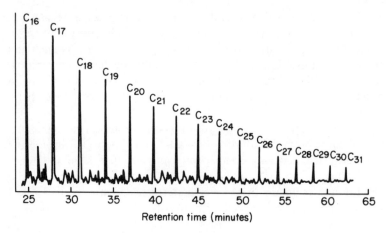

Figure 4 Oil sample from source B: C_{16}–C_{31} segment of chromatogram. Reprinted with permission from Rasmussen.[57] Copyright (1976) American Chemical Society

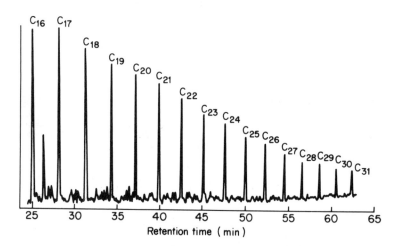

Figure 5 Oil sample from source C: C_{16}–C_{31} segment of chromatogram. Reprinted with permission from Rasmussen.[57] Copyright (1976) American Chemical Society

The Institute of Petroleum[58] in 1970 recommended analytical methods for the identification of the source of pollution by oil of seas, rivers, and beaches. Methods are described for recovering the oil from the sample and for determination of boiling range, vanadium and nickel contents, and wax content. For the determination of boiling range, the hydrocarbon components are separated in the order of their boiling point by means of gas chromatography. For the determination of metals, after incineration of the oil with sulphur and dissolution of the ash, vanadium and nickel are determined by reaction with tungstophosphate and dimethylglyoxime, respectively, and use of X-ray fluorescence

Table 4 G.c. patterns calculated from chromatograms of oil spill and samples of potential sources A, B, C

Peak no.	Oil spill	Source A	Source B	Source C
16.0	100	100	100	100
16.1	7	8	1	1
16.2	3	3	5	3
16.3	6	6	4	2
16.4	1	1	1	1
16.5	4	5	5	1
16.6*	31	31	23	36
16.7	6	7	4	3
16.8*	12	12	8	6
16.9	13	12	10	9
16.10	15	14	13	10
16.11*	13	13	3	1
16.12	1	2	2	1
16.13	1	1	1	1
17.0	100	100	100	100
17.1	2	1	2	1
17.2	2	2	3	2
17.3	3	2	1	2
17.4*	10	10	10	1
17.5	2	2	1	1
17.6	2	2	2	7
17.7	7	8	7	5
17.8	9	9	7	7
17.9*	12	11	9	5
17.10	3	3	3	3
17.11	1	1	1	1
18.0	100	100	100	100
18.1	3	4	5	2
18.2	4	4	6	2
18.3	2	2	2	3
18.4	13	12	13	9
18.5	7	6	4	3
18.6	7	7	7	6
18.7	10	10	8	7
18.8	13	13	11	9
18.9	3	4	2	3
18.10*	8	8	6	12
18.11	2	2	1	1
19.0	100	100	100	100
19.1	3	3	6	5
19.2	3	3	3	3
19.3	3	3	4	2
19.4	1	2	2	1
19.5*	8	9	10	6
19.6	1	2	6	5
19.7	1	2	3	3
19.8	4	5	8	5

*Key areas of comparison.

Table 4 continued overleaf

Table 4 continued

Peak no.	Oil spill	Source A	Source B	Source C
19.9	10	10	9	9
19.10	16	15	13	12
19.11	4	4	3	4
19.12*	10	10	6	8
20.0	100	100	100	100
20.1	3	2	8	5
20.2	1	1	2	2
20.3	6	6	4	4
20.4	1	2	7	2
20.5	9	8	11	7
20.6	6	6	2	2
20.7	6	7	2	2
20.8	7	7	8	6
20.9	7	8	8	8
20.10	7	7	6	8
20.11*	14	15	11	9
20.12	1	1	3	3
20.13	2	3	4	3
21.0	100	100	100	100
21.1	1	2	2	2
21.2	1	2	1	7
21.3*	14	14	18	9
21.4*	11	12	6	1
21.5	12	13	16	9
21.6	9	9	11	10
21.7	2	3	2	1
21.8	4	5	6	4
22.0	100	100	100	100
22.1	3	2	1	2
22.2	5	4	2	10
22.3*	16	16	19	11
22.4	2	2	1	2
22.5	2	2	2	1
22.6	8	7	5	1
22.7	10	11	10	8
22.8*	8	8	5	1
22.9	10	10	12	8
22.10	2	2	2	1
23.0	100	100	100	100
23.1	6	6	4	8
23.2	2	3	5	2
23.3	17	17	18	13
23.4	2	3	11	7
23.5	7	7	7	1
23.6	9	10	10	8
23.7*	13	13	11	7
23.8	2	2	4	2
23.9	2	2	2	2

*Key areas of comparison.

Table 4 continued on next page

Table 4 continued

Peak no.	Oil spill	Source A	Source B	Source C
24.0	100	100	100	100
24.1	4	4	3	4
24.2	1	1	3	2
24.3*	15	15	12	8
24.4	6	6	3	1
24.5	6	7	7	1
24.6	7	8	10	9
24.7	12	12	10	8
25.0	100	100	100	100
25.1	13	14	14	6
25.2*	13	13	8	9
25.3	5	5	2	1
25.4	11	11	10	9
25.5	13	13	12	11
25.6	4	3	3	2
26.0	100	100	100	100
26.1	3	3	4	4
26.2	13	14	13	9
26.3*	13	13	6	11
26.4	5	5	4	1
26.5	7	7	4	5
26.6*	14	14	10	9
26.7	2	2	4	2
27.0	100	100	100	100
27.1*	11	12	18	11
27.2	7	7	8	8
27.3	11	11	15	12
27.4	10	11	8	9
28.0	100	100	100	100
28.1	17	17	18	14
28.2	12	11	6	9
28.3*	2	2	6	11
28.4	10	10	9	9
29.0	100	100	100	100
29.1	5	5	3	4
29.2*	11	12	10	2
29.3	10	10	10	12
29.4	15	15	17	12
29.5	14	14	14	14
29.6	2	3	7	8
30.0	100	100	100	100
30.1	12	11	17	5
30.2	15	15	17	5
30.3*	18	18	13	15
30.4	12	13	9	10
30.5*	12	12	9	5

*Key areas of comparison.

Reprinted with permission from Rasmussen.[57]
Copyright (1976) American Chemical Society.

or spectrophotometry. For the determination of wax content, asphalt-free material is dissolved in hot dichloromethane, the solution is cooled to $\simeq -32\,°C$ and the precipitated wax is separated and weighed.

Various other workers have investigated methods for the identification of oil spills.[59-62] Wilson et al.[59] reviewed approaches based on trace metal (Ni–V) ratio, nitrogen and sulphur content, infrared spectroscopy, and gas chromatography. The fingerprinting of oil spills by gas chromatography has been reported using flame ionization and flame photometric detection.[60,61] Interpretation of the flame ionization chromatograms is normally based on the distribution of the n-paraffin peaks;[59] however, the use of the pristane to phytane ratio has been reported by Blumer and Sass[62] and is thought to be more independent of weathering.

Characterization of oils found on beaches

Ramsdale and Wilkinson[63] have identified petroleum sources of beach pollution by gas chromatography. Samples containing up to 90% of sand or up to 80% of emulsified water were identified, without pretreatment by gas chromatography on one of a pair of matched stainless steel columns (750 mm × 3.2 mm i.d.) fitted with precolumns (100 mm) to retain material of high molecular weight, the second column being used as a blank. The column packing is 5% of silicone E301 on Celite (52–60 mesh), the temperature is programmed at 5° per minute from 50 to 300 °C, nitrogen was used as carrier gas, and twin flame ionization detectors were used.

Adlard et al.[64] improved the method of Ramsdale and Wilkinson[63] by using a S-selective flame photometric detector in parallel with the flame ionization detector. Obtaining two independent chromatograms in this way greatly assists identification of a sample. Evaporative weathering of the oil samples has less effect on the information attainable by flame photometric detection than on that attainable by flame ionization detection. A stainless steel column (1 m × 3 mm i.d.) packed with 3% of OV-1 on AW-DMCS Chromosorb G (85–100 mesh) was used, temperature programmed from 60 to 295 °C per minute with helium (35 ml min^{-1}) as carrier gas, but the utility of the two-detector system is enhanced if it is used in conjunction with a stainless steel capillary column (20 m × 0.25 mm) coated with OV-101 and temperature programmed from 60 to 300 °C at 5° per minute because of the greater detail shown by the chromatograms.

Brunnock et al.[65] have also analysed beach pollutants. They showed that weathered crude oil, crude oil sludge, and fuel oil can be differentiated by the n-paraffin profile as shown by gas chromatography, wax content, wax melting point and asphaltene content. The effects of weathering at sea on crude oil were studied; parameters unaffected by evaporation and exposure are the contents of vanadium, nickel and n-paraffins. The scheme developed for the identification of certain weathered crude oils involves the determination of these constituents, together with the sulphur content of the sample.

McKay[66] has investigated the use of automatic solids injection into the gas chromatographic column in his investigations of hydrocarbon pollution of beach sands.

Characterization of oils by gas chromatography, miscellaneous

Adlard and Matthews[67] applied the flame photometric sulphur detector to pollution identification. A sample of the oil pollutant was submitted to gas chromatography on a stainless steel column (1 m × 3 mm) packed with 3% of OV-1 on AW-DMCS Chromosorb G (85–100 mesh). Helium was used as carrier gas (35 ml min^{-1}) and the column temperature was programmed from 60 to 295 °C at 5° per minute. The column effluent was split between a flame ionization and a flame photometric detector. Adlard and Matthews[67] claim that the origin of oil pollutants can be deduced from the two chromatograms. The method can also be used to measure the degree of weathering of oil samples.

Millard and Arvesen[68] discuss the airborne optical detection of oil on water. They undertook absolute radiometry, differential radiometry, and polarimetry measurements utilizing reflected sunlight over the range 380–950 nm to evaluate methods for detecting oil spills on sea water. Maximum contrast between oil and water was observed at wavelengths less than 400 nm and greater than 600 nm, minimum contrast being in the range 450–500 nm. Oil always appeared brighter than the water, but it was not possible to distinguish one oil from another. These workers commented that differential polarization appeared to be a promising technique.

Garra and Muth[69] characterized crude, semi-refined, and refined oils by gas chromatography. Separation followed by dual-response detection (flame ionization for hydrocarbons and flame photometric detection for S-containing compounds) was used as a basis for identifying oil samples. By examination of chromatograms, it was shown that refinery oils can be artificially weathered so that the source of oil spills can be determined.

Characterization of oils. Gas chromatography–mass spectrometry

In some cases it is necessary to identify unambiguously selected components separated during the gas chromatographic examination of oil spill material. Such methods are needed from the standpoint of the enforcement of pollution control laws. The coupling of a mass spectrometer to the separated components emerging from a gas chromatographic separation column enables such positive identifications to be made.

Albaigés and Albrecht[70] propose that a series of petroleum hydrocarbons of geochemical significance (biological markers) such as C_{20}–C_{40} acyclic isoprenoids and C_{27} steranes and triterpenes are used as passive tags for the characterization of oils in the marine environment. They use mass fragmentography of samples to make evident these series of components without resorting to complex enrichment treatments. They point out that computerized gas chromatography–mass spectrometry permits multiple fingerprinting from the same gas chromatographic run. Hence rapid and effective comparisons between samples and long-term storage of the results for future examination can be carried out.

Usually, Albaigés and Albrecht[70] first deasphaltenized with n-pentane (40 volumes) prior to gas chromatography. When, however, the recovery of the branched plus cyclic alkanes was needed for subsequent analysis the saturated hydrocarbon fraction was isolated by conventional silica-gel adsorption chromatography (eluting solvent n-pentane) and refluxed in iso-octane with 5 Å molecular sieves.

The gas chromatograph (Perkin-Elmer 900) equipped with flame ionization and flame photometric detectors was operated either with 9 ft × ⅛ in. packed columns (1% Dexsil 300 on Gas-Chrom Q 100–120) from 150 to 300 °C at 6° per minute or with 200 ft × 0.02 in. capillary columns (OV-101 or Apiezon L) from 120 to 180 °C at 6° per minute. Mass-fragmentographic analyses were performed on an LKB 9000 S/PDP 11 E 10 computerized g.c.–m.s. system. The jet separator was maintained at 290 °C and spectra were recorded and disc stored at 4 second intervals.

Gas chromatographic profiles of petroleum residues, one example of which is shown in Figure 6, exhibit several characteristic features that have been applied for identification or correlation purposes.

Generally, the most apparent is the n-paraffin distribution (see the upper trace in Figure 6) that has proved to be useful in differentiating the main types of pollutant samples (crude oils, fuel oils, and tank washings)[72] or even types of crude oils,[71] although in this case the method involves the quantification of the previously isolated n-paraffins, therefore lengthening the analysis time.

Another relevant feature of the gas chromatographic profile is the acyclic isoprenoid hydrocarbon pattern that is made evident with capillary columns (see the peaks with an asterisk in the upper trace of Figure 6) or by inclusion of the saturated fraction in 5 Å molecular sieves or in urea (see the middle trace in Figure 6). The predominant peaks usually correspond to the C_{19} (pristane) and C_{20} (phytane) isomers, which ratio serves as an identification parameter,[73] although the series extends to lower and higher homologues.

Finally, the sulphur compounds that are present in minor quantities in petroleum products also exhibit a typical gas chromatographic fingerprint easily obtained by flame photometric detection (see the lower trace in Figure 6). This fingerprint has been introduced to complement the flame ionization detection chromatogram with the aim of resolving the ambiguities or increasing the reliability in the identification of the pollutants.[74]

All the above fingerprints exhibit a different usefulness for characterizing oils. Their variation between crudes and their resistance to the sea weathering processes are not enough, in many cases, for providing the unequivocal identification of the pollutant. The n-paraffins can, apparently, be removed by biodegradation as well as the lower acyclic isoprenoids at respectively slower rates.[75] On the other hand, fractions boiling at up to 300 °C can be lost by

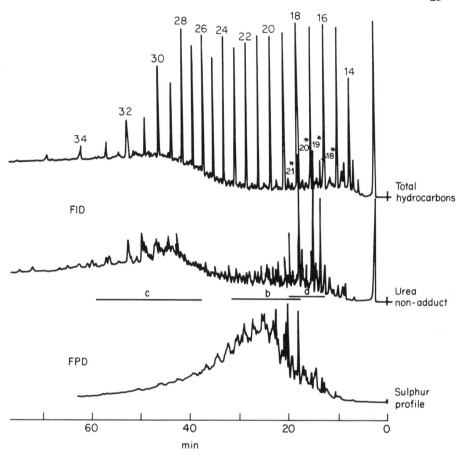

Figure 6 FID and FPD high resolution chromatograms of a crude oil residue (b.p. >220 °C). Numbers above the peaks (with or without asterisk) indicate the number of carbon atoms of the isoprenoid and the n-paraffin hydrocarbons, respectively. Reprinted with permission from Albaigés and Albrecht.[70] Copyright (1979) Gordon and Breach

evaporation affecting both the n-paraffin and the isoprenoid distributions. The flame photometric detector chromatogram is less sensitive to modifications by bacterial metabolism but can also be affected by evaporation, in spite of its higher retention range (Figure 6(b)), as will be shown later. However, the last part of the flame ionization detector chromatogram (Figure 6(c)) appears to be highly promising in overcoming these limitations. In fact, this part corresponds to a hydrocarbon fraction that boils at over 400 °C, so it cannot easily be evaporated under environmental marine conditions. Moreover, it contains a wealth of compounds of geochemical significance, namely isoparaffins and polycyclic alkanes of isoprenoid, sterane and triterpane structure,[76,77] as a result of a complete reduction of precursor isoprenyl alcohols, sterols, and triterpanes, respectively. Therefore, their occurrence and final distribution in crude oils will be related to their particular genetic history, that is to the original sedimentary organic matter and to the processes undergone during its geochemical cycle. In consequence, it was assumed by Albaigés and Albrecht[70] that these factors can provide unique hydrocarbon compositions for each crude oil, by which the unambiguous identification of the samples can be brought about. Besides their geochemical stability, these compounds do, also, remain unaltered after biodegradation[78] being, in this respect, valuable passive tags for characterizing marine pollutants.

The problem is that such components are present frequently at trace levels, as a part of very complex mixtures and can only be recognized after long and tedious enrichment treatments, that are not practical from the standpoint of the routine or monitoring analysis.

Mass fragmentography provides a satisfactory tool for obtaining specific fingerprints for classes and homologous series of compounds, resolved by gas chromatography. In addition, computerized gas chromatography–mass spectrometry allows multiple fingerprinting from the same chromatogram, that is especially important for a quick survey of any compound class in a scanty sample and permits storing the information for further processing or correlation studies. However, to carry out the analyses successfully a precise knowledge of the nature and the gas chromatographic and mass spectrometric behaviour of such compounds is needed. Albaigés et al. have done considerable work in recent years on this topic.[79–82]

In Figures 7 and 8 the proposed formation pathways of the sterane and hopane hydrocarbons, based on field and laboratory results, are shown, to illustrate the different type of compounds that can be found in crude oils depending on their particular genetic history.

In short, in the sterane family two series can be expected; the normal and rearranged steranes, the latter as a consequence of the sedimentary acidic catalytic activity.[83] Cholestane, 24-methylcholestane, and 24-ethylcholestane are generally the basic components. Variations in the stereochemistry of carbons 5, 14, and 20 have also been detected and contribute to the complex pattern exhibited by this family.[84] A similar scheme could explain the degradation of 4-methylsterols to the corresponding methylsteranes, that occur in some samples.[85,86]

The family of the hopanes (Figure 8) originate from two known procariotic source materials, the 3-desoxytriterpanes and the polyhydroxyhopanes, which, by reductive degradation, give rise to the C_{27}–C_{35} members of the series. In this case, the stereochemistry of the C-17 and C-21 in the original compounds is β(H) with one diastereomer at position 22, whilst in petroleum and matured samples, the 17α(H), 21β(H) and 17β(H), 21α(H), with the $22R + 22S$ isomers are found.[81] In addition, two C_{27} members can also be found, the α(H) and the 18α(H)-trisnorhopanes, probably formed by acid-catalysed interconversion.[85]

The general distribution patterns of both the sterane and the hopane series are completed, considering variations in side chain length of the isomeric compounds.

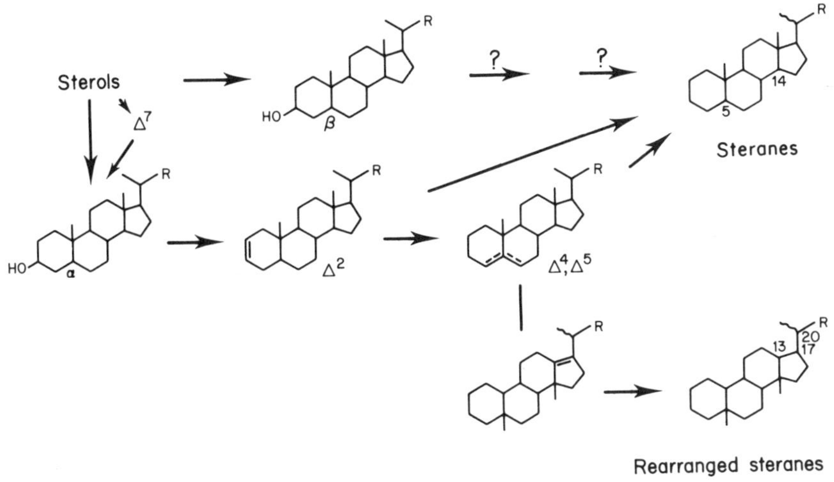

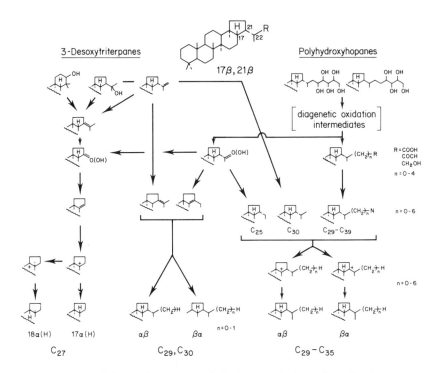

Figure 7 Early transformation routes of sterols in the geological environment, based on laboratory and field results. Reprinted with permission from Albaigés and Albrecht.[70] Copyright (1979) Gordon and Breach

Figure 8 Proposed formation routes of the hopane hydrocarbons in the geological environment, based on laboratory and field results. Reprinted with permission from Albaigés and Albrecht.[70] Copyright (1979) Gordon and Breach

Finally, the acyclic isoprenoid hydrocarbon series extends from 10 to 40 carbon atoms,[79] according to the earlier environmental occurrence of oligoterpenic alcohol derivatives, the oxydoreductive deposition conditions being the major factor influencing the ratio of the corresponding parent and norhydrocarbons derivatives found in geological samples.

The mass spectra of these hydrocarbons exhibit fragmentations that are both characteristic and intense for each family, so they fulfil the ideal requirements for mass fragmentography. The base peak for steranes is 217 (see Figure 9) unless the stereochemistry of the C-14 would be β in which case it is 218.[84] Other characteristic fragment peaks are 151 for the 5β isomers and 259 for the $13\beta(H)$, $17\alpha(H)$ rearranged steranes,[83] so the ratio of the intensities of these peaks can be used for identification purposes. Methylated steranes on ring A (e.g. on C-4) exhibit the peaks corresponding to the above fragmentations but at 14 m/e higher units, therefore the base peak appears at m/e 231 or 232, etc.

Pentacyclic triterpanes are easily distinguished by a base peak at m/e 191 (see Figure 11).[81,82] Several tricyclic diterpanes that have recently been found in petroleum show the base peak at the same nominal mass but are eluted separately from the hopane series in gas chromatography.[86]

Acyclic isoprenoids exhibit prominent ions at m/e 113 + 70n corresponding to the fragmentations induced by the regular side methyl substituents (see Figure 10). To reduce the interferences produced by other isoparaffins the m/e 183 fragmentogram has been found suitable.

The characterization of petroleum pollutants using the above heavy hydrocarbon series was attempted by the mass chromatograms of m/e 183, 191, 217, 231, and 259, which are characteristic for, respectively, acyclic isoprenoids, pentacyclic triterpanes, steranes, 4-methylsteranes, and rearranged steranes. A similar approach has been applied successfully to problems of geochemical correlation of crude oils,[85,87] and to study the chemical transformations in the biodegradation of crude oils.

Albaigés and Albrecht[70] have examined more than 50 crude oils covering most of the commonly used Middle East, African, and Venezuelan oils and those handled in the Mediterranean Sea. Not all provide significant fingerprints for each one of the ions referred, but, in spite of this, relevant differences have been noticed between them. In Figures 10, 11, and 12 are shown examples of individual profiles obtained from petroleum residues of a wide maturity range. Obviously the fingerprinting capability of the method is largely increased when combined mass fragmentographic patterns are used.

Long chain acyclic isoprenoids (Figure 10) are, in fact, rarely found in crude oils, probably because of its lower geochemical stability, that renders their occurrence at very low level concentrations, or because they are particular palaeoecological markers. Nevertheless, they have been found representative of the crude oils produced off-shore in the Mediterranean Spanish coast, oils that, on the other hand, are difficult to differentiate from some of the North African crudes for their very close values of sulphur, nickel, and vanadium concentrations and pristane/phytane ratios. Not only the occurrence but the

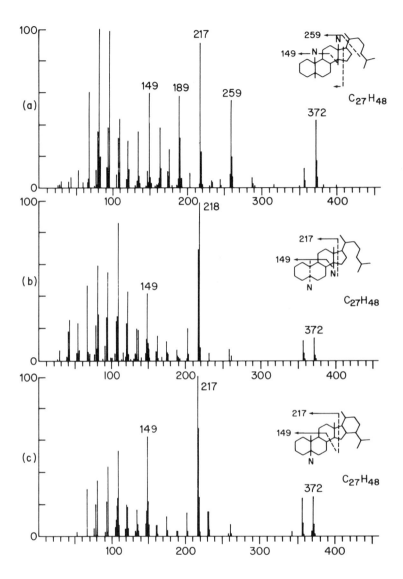

Figure 9 Mass spectra of three isomeric steranes obtained by g.c.–m.s.–c.o.m. of a crude oil extract: (a) 13β,17α-diacholestane; (b) 5α,14β-cholestane; (c) 5α-cholestane. Reprinted with permission from Albaigés and Albrecht.[70] Copyright (1979) Gordon and Breach

internal distribution of the several members of the series can be used as identification criteria for these crudes.

In contrast, pentacyclic triterpanes of the hopane type are ubiquitous in geological samples,[81] and generally constitute the most abundant family of hydrocarbons considered here. Some typical crude oil distributions are presented in Figure 11. The m/e 191 reconstructed fragmentograms show the main

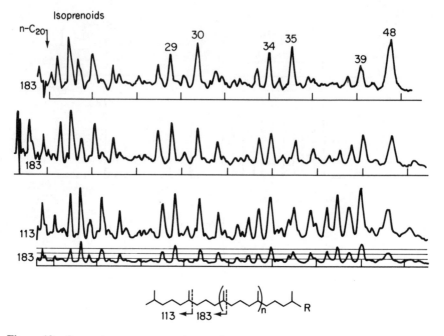

Figure 10 Computer reconstructed mass fragmentograms of long acyclic isoprenoids (m/e 113 and 183). Numbers above the peaks indicate the number of carbon atoms of the individual homologues. Reprinted with permission from Albaigés and Albrecht.[70] Copyright (1979) Gordon and Breach

17α(H), 21β(H)-hopane series whose individual members are identified by the molecular weight fragmentograms (m/e $370 + 14n$). As it has been stated before, two C_{27} isomers are present (18α(H) and 17α(H)-trisnorhopanes, in order of elution) as well as two stereoisomers at C-22 (S and R in order of elution) for each one of the homologues from 31 to 35 carbon atoms. The minor series marked with an asterisk corresponds to the moretane family (17β(H), 21α(H)-hopane) and the prominent peaks eluting before the hopane series in the bottom fragmentograms appear to be the C_{20}–C_{26} tricyclic alkylated diterpanes previously reported.[86] The identification parameters are apparent from these examples, including those that can reflect the organic source matter and the diagenetic and maturation sedimentary conditions, namely the relative concentrations of diterpane, hopane, and moretane hydrocarbon families, the ratios of individual members such as the two C_{27} isomers and the C_{29}–C_{31}, the relative distribution of the higher homologues and the general profile for other unidentified components. Some of these parameters revealed by themselves significant differences between seven Middle East crude oils in a triterpane fingerprinting technique described by Pym et al.[88] Albaigés and Albrecht[70] obtained similar profiles directly from the saturated fraction without problems in sensitivity using the gas chromatographic–mass spectrometric–computer technique. The advantages of the latter are obvious as far as the analysis time

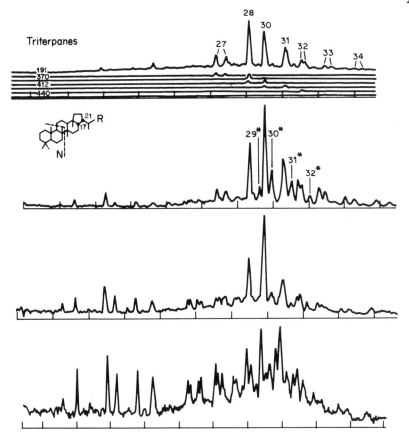

Figure 11 As Figure 10 but for triterpanes (m/e 191). Reprinted with permission from Albaigés and Albrecht.[70] Copyright (1979) Gordon and Breach

is concerned, as well as to the new possibilities offered in identification by the multiparametric profiles.

Sterane type hydrocarbons constitute the most complex family of those considered.[84] More than 70 isomers belonging to the C_{27}–C_{30} desmethyl, 4-methyl, and rearranged types have been detected by Ensminger in geological extracts.[89] They cannot be fully resolved even by high resolution gas chromatography and they elute with triterpanes in the branched + cyclic saturated fraction; however, their characteristic mass spectrometric fragmentations permit them to be easily distinguished by mass fragmentography and although many of them have not been conclusively identified, significant profiles can be obtained for fingerprinting purposes.

Figure 12 shows a sequence of m/e 217 crude oil samples. Molecular weight fragmentograms (m/e 372 + 14n) reveal, in contrast with the triterpane profiles (Figure 11), complex isomeric mixtures. Important differences in the relative distribution of the various members of the series occur depending on the origin

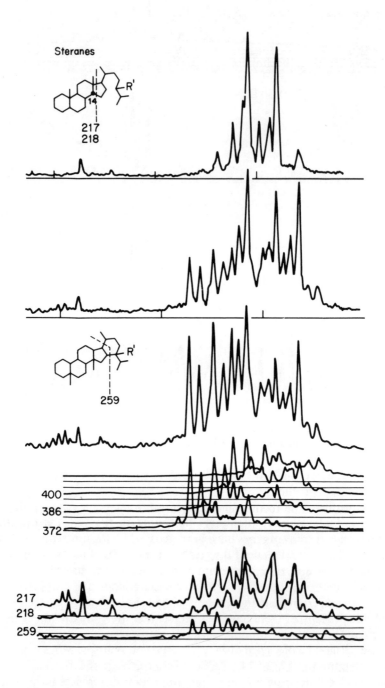

Figure 12 As Figure 10 but for steranes (*m/e* 217, 218, and 259). Reprinted with permission from Albaigés and Albrecht.[70] Copyright (1979) Gordon and Breach

and maturity of the geological sample. Hence, steranic fragmentograms used for identification are those corresponding to m/e 217, 218 and 259 (see the bottom profiles in Figure 12) and to m/e 231, 232, and 273 for the 4-methyl derivatives, respectively. Rearranged steranes (m/e 259) seem to be formed at the same time as the hopane and moretane series, with the two C-20 stereoisomers being formerly present,[89] but later on they disappear with the normal steranes (m/e 217), so they are found at trace level in the matured Middle East and North African crude oils (see Figure 13). The 4-methylsteranes (m/e 231) are generally, much less abundant than their desmethyl isomers, probably because they have originated in some restricted palaeoenvironments. Nevertheless, all of these hydrocarbons are representative of the Spanish offshore crudes, being useful markers in pollution studies from this area.

The most promising results in the application of this fingerprinting method were achieved in the identification of weathered samples where the commonly used hydrocarbon profiles have been severely modified by evaporation or biodegradation. Figure 13 displays the conventional gas chromatographic characterization of Aramco crude oil residues, obtained by distillation (b.p. > 200 °C) and laboratory simulated weathering (3 weeks). From these profiles it can be inferred that the pristane/n-C_{17} and phytane/n-C_{18} ratios are affected by selective biodegradation of linear hydrocarbons. Also, the n-paraffin distribution cannot be used satisfactorily for identification. On the other hand, the flame photometric detector chromatogram has lost, after weathering, an important part of its fingerprinting pattern and is difficult to ascertain in this sample if the relative distribution of the remaining peaks, mainly the triplet eluting on the n-eicosane that can differentiate other Middle East crude oils, have not changed with time. Mass fragmentograms display only appreciable profiles for triterpanes (m/e 191), as to be expected for the majority of the Middle East crude oils. However, they are sufficiently characteristic of the samples and no detectable changes in the fingerprint have been observed, therefore a satisfactory correlation can be brought about.

Mass fragmentography also can afford valuable fingerprints with samples whose flame ionization and flame photometric chromatograms exhibit almost featureless profiles, due to the general occurrence of those hydrocarbons in petroleum and to their environmental stability. Figure 14 shows the profiles displayed by the Venezuelan Laguna crude oil. The triterpane mass fragmentogram (m/e 191) is rather complex but the low retention diterpane series is clearly distinguished, as well as the two C_{27} isomers, that are enhanced in the molecular weight fragmentogram (m/e 370), and the C_{29}–C_{31} homologues. Moreover, the sterane profile exhibits a characteristic predominance of the m/e 218 over the m/e 217 fragmentogram, that is of 14β over the normal isomers. Cholestane and 24-methyl and ethyl derivatives are the three main components. The peak that appears at lower retention time in the fragmentogram has been identified as a pregnane isomer (m/e 288).

Hertz et al.[90] developed an integrated gas chromatographic technique for the characterization of oil spill materials. Dynamic head-space sampling and the complementary analytical techniques of gas chromatography and

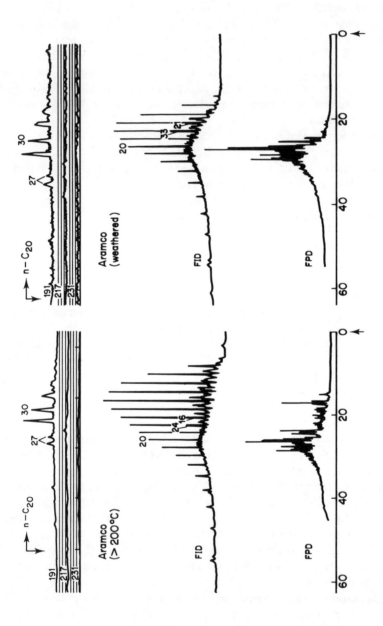

Figure 13 Gas chromatographic and mass fragmentographic fingerprints of Aramco residues obtained by distillation (b.p. >200 °C) and simulated weathering. Pristane/n-C_{17} and phytane/n-C_{18} ratios are indicated. Reprinted with permission from Albaigés and Albrecht.[70] Copyright (1979) Gordon and Breach

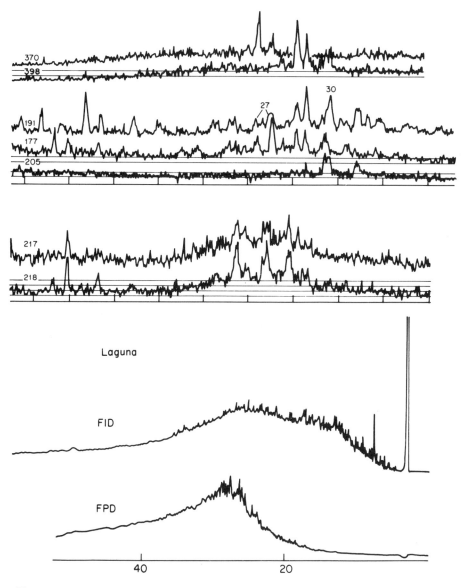

Figure 14 Gas chromatographic and mass fragmentographic characterization of Laguna crude oil (Venezuela). Reprinted with permission from Albaigés and Albrecht.[70] Copyright (1979) Gordon and Breach

coupled-column liquid chromatography are utilized for quantitation of petroleum-containing samples. Gas chromatography–mass spectrometry is employed for identification of individual components in these samples.

Dynamic head-space sampling is the use of a flowing stream of inert gas to purge a sample of volatile organic components and sweep these components onto an appropriate adsorbent.[91,92] Gas chromatography and coupled-column liquid chromatography using an internal and external standard technique were employed for quantitation of petroleum-containing samples. Hertz et al.[90] use gas chromatography–mass spectrometry for identification of individual components. The two quantitative techniques are uniquely complementary in that they utilize different molecular weight regions and compound classes (aliphatic vs. aromatic hydrocarbons) to arrive at a quantitative answer. Furthermore, an indication of the degree of weathering that has occurred in various samples may be determined by examining the n-pentadecane/n-undecane gas chromatographic peak height ratios.

Hertz et al.[90] applied their technique to samples taken at the site of a catastrophic oil spill ($>$50,000 tons of a light Saudi Arabian crude spread along $\sim$100 miles of shoreline). Samples were collected and frozen in the field. These samples along with a sample of oil taken directly from the tanker after the spill (unweathered spill oil) were analysed. Quantitation was facilitated by adding known amounts of a solution of internal standards to the weighed quantities of each sample (1 mg–100 g depending upon the oil content of the sample). The internal standard consisted of a pentane solution of naphthalene and phenanthrene each present at a known concentration (approximately 2 μgμl^{-1}). One port of each flask was connected to a prepurified nitrogen line which directs a stream of gas across the liquid interface at a flow rate of $\sim$180 ml min^{-1}. A second port contained the exit line, which is connected to a 6$\times$0.6 cm stainless steel Swagelok-fitted column packed with Tenax-GC (Applied Science Laboratories, Inc., State College, Pa) used for trapping the purged organics. This column was mounted in a cylindrical jacket which was chilled by a steady stream of cold air. Samples were purged for 4 h, first at room temperature, and then while gradually raising the temperature to 70 °C. At the end of this period, the nitrogen gas was diverted from the flasks directly to the Tenax columns for 2 h to dry the columns. The dried columns were capped and taken to the instrument laboratory for analysis. Following head-space sampling, the less volatile components were removed by pumping the liquid remaining in the flasks through a liquid chromatographic precolumn packed with Bondapak C18 (Waters Associates, Inc., Milford, Mass). The Tenax-GC column from head-space sampling was installed as a precolumn to the analytical column in the gas chromatograph. A heating block was clamped around the precolumn and heated to 375 °C with carrier gas flowing. Just prior to and during this flashing operation, a stream of liquid nitrogen was directed at the head of the analytical column, thus thermally focusing the sample on the column. The oven was temperature programmed, and data were acquired. Gas chromatographic–mass spectrometric data were acquired in an analogous fashion with mass spectra being recorded every 4 s during the course of the entire chromatogram. The Bondapak C18 columns were coupled to an analytical column packed with μBondapak C18 (microparticulate bonded phase l.c. packing material) column, and the organic constituents were elution focused onto the head of the latter using a water–acetonitrile gradient. The gradient was programmed to increase the percentage of acetonitrile in the mobile phase. The effluent from the analytical column was passed through a u.v. photometer (254 nm), and the chromatogram was recorded. Additional analyses were performed by Soxhlet extraction to obtain quantitative comparison values. Portions of each sample were Soxhlet extracted for a minimum of 6 h using diethyl ether.

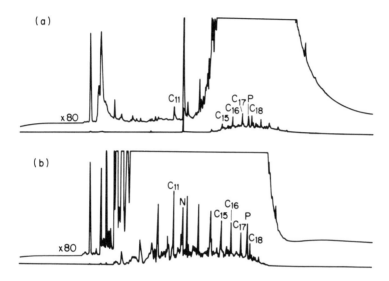

Figure 15 Gas chromatograms from head-space analysis. (a) sediment sample S-1, (b) unweathered spill oil. Peaks labelled N and P are internal standard compounds naphthalene and phenanthrene, respectively, C_{11}, C_{15}, C_{16}, C_{17}, and C_{18} are n-undecane, n-pentadecane, n-hexadecane, n-heptadecane, and n-octadecane, respectively. In both (a) and (b), upper trace is 80 times more sensitive than lower trace. Reprinted with permission from Hertz et al.[90] Copyright (1978) American Chemical Society

Gas chromatograms were quantitated by summing the peak heights of the four aliphatic peaks, n-pentadecane to n-octadecane (Figure 15) and dividing this sum by the product of the peak height of the naphthalene internal standard and the sample weight used. The value so obtained was then divided by the corresponding value from the unweathered spill oil sample. Upon multiplying by 100, the weight per cent oil (wet weight basis) in the sample was obtained (equation 1).

$$\frac{\left(\sum_{n=15}^{18} h_n\right) \cdot WC \cdot hc_{\text{NAP}}}{\left(\sum_{n=15}^{18} hc_n\right) \cdot W \cdot h_{\text{NAP}}} \cdot 100 = \text{wt\% oil in sample} \quad (1)$$

where h_n is peak height of normal aliphatic hydrocarbon of carbon number n in sample chromatogram; hc_n is corresponding peak height from unweathered spill oil sample chromatogram; h_{NAP} is peak height of naphthalene from internal standard in sample chromatogram; hc_{NAP} is corresponding naphthalene peak height in unweathered spill oil sample chromatogram; W is weight of sample analysed; Wc is weight of unweathered spill oil sample used as external standard.

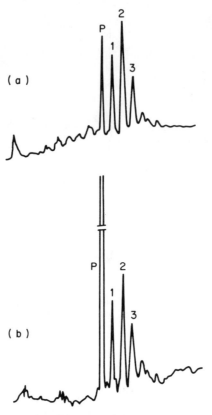

Figure 16 Liquid chromatograms from coupled-column liquid chromatography following head-space sampling. (a) sediment sample S-1, (b) unweathered spill oil. Peak labelled P is internal standard phenanthrene. 1, 2, and 3 are aromatic hydrocarbons used for quantitation. Reprinted with permission from Hertz et al.[90] Copyright (1978) American Chemical Society

An essentially analogous method was used for liquid chromatographic quantitation. The sum of the peak heights of the three aromatic hydrocarbon peaks (numbered in Figure 16) characteristic of the previously head-space sampled unweathered spill oil was referenced to the internal standard phenanthrene (equation 2). Since both the unweathered spill oil and the sample were previously head-space sampled, any loss of phenanthrene does not affect liquid chromatographic quantitation.

$$\frac{\left(\sum_{p=1}^{3} h_p\right) \cdot WC \cdot hc_{PHEN}}{\left(\sum_{p=1}^{3} hc_p\right) \cdot W \cdot h_{PHEN}} \cdot 100 = \text{wt\% oil in sample} \qquad (2)$$

where h_p is peak height of aromatic hydrocarbon of peak number p in sample chromatogram; hc_p is corresponding peak height from unweathered spill oil sample chromatogram; h_{PHEN} is peak height of phenanthrene from internal standard in sample chromatogram; hc_{PHEN} is corresponding phenanthrene peak height in unweathered spill oil sample chromatogram; W is weight of sample analysed; Wc is weight of unweathered spill oil sample used as external standard.

The results from the Karl Fischer titrations were used to correct the total ether-extractable values from Soxhlet extraction to total organic content (equation 3).

$$\left[100 - \left(\frac{W_{dry}}{W} \cdot 100\right)\right] - \%H_2O \qquad (3)$$

$= \text{wt\% extractable organic content of sample}$

where W_{dry} is weight of residue in Soxhlet thimble; W is sample weight; $\%H_2O$ is % water by weight from Karl Fischer titration.

The Karl Fischer data were also used to put the gas and liquid chromatographic results on a dry weight basis.

Table 5 contains data obtained on sediment, tissue, and oil–water emulsion samples collected by Hertz et al.[90] 6 months following the catastrophic oil spill. Figure 15 shows representative gas chromatograms of the head-space sampled sediment S-1 and the unweathered spill oil. Figure 16 presents the corresponding liquid chromatograms. A total ion chromatogram of the unweathered spill oil is presented in Figure 17. All lettered peaks were identified mass spectrometrically, and the identifications are listed in Table 6. As can be seen from Table 6, the major components in the chromatogram of the unweathered spill oil can be determined. This information is valuable in determining the source and chemical and biological consequences of the spill oil examined. Of special interest in this particular petroleum was the sulphur content as witnessed by the presence of the dibenzothiopenes.

In selecting elution peaks for the purpose of quantitation in both the gas and liquid chromatographic procedures, three criteria were of prime importance:

The peaks must be easily identified within the chromatogram. Mass spectrometry may be invaluable in this regard.

These peaks must represent compounds for which the method has high sensitivity.

These peaks must represent compounds which are not unduly affected by weathering or other compound specific losses, e.g. absorption, photodegradation.

Table 5 Hydrocarbon Content of samples obtained from catastrophic oil spill

Percent hydrocarbons by various methods[a]

Sample	Type	Soxhlet extraction corrected for water content Karl Fischer (%)	g.c. (%)	Coupled-column l.c. (%)	Weathering factor (ratio of $n-C_{15}/n-C_{11}$ peak heights)	% water by Karl Fischer
M-1	Oil–water emulsion	67(1)[b]	50 ± 20(4)[b]	80 ± 12(3)[b]	7 ± 7	25
S-1	Sediment	0.2(1)	0.2 ± 0.06(3)	0.7 ± 0.07(2)	60 ± 20	0.8
S-2	Sediment	16(1)	9 ± 3(3)	10 ± 1(2)	30 ± 10	16
S-3	Control sediment	Trace (1)	0.0002 ± 0.00004 (3)	<0.0002 (2)	NA	5
T-1	Tissue	7.7 (1)	0.03 ± 0.01 (3)	1.8 ± 0.2 (2)	NA	76
	Unweathered spill oil	...	...	...	0.6 ± 0.2	0.2

[a]See text for discussion on methods used. [b]Value in parentheses indicates number of separate determinations. Reprinted with permission from Hertz et al.[90] Copyright (1978) American Chemical Society.

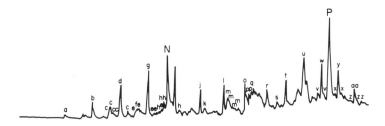

Figure 17 Total ion chromatogram of unweathered spill oil (see Table 6 for identification). Reprinted with permission from Hertz et al.[90] Copyright (1978) American Chemical Society

Table 6 G.c.–m.s. analysis of unweathered spill oil*

$a = n - C_8$	$o = n - C_{15}$
$b = n - C_9$	$p = C_7$-benzene?
$c = C_3$-benzene	$q = C_3$-naphthalene
$d = n - C_{10}$	$r = n - C_{16}$
$e = C_4$-benzene	$s = C_4$-naphthalene
$f = C_{11}$ (not n)	$t = n - C_{17}$
$g = n - C_{11}$	$u = n - C_{18}$
$h = C_5$-benzene	$v = C_1$-dibenzothiophene?
$i = n - C_{12}$	$w = n - C_{19}$
$j = n - C_{13}$	$x = C_6$-naphthalene/C_2-dibenzothiophene
$k = C_1$-naphthalene	$y = n - C_{20}$
$l = n - C_{14}$	$z = C_7$-naphthalene/C_3-dibenzothiophene
$m = C_2$-naphthalene	$aa = n - C_{21}$
$n = C_6$-benzene?	

*Computer-generated total ion chromatogram (Figure 17). Peaks labelled N and P are internal standards naphthalene and phenanthrene, respectively. Identifications followed by '?' are not definite due to incompletely resolved spectra. C_x-naphthalene, C_x-dibenzothiophene = benzene, naphthalene or dibenzothiophene systems containing substituents totalling x carbon atoms. Reprinted with permission from Hertz et al.[90] Copyright (1978) American Chemical Society.

The aliphatic compounds, n-pentadecane to n-octadecane, utilized for quantitation of the gas chromatographic results met these criteria, being readily identified (Figure 15), optimum for head-space sampling and only minimally affected by weathering. The peaks used in the liquid chromatographic analyses also met these criteria.

Liquid chromatographic and gas chromatographic quantitation were facilitated by use of an internal and external standard procedure. The internal standard (naphthalene for gas chromatography, phenanthrene for liquid chromatography) was added to both the unweathered spill oil and the several samples allowing one to correct for differing recoveries during head-space sampling.

Weathering, as defined as the loss of the lower molecular weight petroleum compounds from the sample, is indicated by the ratio n-pentadecane/n-undecane

(Figure 15 and Table 5). The individual weathering numbers (Table 5) in themselves have no absolute meaning, but in a relative sense, provide some insight into the degree of weathering that has occurred.

The generally high standard deviations of the results (standard deviations of the means) in Table 5 can be explained in terms of sample inhomogeneity and the limited amounts of the samples available for analysis.

Smith[93] classified large sets of hydrocarbon oil spectral data by computer into 'correlation sets' for individual classes of compounds. The correlation sets were then used for determining the class to which an unknown compound belongs from its mass spectral parameters. A correlation set is constructed by use of ion series summation, in which a low resolution mass spectrum is expressed as a set of numbers representing the contribution to the total ionization of each of 14 ion series. The technique is particularly valuable in the examination of results from coupled gas chromatography–mass spectrometry of complex organic mixtures. For example, an alkane fraction of a lichen extract gave a spectrogram with 24 peaks (mol. wt. range 212–464) each of which was rapidly classified, generally unambiguously.

Walker et al.[94] studied profiles of hydrocarbons in sediment according to depth in sediment cores collected at Baltimore harbour in Chesapeake Bay, Massachusetts. Gas-liquid chromatography was used to detect hydrocarbons present at different depths in the sediment, while low resolution mass spectrometry was employed to measure concentrations of paraffins, cycloparaffins, aromatics, and polynuclear aromatics. Their data show that the concentrations of total and saturated hydrocarbons decreased with increased depth, and it is commented that identification and quantitation of hydrocarbons in oil-contaminated sediments is required if the fate of these compounds in dredge spoils is to be determined.

Brown and Huffman[95] reported an investigation of the concentration and composition of non-volatile hydrocarbons in Atlantic Ocean and nearby waters. Sea-water samples were taken at depths of 1 and 10 metres and the non-volatile hydrocarbons were identified by mass spectrometric techniques. The results show that the non-volatile hydrocarbons in Atlantic and nearby waters contained aromatics at lower concentrations than would be expected if the source of the hydrocarbons were crude oil or petroleum refinery products. Hydrocarbons appeared to persist in the water to varying degrees, with the most persistent being the cycloparaffins, then isoparaffins, and finally the aromatics.

Walker et al.[96] examined several methods and solvents for use in the extraction of petroleum hydrocarbons from estuarine water and sediments, during an *in situ* study of petroleum degradation in seawater. The use of hexane, benzene, and chloroform as solvents is discussed and compared, and quantitative and qualitative differences were determined by analysis using low resolution computerized mass spectrometry. Using these data, and data obtained following the total recovery of petroleum hydrocarbons, it is concluded that benzene or benzene–methanol azeotrope are the most effective solvents.

Characterization of oils by column, paper, and thin-layer chromatography

Column chromatography

Triems[97,98] classified oils by fractionating the top residue of the oil, freed from components boiling at more than 200 °C, on silica gel with 2,2,4-trimethylpentane, benzene, and acetone as eluants. Three classes of oil were differentiated, viz. paraffinic–naphthenic, naphthenic–aromatic, and aromatic–naphthenic. Each class is subdivided into three groups according to the sulphur content.

Farrington *et al.*[99] used column chromatography and thin-layer chromatography to isolate hydrocarbons, arising from marine contamination, from fish lipids. The hydrocarbon extracts were then examined to select those that can be determined by gas chromatography–mass spectrometry, by combinations of spectrophotometric methods, or by wet chemistry. As a screening method gas chromatography was shown to be fairly accurate and precise for hydrocarbons boiling in the range 287–450 °C and of suitable polarity.

Millson[100] investigated components of sewage sludge and found elementary sulphur in the hydrocarbon fractions eluted from liquid adsorption columns. By using a solid adsorbent such as alumina, silica gel, or Florisil, and heptane as eluant, the sulphur could be separated from weakly adsorbed hydrocarbons, e.g. squalane or biphenyl, but not from more strongly adsorbed hydrocarbons such as phenyldodecane.

Done and Reid[101] applied gel permeation chromatography to the identification of crude oils and products. The technique, which appears more suited to the analysis of crude oils, is based on the separation of oil components in order of their molecular size, for practical purposes their molecular weight.

Oils were dissolved in tetrahydrofuran or toluene and each solvent resulted in a different output profile, and therefore more information for identification. The solutions were pumped through a 24 in. × ⅜ in. column, packed with 6 nm (60 Å) Styragel, and the eluant monitored by a differential refractometer. Elution time was about an hour, although this was improved by staggered injection, and 6 mg samples were required. 'Fingerprints' of the crude oils examined, although claimed to be unique for the 50 oils investigated, fall into several discrete groups. Crude oil residues (b.p. > 525 °C) were excluded by the Styragel employed. While crude oil and weathered samples were found to give similar traces, some changes were introduced by evaporation processes, and it could be difficult to differentiate between similar but weathered crude oils. Heavy fuel oils were more difficult to identify, owing to the variety of fuel oils in use.

Thin-layer and paper chromatography

One of the first reported applications of paper chromatography to the identification of petroleum pollutants was that of Schuldiner,[102] who was able to distinguish between crude oil, marine fuel oil, and crude oil sludge.

Samples were collected in wide-mouthed bottles dragged across the water surface. Ordinary white blotting paper in 3 in. × 8 in. strips was ruled off into rectangles 1.5 in. × 2.5 in., and oil pollutants and suspect samples spotted at the centre. The oils developed radially, although with very viscous oils, non-fluorescing kerosene as a solvent was necessary. Alternatively, heat reduced the viscosity and allowed radial development. For comparison of oils the samples must be treated side-by-side in an identical manner. The chromatograms were evaluated under ultraviolet light, which principally reveals polynuclear aromatics and heterocyclics; characteristic consecutive ring formation, scalloped effects, and ray formation were noted and compared. Crude residual and distillate oils are said to give distinctive patterns, while highly refined products, such as liquid paraffin, have little fluorescence. A classification of the type of oil seems difficult on the reported results, since, on occasions, different products appear to give very similar chromatograms. Changes in viscosity of the pollutant oil due to weathering could affect the chromatogram.

A variation of paper chromatography occasionally referred to as evaporation chromatography was employed by Herd[103] to compare fuel oils suspected of being identical. Paper strips were immersed in an ether solution of the oil and the ether allowed to evaporate, usually overnight. Characteristic bands were revealed under ultraviolet light.

Three varieties of paper chromatography have been described and applied to the identification of petroleum materials in estuarine waters. A 1 cm diameter circle was drawn on chromatographic grade filter paper, and a drop of the sample placed at the centre. Viscous oils often required warming to enhance spreading, and in general this technique is similar to that of Schuldiner[102] described above. The second technique is claimed to have merit for comparing petroleum products. Three concentric circles of 1, 3, and 7 cm diameter were drawn on a 10 cm square of paper and a sample of the oil spotted on the centre circle. A suitable solvent, e.g. benzene, was added slowly until it reached the 3 cm circle, after which the solvent was evaporated. Further development to the 7 cm circle using a more polar solvent, e.g. methanol, gave different effects. More information was achieved using an entirely different solvent system. In this, and the following technique, chromatograms were examined under visible and ultraviolet light. An ascending paper chromatogram was also employed with some success. Samples were placed on a paper sheet 8 in. × 8 in., 1 in. apart and ¾ in. in from one edge and developed in n-heptane, which required about 1–2 hours to rise about 7 in. After evaporation of the solvent, colours, shapes, areas, and distances of travel were compared. If necessary the paper was rechromatographed in a different solvent. Chromatograms from the ascending technique appear easier to compare, but although lubricating oil is markedly different from marine fuel oils, differentiation between more similar products could seldom be confidently achieved.

Thin-layer chromatography has replaced paper chromatography in many applications because of its greater versatility and reproducibility. It has been applied as a rapid method of classifying petroleum and natural and synthetic oils by chromatographing 0.1–1 mg quantities of sample oils.[104] These were spotted onto a silica gel plate and developed with an ascending solvent composed of 70/30 vol/vol chloroform/benzene. After 1 h the solvent was evaporated, and the resulting chromatogram evaluated in ultraviolet light (366 nm) and the fluorescent areas noted. This was followed by spraying the chromatogram with concentrated sulphuric acid and baking it at 120 °C for 15 minutes. Coloured zones formed under visible and ultraviolet light were noted. With this procedure, oils were classified into four groups:

(1) hydrocarbon oils (petroleum products, for example);
(2) synthetic ester oils;
(3) naturally occurring oils;
(4) oils of different composition (silicone oil and low molecular weight polyethylene oxides, for example).

The relatively polar solvent, compared to, say, hexane, provides little separation of the hydrocarbon components of petroleum products, and therefore while they can be distinguished from the more polar oils investigated, insufficient information is obtained for differentiation between petroleum products.

Krieger[105] employed horizontal silica gel plates and hexane as developing solvent. This resulted in an approximately radial chromatogram which was visualized by spraying with 0.03% fluorescein solution and viewing under ultraviolet light. Crude oil and heavy fuel oil, light petroleum, various tar oils, road tar, and olive oil were examined by the technique and some distinction made between them. Natural oil, due to its more polar nature, was not developed, while the remaining products gave generally similar patterns, although in some cases it was possible to match chromatograms of sample and suspect oils.

Normal hexane, a relatively non-polar solvent, was adopted by Lambert[106] to produce more separation between petroleum components after ascending development on a silica gel plate. As many as 8–10 samples were chromatographed on a single plate during about 30 minutes, and identification was achieved by comparison of pollutants with standard petroleum materials. Motor oil, light and heavy fuel oil, various greases, and olive oil were among many oils investigated, initially under ultraviolet light and then after spraying with concentrated sulphuric acid/formaldehyde solution. Well defined coloured zones were obtained with these products with reasonable differentiation between most of the oils examined, although light and heavy fuel oil could be confused. The addition of small amounts of polar solvents, for example pyridine (2–5%), diethyl ether (1%), or acetic acid (0.04–1%) to n-hexane or cyclohexane, is claimed to give improved separation, although in similar work the improvement was slight.[107] The incorporation of various substances in the silica layer, for example picric acid, urea, dimethylformamide, silver nitrate, trinitrobenzene, fluorescein, and caffeine, was studied, but only with the latter three substances was there any improvement. For separation of the components of the more polar natural oils and greases, a more polar solvent, 90/10/2 n-hexane/ethyl acetate/acetic acid, was recommended. Under these conditions, hydrocarbons travel essentially with the solvent front.

An extension of Lambert's technique has been achieved,[107] in which some commercially available petroleum fractions and greases were examined. Gas oil, motor oil, light fuel oil, motor spirit, paraffin, hydrant grease, submersible pump grease, and bitumastic grease were chromatographed on silica gel plates using petroleum ether b.p. 60–80 °C as developing solvent. Chromatograms were evaluated initially under ultraviolet light and chraracteristic colours and patterns noted. Further detail was collected by spraying the plates with 1% formalin in concentrated sulphuric acid and baking them at 60 °C for 16 hours. Examination under visible and ultraviolet light revealed differently coloured areas. Generally, these visualization conditions were sufficient for differentiation, but if there was some ambiguity more information could be obtained from further two-dimensional thin-layer chromatography, in which the plate is rechromatographed after turning it through 90 °C, using a more polar solvent, e.g. 1/9 diethyl ether/petroleum ether b.p. 60–80 °C. In general, less characteristic chromatograms were obtained from the lighter petroleum products than from the heavier products.

Matthews states that thin-layer chromatography is highly suited to the rapid identification of the heavier petroleum and coal tar oils, and their residues,[108] and has developed a scheme of systematic qualitative analysis.[109] The scheme is based on the varying absorption of oils on different solid supports, for example, silica gel, alumina, and kieselguhr G, and on the use of solvents of varying polarity, such as petroleum ether b.p. 40–60 °C, ethanol, toluene, and chloroform. Ascending thin-layer chromatography was adopted for the scheme, the initial step being separation on alumina and development with acetone. Further thin-layer chromatography over an alternative solid support, or development with a solvent of different polarity follows, depending on the distance travelled by the oil (R_f value) and its fluorescence under ultraviolet light (350 nm). In principle, material should be classified as to type from R_f values and its fluorescent patterns, but the author suggests that in view of small variations encountered in R_f values, standards should be used. However, the technique is still of value for comparing pollutant oil with a sample from its potential source. Materials were categorized as:

(1) heavy lubricating oils and greases,
(2) engineering oils and fuels (a wide range of products from kerosene to light lubricating oil),
(3) coal tar products, for example crude tar, anthracene oil, creosotes and pitch,
(4) residual and crude petroleum products,
(5) other fluorescing oils, for example certain natural oils, and finally
(6) non-fluorescing oils, for example liquid paraffin and vegetable oils.

The Matthews[108] procedure is outlined below.

Methods

Materials

Solvents: Petroleum ether 40–60 °C (E(dielectric const. at 20 °C) = 1.9; acetone (E = 20.7); ethanol (industrial methylated spirit 74 over proof) (E = 24.3); toluene and chloroform.

Solid phase: 3×20 and 10×20 cm plates of 250 μm layers using
  (a) 30 g kieselguhr G and 60 ml water slurry,
  (b) 40 g alumina T neutral and 60 ml water slurry,
  (c) 30 g silica gel T and 60 ml water slurry.

All dried overnight. To activate the alumina T, the dry plate is placed in an oven at 105 °C for 30 minutes and allowed to cool in a desiccator. It should be used immediately. $R_t \times 100$ values tended to vary with distance travelled, a standard distance of 5 cm was therefore adopted. To effect sufficient $R_t \times 100$ differences in class 1 oils it was found necessary in the tertiary system to use 8 cm silica gel G. The ascending technique was used, using fresh solvents. The chromatograms were viewed under ultraviolet radiation of 350 nm (and 254 nm). As this kind of thin-layer chromatography seems to give small $R_t \times 100$ variations between plate batches, the use of standards is strongly recommended.

A start line is marked each side of plate, about 2 cm from plate edge. A 0.5 cm score is made across the plate, so that the solvent will ascend 5 cm (8 cm in the case of silica gel) before stopping. A drop of extracted oil is spotted with a melting point tube. Extremely viscous oils (e.g. heavy gear oils) should be warmed or a strong chloroform solution should be used, spotting a little at a time. Weak chloroform solutions lead to $R_t \times 100$ variation and this is not desirable. For the band technique, a toluene solution is prepared (1 part by weight of sample + 10 parts by volume of toluene). Five spots are spotted on to give a band about 1.5 cm wide. It is important that the toluene or chloroform has evaporated before proceeding. Appropriate standards are used in the secondary and tertiary systems, being treated in the same way as the sample.

The procedure given in Table 7 is then followed, fresh solvent being allowed to ascend. The result is then examined under ultraviolet light at 350 nm (and 254 nm) being interpreted by Table 7.

Where results differ slightly from the information given in the tables, an assessment should be made of the nature of the oil based on the principles of Table 7. Clean-up procedures and separation techniques may now be applied. The thin-layer chromatography procedure is then repeated on the oil product from such procedures. Differences in the patterns obtained by the two procedures lead to identification of various oils present.

Rubelt[110] used thin-layer chromatography to identify petroleum distillates, particularly diesel and heating oil. Separation was performed on silica gel plates using n-hexane as developing solvent. It was found that spraying with 0.03% aqueous fluorescein solution and viewing under ultraviolet light produced similar chromatograms in the case of diesel and heating oil, but spraying with

Table 7 Systematic scheme for differentiation of oils. Fluorescence at 350 nm

System 1	Result	System 2	Result	System 3	Result	Conclusion
	Class I $R_t \times 100 = 0$ Blue streak to front	(a) Take up in pet.ether (b) Flame test	(1)(a) Partially insoluble (b) Flame test probably positive			Lubricating greases
			(2)(a) Soluble (b) Flame test negative	Silica gel G 8 cm/acetone: pet.ether 40–60 °C 4:1	$R_t \times 100 = 0$ $R_t \times 100 > 0$	Petroleums. Heavy lube oils, e.g. motor oils. As R_t decreases oil viscosity increases
	Class II $R_t \times 100 > 0$ Various shades of blue		(a) $R_t \times 100$ low (b) $R_t \times 100$ medium	Use a system specific for a specific oil	Specific for the oil	(a) Heavy engineering oils (b) Light engineering oils
		Alumina T/ethanol (I.M.S.)	(c) $R_t \times 100 = 60$–70 long pale blue spot	Use g.c. if further differentiation required		Diesel fuel oils
			(d) Diffuse at solvent front very pale blue	Use g.c. if further differentiation required		Solvent fuels
	Special blue-green diffuse $R_t \times 100 = 70$–100		Pale green colour $R_t \times 100 = 20$ blue-green-brown head diffuse at at front	Band technique activated alumina T/pet.ether 40–60 °C	Banded	Anthracene oil

Alumina T/acetone		Band technique activated alumina T/pet.ether 40–60 °C		Band technique Alumina T/acetone		
	Class III orange or brown head at front and brown streak	Banded most colours present		Blue band becomes increasingly yellow. Orange head becomes increasingly red with refinement	Crude coal tar	
					Refined coal tar	
					Coal tar pitch	
		Banded			Coal tar (characterized by cold sequence)	
	Class IV yellow brown streak and head sometimes a little blue	Band technique kieselguhr G/acetone	Dull yellow-brown head		Lake asphalts	
			Bright yellow band		Bitumens	
			Blue-yellow head virtually no band		Crude oils	
			Blue-yellow head and band	Band technique alumina T/acetone	Yellow increases blue and purple decrease in band and head as oil gets heavier	Light fuel oil
						Middle fuel oil
						Heavy fuel oil
						Other fluorescing oils
	Class V Characteristics do not fit standard patterns			Special techniques		
	Class VI No fluorescences at 356 or 254 nm				Non-fluorescing oils	

Reprinted with permission from Matthews[109] Copyright (1970) Blackwell Scientific Publications Ltd.

conc. sulphuric acid/formaldehyde reagent allowed more differentiation. The paraffins, which travelled with the solvent front, were scraped off, eluted, and rechromatographed on a silica gel plate impregnated with urea. Normal paraffins formed inclusion compounds with urea and were retained more than the branched and cyclic compounds which travel with the solvent front. Although the separation was not entirely quantitative, the n-paraffins could be eluted from the layer after the complex had been destroyed with hot water or carbon disulphide, the n-paraffins could be identified by gas chromatography. A similar technique has been employed in the study of naturally occurring saturated hydrocarbons.[111]

Silica gel in conjunction with cyclohexane as solvent[112] was found to be the most suitable absorbent for the separation of saturated from aromatic hydrocarbons in the gas oil to lubricating oil range. Better separation was claimed if the cyclohexane is saturated with less than 0.5% dimethyl sulphoxide. Under the conditions used, paraffins travelled with the solvent front, followed respectively by the monoaromatic, diaromatics, polynuclear aromatics, and polar substances, which generally remained on the starting point. Chromatograms can be revealed by ultraviolet light in the normal manner to show aromatic and heterocyclic compounds, and then spraying with dichromate/sulphuric acid reagent to reveal all components, including saturated hydrocarbons.[112] Rhodamine B in fluorescein gives an absorbing spot on a fluorescing background under ultraviolet light at 350 or 254 nm. A solution of phosphorus pentachloride in carbon tetrachloride gave some classification of hydrocarbon type in lubricating oils in the following manner:[113]

(1) Paraffins   light brown
(2) Monoaromatics   wine red
(3) Diaromatics   dark blue-green
(4) Polar aromatics   dark brown

Lubricating oils, particularly used motor oil, are difficult to identify by most techniques. A modern lubricant could contain a variety of additives in addition to the base oil, for example an antioxidant, a detergent inhibitor, an antiwear agent, an ashless dispersant, a viscosity index improver, a pour-point depressant and a rust inhibitor. Therefore thin-layer chromatography of the additives, rather than the base oils, could be a more specific method of identification.[109,110] With used oils the degraded additives could be expected to be even more specific when comparing pollutant and suspect oils, since they are dependent on the use of the oil.

Sauer and Fitzgerald[114] have described thin-layer chromatographic technique for the identification of water-borne petroleum oils. Aromatic and polar compounds are removed from the sample by liquid/liquid extraction with acidified methanol, the extract is chromatographed on a silica gel thin-layer plate, and the separated components are detected by their fluorescence under long- and short-wave ultraviolet light. Unsaturated non-fluorescing compounds are detected by iodine staining.

The preliminary work of Sauer and Fitzgerald[114] indicated that the resolving power of thin-layer chromatography to separate fluorescing compounds found in oils was poor when the oil, or a dilution thereof, was spotted directly on to the plate; presumably due to the overwhelming complexity of the mixture of compounds generally present in oils. Furthermore, it revealed that only those compounds capable of at least some resistance to chemical alteration due to weathering should be considered. In addition, a simple visualization process such as black light illumination was considered more desirable than more involved techniques such as sulphuric acid charring. For these reasons Sauer and Fitzgerald[114] developed a preliminary fractionation of the starting material: polycyclic aromatic compounds, hetero and polar compounds (N, O, and S containing) were isolated from oils and chromatographed. Although these compound types have some solubility in water (and could diffuse into the aqueous layer from an oil spill on water)[115] and are subject to oxidative loss of fluorescence, they have provided a satisfactory thin-layer chromatographic fingerprint of oils. This method is described below.

Method

One ml of oil sample is placed into a 15 ml centrifuge tube with 1 g of colour-tagged anhydrous calcium sulphate. The mixture is then shaken on a vortex mixer for 1 min. If the colour indicator is not visible at the end of 1 min, an additional gram of calcium sulphate is added and the sample is remixed. One ml of 0.4% glacial acetic acid/methanol (by volume) is added, and the mixture is again shaken for 3 min or until well emulsified. The mixture is then centrifuged for 10 min at 2500 rpm. The methanolic phase is withdrawn and placed in a stoppered 5 ml glass vial. If the sample is a light distillate oil (gasoline, kerosene, jet fuel of a No. 1 fuel oil), it is necessary to concentrate the methanol extract. This is accomplished by placing the vial containing the extract on a hot plate at 65 °C with a Pasteur pipette suspended about 1 cm above the liquid surface. A vacuum is applied to the pipette until fine oil droplets precipitate out of solution.

Five microlitres of the methanolic extract is spotted along the lower edge of an unactivated 20×20 cm t.l.c. plate (Adsorbosil 1 or 5, Applied Science Laboratory). The plate is developed horizontally in an unsaturated atmosphere. To achieve reproducibility of R_f values from laboratory to laboratory, it is necessary to standardize the water content of the absorbent. This is done by placing the plate adsorbent side down on the developing chamber and allowing it to precondition for 45 min over a glass tray containing 65% sulphuric acid/water (by volume). This results in uniformly deactivated plates. Preconditioning is stopped by sliding a stainless steel sheet between the sulphuric acid tray and the face of the plate. The sulphuric acid tray is then removed and replaced with a tray containing 10% acetic acid/methanol (by volume). An area 20 cm wide and 2 cm high on the plate (located just above the spotted sample) is preconditioned by exposing the thin-layer plate to the methanol vapour

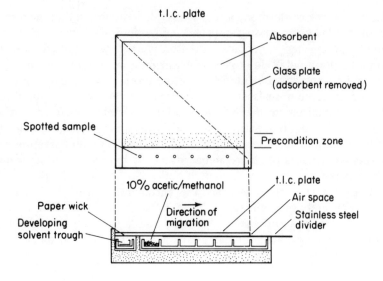

Figure 18 Longitudinal section of developing chamber. Reprinted with permission from Sauer and Fitzgerald.[114] Copyright (1976) American Chemical Society

by sliding out the stainless steel divider 2 cm (Figure 18). After 5 min the methanolic conditioning is stopped. This procedure establishes a methanol deactivated zone 2 cm wide across the bottom of the plate.

The solvent reservoir is filled with 4% methanol/hexane (by volume), and the development is started when the solvent wick is brought in contact with the adsorbent. Development is allowed to continue for 45 min (about 15 cm solvent migration). The plate is then removed from the developing chamber and allowed to air dry before it is visualized under long (365 nm) and short (254 nm) wavelength ultraviolet light. After u.v. light inspection the plate is exposed to iodine vapour in an airtight bell jar and then reinspected. The chromatograms were photographed with Kodacrome 25 colour film in a Pentax 35 mm single-lens reflex camera. The fluorescence on the plate was recorded using multiple exposure times 1.5, 2.0 and 2.5 s at f-1.4. After iodine staining, the chromatograms are rephotographed using transmitted daylight and adjusting the f-stop and shutter speed setting accordingly.

This method was used primarily as a means of assessing a match or mismatch between an oil spill sample and suspect oil samples solely by visual examination of their thin-layer chromatograms under u.v. light and iodine staining. Generally, black light inspection yields more information than iodine staining, and this sole visualization technique is generally conclusive. However, some oils, such as hydraulic fluids and some lubricating oils, contain an appreciably lower aromatic hydrocarbon content, while generally possessing unsaturated stabilizers and antioxidants.[116] Visualization of these non-fluorescing additives is accomplished with the iodine staining technique.

The acidified methanol extract of petroleum oils contains not only light aromatic hydrocarbons and unsaturated, non-fluorescing additives, but also very heavy, polar compounds. As a result, the use of a relatively non-polar developing solvent leaves many polar fluorescing compounds at the starting position, yet provides good separation of the lighter (1-3 fused ring) aromatics. Conversely, the use of a strong eluting solvent provides good separation of all the heavy polar components but runs all the light and medium aromatics together in a brightly fluorescing band at the solvent front. This situation necessitated the use of the methanolic preconditioning gradient across the face of this plate. Furthermore, because all components of the extract are soluble in the preconditioning solution, development of the plate begins with the entire sample migrating quickly across the methanol preconditioned zone and concentrating in a narrow band, rather than a circular spot, at the far edge of this zone. The components which separate out on the plate therefore retain this band shape; consequently, close-running compounds are much more highly resolved. However, actual chromatographic separation begins to occur only at the far edge of the methanol preconditioned zone where a steep methanol gradient in the adsorbent layer has been established (Figure 19).

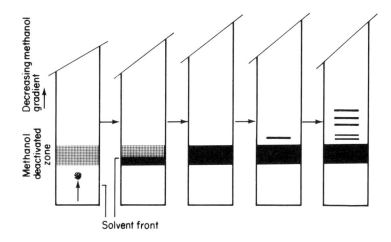

Figure 19 Sequence of sample development utilizing vapour preconditioning and preconditioning zone. Reprinted with permission from Sauer and Fitzgerald.[114] Copyright (1976) American Chemical Society

Brown et al.[117] have described a rapid field method for detecting down to 2 μg of oil in sediments associated with marine oil spills. The method was employed in connection with the Argo Merchant oil spill off Nantuckett in December 1976.

Method

Apparatus and reagents

Ultraviolet light (either long, short or mixed wavelength u.v.) is satisfactory; 30 ml glass beaker with pouring spout; 5 ml glass beaker for measuring sediment sample; 2 ml pipette; 5 ml glass vials with metal or foil-lined caps; Corning glass disposable micro-sampling pipettes (25 μl); 30 ml jar (for developing chamber) fitted with metal or foil-lined cap; 7.5×2.5 cm strips of I.T.L.C., type SA chromatographic paper; pesticide quality hexane; developing solvent composed of 35% petroleum ether (certified A.C.S., boiling range 38.3–48.8 °C) and 65% benzene (certified A.C.S., boiling point 80.1 °C); sodium sulphate (anhydrous, granular, Reagent, A.C.S.); hexane containing a known quantity of oil (hexane containing 160μg of Empire Mix crude oil per ml will suffice).

Procedure

Carefully fill the 5 ml sample-measuring beaker with sediment, (pour off excess water and replace with sediment). With a spatula, transfer the sediment sample to the 30 ml beaker. Add approximately 1 g of sodium sulphate to the sediment sample and mix with the spatula. Pipette 2 ml of hexane into the 5 ml sample-measuring beaker, stir with the spatula and pour into the 30 ml beaker. Mix sediment and hexane thoroughly for 1 min (care should be taken to treat every sample exactly the same in this step). Pour the hexane into the 5 ml vial and allow the hexane volume to reduce approximately 0.5 ml. (Note: approximately 1.5 ml of hexane should be recovered after extraction.) Using a micro-sampling pipette, place 25 μl of the concentrated hexane extract on the active side of the I.T.L.C., type SA chromatographic paper strip, approximately 1.5 cm above the bottom of the paper. Allow the spot to dry thoroughly. Hold the paper strip (with the sample spot down) in the developing jar such that the paper is approximately 1 cm below the developing solvent (35% petroleum ether/65% benzene) for 45–60 s. (This should allow the front to ascend approximately 2.5 cm.) Remove the strip from the developing solvent, allow the strip to dry and view under ultraviolet light. Repeat the above steps with a solution of hexane containing a known concentration of oil. (This will serve as a standard for comparing to unknown samples.)

Interpretation of results

The presence of a blue fluorescent spot on the developed chromatogram is indicative of the presence of oil. The greater the intensity of the fluorescence, the greater the quantity of oil. Comparison of intensity of fluorescence between samples will aid in identifying areas containing the greatest concentrations of oil.

Whittle[118] has described a thin-layer chromatographic method for the identification of hydrocarbon marker dyes in oil polluted waters.

Characterization of oils by infrared and Raman spectroscopy

In the 7–11 μm 'fingerprint' region, C–O absorptions in particular can occur in petroleum products due to contamination by naturally occurring organic and hydrocarbon degradation products. These absorptions can seriously interfere, and unlike gas chromatography which is a purification procedure in itself, a prior clean-up is often required. In contrast to gas chromatography, infrared analysis gives little indication of the boiling range of the sample, and therefore a rapid classification of the type of material is not possible unless much experience has been obtained. The spectrum reveals the presence and relative amounts of groups of compounds, such as aliphatic, olefinic, and aromatic hydrocarbons in petroleum products. For example, it is possible to differentiate between premium and regular grade paraffin by examination of the increased absorptions at 12.43 and 6.23 μm in regular grade paraffin owing to higher aromatic hydrocarbon content.[119]

Rosen[120] used infrared spectroscopy to gain qualitative information on petroleum refinery wastes causing taste and odour problems in receiving surface waters. The organic matter in solution was collected on activated carbon from large volumes of water (up to 20,000 gal). Organic material was recovered and separated into neutral, basic and acidic fractions. The neutral fraction was further separated by elution from a silica gel column into aliphatic, aromatic, and oxygenated fractions. Spectra prepared between 5 and 10 μm allowed a tentative identification of refinery waste. A similar technique was employed by Voege[121] who established the presence of light lubricating oil, alkyd resin, and aliphatic fatty acids in a pollution sample. In both procedures the evidence tends to be presumptive rather than conclusive, and since the method is very time-consuming its application is limited. The efficient removal of oxygenated matter using a chromatographic column appears to be a critical factor. Unfortunately it is not easy to decide when this pretreatment has been successful with unknown oils.

Gross pollution is easier to investigate by infrared spectroscopy and it is claimed to give more rewarding results, since a simple liquid-liquid extraction followed by drying can be sufficient pretreatment.[121] Although many articles recommend the use of infrared analysis for identification of organic pollutants, few deal in detail with petroleum oils. Cole[122] used infrared in conjunction with gas chromatography to indentify steam-cracked naphtha and powerformate occurring in refinery waste waters. Gas chromatography was often found to be insufficient for these materials, but infrared analysis immediately revealed the higher olefin content of the steam-cracked naphtha, i.e. absorption bands at 6.0 and 11 μm. The aromatic content, indicated by bands at 6.2, 12, and 15 μm was also distinctive. Fuel oil was identified by comparison of pollutant and suspect oils and shown to be distinguishable from motor oil by examination of spectra recorded between 2.5 and 15 μm. The authors concluded that infrared spectroscopy is not as specific as might be expected, and supporting data is required.[123]

A more systematic approach is that of Kawahara.[124-127] The method is more applicable to essentially non-volatile petroleum products, and the author states that useful information cannot be obtained from infrared spectroscopy applied to volatile products such as naphthas and gasolines, unless on-the-spot collection of the sample is made immediately following a spill.[124] Heavy residual fuel oils and asphalts are not amenable to gas chromatography and give similar infrared spectra. However, a differentiation can be made by comparing certain absorption intensities.[126] Samples were extracted with chloroform, filtered, dried, and the solvent evaporated off at 100 °C for a few minutes using an infrared lamp. A rock salt smear was prepared from the residue in a little chloroform, and the final traces of solvent removed using the infrared lamp. The method, which in effect compares the paraffinic and aromatic nature of the sample, involves calculation of the following absorption intensity ratios:

$$\frac{13.88\,\mu m \text{ polymethylene chain}}{7.27\,\mu m \text{ methyl groups}} \qquad \frac{6.25\,\mu m \text{ aromatic C-C}}{7.27\,\mu m \text{ methyl groups}}$$

$$\frac{3.28\,\mu m \text{ aromatic C-H}}{3.42\,\mu m \text{ aliphatic C-H}} \qquad \frac{12.34\,\mu m \text{ aromatic rings}}{13.88\,\mu m \text{ polymethylene chain}}$$

$$\frac{12.34\,\mu m \text{ aromatic rings}}{7.27\,\mu m \text{ methyl groups}} \qquad \frac{6.25\,\mu m \text{ aromatic C-C}}{13.88\,\mu m \text{ polymethylene chain}}$$

Peaks observed at 5.90 μm and 8.70 μm were thought to reflect oxidative effects on the asphaltic material, while asphaltic sulphoxide and sulphone were tentatively inferred by bands at 9.76, 8.66, and 7.72 μm. The $\frac{12.34}{13.88}$, $\frac{12.34}{7.27}$, and $\frac{6.25}{13.88}$ μm ratios tended to show the greatest difference between different samples. When the ratios $\frac{12.34}{7.27}$ versus $\frac{810}{720}$ μm were plotted graphically, the intermediate fuel oils behaved similarly.[127] Weathering caused fuel oils to fall below the curve although with asphalts the effect was significantly less. Since no prior purification was employed the method relies on an uncontaminated, unweathered sample of oil being available. Mattson and Mark[128] reported some criticism of Kawahara's technique. They claim that evaporation of the solvent chloroform by infrared heating removes volatiles and causes large changes in the ratios. An oil sample was shown to suffer such alteration by the infrared source during repeated analysis. The absorption of all bands decreased non-uniformly between 20 and 100% over a period of 30 minutes. They propose the application of internal reflection spectrometry as a rapid, direct, qualitative technique requiring no sample retreatment. Unlike infrared spectroscopy which involves transmission of infrared radiation through a homogeneous transparent solution, infrared reflection spectrometry is based on attenuated reflection of the infrared radiation between the sample layer and a suitable, highly polished crystal surface. The attenuation of the radiation passing through a cell is monitored and recorded in the same manner as transmission infrared spectroscopy, although the resulting spectrum shows absorptions at slightly shifted wavelengths.[129] In contrast to infrared spectroscopy there is no decrease in relative sensitivity in the lower energy region of the spectrum, and since no solvent is required, no part of the spectrum contains solvent absorptions. Oil samples contaminated with sand, sediment, and other solid substances have been analysed directly, after being placed between 0.5 mm 23 reflection crystals. Crude oils, which were relatively uncontaminated and needed less sensitivity, were smeared on a 2 mm 5 reflection crystal. The technique has been used to differentiate between crude oils from natural marine seepage, and accidental leaks from a drilling platform. The technique overcomes some of the faults of infrared spectroscopy, but is still affected by weathering and contamination of samples by other organic matter. The authors suggested the following absorption bands as being important in petroleum product identification:

3.23–2.78 μm	water
3.28	aromatic CH
3.39	–CH_3
3.42	>CH_2
3.51	>CH_2
5.88	>C=O
6.25	aromatic C–C
6.90 μm	>CH_2
7.27	–CH_3
9.71	>S=O, PO_4?
11.63	aromatic CH
13.60	aromatic CH
13.98	long chain–CH_2–

Kawahara (Kawahara and Ballinger,[130] Kawahara[131]) has used his method to characterize a number of known and unknown petroleum samples. All of these studies used the normal transmission method to obtain infrared spectra; however, the feasibility of using internal reflection to obtain infrared spectra has been demonstrated by several groups (Mattson et al.,[132] Mark et al.,[133] and Baier[134]). The advantage of the latter method is that chemical extraction of petroleum from such as sand and water is unnecessary.

Mattson[135] has also reported on investigations of samples from 40 oil sources and has shown that there exist sufficient differences between the fingerprints by infrared spectroscopy of oils from various sources such as oil spills for the technique to be used to identify them.

Powell et al.,[136] has described a near infrared method for the determination of total hydrogen bonded to carbon which he considers should be applicable to the characterization of oil in polluted river waters. The demonstration is based on the integrated absorption of the first overtone of the C–H stretching band at 1680–1785 nm, which is rectilinearly related to the concentration of C-bonded H for the six hydrocarbons studied (as 0.01 M to 0.1 M solution in carbon tetrachloride).

Pierre[137] has reported an exploratory study of the characterization of surface oil slicks by infrared reflectance spectroscopy. A double beam spectrophotometer was modified for studying the reflectance spectra (at angles of incidence of 45°, 60°, and 70°) of oil layers (20–30 μm thick) on the surface of water using pure water as reference.

Ahmadjian and Brown[138] have used laser raman spectroscopy to identify petroleums.

Lynch and Brown[139] have obtained infrared spectra for over 50 samples of crude oils, fuel oils, and other petroleum products. These were characteristic of each sample. They showed that computer analysis of absorptivities can be used to match unknown samples with those already known. Lynch and Brown are of the opinion that with appropriate data analysis infrared spectroscopy can be used to identify the type and source of a reasonably large number of petroleum samples.

The method of Lynch and Brown[139] is an extension of the ones suggested by Kawahara[140] and Mattson;[141] however, they used additional low frequency bands and higher instrument resolution. Furthermore, they developed a new method for numerical analysis of the spectral data, which utilizes a digital computer to match an unknown petroleum sample with the correct known. These workers used a high resolution instrument (Perkin Elmer Model 521). A complete spectrum from 600 to 4000 cm^{-1} of a typical sample of crude oil from Ecuador is shown in Figure 20. The sample thickness was 0.1 mm and, at this thickness, the strong hydrocarbon bands at 1375, ~1450, and 2800–3000 cm^{-1} are completely absorbing. These bands are similar in the spectra of all of the samples of hydrocarbons and they were not used in the analysis.

The analysis of Lynch and Brown,[139] both by visual inspection and on a computer (discussed below), is confined to bands from 650 to 1200 cm^{-1}. Other investigators have used bands in this region, e.g. Kawahara[140] used bands at 720 and 810 cm^{-1} whereas Mattson[141] used bands at 725, 747, 814,

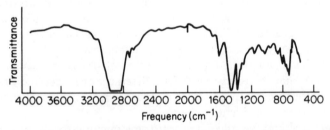

Figure 20 Infrared spectrum of Ecuadorian crude oil using NaCl cell with thickness of 0.1 mm. Reprinted with permission from Lynch and Brown.[139] Copyright (1973) American Chemical Society

874, and 1034 cm^{-1}. Spectra for these previous investigations were measured with lower resolution. Using a high resolution instrument Lynch and Brown[139] observed many sharp bands between 650 and 1200 cm^{-1}; this is especially true in the 700–900 cm^{-1} region where they observed as many as 19 bands in spectra of crude oils and the lighter distillates. The intensities and the number of bands between 650 and 1200 cm^{-1} are extremely characteristic of petroleum samples, and they provide a unique fingerprint of each individual sample. As will be shown, this fingerprint can be used to match an unknown sample to the correct known rapidly and unambiguously.

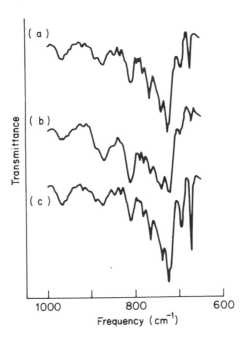

Figure 21 Infrared spectra in 650–1000 cm^{-1} region of crude oils from Nigeria (a), Ecuador (b), and Rocky Mountains (c). All spectra were measured with samples in 0.1 mm NaCl cell. Reprinted with permission from Lynch and Brown.[139] Copyright (1973) American Chemical Society

Spectra from 650 to 1000 cm^{-1} of three samples of crude oils from different parts of the world are shown in Figure 21. Sample (a) is from Nigeria, sample (b) from Ecuador, and sample (c) from the Rocky Mountains. The most noticeable difference between the spectra are the bands at 670 and 695 cm^{-1}; however, the band at 670 cm^{-1} is due to a highly volatile compound and it disappears rapidly from samples exposed to the atmosphere. The band at 695 cm^{-1}, apparently due to aromatics, decreases in spectra of samples left on the surface of the water, but it is useful in initial analyses. For the spectra shown in Figure 21 the doublet at 720–725, the shoulder at 745, the band at 765, the two weak bands at ~790, and the broad band at 870 cm^{-1} are very useful in visual identification of the samples.

Spectra from 650 to 1000 cm^{-1} of three samples from the same part of the world, North Africa, are shown in Figure 22. Spectra (a) and (b) are almost identical and can be distinguished only by differences in the relative intensities of the bands. Spectrum (c) is similar to the other two, but bands at 670, 695, 725, and 740 cm^{-1} have significantly different intensities. It should be noted that the band contours in the spectra of all three samples are almost identical. Samples from the same part of the world all have similar band contours between 650 and 1000 cm^{-1}, and the band contours can be used to classify a sample as to its general geographic origin. However, samples from the same general location can be distinguished from each other by comparing relative intensities of a selected number of bands.

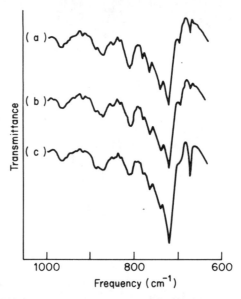

Figure 22 Infrared spectra in 650–1000 cm^{-1} region of crude oils from North Africa: (a) Sertica, (b) Essider, and (c) Amna. All spectra were measured with samples in 0.1 mm NaCl cell. Reprinted with permission from Lynch and Brown[139]

A comparison of the spectra of a crude oil and three of its distillates is shown in Figure 23. Spectrum (a) is of Ecuadorian crude oil, spectrum (b) of its kerosene fraction, spectrum (c) of its No. 2 fuel fraction, and spectrum (d) of the high boiling residual fraction. The kerosene and the No. 2 fuel have more and sharper bands than the crude they come from. Kerosenes and No. 2 fuel oils usually have the same bands in the 650–1000 cm^{-1} region, but the relative intensities of the bands are considerably different from one sample to another. As expected the residual distillate has fewer broader bands than its parent crude or the other

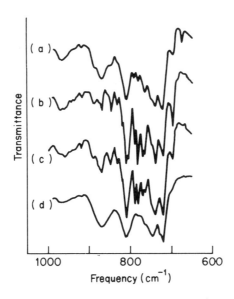

Figure 23 Infrared spectra in 650–1000 cm^{-1} region of Ecuadorian crude oil and its distillates; (a) crude; (b) kerosene distillate; (c) No. 2 fuel oil; and (d) residual distillate. All spectra were measured with samples in 0.1 mm NaCl cell. Reprinted with permission from Lynch and Brown.[139] Copyright (1973) American Chemical Society

two distillates. In general it can be said that by cursory inspection of the spectra, petroleum samples can be classified into three categories: crudes, kerosene/No. 2 fuel, or residual distillates.

To test the capabilities of infrared spectroscopy to provide for a unique identification of petroleum Lynch and Brown[139] examined unknown (to them) mixtures of crude oils from the known samples. Close visual inspection of the spectrum of this unknown sample suggested the identity of the two components. They tried to reproduce the spectrum of the unknown crude by mixing various amounts of these two crudes together. Finally, it was concluded that the unknown sample contained Ecuadorian and Nigerian crudes in a ratio between 2:1 and 1:1. The actual unknown sample did contain Ecuadorian and Nigerian crudes in the ratio of 3:2 thus, they correctly identified the two components and bracketed their concentrations.

Lynch and Brown[139] investigated the applicability of computer analysis of infrared spectral data on unknown oil samples to the identification of the origin of the oil. The first stage of this work was to measure spectra of known samples and to catalogue the spectral data so that unknowns can be identified rapidly and conveniently.

The first problem was to obtain an accurate estimation of L_0 at low frequencies when cells with NaCl windows were used. For a 5 mm thick NaCl window, the intensity of light transmitted through the window decreases about 2% from 800 to 700 cm^{-1}, and then it falls off rapidly below 700 cm^{-1}. They calculated the absorbances for a 5 mm NaCl window in 10 cm^{-1} increments between 610 and 820 cm^{-1}. The calculated absorbances were then divided by the thickness of the windows to give absorptivities, and these were fitted by least squares to a cubic equation. For any given spectrum of a petroleum sample in which NaCl windows are used, the thickness of the NaCl windows and the per cent transmittance at 650 cm^{-1} were read into a computer program. The program then calculates the background from 610 to 1200 cm^{-1} assuming that it obeys the cubic equation between 610 and 820 cm^{-1}, and that it is level above 820 cm^{-1}. This method was checked by comparing spectra measured using NaCl windows with those using AgCl windows (constant I_0 in this region), and the absorptivities agreed within 1%.

Once the background I_0 is determined for a given sample, the per cent transmittance of each of the bands is read into the computer program and the absorbances are calculated. For each of the known samples the absorbances are divided by the sample thickness to give pseudoabsorptivities. These are stored in a computer file so that they may be compared with any unknown at any later date.

To identify an unknown sample, the same information is read into the computer, i.e. the per cent transmittance of the background and per cent transmittance for each of the selected bands. If a band is missing in a spectrum of an unknown or known, it is given a zero per cent transmittance and is eliminated from the analysis. Absorbances are calculated for each of the unknown bands, and these are converted to pseudoabsorptivities by dividing by the estimated or known thickness of the sample.

The next process in the analysis is to take the ratio of the absorptivities of each of the known samples to those of the unknown. If the samples are the same, the ratios for all of the bands would be identical and, if they have the same thickness, the ratios would all be 1.0. Since the exact thickness of many of the unknown samples is not known they devised a method for eliminating the sample thickness from the analysis. The ratios for each known to unknown are averaged, and then the ratios are divided by the average. The ratio of the thickness of the known sample to the thickness of the unknown is a constant for any known, and all ratios including the average contain this constant. Thus, when each ratio is divided by the average the constant cancels, and the problem of estimating the thickness is eliminated. Furthermore, the new average ratio becomes 1.0 so that deviations are compared with the ideal ratio. The ratios are then listed and the differences between the ratios and 1.0 are determined. Finally the program lists the number of bands with a difference (ratio-average) within 0.05, 0.10, 0.25, and 0.50.

An outline of the method of analysis and a list of the frequencies of all of the bands used are given in Table 8. Lynch and Brown[139] used a total of 21 bands in the analysis, but rarely did a spectrum of a sample, known or unknown, have all 21 bands.

Lynch and Brown[139] quote several examples of the application of their technique to the identification of oil spill materials. They claim that it provides an unambiguous identification for petroleum products, eliminating the need

Table 8 Outline of computer analysis and listing of frequencies of bands used in analysis

Computer analysis
Absorptivities of known samples calculated and stored in a computer file.
Absorptivities of unknown samples calculated.
Ratio of absorptivities of each known to those of the unknown calculated.
Average ratio for each known to unknown calculated.
Ratios divided by average to make average 1.0.
Difference between ratios and 1.0 (the average) calculated.
Listing of the number of bands for each known with a difference (ratio-average) less than ± 0.05, ± 0.25, and ± 0.50.

Frequencies (cm^{-1}) of bands used in analysis to nearest 5 cm^{-1}

695	790	890
720	805	915
725	810	955
740	820	1020
765	835	1070
770	845	1145
780	870	1160

Reprinted with permission from Lynch and Brown.[139] Copyright (1973) American Chemical Society.

for adding tracer materials to petroleum products or for other methods of analysis. Ideally, for the method to be used on a worldwide scale, infrared spectra of all possible petroleum samples would have to be measured and the absorptivities of the selected bands stored in a computer file. This could all be accomplished within a relatively short period of time. In addition, to identify samples taken from natural waters, the effects of weathering on the infrared spectra of samples would have to be known.

Bogatie[142] applied infrared spectroscopy to the rapid identification of oil and grease spills from pulp and paper mills. In the case of oils and greases classical infrared methods are most useful in differentiating between broad groups or families of oils and greases. For example, Figure 24 shows infrared spectra for three major groups; (a) tall oil, (b) petroleum fuel oil, and (c) a non-petroleum product. Tall oil shows a strong band at 1700 cm^{-1} representing carboxylic acid. The petroleum product has no 1700 cm^{-1} band because it does not contain carboxylic acid groups. The non-petroleum hydraulic fluid has a spectrum much different from either tall oil or petroleum, especially in the carbon–hydrogen bonding region (2950 cm^{-1}), which shows only a fraction of the carbon–hydrogen content found in the other two oils. Figure 25 serves as another example of the classical method. Tall oil; a carboxylic acid with a strong band at 1700 cm^{-1}, is clearly different from tall oil soap, a sodium salt, which has characteristic bands at 1600 and 1450 cm^{-1}. Thus, the classical method is most useful in distinguishing between broad groups or families of oil and grease.

Unfortunately, the use of the classical method in distinguishing between oils or greases within common groups or families was found to be inadequate. Petroleum oils and grease are derived from a common base and thus have very similar infrared spectra. Differences in spectra result from only small changes

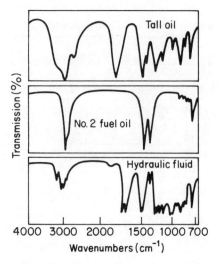

Figure 24 Infrared spectra for three groups of oils. Reprinted with permission from Bogatie.[142] Copyright (1974) Technical Association of the Pulp and Paper Industry

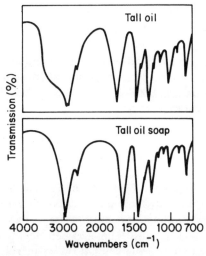

Figure 25 Infrared spectra for tall oil and tall oil soap. Reprinted with permission from Bogatie.[142] Copyright (1974) Technical Association of the Pulp and Paper Industry

in molecular structure or chemical additives. This is illustrated in Figure 26 which shows spectra for fuel oil, turbine lube oil, and grease, all petroleum products. Bands are nearly identical except for minor differences in the 650–1000 cm^{-1} region, located at the right-hand side of the spectrogram. Figure 27 gives a magnification of this area. Even more difficult is the differentiation between No. 2 and No. 6 fuel oil, as is evident in Figure 28. Again one can observe only small differences in the 650–1000 cm^{-1} region, which become more distinct when magnified, as shown in Figure 29.

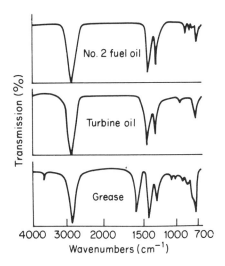

Figure 26 Infrared spectra for three petroleum products; No. 2 fuel oil, turbine oil and grease. Reprinted with permission from Bogatie.[142] Copyright (1974) Technical Association of the Pulp and Paper Industry

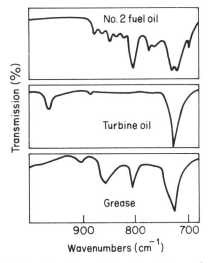

Figure 27 Magnification of the 650–1000 cm^{-1} area shown in Figure 26. Reprinted with permission from Bogatie.[142] Copyright (1974) Technical Association of the Pulp and Paper Industry

An accomplished infrared spectroscopist would normally be able to perform the above identifications. However, a less complicated, yet reliable and fast method was desired by Bogatie.[142] The ratio of infrared absorbance technique was found to be suitable.[143,144] Even though the bands for each oil in Figures 26–29 are at or near the same wave number, the intensities or band absorbances

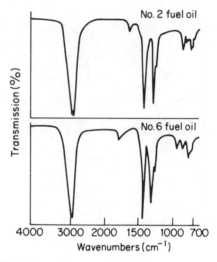

Figure 28 Infrared spectra for No. 2 fuel oil and No. 6 fuel oil. Reprinted with permission from Bogatie.[142] Copyright (1974) Technical Association of the Pulp and Paper Industry

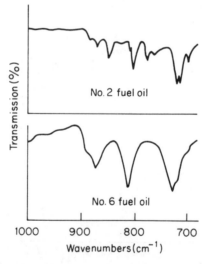

Figure 29 Magnification of the 650–1000 cm^{-1} area shown in Figure 28. Reprinted with permission from Bogatie.[142] Copyright (1974) Technical Association of the Pulp and Paper Industry

are different. When compared to other bands in the spectrum, these differences, expressed as ratios of absorbances, are characteristic for each oil and grease. The ratios of the absorbance of one band to another band can be used in identifying a particular compound, as discussed below.

Bogatie[142] determined absorbances by the baseline technique (Figure 30). The

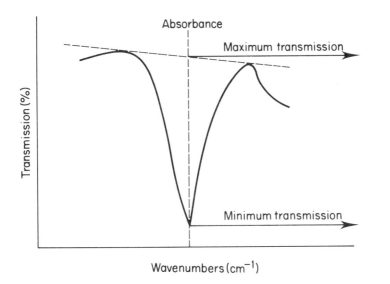

Figure 30 Absorbance determination using baseline technique. Reprinted with permission from Bogatie.[142] Copyright (1974) Technical Association of the Pulp and Paper Industry

length of the vertical line which intersects the line that is tangential to the two proximate inflections represents the absorbance or depth of the absorption. Per cent transmission values are obtained at the maximum and minimum points and then converted to absorbance values by using a transmission–absorbance conversion table. The difference between the absorbance values for the maximum and minimum points is then the absorbance of the particular peak.

The absorbance of any given peak at a specific wavelength represents the number of functional groups absorbing the infrared radiation at that wavelength. Common functional groups are O–H, N–H, C–H, C=C, C≡C, straight and cyclic chain hydrocarbons.[145] The degree of absorbance for these wavelengths is similar for all oils and greases. However, functional groups causing peaks or absorption bands at other wavelengths may give varying degrees of absorbance.

When the absorbance values of these bands are ratioed to the 2960, 1375 and 720 wavenumbers (cm^{-1}), at least one ratio will stand out as being distinctive for a specific oil or grease.

In some cases, chemical additives may cause characteristic absorption bands which will distinguish a particular product from the other oils or greases. Gear oil was found to have such bands at 1540 and 1750 cm^{-1}. Many of the greases were found to have a band at 3700 cm^{-1} being caused by the presence of an aluminium stearate additive.

To find a characteristic ratio which would help to distinguish between various oils and greases, all possible ratios, shown in Figure 31, were calculated for each product, using the frequency bands listed in Table 9. After considering the values of all possible ratios, there was at least one ratio that was characteristic

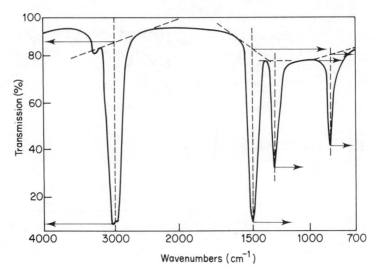

Figure 31 Absorbance determination for total spectrum. Reprtined with permission from Bogatie[142]. Copyright (1974) Technical Association of the Pulp and Paper Industry

for any given oil and grease. A total of 11 ratios listed in Table 10 was found to characterize each of the 15 oil and grease products studied by Bogatie.[142] Consequently, $11 \times 15 = 165$ ratio values were obtained, as listed in Table 11. When analysing the spectrum of an unidentified oil, all 11 ratios must be calculated and compared to the ratios used in Table 11 to determine the identity of the oil or grease. At first glance, a computer analysis would appear as the only way to solve 15 equations with 11 unknowns. However, a more simple graphical method was developed and is described by Bogatie.[142]

Whittle et al.[146] point out that following the work of Simard et al.,[147] the estimation of oil in water by extraction into carbon tetrachloride and

Table 9 Significant wavenumbers (cm^{-1})

680	860	1600	1780
720	1375	1700	2960
810	1540	1750	3700

Reprinted with permission from Bagatie.[142] Copyright (1974) Technical Association of the Pulp and Paper Industry.

Table 10 Significant ratios

$2960 \, cm^{-1}$	$1375 \, cm^{-1}$	$720 \, cm^{-1}$	$1750 \, cm^{-1}$
3700/2960	1600/1375	860/720	1540/1750
1780/2960	810/1375	810/720	
1750/2960	720/1375		
1700/2960			
680/2960			

Reprinted with permission from Bagatie.[142] Copyright (1974) Technical Association of the Pulp and Paper Industry.

Table 11 Ratio calculation results

	720/1375	810/1375	680/2960	1700/2960	860/720	810/720	3700/2960	1780/2960	1600/1375	1750/2960	1540/1750
Fuel oils											
No. 2	0.175	0.217	0.006	0.003	0.177	1.241	0.001	0.001	0.431	0.001	0.001
No. 6	0.118	0.126	0.002	0.016	0.520	1.068	0.001	0.001	0.125	0.001	0.001
Tall oils											
	1.472	0.512	0.017	1.005	0.174	0.393	0.001	0.001	0.448	0.001	0.001
Hydraulic fluid											
	0.141	1.600	0.857	0.006	(20.0)	(11.3)	0.001	0.001	4.721	0.001	0.001
Lube oils											
Paper machine	0.223	0.024	0.001	0.001	0.001	0.108	0.001	0.005	0.001	0.001	0.001
Wormgear	0.295	0.066	0.001	0.001	0.215	0.223	0.001	0.001	0.098	0.105	0.001
Turbine	0.180	0.010	0.001	0.003	0.001	0.056	0.001	0.001	0.018	0.001	0.001
Pump	0.254	0.009	0.001	0.001	0.058	0.035	0.008	0.001	0.013	0.001	0.001
Gear	0.304	0.020	0.001	0.001	0.007	0.068	0.001	0.001	0.021	0.126	0.179
New motor oil	0.128	0.031	0.001	0.029	0.107	0.243	0.001	0.001	0.040	0.001	0.001
Used motor oil	0.144	0.016	0.001	0.226	0.079	0.114	0.001	0.001	0.054	0.001	0.001
Grease											
Pump mill	0.356	0.024	0.001	0.001	0.001	0.067	0.061	0.133	0.023	0.001	0.001
Woodyard	0.303	0.059	0.001	0.001	0.001	0.194	0.010	0.001	1.129	0.001	0.001
Crane	0.327	0.077	0.001	0.001	2.07	0.235	0.040	0.001	0.778	0.001	0.001
Silicone	0.002	0.742	0.001	0.001	0.001	(320.0)	0.001	0.001	0.001	0.001	0.001

Reprinted with permission from Bagatie.[142] Copyright (1974) Technical Association of the Pulp and Paper Industry.

measurement of the infrared absorbance in the region 3400 cm^{-1} (2.94 μm) to 2600 cm^{-1} (3.85 μm) has become a widely used method. Standard methods based on this procedure have been published by many authorities and these include Stichting Concawe,[148] the United States of America Environmental Protection Agency[149] and the American Petroleum Institute (API-733-58).[150] Coles et al.[151] reviewed these methods and used the less toxic Freon 113 (1,1,2-trichloro-1,2,2-trifluorethane) as solvent, as did Greenfield.[152] Whittle et al.[146] set out to find a method of calibration which does not require a standard oil. They examined two of the four methods of calculation described in the literature i.e.

(1) Simard et al.[147] using the sum of the absorbances et 2925, 2860, and 2960 cm^{-1} which they attributed to CH_2, CH_3, and CH groups respectively. This method gave some compensation for variations in oil composition compared to an absorbance reading at a single wavelength.
(2) Coles et al.[151] also using the sum of three absorbances but they used 2930, 2960, and 3030 cm^{-1}, attributing the 3030 cm^{-1} peak to CH aromatic groups, the 2960 cm^{-1} peak to CH_3 and the 2930 cm^{-1} peak to CH_2
(3) API-733-58[150] using the sum of the 2930 and 3030 cm^{-1} peaks and the authors[149,152] who use only a single absorbance reading at 2930 cm^{-1}.

According to Coles et al.[151] the single most critical choice affecting the accuracy of the method is that of an oil for calibration, and where possible calibration standards should be prepared from a sample of the oil actually being measured. When this is not possible the use of a synthetic oil, a mixture of n-hexadecane (37.5%), iso-octane (37.5%) and benzene (25%), suggested by Simard et al.,[147] appears to have been generally accepted. API-733-58[150] quotes the accuracy of the method as ±10% using the authentic oil for calibration and ±20% if the synthetic oil is used, except when oils of high aromatic content are determined. Coles et al.[151] commented on the poor accuracy achieved when determining gasoline and suggested a gasoline standard for the calibration of the paraffinic components and a mixture of benzene, toluene, and xylenes to calibrate the aromatic components. The methods of calculation described by Whittle et al.[146] are based on measurements made on individual standard solutions of pure compounds, e.g. n-hexadecane, toluene, etc., in carbon tetrachloride, and takes into account the variations in composition of different oils.

Method 1

The first method of calculation involved equating the absorption of each of the three common groups found in petroleum, i.e. the CH_2 group at 2930 cm^{-1}, the CH_3 groups at 2960 cm^{-1} and the CH aromatic group at 3030 cm^{-1}, to concentration as follows:

$$c = x.A_{2930} + y.A_{2960} + z.A_{3030} \tag{4}$$

where A is absorption of the subscripted peak, c is concentration in mg l^{-1}, x, y, z are coefficients related to the absorptivities of the respective groups. By choosing three suitable compounds, each one rich in one of the groups and making the appropriate absorbance readings, then these equations can be solved for x, y, and z.

In practice the aromatic peak is much weaker than the other two peaks, so z has a greater value than x or y. Therefore since the accurate measurement of A_{3030} is important the absorbance due to the shoulders of the CH_2 and CH_3 peaks must be corrected for. Whittle et al.[146] found that a suitable correction factor F can be calculated from the n-hexadecane absorbances by basing the correction on the absorbance at 2930 cm^{-1} and letting $F = A_{2930}/A_{3030}$. Equation (4) now becomes

$$c = x.A_{2930} + y.A_{2960} + z.(A_{3030} - A_{2930}/F) \qquad (5)$$

Table 12 gives the absorbances of solutions of n-hexadecane, pristane, and toluene and substituting these values in equation (5) gives: For n-hexadecane

$$160 = x.0.5225 + y.0.2249 + z.\,(0.0151 - 0.5225/F) \qquad (6)$$

For pristane

$$160 = x.0.3537 + y.0.3249 + z.(0.0137 - 0.3537/F) \qquad (7)$$

For toluene

$$160 = x.0.0407 + y.0.0286 + z.(0.0748 - 0.0407/F) \qquad (8)$$

For pure n-hexadecane and pristane the aromatic CH absorption should be zero, the small recorded measurements at 3030 cm^{-1} being due to the shoulder of the CH_2 and CH_3 peaks. For n-hexadecane the term $z.\,(0.0151 - 0.5225/F)$ equals zero (from the definition of F above) and it is assumed that the term for pristane also approximates to zero. Equations (6) and (7) can then be solved for x and y giving $x = 177.4$ and $y = 299.4$. Substituting the values for x and y in equation (8) gives $z = 1960$.

For oil solutions the concentrations can be calculated from equation (5).

Table 12 Mean values of absorbances for compounds (corrected for blanks)

n-Hexadecane	0.5225	0.2249	0.0151
Pristane	0.3537	0.3249	0.0137
Iso-octane	0.1973	0.4420	0.0217
Benzene			0.1190
Toluene	0.0407	0.0286	0.0748
Ethyl benzene	0.0789	0.1343	0.0755
Blank	0.0050	0.0042	0.0020

Pye Unicam SP 1200
Reprinted with permission from Whittle et al.[146] Copyright (1980) Royal Society of Chemistry.

Method 2

The second procedure is based on calculating an 'absorptivity' for each group, related to the 'molar concentration' of that group, from the appropriate absorbance of a solution of a compound rich in that group.'

$$a = \frac{A \cdot M}{C \cdot N} \tag{9}$$

a = absorptivity (mg·mole)$^{-1}$ 1 cm^{-1};
A = absorbance of the appropriate group in a 1 cm cell;
M = relative molecular mass of the compound used;
c = concentration in mg l^{-1};
N = number of groups in one molecule of the compound used.

For example n-hexadecane is chosen as the compound rich in CH$_2$. From Table 12 $A_{2930} = 0.5225$

$$a_{CH_2} = \frac{0.5225 \cdot 226.45}{160 \cdot 14}$$

$$= 0.05283 \text{ (mg mole)}^{-1} \text{ 1 cm}^{-1}$$

Similarly a_{CH_2} and a_{CH} can be calculated from iso-octane and benzene.

For oil solutions the concentration is calculated using the equation:

$$c = \frac{A_{2930} \times m_{CH_2}}{a_{CH_2}} \times \frac{A_{2960} \times m_{CH_3}}{a_{CH_3}} + \frac{(A_{3030} - A_{2930}/F) \times m_{CH}}{a_{CH}}$$

where m is relative molecular mass of the subscripted 'group' and F is correction factor as calculated in Method 1.

The factors obtained for Method 1 are shown in Table 13 which also shows the 'absorptivities' of each group calculated from the appropriate absorbance of the compound with the highest proportion of each group.

Table 13 Factors obtained for Method 1 (Pye Unicam SP1200)

Method		Factors	
	x	y	z
1A (n-hexadecane, pristane, toluene)	177.4	299.4	1960
1B (n-hexadecane, pristane, ethyl benzene)	177.4	299.4	1445
1C (n-hexadecane, iso-octane, benzene)	179.0	295.4	1344

Absorptivities Group	Compound	a((mg mole (group))$^{-1}$ 1 cm^{-1}
CH$_2$	n-hexadecane	0.05283
CH$_3$	Iso-octane	0.06025
CH (aromatic)	Benzene	0.00968

Reprinted with permission from Whittle et al.[146] Copyright (1980) Royal Society of Chemistry.

Table 14 Real oil solutions. Results show mean 'recoveries', i.e. (mean calculated concentration/true concentration) × 100

Oil	Method* 1A	1B	1C
White spirit	108.1	104.0	102.8
Premium kerosene	106.9	103.7	102.8
Regular kerosene	108.0	104.1	103.0
Petrol 3 star	102.2	93.6	91.5
Gas oil 1	95.3	92.6	91.9
Gas oil 2	96.1	93.5	92.8
Lube oil 20w	88.9	87.1	86.6
Lube oil 20–50w	87.7	85.7	85.1
Residual fuel oil	89.3	87.1	86.6
North sea crude	107.6	104.9	104.0
Light Nigerian crude	101.1	97.7	96.8
Kuwait crude	101.0	100.0	99.5
Range	87.7–108.1	85.7–104.9	85.1–104.0
Range %	20.4	19.2	18.9
Mean	99.4	96.2	95.3

*Key see table 13.
All samples examined as 160 mg l^{-1} solutions in carbon tetrachloride, except gas oil 20–200 mg l^{-1}.
Reprinted with permission from Whittle et al.[146] Copyright (1980) Royal Society of Chemistry.

Table 15 Linearity

Concentration (mg l^{-1})	Mean (mg l^{-1})	Standard deviation ($n=f$)
20	20.32	2.16
40	40.76	1.76
80	77.56	2.00
120	116.32	1.64
160	151.84	1.52
200	187.72	2.64

Slope of linear least squares fit* to 36 points = 0.929.
Intercept = 3.08.
Overall standard deviation = 2.16.
*Based on the function $\Sigma_1^n (y_i - a_o - a_r x_i)$, i.e. unweighted.
Reprinted with permission from Whittle et al.[146] Copyright (1980) Royal Society of Chemistry.

The accuracy of the new methods of calculation are shown in Table 14. The important criteria of the results is considered to be the range of the recoveries and the range values.

Whittle et al.[146] selected Method 1A, based on n-hexadecane, pristane, and toluene, for the studies on linearity and the applicability of other instruments. This method gave the least bias.

To test the linearity of Method 1A, solutions of gas oil (20–200 mg l^{-1}) in carbon tetrachloride were prepared and six sets of absorbance measurements were recorded for each concentration and the results, corrected for blanks, are shown in Table 15. The slope is less than one because of the bias for the particular oil used and the small intercept may be due to errors in measuring the blank.

The factors x, y, z, and F used in Method 1, vary between instruments due to factors such as resolution and recorder response, and need to be experimentally determined for each instrument. Table 16 shows the factors for Method 1A as determined for the Pye Unicam SP1200 and the Perkin Elmer 297. The oils selected for the evaluation of the Method 1A on the Perkin Elmer 297 were those which gave the greatest positive and negative bias and two types of oil in between. With the exception of the gas oil both instruments give very close results.

Table 16 Comparison of instruments (Method 1A)

(a) Factors

	P.E.297	SP1200
x	115.8	177.4
y	308.1	299.4
z	1582	1960
F	137	34.6

(b) Real oil solutions

	Mean 'recoveries' (%)	
Oil	P.E.297	SP1200
White spirit	108.4	108.1
Petrol 3 star	101.1	102.2
Gas oil 1	90.1	95.3
Lube oil 20–50w	86.7	87.7

Reprinted with permission from Whittle et al.[146] Copyright (1980) Royal Society of Chemistry.

Characterization of oils by ultraviolet spectroscopy

The absorption of ultraviolet radiation is much less useful than fluorescence techniques. Spectra of materials as complex as petroleum products are insufficiently detailed for identification purposes.

Levy[153] has identified petroleum products in a marine environment by ultraviolet spectroscopy. The sample (down to 0.06 mg) in hexane is filtered through Whatman No. 42 paper, the filtrate is diluted to 100 ml and the extinction is measured at 256 nm. The dilution is adjusted, if necessary, to bring the extinction within the range 0.4–0.8, and the spectrum relative to hexane is then recorded over the range 350–210 nm. The absorption at $\simeq 256$ nm and $\simeq 228$ nm is measured, the ratios of these values is independent of concentration of the test solution and can be used for distinguishing between samples of residual fuel oils and lubricating oils even after weathering.

Characterization of oils by fluorescence techniques

Petroleum products contain many fluorescing components, e.g. aromatic hydrocarbons, polycyclic aromatic hydrocarbons, and various heterocyclic compounds. The use of improved technique and instrumentation has led to the use of this technique for the identification of crude and residual oil pollutants in a marine environment,[154,155] and of motor oils and related petroleum products.[156-159]

Fluorescence emission spectra obtained by irradiating a cyclohexane solution at 340 nm are similar for many crude and semi-refined oils (residual fuels). A maximum emission occurs at about 386 nm with slight shoulders at 405 and 440 nm.[155] The intensity of the maximum and its shoulders was different for each oil examined, and the ratio of the intensities 440/386 nm was a further identification parameter. The emission spectra are also dependent on the oil concentration and each oil reacted differently to dilution, which is a third parameter for identification. Evaporation of volatiles in simulated weathering experiments demonstrated that loss of 28–39 wt % had little effect. The effect of sunlight and general exposure on a film of oil during 4 days had very serious effects on the intensities and general shape of many spectra. Immediate sampling is therefore recommended.

Freegarde[154] demonstrated that emission spectra from crude oils in solution at 77 K and excitation at 250 nm, contained much more detail. A mixture of fluorescent compounds, such as those in most petroleum materials, gave a different emission spectrum at different excitation (irradiation) wavelengths. An identification technique was developed based on the construction of a contour map of the oil examined, formed by scanning the emission spectrum at 20 excitation wavelengths, and connecting contours of equal fluorescent intensity on a graph of emission wavelength versus excitation wavelength. Although in the comparison of fresh oils the method could be sensitive, weathering effects similar to those discussed above[155] would hinder identification. The spectra could be obtained fairly quickly but construction of the contour map seems much more time-consuming.

Synchronous excitation of fluorescence emission is a technique recommended for the rapid identification of motor oils and related petroleum materials.[156-158] Samples of oil in cyclohexane solution are excited at a wavelength which synchronously trailed the plotted emission, usually by about 25 nm. The differences between unused motor oils were found to be quantitative rather than qualitative variations, which accords with the findings of Parker and Barnes.[159] Used motor oils varied widely in both aspects, and in this area the technique is promising. Petrols vary distinguishably with octane number and different makes of high octane petrol tended to vary considerably. As volatility of the product decreases, synchronously excited emission at higher wavelengths predominates. Paraffin, TVO, and diesel fuel give spectra seemingly containing less useful detail. Substantial differences existed in the spectra of different types of lubricants, motor oils, gear oils, and cutting oil. Differentiations between coal tar pitches may be possible, although bitumens may be less amenable to the technique.

Wakeham[160] has discussed the application of synchronous fluorescence spectroscopy to the characterization of indigenous and petroleum derived hydrocarbons in lacustrine sediments. The author reports a comparison, using standard oils, of conventional fluorescence emission spectra and spectra produced by synchronously scanning both excitation and emission monochromators. Graphically presented, results demonstrate the increased resolution obtained using the synchronous technique.

The US Environmental Protection Agency[161] has reviewed publications on the identification of petroleum oils in water using fluorescence techniques.

Characterization of oils by metals analysis

Determination of metals can assist in the characterization of crude oil spillages. Distilled products should not contain these elements. In particular, the ratio of vanadium to nickel in an oil can be characteristic of its source of origin. The ratio of trace elements particularly V/Ni have been shown to be unaffected by prolonged weathering effects.

Neutron activation analysis

Neutron activation analysis techniques have been reported in several papers dealing with the determination of trace elements in crude oil and petroleum products, containing crude oil distillation residues, e.g. marine fuel oil.[162,163]

Bryan et al.[164] developed neutron activation techniques for the identification of metals in oil slicks. Determinations before and after exposure of oils to wave action, elevated temperature, ultraviolet radiation, and bacterial attack showed that the concentration of most of the elements were only slightly affected by a minimum exposure, after which they were stable to further exposure. Each trace element concentration provided an evidence point for comparison of oil samples, and it was found that more than enough stable evidence points could be determined to distinguish between different oil samples. Determinations of the ^{34}S to ^{32}S ratios also provided a useful evidence point.

Lukens et al.[165] discuss procedures in the comparison of slicks with suspected source oils. In particular, neutron activation analysis was used to determine qualitative and quantitative trace-element patterns in a wide variety of oils, and efforts were made to develop an objective means of comparing these patterns. Nearly 300 trace-element patterns were accumulated, by means of which it can be ascertained, with almost complete certainty, whether or not two samples are of the same oil.

Lukens[166] has also published an instruction manual for oil slick identification by trace-element patterns measured by neutron activation analysis. The following steps are described: acquisition of the oil slick sample and of oil samples from suspected sources of the slick; preparation of samples; irradiation of samples and comparator standards under identical conditions; activity measurements on irradiated samples and standards by γ-ray spectrometry; processing of the spectrometric data in order to identify the trace elements and to determine their concentrations; objective comparison of trace-element patterns of each suspect oil with the slick oil; and estimation of match between suspect and slick oils.

Flaherty and Eldridge[167] have discussed the application of neutron activation analysis to the determination of vanadium in oil. The oil sample and standard solution (each 3 g) were irradiated for 10 min in a neutron flux of $\simeq 10^8$ neutrons cm^{-2} S^{-1}, and the γ-ray spectrum was monitored with a NaI(Tl) detector. The maximum deviations of the results over several months were within $\pm 7\%$ for the range 20–2000 ppm of vanadium.

Synthetic complexes of manganese, mercury, indium, and rhenium with asphaltenes and phthalocyanines, not normally found in petroleum, have been suggested as tracers for crude oils. These were then determined by neutron activation analyses.[168]

X-ray fluorescence spectrometry

X-ray fluorescence has been employed to determine vanadium and nickel levels in oils.[169] Louis[170] has explained the principles and operation of this technique to the examination of mineral oils and light petroleum fractions. The method is rapid, pretreatment of samples is avoided, and improved detection limits are obtained.

Kubo et al.[171] have investigated the application of energy dispersive X-ray fluorescence spectrometry to the determination of vanadium, iron, nickel, molybdenum, arsenic, selenium, and zinc in fuel oils and shale oils.

Atomic absorption spectrometry

Atomic absorption spectrometry has been used to determine vanadium and nickel.[172,173] Stuart and Branch[172] found that the ratio of vanadium to nickel in oils remains constant during evaporation of lighter fractions, and the value of this ratio as determined by atomic absorption spectrophotometry of the oil diluted with isobutyl methyl ketone can be used to compare oils and to establish their origin.

γ-Ray spectrometry

γ-Ray spectrometry has been examined[174] and found to compare favourably with conventional chemical techniques for determining metals in oils, but required less time and effort. However, as a computer and an atomic reactor are required the method is instrumentally complicated.

Oscillographic polarography

Budnikov et al.[175] determined vanadium and nickel in benzene solutions of bituminous oils by this technique.

Miscellaneous

Lieberman[176] discussed the application of various techniques of metals analysis to oil pollution source identification. These workers subjected crude oils, residual fuel oils, and distillates fuel oil to simulated weathering for 10 to 21 days, at 55 °F and 80 °F, at high and low rates of salt water washing. 'Weathered' and 'unweathered' samples were analysed by low voltage m.s. (for polycyclic aromatics), high voltage m.s. (naphthenes), gas chromatography (n-alkanes), emission spectroscopy (Ni and V), X-ray spectrometry (total S) and the Kjeldahl technique (total N). Several compound indices were sufficiently unaffected by simulated weathering to allow discrimination between like and unlike pairs of oils. Discriminant-function analysis was used to select the best compound indices for the oils under examination. By using these indices, weathered and unweathered samples were correctly paired with high statistical confidence.

Determination in water

Various methods have been developed for the determination of oils in waters. These range from gravimetric and turbidmetric methods to methods involving techniques such as direct gas chromatography, gas stripping methods, and headspace analysis coupled with gas chromatography, infrared spectroscopy, thin-layer chromatography, and ultraviolet and fluorescence techniques. These methods are described separately below.

The standard ABCM gravimetric procedure[177] for determining oil in effluents involves treating the sample with a magnesia floc produced *in situ* on to which the oil is absorbed. The floc is removed, destroyed with hydrochloric acid, and the liberated oil taken up into the petroleum ether or occasionally chloroform. After careful removal of the solvent, the residual oil is determined gravimetrically. The method is applicable to only the less volatile oil products, and includes any material which is absorbed into the floc and is soluble in the organic solvent. It determines 'non-volatile ether-extractable material'. Several authorities have expressed dissatisfaction with the procedure and raised doubts as to its accuracy and precision in the lower concentration ranges, i.e. < 10 mg l^{-1}. Milner[178] found the method reasonably accurate and reproducible down to about 3 mg l^{-1} if an alumina floc and more particularly, a standard cooling and weighing system, were used. Determinations on samples containing soluble oils, for example, cutting oils, gave erratic results. A standard procedure for volatile liquids immiscible with water based on the turbidity produced when an acetone solution of the oil is diluted with water[179] is also non-specific and tends to be inaccurate below about 5 mg l^{-1}.

In general, methods based on direct solvent extraction followed by evaporation and gravimetric determination of the residual oil, suffer similar disadvantages.[180-182] Recoveries of hydrocarbon oils by solvent extractions are not always quantitative.

Taras and Blum[183] have pointed out that the scope of the standard method of the American Public Health Association for the determination of grease can be extended to include emulsifying oils by saturating the acidified sample with sodium chloride before filtration. Trichlorotrifluoroethane and hexane are equally effective as extracting solvents, but the former is preferred on account of its non-inflammability and greater density. The most difficult oil sample studied was $\simeq 85\%$ recoverable by the modified method.

In view of the unsatisfactory situation regarding gravimetric methods for the determination of oil in water it is recommended that alternate more specific, techniques are considered as discussed below.

Gas chromatographic methods

Gas chromatography is limited to sufficiently volatile materials. Since petroleum products are readily classified as to type by this technique, the method is highly specific and although substances with similar retention could interfere, the chromatogram profile enables many significant interferences to be noticed and discounted.

Gas chromatography has been used to estimate concentrations of volatile petroleum materials in ground water.[184] One gallon samples of well water are solvent extracted. Iso-octane was employed when a low boiling petroleum solvent was thought to be the pollutant. A quantitative determination of 20 mg l^{-1} was achieved using *n*-octane as an internal standard.

Direct injection of a petrol in water solution, 10 mg l^{-1}, was found impractical due to background signal or unknown interference peaks.[185] McAucliffe[186] found various impurity peaks in the direct injection of aqueous solutions of hydrocarbons, which limited sensitivity to about 1 mg^{-1} of individual hydrocarbons. He employed a 12 ft × ¼ in. column containing 25% SE-30 on firebrick and a precolumn containing ascarite to absorb the water and improve the sensitivity. Other workers have used similar precolumns to remove water.

Two disadvantages of quantitative gas chromatographic analysis of low concentrations are that if direct aqueous injection is employed, then background 'noise' considerably affects sensitivity, and if a solvent is used for extraction of the oil, impurities in the solvent can be very significant. Head-space analysis and degassing techniques, discussed later, avoid these disadvantages, but are usually applicable only to the more volatile petroleum products.

A very practical description of the application of gas chromatography to quantitative analysis of petroleum products in aqueous and soil samples is contained in the report by CONCAWE.[187] Well detailed procedures are given for three hydrocarbon mixtures:

(1) b.p. 150–450 °C,
(2) b.p. 100–450 °C, and
(3) b.p. up to 270 °C.

Methods (1) and (2) which determine oils down to 0.1 mg total hydrocarbons per litre, are based on solvent extraction with carbon tetrachloride and pentane respectively, and involve concentration of the extract to small volume before injection. Method (3) enables direct injection of the aqueous samples, but the limit is then about 5 mg l^{-1} of total hydrocarbons. A reference mixture of known hydrocarbons or an oil similar in boiling range and nature to the oil pollutant is used for calibration, and the areas of peaks on the chromatograms are summated using an integrator.

Compounds of petrol, and of diesel and gas oil (b.p. 174–349 °C) in aqueous solution, which were found to be essentially aromatic hydrocarbons up to about C_9, have been determined by gas chromatography.[185] The components were extracted into nitrobenzene and separated on a column containing polyethylene glycol. The technique determines individual hydrocarbons down to about 0.1 mg l^{-1} with ±0.1 mg l^{-1} precision, although at >0.5 mg l^{-1} the precision is reported as being ±5% for major components and ±10% for others.

Jeltes[188,189] has described a gas chromatographic method for the determination of mineral oil in water in which a true solution of diesel fuel and gas oil in water were analysed by isothermal gas chromatography at 110 °C with polyoxyethylene glycol 1500 as stationary phase and Chromosorb W as support. For two-phase oil–water systems, the oil was extracted with carbon tetrachloride and the extracts were analysed by temperature programmed gas chromatography with SE-30 as stationary phase by a method similar to that of Beynon.[187] Components of mineral oils, i.e. water-soluble types boiling below 300 °C, could be identified and determined, and it is possible sometimes to establish the origin of the oil.

Goma and Durand[190] reported that the sampling errors caused in the injection of water–hydrocarbon mixture on to a gas chromatographic column can be overcome by the addition of acetone for solubilization only when the hydrocarbon chain length is less than 12. Subjection of the sample to ultrasonic vibration (20 kHz, 100 W) gives a homogenous emulsion which must be used immediately, but it is more satisfactory to add a surface-active agent (40% of the amount of hydrocarbon if polysorbate 80 is used) and then disperse the sample with ultrasonic vibration. The emulsified sample is injected directly on to a column (4 ft × 0.25 in.) of 10% of SE-30 on Chromosorb W HMDS (80–100 mesh) operated at 170–210 °C, with hydrogen flame ionization detection. Results are given for the hydrocarbons from C_{11} to C_{18}.

Lure et al.[191] have described a method which involves extraction of the sample (≃500 ml, containing less than 200 mg of hydrocarbons) with hexane concentration of the dried extract to 1 ml and gas chromatography of an aliquot (less than 2 μl) of this solution. For lower levels of hydrocarbons, 2 litres of sample is allowed to percolate through a column (13 cm × 1 cm) on activated carbon, the carbon is dried in air, then the hydrocarbons are extracted with chloroform and the extract is concentrated for gas chromatography. The gas chromatography is carried out on a column (3–4 m × 4–5 mm) packed with 10% of SE-30 or 20% of 3,3′-oxydipropionitrile supported on Chromosorb with helium as carrier gas and a katharometer detector.

Jeltes and Van Tonkelaar[192] compared gas chromatographic and infrared methods for the determination of dissolved mineral oil in water. They saturated various petroleum fractions with water by shaking for 4 min, then, after 2 days, the clear aqueous phases were extracted with carbon tetrachloride. The polar compounds were removed from the carbon tetrachloride extracts by shaking with Florisil and decanting after the Florisil had settled. Thes extracts were analysed by infrared and gas chromatographic methods, both before and after treatment with Florisil, and also after the two-phase systems had been exposed for 8 weeks to light and air. Gas–liquid chromatography provided the most suitable method for the qualitative and quantitative determination of hydrocarbons; the results were unaffected by Florisil treatment or exposure to light and air. Infrared analysis revealed the presence of dissolved polar compounds.

Desbaumes and Imhoff[194] have described a method for the determination of volatile hydrocarbons and their halogenated derivatives in water.

Bridie et al.[193] have studied the solvent extraction of hydrocarbons and their oxidative products from oxidized and non-oxidized kerosene–water mixtures, using pentane, chloroform and carbon tetrachloride. Extracts are treated with Florisil to remove non-hydrocarbons before analysis by temperature programmed gas chromatography. From the results reported it is concluded that, although each of the solvents extracts the same amount of hydrocarbons, pentane extracts the smallest amount of non-hydrocarbons. Florisil effectively removes non-hydrocarbons from pentane extracts, but also removes 10–25% of the aromatic hydrocarbons. However, as the other solvents are less susceptible than pentane to treatment with Florisil, pentane is considered by these workers to be the most suitable solvent for use in determining oil in water.

Morgan[195] has described a gas chromatographic method for the determination of Bunker C fuel oil in marine organisms at the 0.5 mg kg^{-1} level. Pentane–methanol extraction of tissues, using a blender, is followed by adsorption chromatography. After evaporation, the sample is analysed. To determine weathered Bunker C fuel oil, additional detection methods may be necessary.

Determination of oils in sewage

The UK Department of the Environment standard methods[196] involving carbon tetrachloride extraction for the determination of extractable hydrocarbon oils and greases in sewage have been published. The first method consists of solvent extraction using carbon tetrachloride followed by gravimetric determination of the extractable material. The second method partially fractionates this material according to the type of material using chromatographic techniques. Separation on a silica gel column with an upper layer of Florisil is used to remove aromatic and polar compounds, respectively, while the eluate contains the aliphatic hydrocarbon ingredients.

Head-space analysis

May et al.[197] have described a gas chromatographic method for analysing hydrocarbons in marine sediments and sea water which is sensitive at the submicrogram per kilogram level. Dynamic head-space sampling for volatile hydrocarbon components, followed by coupled-column liquid chromatography for analysing the non-volatile components, requires minimal sample handling, thus reducing the risk of sample component loss and/or sample contamination. The volatile components are concentrated on a Tenax gas chromatographic precolumn and determined by gas chromatography or gas chromatography–mass spectrometry.

Khazal et al.[198] and Drozd and Novák[199] examined and compared the methods of head-space-gas and liquid-extraction analysis, comprising the gas chromatography of samples of the gaseous or liquid-extract phases withdrawn from closed equilibrated systems and involving standard-addition quantitation, for the determination of trace amounts of hydrocarbons in water. The liquid-extraction method[200] (see Figure 32) is more accurate but it yields chromatograms with an interfering background due to the liquid extractant. The sensitivity of determination of volatile hydrocarbons in water is roughly the same for each method, concentration amenable to reliable determination amounting to tens of $\mu g\, l^{-1}$ on a packed column with a flame ionization detector.

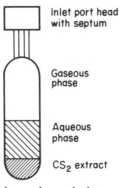

Figure 32 Representation of the system employed in the liquid extraction method. Reprinted with permission from Khazal et al.[198] Copyright (1978) Elsevier Science Publisher B.V.

These workers showed that standard addition technique is suitable for quantitative determination of trace amounts of hydrocarbons in water, using both head-space analysis and liquid-extraction techniques.

With both methods, the chromatographic analyses were carried out at a sensitivity attenuation 1/16; glass column (180 cm × 3 mm i.d.) packed with 8.14 g of 10% (w/w) Apiezon K on Chromaton N (0.2–0.25 mm); column temperature 80 °C; nitrogen carrier gas; flow rates of 26, 29.4, and 200 ml min^{-1}, for nitrogen, hydrogen, and air, respectively. The Chromaton N and Apiezon K were products of Lachema and AEI (Manchester, Great Britain) respectively.

A mass balance of solute in the system leads to the following relations[201] for the head-space-gas and liquid-extraction methods of analysis, respectively:

$$W_i = W_{iG}\left(1 + K_{LG} \cdot \frac{V_L}{V_G}\right) = \frac{W_s - w_i}{(A_i' v_G / A_i v_G') - 1} \tag{10}$$

$$W_i = W_{ie}\left(1 + \frac{V_p}{K_{ep} V_e}\right) = \frac{W_s}{(A_i' v_e / A_i v_e') - 1} \tag{11}$$

where W_i is the initial weight of solute i in either system (amount to be determined). W_{iG} is the weight of the solute in the gaseous phase, V_L and V_G are the volumes of the aqueous and the gaseous phase and K_{LG} is the distribution constant of the solute, defined as the ratio of solute concentrations in the liquid and the gaseous phase, W_{ie} is the weight of the solute in the liquid extract, V_p and V_e are the volumes of the parent liquid and the extract, K_{ep} is the distribution constant of solute, defined as the ratio of its concentrations in the extract and parent liquid, W_s is the weight of standard added to the system, w_1 is the weight of the solute taken out of the system in sampling the gaseous phase for the first analysis, A_i and A'_i are the solute peak areas in the chromatograms obtained in the first and in the second analyses, v_G and v'_G are the volumes of the gaseous phase and v_e and v'_e are the volumes of the extract used in the gas chromatograph in the first and in the second analyses, respectively. If $v_G = v'_g$ and $v_e = v'_e$, in the present case, the expressions are simplified accordingly; all the results presented have been calculated from peak heights.

The expression $1 + (V_p/K_{ep}V_e)$ in equation (11) (a system factor) applies to a system comprising only the parent liquid and the liquid extract. However, the system used by Khazal et al.[198] comprised also the gaseous phase so that the situation has to be expressed by

$$W_i = W_{ie}\left(1 + \frac{V_p}{K_{ep}V_e} + \frac{V_{Ge}}{K_{eG}V_e}\right) \tag{12}$$

In this case V_{Ge} and K_{eG} are the volume of the gaseous phase in the ternary parent liquid-extract–gas system and the distribution constant of the solute, defined by the ratio of solute concentrations in the extract and the gaseous phase. The last expression in equation (11) remains unaltered, i.e. the calculation of the results is independent of the form of the system factor.[201]

In all the measurements the values of W_s were equal to those of W_1, and the values of w_1 were determined by external calibration. Table 17 shows the results of the head-space analysis, while the results of liquid-extraction analysis are given in Table 18. In both kinds of analysis, chromatograms of comparable sizes were recorded at a fixed detector-sensitivity setting. It is apparent from the data in Tables 17 and 18 that the head-space-gas method of analysis is less accurate than its liquid-extraction analogue. On the other hand, a direct comparison of the data suggests that the head-space-gas method is more sensitive. Supposing the entire amounts of solutes (values of w_i) had originally been present in the aqueous phase only, as would be the case in an actual analysis of water, it is possible to infer from the data in Tables 17 and 18 that concentrations of tenths and tens of $\mu g\ ml^{-1}$ of hydrocarbons in water were determined by the head-space-gas and liquid-extraction standard-addition methods, respectively. This sensitivity is still insufficient for many applications in modern water pollution control; often it is necessary to determine hydrocarbons in water in concentrations of tenths of $\mu g\ l^{-1}$ or less. This sensitivity could theoretically be attained by employing capillary gas

Table 17 Results obtained by the head-space-gas–standard-addition method

Solute	$W_t(\mu g)$ Given	$W_t(\mu g)$ Found	Error μg	Error %	$S/\sqrt{n}$	t_{exptl}
Hexane	2.02	1.88	−0.14	6.9	0.08	1.75
Benzene	4.25	4.23	−0.02	0.5	0.20	0.10
2,4-Dimethylhexane	4.42	4.57	0.15	3.4	0.25	0.60
Octane	5.67	6.43	0.76	13.4	0.43	1.77
l-Nonene	7.12	8.36	1.24	17.4	0.63	1.97

$S\sqrt{n}$ = standard deviation of the average, n = number of determinations (9),
t_{exptl} = experimental student coefficient; t_{crit} = 2.26.
Reprinted with permission from Khazal et al.[198] Copyright (1978) Elsevier Science Publishers B.V.

Table 18 Results obtained by the liquid-extraction–standard-addition method

Solute	$W_t(\mu g)$ Given	$W_t(\mu g)$ Found	Error μg	Error %	$S/\sqrt{n}$	t_{exptl}
Benzene	60.8	57.2	−3.6	5.9	2.8	1.28
2,4-Dimethylhexane	63.1	60.5	−2.6	4.1	2.4	1.08
Octane	81.0	80.5	−0.5	0.6	3.6	0.14
l-nonene	101.7	97.7	−4.0	3.9	5.0	0.80

For the meaning of symbols see Table 17; $n = 10$, $t_{\text{crit}} = 2.22$.
Reprinted with permission from Khazal et al.[198] Copyright (1978) Elsevier Science Publishers B.V.

chromatography with unsplit sample injection and utilizing the reserve (about an order of magnitude) in detector sensitivity.

A great advantage of the head-space-gas method over the liquid-extraction method is that it is not necessary to introduce into the system any substance that might interfere with the analysis; with the liquid-extraction method the background chromatogram of the extractant may be a source of serious difficulties. This situation is apparent from the chromatograms in Figures 33 and 34. While the chromatogram of a head-space-gas sample is free from artefact peaks, in the chromatogram of a sample of the liquid extract the peak of hexane is completely obscured by the background response of carbon disulphide. In addition, the peak of benzene had to be corrected for a blank value due to an impurity present in the carbon disulphide. With larger samples of the extract and lower concentrations of the solutes the situation would obviously be worsened accordingly.

Drozd et al.[202] examined the reliability and reproducibility of qualitative and quantitative head-space analyses of parts per billion of various aliphatic and aromatic hydrocarbons in water using capillary column gas chromatography utilizing a simple all-glass splitless sample injection system. They examined the suitability of the standard addition method for quantitative head-space gas analysis for concentrations in the condensed phase varying from units to hundreds of parts per billion.

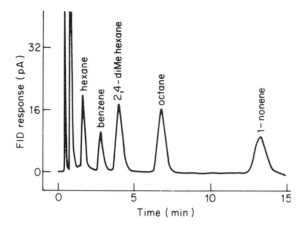

Figure 33 Chromatogram of a 1 ml sample of the head-space gas. FID, sensitivity attenuation 1/16. Reprinted with permission from Khazal et al.[198] Copyright (1978) Elsevier Science Publishers B.V.

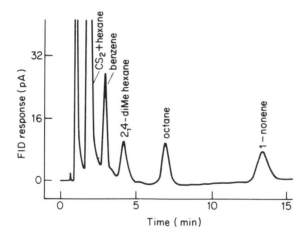

Figure 34 Chromatogram of a 1μl sample of the CS$_2$ extract. Details as in Figure 33. Reprinted with permission from Khazal et al.[198] Copyright (1978) Elsevier Science Publishers B.V.

Although 10 min appeared to be sufficient for equilibration to be achieved in the head-space method, an equilibration time of 30 min at 40 °C was chosen for practical reasons. The equilibration device (Figure 35) consists of a 100 ml glass vessel with a flat bottom, provided with a jacket, through which thermostated water is pumped with an ultrathermostat. Thick-walled capillary tube at the top ends in a flange on which a rubber septum is clamped with an aluminium cap.

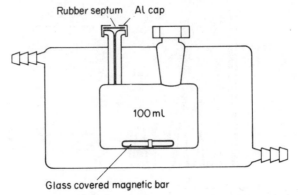

Figure 35 Equilibration device. Reprinted with permission from Drozd et al.[202] Copyright (1978) Elsevier Science Publishers B.V.

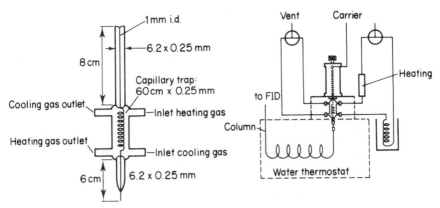

Figure 36 Schematic diagram of the all-glass splitless injection system. Reprinted with permission from Drozd et al.[202] Copyright (1978) Elsevier Science Publishers B.V.

The head-space-gas samples were analysed on a gas chromatograph equipped with a splitless sampling system and capillary columns. The column temperature could be controlled to within ±0.02 °C while the inlet pressure variations were within ±0.002 atm during each analysis. The compounds were detected with a flame ionization detector (10^{-11} A f.s.d). Measurements of retention times and peak areas were effected with a digitizer computer system (sampling rate two per second) with subsequent off-line data processing using paper tape.

The splitless sample introduction system is shown schematically in Figure 36. The all-glass part between the septum and the column (Figure 36(b)) consists of a liner, a capillary trap, and a second liner that is tapered to match the outside diameter of the column. The capillary trap, coated with a thin film of OV-101 is surrounded by a jacket provided with four fittings to permit the transport of the cooled or heated gas during operation. The pathway of the auxiliary gas flows is controlled by two three-way metal valves. The auxiliary gas lines are made of thin-walled copper tubing (o.d. 3 mm, i.d. 2 mm). They were kept as short as possible and insulated to prevent excessive heat losses. The coiled part of the cooling line was placed in a dry ice–ethanol bath at about $-70\,°C$. In the heating line the tube was partly coiled around a massive aluminium cylinder (2 m × 2 cm o.d.) provided with two cartridge heaters. On applying a flow of dry nitrogen of *ca.* 6 l min^{-1} a temperature between -50 and $-60\,°C$ was obtained in the centre of the jacket during cooling. After switching to heating, this temperature increased to above 150 °C within 20 seconds. The head-space-gas samples were taken from the equilibration system with a gas-tight syringe provided with a stainless steel piston which was preheated to about 60 °C.

Capillary columns were used:

(1) Column 1 — a stainless steel column (10 m × 0.25 mm i.d.) using a 10% solution of squalane in *n*-hexane.
(2) Column 2 — an HCl-etched[203] alkali-glass column (73 m × 0.25 mm i.d.), deactivated according to Blomberg,[204] was coated dynamically with a 60% solution of squalane in *n*-hexane using a slightly modified mercury plug method as reported by Schomburg and Husmann.[205]
(3) Column 3 — this was a Duran glass column (29 m × 0.25 mm i.d.) provided with a layer of barium carbonate and deactivated according to Grob and coworkers,[206,207] and coated as described by Bouche and Verzele[208] with a 0.6% solution of squalane in *n*-hexane.

A high separation efficiency and long-term stability with respect to solute retention, preferably retention indices, are the basic requirements of a column for the reliable routine analysis of complex mixtures of hydrocarbons.[209] The resistance of the column to the high concentration of water vapour is a serious additional complication in the head-space-gas analysis of organic compounds in water.

In Table 19 three different types of squalane-coated capillary columns are compared in terms of their efficiency, reproducibility of retention indices, and resistance to water vapour. These results were obtained with a split injection system (splitting ratio 1:400) with synthetic mixtures of hydrocarbons using the retention time of methane as the 'dead time'. Columns 1 (stainless steel) and 3 (Duran glass, covered in advance with a layer of barium carbonate) are acceptable.

In Table 20 retention indices of five aromatic hydrocarbons based on 'dead time' measurements with a splitter are compared with those measured during head-space-gas analysis either measured directly or calculated as discussed above. Correct results can be achieved by employing 'dead times' calculated by the extrapolation method. Retention indices based on the direct measurement of the 'dead time' differ markedly for solutes at the beginning of the

Table 19 Column Properties

	Column 1						Column 2						Column 3		
	Freshly prepared			After 2 months of head-space analysis			Freshly prepared			After 2 weeks of head-space analysis			Freshly prepared		
	I_{sq}^{70}	k	N/m	I_{sq}^{70}	k	N/m	I_{sq}^{70}	k	N/m	I_{sq}^{70}	k	N/m	I_{sq}^{70}	k	N/m
Benzene	641.1			641.1			642.3			653.4			642.0		
Toluene	749.2			749.4			749.2			759.3			749.3		
Ethylbenzene	839.0			839.2			838.8			848.7			839.4		
m-Xylene	855.3			855.6			855.3			866.0			855.6		
o-Xylene	874.3			874.5			874.4			885.9			874.6		
n-Nonane		18.6	2900		12.8	2300		5.5	2400		4.9	2400		5.8	2500

N = number of theoretical plates.
Reprinted with permission from Drozd et al.[202] Copyright (1978) Elsevier Science Publishers B.V.

Table 20 Effect of the determination of the 'dead time' on retention indices of aromatic hydrocarbons on a squalane capillary column (column 3)

Compound	I^{70}_{sq}		
	(a)	(b)	(c)
Benzene	641.1	643.3	641.1
Toluene	749.2	749.8	749.0
Ethylbenzene	839.0	839.1	838.8
m-Xylene	855.3	855.5	855.2
o-Xylene	874.3	874.4	874.4
n	3	6	6

I^{70}_{sq} values determined (a) directly with split injection, (b) directly with the splitless injection system and (c) calculated by logarithmic extrapolation. n = number of measurements.
Reprinted with permission from Drozd et al.[202] Copyright (1978) Elsevier Science Publishers B.V.

chromatogram. Compounds with a larger retention time are less sensitive to the method used.

From the mass balance of a substance i, it follows that the total weight of this substance in the sample (W_i) can be calculated according to the equation

$$W_i = \frac{W_s - w_i}{(A_i' v_g A_i v_g') - 1} \quad (13)$$

It is assumed that the distribution coefficient of a compound i in the gas-liquid system as well as the volume of both phases are not influenced by the addition of the standard. The weight of substance i in the head-space-gas sample (w_i) is usually negligibly small compared with the amount of the added standard (W_i). The peak-area ratio after and before addition of the standard (A_i'/A_i) can be replaced with the ratio of the corresponding peak heights for symmetrical peaks. This is an advantage for concentrations near the detection limit, as will be shown below.

Head-space-gas samples (1 ml) were taken from the same model samples after equilibration times of 10 min, 30 min, 2 h and 2.5 h. The concentrations in the condensed phase varied between 25 and 100 ppb. The representative chromatogram in Figure 37 shows that an equilibration time of 10 min is sufficient for equilibrium to be achieved. It appeared that a detection limit of tenths of a part per billion could be obtained under the proposed experimental conditions (detector sensitivity setting 10^{-11} A f.s.d. and a noise level of 10^{-14} A) without a serious decrease in quantitative reliability.

The results of the determination of hydrocarbons in model water samples by using the standard addition method are presented in Table 21. Each of the samples contained nine hydrocarbons with a concentration varying from less than unity to hundreds of parts per billion in the condensed phase. In the calculation, peak heights measured manually were employed instead of integrator- or computer-measured peak areas for reasons of simplicity and

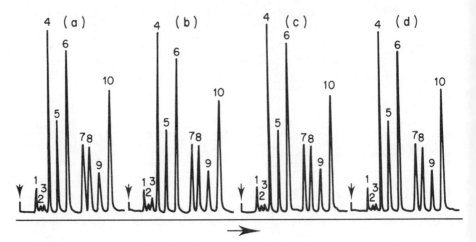

Figure 37 Chromatograms of 1 ml head-space gas samples, after an equilibration time of (a) 10 min, (b) 30 min, (c) 2 h and (d) 2.5 h at 40 °C. Peaks; 1, acetone (10 ppm in the liquid phase); 2, n-hexane (25 ppb); 3, benzene (46 ppb); 4, n-heptane (27 ppb); 5, toluene (57 ppb); 6, n-octane (42 ppb); 7, ethylbenzene (79 ppb); 8, m-xylene (78 ppb); 9, o-xylene (69 ppb); 10, n-nonane (44 ppb). Column 1, carrier gas N_2, inlet pressure 0.11 atm, temperature 70 °C. Reprinted with permission from Drozd et al.[202] Copyright (1978) Elsevier Science Publishers B.V.

because more reliable results were obtained at low concentrations owing to a low signal-to-noise ratio.

At the highest concentrations studied (hundreds of parts per billion), where peak asymmetry caused by overloading of the column was observed, peak-area measurements offered better results. In all other instances the reliability was either similar or only slightly improved when computer-measured peak areas were used.

Various other workers[210,211] have studied the application of head-space analysis to the determination of hydrocarbons in water. McAucliffe[212] determined dissolved individual hydrocarbons in 5 ml aqueous samples by injecting up to 5 ml of the head space. For petroleum oils which contain numerous hydrocarbons, very much larger aqueous samples are required. The percentage of hydrocarbons in the gaseous phase, after water containing the hydrocarbons in solution was equilibrated with an equal volume of gas, was found to be 96.7–99.2% for most C_3–C_8 alkanes. In the case of benzene and toluene the values were 18.5 and 21.0%, respectively, indicating that the lower aromatic hydrocarbons may be less amenable to the technique.

Gas stripping methods

Swinnerton and Linnenbom[213] were the first to examine the applicability of gas stripping methods to the determination of hydrocarbons in water. They

Table 21 Results obtained by head-space-gas analysis of hydrocarbons in model water samples with the standard addition method

Solute	Mean W_i (µg) Given	Mean W_i (µg) Found	Difference (found − given) (%)	s.d. (w_i·found) (%)
n-heptane	0.179	0.183	2.2	10.2
	0.358	0.337	−6.0	14.7
	4.476	4.537	7.6	9.1
	8.952	9.247	3.3	8.6
n-Octane	0.281	0.267	−4.7	19.4
	0.561	0.549	−2.2	12.6
	7.016	6.669	−4.9	19.6
	14.03	13.16	−6.2	16.1
n-Nonane	0.292	0.321	10.2	24.1
	0.584	0.598	2.4	15.3
	7.304	6.732	−7.8	18.6
	14.07	12.86	−11.9	14.4
Toluene	0.378	0.358	−5.4	13.7
	0.756	0.722	−4.5	13.7
	9.448	8.612	−8.8	15.0
	18.90	16.82	−10.9	14.2
Ethylbenzene	0.524	0.481	−8.2	22.7
	1.047	0.994	−5.1	11.0
	13.09	10.86	−17.0	21.4
	26.18	22.02	−15.9	19.8
m-Xylene	0.523	0.478	−8.6	18.6
	1.046	0.969	−7.4	20.5
	13.07	11.22	−14.1	14.2
	26.15	22.50	−13.9	24.5
o-Xylene	0.463	0.486	5.1	26.7
	0.925	0.819	−11.5	14.0
	11.57	10.06	−13.1	19.1
	23.14	17.36	−25.0	12.6

Number of measurements: 10
Reprinted with permission from Drozd et al.[202] Copyright (1978) Elsevier Science Publishers B.V.

determined C_1–C_6 hydrocarbons by stripping them from water with a stream of helium.

After gas stripping the hydrocarbons can be either passed direct to a gas chromatograph or, to increase sensitivity, trapped in a cold trap then released into the gas chromatograph. Alternatively, the stripped hydrocarbons can be trapped in, for example, active carbon, then released into the gas chromatography. This method offers the possibility of determining trace amounts of organic compounds in water even below the ppt level (1 part in 10^{12}, w/w), particularly for the more volatile compounds.[214] Many factors, such as interference by artefacts because of impurities in the stripping gas, the large amount of water passing the trap, adsorption of less volatile compounds in drying filters, the selection of sorbents and the adsorption and desorption efficiency, are serious drawbacks of the method, particularly for quantitative analyses. However, Grob and co-workers[215-218] recently reported an impressive improvement of the method by using a closed loop system, provided with a small-volume effective charcoal filter, but several precautions are necessary when working at such low concentrations. The complicated procedure and the sophisticated equipment required result in many more or less unknown factors and a semiquantitative analysis. In view of the absolute amounts of pollutants involved, their overall results were excellent.

Examples of the various types of methods are discussed below.

Desbaumes and Imhoff[219] swept volatile hydrocarbons and their halogenated derivatives from water contained in a heated metallic column, by a current of purified air and their concentrations were determined with a flame ionization detector. The condensed vapours are analysed qualitatively by gas chromatography. Details and diagrams of the equipment are given and the operating procedure is described. Samples must be stored only in glass or stainless steel containers. Substantial losses may still occur if the storage time is less than 10 hours.

Novák et al.[220] have analysed for simple hydrocarbons in drinking water. The hydrocarbons were extracted by bubbling an inert gas through the water, and the compounds were separated and identified by combined gas chromatography–mass spectrometry. The glass gas chromatographic column (5 m × 2.5 m) was packed with 10% of Carbowax 20 m on Chromosorb W AW and temperature programmed from 60 to 200 °C at 4° per minute, with helium as carrier ($\simeq 20$ ml min^{-1}). The column effluent, after removal of the helium in a separator, was introduced directly into the ion source of the mass spectrometer. Compounds at concentrations down to 0.1 or 1 part per 10^9 of water could be determined. The detection limit was considerably better than those obtained with methods based on solvent extraction and chromatographic or spectral analysis, but compounds with boiling point considerably greater than 100 °C could not be determined.

Polak and Lu[221] have described a gas stripping method for the determination of the total amount of volatile but slightly soluble organic materials dissolved in water from oil and oil products. Helium is bubbled through a sample of the aqueous liquid and the gas carries the organic vapours directly to a flame ionization detector. The detector response plotted against time gives an exponential curve, from which the amount of organic material is derived with the aid of an electronic digital integrator. A detector-response factor is required and this is determined with samples prepared by saturating water with hexane or with benzene.

Kaiser[222] has described a sensitive degrassing technique for trace hydrocarbons in which volatile hydrocarbons up to C_{12} (probably higher if the aqueous sample is warmed) are removed from aqueous solution at 20 °C by a stream of dry nitrogen during 2–10 minutes, and passed into a gas chromatographic column cooled in liquid nitrogen. After the dosing period was completed, the column temperature was programmed at a rate of 7.5 °C min^{-1}, and the hydrocarbons eluted and detected in the usual manner.

Substances can be removed from water, other liquids, and solids without any further preparation. The detection limit achieved for individual hydrocarbons in water was 10^{-2} ppb (10^{-9} wt %).

Colenutt and Thorburn[223] have recently described a technique for routine analysis of organic compounds in water, in the ppb concentration range, by gas chromatography preceded by a concentration step by gas stripping. Nitrogen was purified over active carbon, passed through the water sample and then through an active carbon trap where the organic compounds were concentrated. The method has been used on samples of streams and lakes, and at the same time rain water was collected and analysed. The results show that the influence of rainfall on the pollutants concentration is significant so that the highest level of water pollution occurs in areas of highest atmospheric pollution.

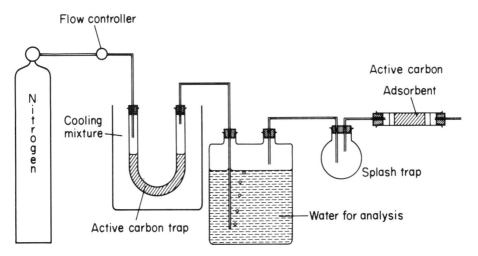

Figure 38 Diagram of the gas stripping apparatus. Reprinted with permission from Colenutt and Thorburn.[223] Copyright (1980) Gordon and Brack

The gas stripping technique used by these workers is illustrated diagrammatically in Figure 38. Nitrogen from a gas cylinder was purified by passage over active carbon and thence passed through the water samples at a flow rate of up to 1 l min^{-1}. Typically 1–5 l of water were used for the analysis, the sample being contained in a glass flask whose size closely matched the volume of water. This avoided any problems of contamination through the introduction of contaminated laboratory air. To avoid the use of grease lubrication Teflon sleeves were used in the ground glass joints. After flowing through the water the gas was passed through an active carbon trap where the organic compounds were concentrated. The trap consisted of a silica glass tube about 80 mm long, 6 mm outer diameter, and 3 mm bore, drawn to a point at one end. Sutcliff Speakman active carbon of 100/120 mesh was held in the trap by plugs of glass fibre yarn.

Compounds adsorbed on the trap were desorbed with carbon disulphide. The solvent was introduced into the trap by Teflon. Pressure applied on the bulb of the pipette was used to drive the solvent back and forth over the carbon. The carbon disulphide was eventually collected in calibrated tubes.

The time required for complete stripping of compounds from aqueous solution depends upon several factors. A time of 2 hours for normal paraffins up to n-hexadecane was found to be sufficient.

Carbon disulphide is an excellent solvent in the washing of active carbon traps as the data in Table 22 show. These recoveries were obtained when measured amounts of the organic compound were added to a carbon tap, and subsequently desorbed in the manner described. An additional advantage of carbon disulphide is the relatively low sensitivity of the flame ionization detector for this solvent.

Table 22 Percentage recovery of selected compounds from spiked carbon traps after washing with carbon disulphide.

n-Octane	95	98
n-Nonane	95	98
n-Decane	95	98
n-Undecane	97	99
n-Dodecane	98	99
n-Tetradecane	98	99
Benzene	96	99
Toluene	97	99
e-Xylene	97	99

Reprinted with permission from Colenutt and Thorburn.[223] Copyright (1980) Gordon and Brack.

The system used by Colenutt and Thornburn[223] was a Perkin Elmer F11 chromatograph fitted with a flame ionization detector. By following the movement of particular peaks with a change in the nature of the stationary phase and the injection of standard compounds it was possible to identify tentatively a number of compounds in the mixture.

Confirmation of peak identity was achieved by mass spectrometry. The system used was a Perkin Elmer 990 chromatograph linked to a Hitachi Perkin Elmer RMS4 mass spectrometer. Using this system the mass spectrum of each peak was recorded after elution of the compound from the column, and the substance identified.

Routine injections were between 1 and 10 μl of the extract from the carbon traps. No concentration of the carbon disulphide solution was necessary, or indeed desirable.

Colenutt and Thorburn[223] applied their technique to various synthetic and actual samples of hydrocarbons in water. Synthetic solutions of 10 ppb n-alkanes from n-octane to n-hexadecane prepared by adding acetone solutions of the hydrocarbon to distilled water were put through the procedure and the gas

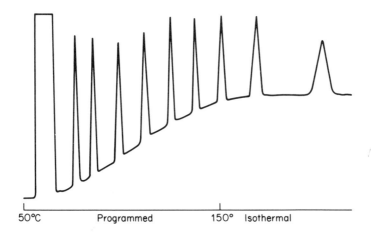

Figure 39 Chromatogram of the C_8–C_{16} *n*-alkanes stripped from a nominal 10 ppb solution. Analytical conditions: 5% SE 30 coated on acid washed silanized Chromosorb W. Nitrogen carrier gas at a flow rate of 30 ml min^{-1}. The hydrocarbons are eluted in order of increasing molecular weight. Reprinted with permission from Colenutt and Thorburn.[223] Copyright (1980) Gordon and Brack

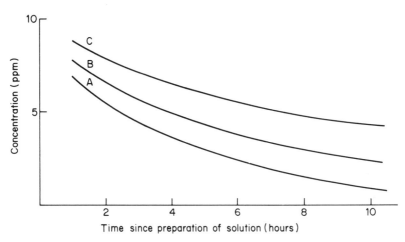

Figure 40 Decrease in hydrocarbon concentration of a nominal 10 ppb solution A, *n*-octane; B, *n*-nonane; C, *n*-decane. Reprinted with permission from Colenutt and Thorburn.[223] Copyright (1980) Gordon and Brack

chromatograms of the carbon disulphide extracts are shown in Figure 39. If the solution was analysed almost immediately after preparation a value close to the nominal 10 ppb for each component was obtained. However, if the aqueous sample was left exposed in an open laboratory for any length of time, the concentration of the lower molecular weight compounds decreased, as

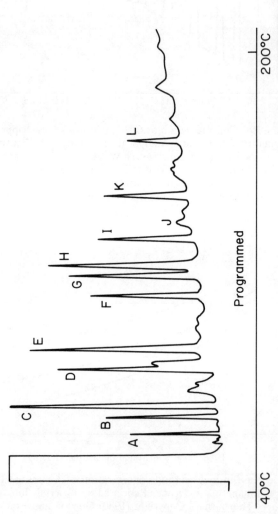

Figure 41 Chromatogram of Welsh Harp water sample. Analytical conditions: 10% Carbowax 20M coated on acid washed silanized Chromosorb W. Nitrogen carrier gas at a flow rate of 30 ml min^{-1}. A, n-octane; B, n-nonane; C, benzene; D, n-decane; E, toluene; F, n-undecane; G, ethyl benzene; H, xylenes; I, n-dodecane; J, n-propyl benzene; K, n-tridecane; L, n-tetradecane. Reprinted with permission from Colenutt and Thorburn.[223] Copyright (1980) Gordon and Brack

Figure 40 shows. Thus the concentrations of the lower alkanes are somewhat suspect in that the evaporation effect prior to sampling is unknown.

In Figure 41 is shown a typical gas chromatogram obtained by Colenutt and Thorburn[223] for the extract of a river water sample showing the presence of aliphatic and aromatic hydrocarbons at the 1–10 ppb level. These workers found up to 10 times greater concentrations of hydrocarbons in rain water to that found in river water.

Wasik[224] has used an electrolytic stripping cell to determine hydrocarbons in sea water. Dissolved hydrocarbons in a known quantity of sea water were equilibrated with hydrogen bubbles, evolved electrolytically from a gold electrode, rising through a cylindrical cell. In an upper head-space compartment of the cell the hydrocarbon concentration is determined by gas chromatography. The major advantages of this cell are that the hydrocarbons in the upper compartment are in equilibrium with the hydrocarbons in solution, and that the hydrogen used as an extracting solvent does not introduce impurities into samples.

Figure 42 shows the layout of the stripping cell. The large compartment (A) may have a volume of 1–3 litres. The small compartment (B) has a volume of approximately 200 ml. The two compartments are connected by a 35 mm o.d. glass tube with a 30 mm o.d. coarse porosity glass frit at one end. On compartment (B) side of the frit is a silicic acid plug. This prevents flow of liquid between the two compartments and provides a low resistance for the flow of electrical current. The two gold electrodes (1 mm o.d. wire) are held in the respective compartments by means of septums held in ¼ in. × ¼ in. union fittings. The electrode in the large compartment (50 cm) has a spiral configuration to provide a large surface area. The supporting electrolyte in the small compartment is 5% sulphuric acid solution. The stripping cell was in a constant temperature water bath controlled to ±0.005 °C. The hydrogen

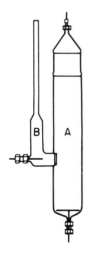

Figure 42 Electrolytic stripping cell. Reprinted with permission from Wasik.[224] Copyright (1974) Preston Publications Inc.

head-space was sampled by a gas sampling valve thermostated at 100 °C. The 1/16 in. o.d. stainless steel tubing connecting the stripping cell with the gas sampling valve was heated to 100 °C by a heating tape. The chromatographic column was a 15.24 m × 0.508 mm i.d. SCOT column was prepared with finely ground diatomaceous earth on a fumed silica support and coated with a mixture of *m-bis*(*m*-phenoxyphenoxy)benzene and Apiezon L. The column was operated at 80 °C. The effluent was monitored by a hydrogen flame detector.

A constant current device provided the electrical current to the stripping cell. The volume of hydrogen, V, bubbled through the large compartment was calculated from equation (14):

$$V = it/F \cdot 22.414/2 \cdot T_e/273.15 \qquad (14)$$

where i is the current, t the time in seconds, F the Faraday constant and T, the temperature of the stripping cell.

Wasik[224] used this method to successfully determine ppb of gasoline in sea water. He found that a convenient method for concentration of the hydrocarbons is to recycle the hydrogen stream containing the hydrocarbons many times over a small amount of charcoal (2.3 mg). Figure 43 shows a schematic drawing of the apparatus. The hydrogen from the stripping cell is allowed to enter a flow circuit composed of a filter tube containing the charcoal, a stainless steel bellows pump, and the connecting stainless steel tubing. The pump increase the flow in the circuit to ten times the flow of hydrogen entering and leaving the circuit. The flow circuit is enclosed in an air bath

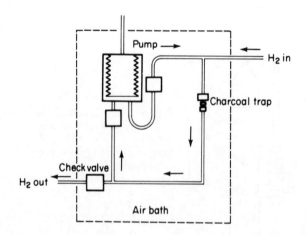

Figure 43 Apparatus for absorbing hydrocarbons on charcoal. Reprinted with permission from Wasik.[224] Copyright (1974). Preston Publications Inc.

whose temperature is greater than that of the stripping cell in order to prevent water vapour from condensing in the flow circuit. The bellows pump is operated by pulsating air pressure. The flow can be controlled either by regulating the pressure limits of the pulsating air or by controlling the time between air pulses. The filter holder is depicted in more detail in Figure 44. The charcoal is sandwiched between two ⅛ in. diameter circular pieces of stainless steel screen (325 × 2300 mesh) (A) positioned on the inside shoulder of a ⅛ in. × 1/16 in. reducing union (B). The charcoal, while still in the filter tube, was extracted three times with 5 μl of carbon disulphide. A 2–3 μl aliquot of this solution was then injected into a SCOT capillary column.

Figure 44 Charcoal filter holder made for ⅛ in. × 1/16 in. stainless steel reducing. Reprinted with permission from Wasik.[224] Copyright (1974). Preston Publications Inc.

Infrared spectroscopy

Infrared spectroscopy is generally accepted as an excellent technique for determining petroleum products of the order of 1 mg or less. Invariably, absorption intensities of C–H vibrations in aliphatic hydrocarbons are measured, and related to the quantity of the oil present. However, Jeltes[225-227] points out that when in very low concentrations the substances dealt with are in true solution and are aromatic hydrocarbons, infrared spectroscopy is not so suitable. In contrast to gas chromatography, the techniques are applicable to non-volatile petroleum products.

The general infrared spectroscopic technique was published as early as 1951 by Simard *et al.*,[228] the oil being extracted from a 1 litre sample into carbon tetrachloride and the spectra recorded between 3.2 and 3.6 μm. Carbon tetrachloride is suitable because it contains no C–H bonds. The intensities of absorptions due mainly to saturated hydrocarbons at 3.38, 3.42, and 3.50 μm were determined, and the amount of oil measured down to about

0.1 mg l^{-1} using a standard solution of the oil if possible, or a solution of 37.5% iso-octane, 37.5% cetane and 25% benzene for calibration. Other workers have used this procedure with only slight modifications for the determination of petroleum products in water or soil.[229-234] Similar absorption intensities were measured, i.e. 3.0–3.5 µm,[230] 3.2–3.6 µm,[231] 3.3–3.6 µm,[229,232] 3.39, 3.42, and 3.51 µm,[234] 3.25–3.72 µm.[233] Various other workers have discussed infrared methods.[239-244]

Should any organic matter, such as fatty acids, glycerides, chemical and biochemical oxidation products of petroleum oils, be coextracted into the organic solvent, it can seriously interfere with the determination. To overcome this problem, a prior separation stage has been introduced, involving percolation of the carbon tetrachloride extract through a bed of alumina or Florisil.[235-237,239] This method is preferred by Hughes *et al.*[238] who used a modified impeller and a sample bottle immersed in a 30 kHz ultrasonic cleaning bath for dispersion of the carbon tetrachloride solvent.

Polar materials, for example, carboxylic acids, esters, ketones, phenols, amines, are strongly absorbed, whereas weakly polar hydrocarbons are eluted preferentially and examined by infrared spectroscopy. Heterocyclic compounds in petroleum products may also be strongly retained. Lindgreen[235] demonstrated that relatively massive contamination by ethanol, acetone, acetic acid, formic acid, oleic acid, vanillin and margarine have little effect on the method, although without prior separation, serious interference occurred.

Beynon[237] describes the technique with excellent practical details. A 3.5–5 litre pollution sample was extracted with carbon tetrachloride, and a portion of the extract dried, and then eluted from a Florisil column. The eluant and a blank are examined by infrared analysis and the intensities measured at 3.38 µm, and if necessary 3.42 and 3.50 µm. A lower limit of 0.1 mg kg^{-1} is quoted with undetermined precision.

A variation in the above technique is the formation of an alumina floc *in situ*, on to which the oil is absorbed.[245] This is filtered off, extracted with carbon tetrachloride, and the spectra produced in the usual manner.

Hellmann[246] investigated the possibilities and limits of infrared spectroscopy for the determination of mineral and fuel oil in surface waters. The method is applicable to concentrations down to about 0.01–0.2 mg l^{-1}. It was used to detect traces of fuel or mineral oil and to determine the dispersion of oil and its emulsions in water, the biochemical decomposition of mixtures of mineral oil products and the effectiveness of various oil-binding agents and emulsifiers. With oil-polluted samples of soil and ground water errors were caused by the presence of non-polar hydrocarbons. It was also found that aromatic compounds could not be assessed by infrared spectrophotometry.

Hellmann and Zehle[247] determined microgram amounts of alkanes in water by extracting one litre of water with as little as 3 ml of carbon tetrachloride. This gives a detection limit of 1–2 µg, with an error of 15–20%.

Gruenfeld[248] compared the relative extraction efficiencies of carbon tetrachloride and trichlorotrifluoroethane in the extraction of oils from water and the effects of adding sulphuric acid and sodium chloride were examined by extracting 1 litre synthetic samples (prepared by emulsifying each of four test oils with tap water) with successive 25 ml portions of each solvent. The amount of oil in each extract was determined by measuring the extinction of 2930 cm^{-1}. Although the solvents were almost equally effective in extracting the oil trichlorotrifluoroethane is recommended because of its lower toxicity. Sulphuric acid and sodium chloride improved the extraction efficiency, and complete separation of the oil was achieved with four extractions when 5 g of sodium chloride and 5 ml of 50% sulphuric acid were added.

Mallevialle[249] carried out a systematic study of the factors governing the determination of hydrocarbons in water by extraction with carbon tetrachloride, Florisil chromatography, and measurement of infrared adsorption. The method, described below, was unsuitable for aromatic hydrocarbons.

Mallevialle[249] standardized on extracting 1 litre of water, adjusted to pH 3 with 10 ml of carbon tetrachloride. The water was stirred mechanically for 30 min (with a 100 stroke a min stirring machine mechanism with a backward and forward motion over a 6 cm stroke).

Efficiency ranges from 65 to 95% for the first extraction and from 95 to 100% for the second; these variations can be explained by the fact that hydrocarbons are in the state of pseudosolutions or microemulsions according to their nature and the presence of surface-active elements. Almost the whole of the hydrocarbons has passed to the organic phase after two extractions from volumes of water of from 5 to 1 litre.

Problems can arise during clarification through the formation of an emulsion at the interface of the two liquids. For example, in the case of water containing a large amount of humic matter, it will be better to increase the ion strength of the solution (sodium chloride) and to use a less vogorous stirring system for a longer time. In extreme cases the organic phase and the emulsion can be centrifuged to improve separation.

The carbon tetrachloride was then dried by passing it through a small column of sodium sulphate because the presence of water would affect the infrared absorption measurements.

The infrared spectra (KBr disc) of mineral hydrocarbons are of the type illustrated in Figure 45. The peaks of high intensity due to CH, CH_2 and CH_3 in the 2800–3100 cm^{-1} range are characteristic of the aliphatic chains and can therefore be used to measure hydrocarbon content. Measurements were made in infrasil cells with a 10, 20, and 50 mm optical path. Mallevialle[249] used a reference mixture as follows: 37.5% trimethylpentane; 37.5% cetane; 25% benzene.

Figure 46 gives details of the peaks due to the CH, CH_2, and CH_3 groups for the standard mixture, gasoline, and all the heavy hydrocarbons. Mallevialle[249] made a comparative study of the various possibilities on numerous samples from a variety of sources.

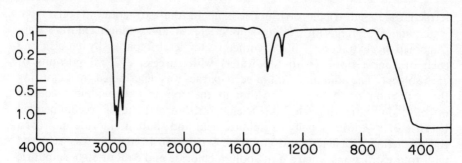

Figure 45 Infrared spectrum of crude oil (film on KBr window). Reprinted with permission from Mallevialle.[249] Copyright (1974) Pergamon Press Ltd

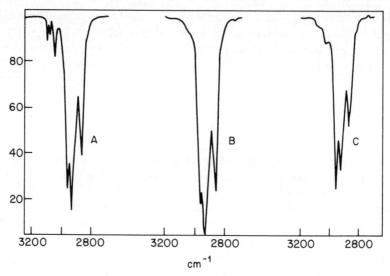

Figure 46 Infrared spectra over the range 3200–2800 cm^{-1}. A, standard mixture; B, crude oil; C, petroleum. Reprinted with permission from Mallevialle.[249] Copyright (1974) Pergamon Press Ltd

The peaks over the range 3000–3100 cm^{-1} are due to the vibrations of the CH groups, particularly of the aromatics; as the intensity of these peaks is much weaker, the aromatics are normally measured by fluorescence or by u.v. absorption. The peaks at 2962 and 2872 ($\pm$10) cm^{-1} correspond to the CH$_2$. In Figure 47 are shown five standard curves:

A: sum of absorptions at 2926 and 2853: $CH_{2as} + CH_{2s}$;
B: sum of absorptions at 3040, 2962, 2926: $CH_{2as}\ CH_{3as}$;
C: sum of absorptions at 2962 and 2926: $CH_{3as} + CH_{2as}$;
D: sum of absorptions at 2962 and 2926 and 2853: $CH_{2as} + CH_{3as} + CH_{2s}$;
E: sum of absorptions at 2962 and 2926 but using coefficients allowing for the specific absorption differences of the CH$_3$ and CH$_2$ groups:

$1.4(0.12 E_1 + 0.19 E_2)$ where E_1 is absorption due to the CH_3 and E_2 is absorption due to the CH_2. The linearity of the points obtained is satisfactory in each case.

The types of Florisil (100–200 mesh) available commercially were not absolutely pure because they release organics which absorb over the infrared range studied; in practice, this interference can be removed in the case of a 2 g column of Florisil (50 × 100 mm) by running 20 ml of carbon tetrachloride through the column. The extract of not less than 15 ml is then run through the column and the first 10 ml of eluate are ignored: this column can be reused several times after washing with carbon tetrachloride.

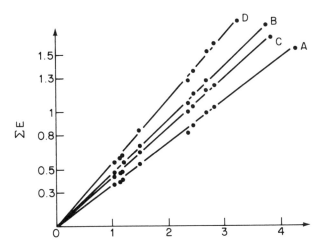

Figure 47 Milligram standard in 10 ml carbon tetrachloride. Reprinted with permission from Mallevialle.[249] Copyright (1974) Pergamon Press Ltd

Using this method Mallevialle[249] was able to measure as little as 0.1–0.2 mg l^{-1} by means of two extractions with 10 ml of carbon tetrachloride from 5 litres of water (optical path: 10 mm) with an accuracy of ±0.05 mg l^{-1}. The same result could also be obtained with 1 litre of water by increasing the optical path to 50 mm.

The error is considerably reduced (<15%) when working with oil contents exceeding 1 mg l^{-1}.

In Table 23 are shown results obtained using the five infrared calibration curves (Figure 47) in the analysis of known synthetic solutions of various refinery distillation fractions and commercial hydrocarbons. These results seem to suggest that calibration method D is the most suitable for all types of hydrocarbons.

Martin and Geyer[250] also studied infrared spectroscopy as a means of estimating mineral oils in water. The method was unsatisfactory in the case of waste waters containing emulsifying agents, which must be removed, together with other polar compounds, by extraction with carbon tetrachloride and

Table 23

mg of hydrocarbon	mg	Calculated mg of hydrocarbon				
		A	B	C	D	E
Esso HDMS	11.2	13.35	10.25	11.5	11.8	
Antar 20 W40	10.0	12.15	9.8	10.6	11	10.8
Antar Z 10	9.6	10.65	9.1	9.8	9.7	10
Mobiloil 10 W50	7	9.2	7.5	8.1	8.2	8.3
Diesel, motor	8.2	9.4	8.05	8.6	8.8	8.7
Car gasoline	9.6	9.2	10.4	10.8	9.6	8.6
Crude oil	10.2	10.4	10	9.8	9.75	9.9
Slop	6.4	6.25	5.6	5.8	5.9	5.75
Medium diesel	15	17.1	13.2	14.6	15.2	15.2
Light diesel	10	11.05	9.5	10.1	10.2	10.2
Kerosene	8.2	8.4	7.95	8.35	8.2	8.2
Heavy gasoline	6.9	6.9	7.6	8.1	7.6	7.7

Reprinted with permission from Mallevialle.[249] Copyright (1974) Pergamon Press Ltd.

treatment of this extract with an adsorption agent. They compared results obtained by shaking with Florisil or by passing through a chromatographic column of aluminium oxide or Florisil. For some waste waters, containing phenol, only the aluminium oxide column can be used. It is recommended that the Florisil used for adsorption of polar compounds should be replaced by aluminium oxide.

Geyer et al.[251] also tested the feasibility of infrared methods in the case of water containing either toluene, trichloroethylene, or methylene chloride. The characteristic absorption bands of these substances in a carbon tetrachloride extract, following clean-up on an alumina column, were illustrated in the presence or absence of mineral oil. Toluene and oil could only be distinguished with difficulty, while trichloroethylene did not interfere with the quantitative determination of mineral oil; qualitative separation only was possible in the presence of methylene chloride.

Götz[252] has pointed out that the determination of hydrocarbons in water and sludge requires a preliminary treatment for removal of fats and fatty acids which interfere with the estimation. The efficiency of column chromatography for this purpose using Kieselgel, Florisil, and aluminium oxide as adsorbents, was tested by these workers.

Ahmed et al.[253] investigated the precision and accuracy of infrared spectroscopic methods for determining oil in water using two techniques of oil preconcentration, namely carbon tetrachloride extraction and polyurethane foam adsorption.

Method

Carbon tetrachloride extraction All samples were preserved with 3 ml of concentrated sulphuric acid, refrigerated, and extracted within 30 hours after being obtained. Once in the laboratory, the sample was poured into a 2 litre separatory funnel and 30 ml of carbon tetrachloride was used to wash the bottle from which the sample was removed. This same carbon tetrachloride was then used to extract the water sample. After 30 seconds agitation and a 3 min settling period, the non-aqueous phase was drained through a funnel containing about 30 g of anhydrous sodium sulphate over a glass wool plug and collected in a 100 ml volumetric flask. This rinsing and extraction procedure was repeated twice more. The sodium sulphate was rinsed with 5 ml of carbon tetrachloride which was added to the extracts and the volumetric flask was made up to 100.0 ml.

Infrared spectra were taken of these extracts in a 10.0 mm cell with sodium chloride windows. Solutions were transferred via a Pasteur pipette, washing the cell twice with the solution to be observed before filling. Spectra were scanned from $3400 \, cm^{-1}$ to $2500 \, cm^{-1}$ using 5–10 times expansion of the per cent transmission scale. The C–H stretching band at $2930 \, cm^{-1}$ was used for analysis. The absorptivity at this wavelength as calibrated with a motor oil (SAE 30 weight) was $3.39 \, l \, g^{-1} \, cm^{-1}$. A typical procedural blank corresponded to an oil concentration of less than $0.010 \, mg \, l^{-1}$.

Method

Polyurethane foam adsorption The foam adsorption method of Schatzberg and Jackson[254] was used. 10–50 litres of sea water more passed through a foam dsc (8 cm diameter, 4 cm thick, 100 pores/in.) and then the retained oil was extracted in a Soxhlet apparatus with carbon tetrachloride. This extract was measured as described above. It was necessary to clean new discs by Soxhlet extraction with carbon tetrachloride for 4–6 hours to reduce blank analyses to acceptable levels. Recoveries of known concentrations of oil were greater than 85% for those concentrations above $5 \, mg \, l^{-1}$.

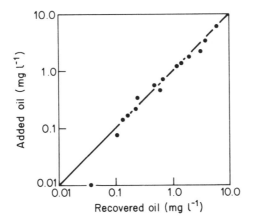

Figure 48 Oil concentration as determined by infrared spectrometric technique *vs.* known concentration

The results in Figure 48 obtained on synthetic solutions of oil in carbon tetrachloride show that the average recovery is 101% and the average error is 13%. Therefore, the method itself seems quite reliable since the quantity of oil in the sample bottle can be determined with a relative standard deviation of less than 15% for concentrations exceeding about 0.05 mg l^{-1}.

Replication experiments on samples taken in Boston Harbour gave relative standard deviations for each set of replicates from 46 to 130% with the average being 75%; this variability is far in excess of the variability of the analytical procedure itself (15%). This error must, therefore, be associated with sampling from a heterogeneous system such as oil in water.

Comparative data for the extraction and foam adsorption preconcentration methods are given in Table 24. In all cases, the oil concentration resulting from the foam technique is the lower of the two values; in fact, it is low by a factor of five on the average. Although there are sampling errors associated with both techniques it is clear that the foam technique has a much lower collection efficiency than liquid-liquid extraction and is, therefore, a less useful procedure.

Table 24 Comparison of oil concentrations determined using different preconcentration techniques (surface sample only)

CCl$_4$ extraction (mg l^{-1})	Foam adsorption (mg l^{-1})	Ratio
84.1	7.2	11.7
0.10	0.08	1.25
0.21	0.03	7.0
0.283	0.19	1.5
0.239	0.07	3.4
	Av.	5.0

Adsorption on to molecular sieve 5A is another technique that has been examined for the preconcentration of oil in water samples. Uchiyama[255] gives details of a procedure for the separation and determination of mineral oil, animal oil, and vegetable oil in water. After extraction with carbon tetrachloride, the extract is treated with molecular sieve 5A, on which animal and vegetable oils are adsorbed. The oil is then determined by infrared analysis. Results of this procedure are given below.

A flow sheet of the analytical procedure is shown in Figure 49. 10 g sodium chloride and 10 ml carbon tetrachloride were added to 350 ml sample water in a 500 ml separable funnel. The funnel was shaken for 5 min to separate the carbon tetrachloride layer. After shaking the carbon tetrachloride layer with 1 g sodium sulphate anhydrous, the carbon tetrachloride layer was filtered with a filter paper. After adjusting the volume to 10 ml with carbon tetrachloride, the solution may be used for infrared analysis (determining the concentration of total oil). After analysing the concentration of total oil, the carbon tetrachloride solution was mixed with 3 g molecular sieve 5A 1/16 in. pellets heated to 400/500 °C for 3 hours then cooled prior to use. After standing the

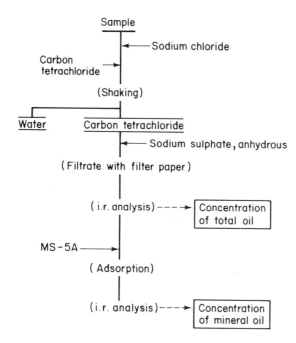

Figure 49 Flow sheet of MS-5A method. Reprinted with permission from Uchiyama.[255] Copyright (1978) Pergamon Press Ltd

solution with molecular sieve 5A at room temperature for 1.5 h with occasional shaking, the solution was used for infrared analysis using 10 mm quartz cells (determining the concentration of mineral oil).

The infrared spectra of oils is the type illustrated in Figure 50. The peaks of high intensity due to CH, CH_2, and CH_3 in the 2800–3100 cm^{-1} range are characteristic of the aliphatic chains and can therefore be used to measure oil content. The peak of 2950 cm^{-1}, the peak of the highest intensity, was used by Uchiyama.[255]

Figure 51 shows the relationship between the adsorption time and the quantity of non-adsorbed oil for various types of oils. Mineral oils were not adsorbed on molecular sieve or the rate of adsorption was very slow. The rate of adsorption of animal and vegetable oils was very fast and they were adsorbed perfectly on sieves in about one hour. Other animal and vegetable oils were the same as lard and soy bean oil. The difference in this adsorption rate is very useful for separating mineral oils from animal and vegetable oils.

Figure 52 shows the adsorptive capacity of various oils on molecular sieve (3 g) in 10 ml carbon tetrachloride. The adsorptive capacity of the sieves was 2 mg oil 3 g^{-1} sieves. This shows that if the quantity of oil was higher than 2 mg, the carbon tetrachloride solution of oils should be diluted before molecular sieve treatment.

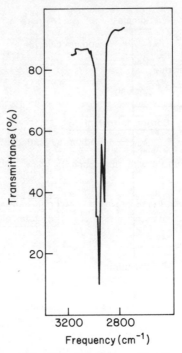

Figure 50 The infrared spectrum of oil. Reprinted with permission from Uchiyama.[255] Copyright (1978) Pergamon Press Ltd

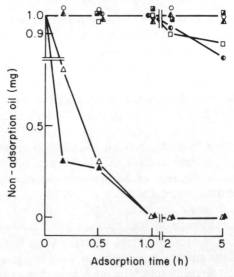

Figure 51 The rate of absorption of oils on MS-5A: ○, heavy oil; □, pump oil; ◐, machine oil; ▲, engine oil; ◼, spindle oil; △, lard; ▲, soy bean oil. Reprinted with permission from Uchiyama.[255] Copyright (1978) Pergamon Press Ltd

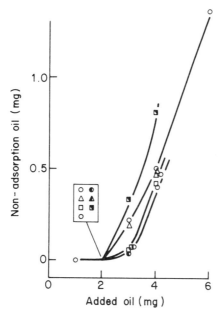

Figure 52 The absorptive capacity of MS-5A; ○, beef tallow; △, lard; ◼, whale oil; ☐, rape seed oil; ⊘, cotton seed oil; ●, soy bean oil; ▲, olive oil. Reprinted with permission from Uchiyama.[255] Copyright (1978) Pergamon Press Ltd

The standard curves for various oils are shown in Figure 53. The linearity of the points obtained was satisfactory in each oil. In most cases where the exact nature of the pollution is unknown, the kind of oil used as the standard material is very important. Four chemical reagents were studied as the standard materials. From Figure 53 oleic acid seems to be the most suitable for all types of oils.

Table 25 shows the satisfactory results obtained using this procedure in determinations on river and waste water samples spiked with various oils (methods calibrated against oleic acid).

Carsin[256] has carried out a literature review of the sampling of hydrocarbons from polluted water. He reviewed sampling equipment used in the determination of hydrocarbon contaminants in natural waters and a new type of sampler in which a floating collector is employed, and an aspirator pump draws the contaminated water through a bed of oil absorbent polyurethane foam. Problems of operation of the sampling apparatus, including the behaviour at sea under different weather conditions are discussed. The recovery and subsequent analysis of the oil fraction by infrared spectrometry are also described.

Hellmann[257] has shown that certain plastics are permeable to hydrocarbons and may therefore be unsuitable to contain samples of water for analysis. A comparative test by infrared spectrophotometry showed that the hydrocarbon

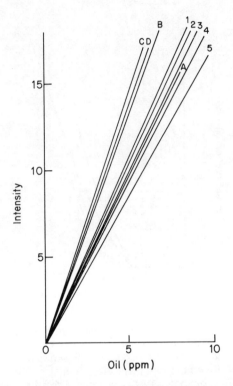

Figure 53 The standard curves: 1, engine oil, pump oil; 2, lard, beef tallow, whale oil; 3, machine oil, olive oil; 4, rape seed oil, cotton seed oil; 5, heavy oil, bean oil; A, oleic acid; B, decane; C, palmitic acid; D, stearic acid. Reprinted with permission from Uchiyama.[255] Copyright (1978) Pergamon Press Ltd

content (initially 1–6 mg l^{-1}) of oil-contaminated water fell to about 50% of the initial value after 12 h in a plastic container. For very volatile compounds, the loss in the same time may be as much as 90%.

Mark[258] has described an infrared method for the determination of the oil content of marine sediments. He showed that the magnitude of the CH_2 stretching band at 2925 cm^{-1}, normally used to determine oil in a sediment, is enhanced when biological matter is also present. The concentration of this material can generally be estimated from the magnitude of the protein-NH band at 1650 cm^{-1} with use of a calculated correction to the total absorption at 2925 cm^{-1}, but the oil must contribute less than 10% to the total absorption at 2925 cm^{-1}. It is desirable, however, that the nature of the organic matter be determined by means of a study of the complete infrared spectrum.

Thin-layer chromatography

The great advantage of thin-layer chromatography is its simplicity and cheapness. Hydrocarbon components of petroleum products can be separated very

Table 25 Analytical results of oils in river waters and waste waters

Water*	Mineral oil added (mg)				Vegetable oil added (mg)				Animal oil added (mg)			Total oil (mg)		Mineral oil (mg)	
	machine	pump	heavy	engine	spindle	rape seed	cotton	olive	soy bean	beef tallow	lard	whale	added	detected	added	detected
A	0.5	0.5	0.5			0.5	0.5			0.5	0.5		3.5	3.32	1.5	1.48
B			0.5	0.5	0.5			0.5	0.5		0.5	0.5	3.5	3.28	1.5	1.58
C	0.5	0.5	0.5		0.5	0.5	0.5			0.5	0.5		3.5	3.36	1.5	1.46
D			0.5	0.5	0.5			0.5	0.5		0.5	0.5	3.5	3.32	1.5	1.46

*A, water from river A; B, water from river B; C, effluent of the sewage treatment works C; D, effluent of sewage treatment works D. Reprinted with permission from Uchiyama.[255] Copyright (1978) Pergamon Press Ltd.

efficiently from polar contaminants on the thin-layer chromatography plate. In all quantitative methods, a sample oil is compared to a range of known concentrations of an identical or similar oil, or hydrocarbon mixture.

Many of the thin-layer chromatographic procedures for qualitative analysis of petroleum products can be used for semiquantitative analysis. However, a more useful approach is that of Geobgen,[259] who employed the common ascending thin-layer technique on silica gel plates, using n-hexane as the developing solvent. One litre waste water samples were extracted directly with carbon tetrachloride, and a portion of the extract chromatographed during about 15 minutes. A non-polar developing solvent such as n-hexane will effect a separation between saturated hydrocarbons, which tended to travel near the solvent front, and the more strongly absorbed aromatic hydrocarbons. The saturated hydrocarbon areas were evaluated as quickly fading yellow areas upon a blue background after spraying with bromothymol blue solution. Aromatic hydrocarbons were visualized by examination of their fluorescence under ultraviolet light. However, saturated hydrocarbon areas were used for quantitative determinations, being compared to a range of known concentrations of a standard solution containing 1 g eicosan, 0.1 g phenanthrene, and 1 g paraffin oil. Geobgen[259] claimed that such a standard solution is satisfactory, since experience has shown that virtually all waste water samples contain less than 5–10% aromatic hydrocarbons. With this procedure concentrations of <5 mg l^{-1} can be determined, giving a lower limit. Overloading of the plates can cause the polar material contaminating the oil sample, especially in sewage samples, to tail into the saturated paraffin area and cause interference.

Channel thin-layer chromatography eliminates most of the horizontal spreading of a vertically moving component on the plate, and thereby increases the sensitivity. It has been applied to analysis of oil pollutants.[260] In this modification, two parallel scratches are made 2 mm apart on a silica gel coated plate. The sample is applied so that it develops between these scratches with carbon tetrachloride. This permits the thin-layer chromatography of smaller quantities and the resulting rectangular spots are easily measured. The procedure is limited to oils that boil above 150 °C. One litre samples are extracted with only 2 ml carbon tetrachloride, and loss of any solvent into the water is claimed not to affect the subsequent determination. Therefore exact separation is unnecessary, and the separated extract is chromatographed. On exposure to iodine crystals, brown rectangular areas appear which are easily measured. The developing solvent used, again carbon tetrachloride, probably does not separate saturated and lower aromatic hydrocarbons significantly, and therefore the composition of the oil is less important. A non-volatile gas oil is proposed as a standard hydrocarbon material, and various concentrations are chromatographed to calibrate the plate in terms of area of spot (mm^2) produced by 0.1 mg of the gas oil. Determination of petroleum oils in the range 1–15 mg l^{-1} are possible. A narrower width between scratches has been used in other analytical work to give greater sensitivity.[261]

Channel thin-layer chromatography has also been used by Berthold.[262]

During development, the samples are confined to narrow bands (2 mm × 50 mm), formed by removal of parallel thin lines of the adsorbent. This ensures that the breadth of the spots is constant, and thus the amount of material in the spot can be calculated, after calibration, from the length of the spot. Five such bands can be accommodated on a normal plate. The technique was applied to trace analysis for hydrocarbons in water.

A microcircular thin-layer chromatographic technique was developed by Koppe and Muhle[263] for the detection and determination of dissolved hydrocarbons in natural waters. Silica gel plates were employed and developed in a horizontal position. Three litre aqueous samples were extracted with 12 ml carbon tetrachloride and the extracted organic layer chromatographed using carbon tetrachloride as the initial developing solvent. Circular patterns were produced on spraying with 0.5% phosphomolybdic acid in butanol solution, the non-polar components of the oil forming blue-grey peripheries, while weakly polar components produced violet-grey colours in the solvent field. Ethanol was applied as a second developing solvent, which caused the non-volatile hydrocarbon components to concentrate on the periphery. Perhydropyrene was found to be a suitable comparison standard. Under the specified conditions, the lower detection limit is claimed to be 0.003 mg l^{-1} for polynuclear aromatics such as phenanthrene, but for oils containing a considerable quantity of saturated hydrocarbons, the limit is 0.1 mg l^{-1}. A photobromination process was developed for the more volatile hydrocarbons down to n-hexane, which are not determined by the above procedure. The carbon tetrachloride extract was brominated under ultraviolet light by using bromine. Saturated compounds from polybrominated substitution compounds of considerably less volatility. These were chromatographed in a similar manner, using ethanol as the only developing solvent. Ammoniacal silver nitrate was used as a visualization agent, yielding basically grey areas after spraying. This modification enabled the detection of 0.01 mg l^{-1} decalin. Detection of petroleum oils containing numerous components would be many times less sensitive.

Semenov et al.[264] determined small amounts of petroleum products in chloroform extracts of natural water by extracting the sample (200–500 ml) followed by thin-layer chromatography on alumina. He developed the chromatogram with light petroleum–carbon tetrachloride–acetic acid (35:15:1), and examined the plate in u.v. radiation; the petroleum products exhibit three zones (pale blue, yellow, and brown). Each zone is then extracted with chloroform, the fluorescence of the extracts measured and the results referred to a calibration graph. The sensitivity is 0.1 mg l^{-1}. The infrared and fluorescence spectra of the zone obtained with various petroleum products are discussed.

Goretti et al.[265] discussed the thin-layer chromatographic determination of the constituents of ether extracts of industrial waste waters. A light petroleum extract of the sample is evaporated and the residue is weighed, dissolved in ethyl ether (5–50 μg) and applied to silica gel in a lane formed by scratching two lines 3 mm apart along the plate and widening them beyond the width of the applied spot at the start. The chromatogram is developed with chloroform–benzene (1:3) for 13 cm at 18 °C in 75 min. Separate controls are prepared for various amounts of vegetable oils (mixtures of olive and seed oils) and mineral oils (petroleum hydrocarbons containing 60–70% of naphthenes). The materials are identified by the fact that mineral oils travel with the solvent front, vegetable oils show an R_F of $\simeq 0.5$ and other compounds (e.g. phenols, acids, and alcohols) remain at the start. For determination of the constituents, the spot lengths are compared with those of the appropriate controls, the relationship between spot length and the amount of applied material being rectilinear up to 8 µg. The materials in the respective spots were analysed and identified by extraction followed by gas chromatography.

Hunter[266] discussed the quantitation of environmental hydrocarbons using thin-layer chromatography and compared the relative effectiveness of gravimetric and densitometric evaluations of the developed plate. The method involves the use of silica gel adsorbent, which is capable of separating samples into saturates (alkanes) and unsaturates (mainly aromatics) at sensitivities to 0.5 µg. Comparison of the data obtained by gravimetric and densitometric methods indicates that densitometry is less accurate owing to variable hydrocarbon response, and it is concluded that the need for repeated checks and calibration make this method only marginally more convenient than the more accurate gravimetric procedure.

Paper chromatography

Sinel'nikov[267] determined bitumoids in open reservoir water using paper chromatography. The water sample is extracted repeatedly with chloroform at pH 7.0 and then at pH 3.0. The combined extracts are evaporated at 40 °C to a small volume and the bitumoids are concentrated in the zone of capillary rise on a strip of chromatographic paper and are separated by dipping the strip into 5 ml of 70% ethanol and allowing a chromatogram to develop for 12 hours. After drying, the separate bitumoid fractions are cut out of the paper and extracted with 10 ml of chloroform and the fluorescence of the extract is measured.

Fluorescence techniques

Fluorescence techniques, when they are applicable, are extremely sensitive. The predominant fluorescent substances in petroleum products are polycyclic aromatic hydrocarbons and heterocyclic compounds, and therefore fluorescence should be dependent on the type of oil being examined. This has been found to be an important factor,[268] and variations in fluorescence intensities of up to eight times have been found within a small number of lubricating oils examined.[269] Therefore most fluorescent techniques for determining oil in water with useful accuracy, demand a sample of the polluting oil, in order that the fluorescence characteristics of the oil can be evaluated. Saturated hydrocarbons do not show any significant fluorescence. In rivers, lakes, and sea water in the vicinity of estuaries, the background fluorescence of naturally occurring organic matter and sewage effluent can be very high, and cause serious interference, while in ground water and in distant sea water it is usually less significant.

Most published procedures deal with the determination of petroleum products in industrial effluents from flotation processes, boilers, refineries etc.[270-281] Usually the oil is extracted from the aqueous sample with a small volume of organic solvent, for example, benzene, petroleum ether,[270,276,277] chloromethane, diethylether, or toluene,[279] or non-fluorescing gasoline.[276] In all these methods, a sample of the polluting oil was available and hence calibration was possible. Often the fluorescence level of the blank was that of the incoming process or boiler feed water. The intensity of fluorescence was determined in commercial fluorimeters, and concentration limits of $01.5-0.2$ mg l^{-1},[274] $100-300$ mg l^{-1} with $\pm 0.7\%$ error,[270] and a lower detetion limit of 0.5 μg l^{-1} [271] have been reported.

Fluorescence techniques have been used occasionally to detect the presence of a volatile petroleum product in polluted water without any accurate quantitative reading, in relation to a known incident of pollution, where the oil and its fluorescent characteristics could be ascertained. Under these limitations the technique could be sensitive and rapid.

Parker and Barnes[269] employed a spectrofluorimeter to determine lubricating oil mist in air, and demonstrated several features of importance regarding oils in water. They examined back-axle oil, commercial engine oils, bicycle oil, and high grade engine oils by excitation of the oil in cyclohexane solution with ultraviolet light at 248 nm or 286 nm, and then measured the resulting fluorescence intensity at 357 nm. Even within the small group of oils examined the fluorescence intensities depended considerably on the nature of the oil. The measurement of fluorescence emission and excitation spectra gave some qualitative information on the lubricating oil, but the authors reached the conclusion that nothing more than a rough quantitative estimate of oil content is provided, unless the oil itself is available for calibration, in which case 0.06 μg l^{-1} air can be determined.

Freegarde et al.[282] extended the technique to crude oil in sea water, which was extracted by cyclohexane and the fluorescence emission spectra measured. The sensitivity was limited by the blank, but since it was found that little plankton or sewage pollution was present, the minimum detectable oil content was about 0.001 mg l^{-1}.

Nietsch[283-286] and Leoy[287] applied simple fluorescence techniques to the detection of petroleum products in natural waters. Because of the relative purity of the water he was able to detect seepage of the petroleum products into spring water down to about 0.001 μg l^{-1} in favourable circumstances. However, a more realistic level would be about 1 mg l^{-1} with a large undetermined error. The fluorescence of the polluting oil, which could be extracted directly with ether or petroleum ether, or absorbed on to magnesia, was compared to that of solutions of a standard motor oil in water. Only estimates of the oil content are permissible, a limitation that has led to an evaluation of the application of fluorescence to accurate quantitative determination of petroleum oils in water.[288] The authors realized that if any accuracy was to be achieved the oil pollutant must be known to enable its fluorescence characteristics to be ascertained. Aqueous emulsions of various oils were ultrasonically obtained, and their characteristic spectra obtained. The wavelengths of maximum intensity produced by excitation at a suitable wavelength (365 nm) were neither characteristic nor reproducible. However, qualitative classification of the oil is claimed as feasible through measurement of three parameters:

(1) the value of total intensity,
(2) the width of the curve in nm at half its height, and
(3) the wavelength corresponding to the centre point of this width.

For crude oil these values were found to be (1) 21 units, (2) 72 nm, and (3) 494 nm, and (1) 180 units, (2) 50 nm, and (3) 484 nm for a commercial oil. The qualitative data then allows accurate quantitative calibration of the fluorescence and $1 \, mg \, l^{-1} \pm 10\%$ is claimed as the lower limit. Unfortunately, practical application to natural waters is complicated, for the authors conclude by reporting that other fluorescing materials present interfere by 'overshadowing' the fluorescence of the oil. Also, at lower concentrations, the sensitivity of the method is insufficient. Therefore chemical separation and enrichment procedures are recommended before the method may become applicable. Paper chromatography has been used as a separation technique before application of quantitative fluorescence analysis, to determine bitumens in water.[289]

Combinations of analytical techniques

Wade and Quinn[290] measured the hydrocarbon content of sea surface and subsurface samples. Hydrocarbons were extracted from the samples and analysed by thin-layer and gas-liquid chromatography. The hydrocarbon content of the surface microlayer samples ranged from 14 to 599 $\mu g \, l^{-1}$ with an average of 155 $\mu g \, l^{-1}$, and the concentration in the subsurface samples ranged from 13 to 239 $\mu g \, l^{-1}$ and averaged 73 $\mu g \, l^{-1}$. Several isolated hydrocarbon fractions were analysed by infrared spectrometry and each fraction was found to contain a minimum of 95% hydrocarbon material, including both alkanes and aromatics.

Golden[291] described a procedure which separates hydrocarbons into three groups: C_5–C_{10} aliphatic and short chain aromatic hydrocarbons, C_{10}–C_{32} aliphatic hydrocarbons, and polycyclic hydrocarbons such as 3,4-benzopyrene. Nitrobenzene was used to extract the first group which are then identified by gas chromatography. 'Heavy' hydrocarbons are extracted with carbon tetrachloride, passed through Florisil, and analysed by infrared spectroscopy, gas chromatography, and by weighing. Polycyclic hydrocarbons are extracted with cyclohexane and determined by thin-layer chromatography and ultraviolet spectrography.

Liu[292] measured hydrocarbon contents in anaerobically digested chemical sewage sludges. Infrared, gas chromatographic, mass spectroscopic, and biological assay techniques were used. Levels of petroleum hydrocarbons ranged from 434 to 7580 mg l^{-1}. A simple extraction method is described for the measurement of total lipids which tended to be associated with solid and particulate material.

Miscellaneous techniques

Zsolnay and Kiel[293] have used flow calorimetry to determine total hydrocarbons in sea water. In this method the sea water (1 litre) was extracted with trichlorotrifluorethane (10 ml) and the extract was concentrated, first in a vacuum desiccator, then with a stream of nitrogen, to 100 μl. A 50 μl portion of this solution was injected into a stainless steel column (5 cm × 1.8 mm) packed with silica gel (0.063–0.2 mm) deactivated with 10% of water. Elution was effected, under pressure of helium, with trichlorotrifluoroethane at 5.2 ml per hour and the eluate passed through the calorimeter. In this the solution flowed over a reference thermistor and thence over a detector thermistor. The latter was embedded in porous glass beads on which the solutes were adsorbed with evolutions of heat. The difference in temperature between the two thermistors was recorded. Each solute first displaced the adsorbed solvent molecules, giving a desorption peak (negative); this was followed by an adsorption peak (positive). The area of the desorption peak was proportional to the amount of solute present. All hydrocarbons were eluted as one peak, the silica gel column removing the non-hydrocarbon solutes. Responses to unsaturated hydrocarbons were less than those to saturated ones; if the peak area were calculated as nonodecane, an error of 25% could be introduced. The limit of detection was about 4 μg l^{-1}.

The turbidimetric approach to the determination of oil is illustrated by the work of Witmer and Goilan[294] on a method for determining oil in ship ballast water. Samples were collected in square-shaped sample bottles (50 ml), which had been wetted with a surfactant (Triton X-100) to prevent oil plating out on the walls, and were ultrasonically emulsified, with cooling. The light transmittance of each sample was then measured in a turbidity meter, with use of oil-free ballast water as a reference, and the apparent oil content was determined from a calibration graph for the specific oil type in question. The oil-droplet size distribution was determined by photomicrography of the sample as it is passed through a flow cell (0.25 in. × 0.375 in.).

Lee and Walden[295] have described a photometric method for determining traces of kerosene in effluents. The method, which is based on the miscibility of kerosene with acetone and its relative insolubility in water, can be applied to samples containing up to 10 mg of kerosene per litre. To the sample (1 litre) is added activated carbon (0.2 g). The sample is stirred for 5 minutes then filtered through a glass-wool plug in the stem of a funnel. Excess of water is removed from the stem by suction with a water pump for 5 min. The carbon column is washed with acetone (5 × 3 ml). The combined washings are evaporated to 2 ml and diluted to 10 ml with acid lauryl sulphate solution (1 ml of conc. sulphuric acid and 1 g of Na lauryl sulphate per litre cooled to 10 °C before use) to produce a stable turbidity. The extinction is measured at 550 nm and the kerosene concentration obtained from a calibration graph.

Freegarde et al.[296] have reviewed methods of identifying and determining oils in sea water. They describe a method for determining down to 0.001 mg l^{-1} of crude oil in sea water.

AROMATICS

Ultraviolet spectroscopy is a sensitive technique for quantitatively determining aromatic hydrocarbons and various heterocyclic compounds, in water or organic solvents. Saturated hydrocarbons such as paraffins generally do not have any significant absorption. The ultraviolet spectroscopic technique is very often neither accurate nor specific since it has been demonstrated that although the general shape of spectra of oils are similar, the intensity of the maximum absorption varies according to the type of oil.[297] If the polluting oil is known, and a sample is available for calibration, then an accurate sensitive method can sometimes be developed. Unfortunately the background absorption of river water organics is often high. A river sample was found to produce seven times more absorption than water containing 100 mg l^{-1} benzene, while phenol and pyridine gave 4 and 25 times more absorption respectively.[298] A separation procedure based on the passage of an extract through a column of absorbent, e.g. silica gel, alumina, or Fluorisil, would elute saturated hydrocarbons almost quantitatively. However, the absorbing portion of petroleum oils, i.e. aromatic and polycyclic aromatic hydrocarbons, and heterocyclic compounds, would be difficult to separate positively in a quantitative manner. This probably also applies to fluorescence. However, in the cases of ground water, sea water distant from estuaries, and possibly very clean rivers, which can all have a much lower background absorption, the technique is more applicable.

Many reported determinations of oils involving ultraviolet analysis are concerned with industrial waste water, and often the blank sample is the incoming river water. Saltzman[298] devised a differential photometric monitoring method that continuously determined aromatic hydrocarbons in water. The incoming river water was compared to the plant effluent in a dual-beam analyser with two sample cells. The accuracy of the method is greatly dependent on the aromatic contamination being known, constant composition, and also on the strong background river water absorption remaining constant during the plant process. In most of the published procedures[299-309] the actual pollutant was known and available for calibration. The aqueous sample has been extracted with carbon tetrachloride,[300,306] hexane,[307,308] ether,[301,303] ether/octane,[305] or octane[302] and the absorption of the solution examined at either 220 nm, 240–280 nm,[307] 260 nm,[306] about 300 nm,[301] or 360 nm.[300] The limits of the procedure are reported as 0.5–50 mg l^{-1},[303] 0.1–200 mg l^{-1},[299]

>0.5 mg l^{-1},[308] 0.03–50 mg l^{-1},[305] 1–1000 mg l^{-1},[304] 13.5 mg l^{-1},[300] and 2.5–2000 mg l^{-1}.[302] Moneva and Angelieva[301] determined the level of petroleum products in waterways by carbon tetrachloride extraction and absorption at 262 nm. Using appropriate calibration curves, a sensitivity limit of about 0.2 mg l^{-1} with 1–2% error was found. The technique was applied to river water in which 0.8–16.0 mg l^{-1} of petroleum products was determined.

Reisus[308] evaluated ultraviolet spectroscopy and its application to the determination of diesel fuel oil in natural waters, based on liquid-liquid extraction with n-hexane and absorption of ultraviolet light at 220 nm. Detection of 0.01 mg l^{11} of diesel fuel was feasible and the errors in higher concentration ranges were calculated as being:

mg l^{-1}	Percentage error
>5	±5
1–5	±10
0.5–1	±20
<0.5	±50

On the basis of his findings, the author concludes that naturally occurring substances interfere, especially phenolic substances which are coextracted into the hexane, and also that the absorption varies with the type of oil. Therefore, the procedure is inaccurate for pollution by unknown oils or oil mixtures, and also when hexane-soluble neutral organic compounds are present, which absorb in a similar range.

Recently Hennig[311] has applied ultraviolet spectroscopy to the determination of aromatic constituents of residual fuel oil in sediment samples. Examination of the ultraviolet spectra of samples of an oil pollutant from a beach and crude oil, at various concentrations, revealed strong absorption maxima at approximately 228 nm and 256 nm. The ratio of the peak heights at these wavelengths is constant for a particular oil, and is independent of concentration. These permit quantitative analysis of sediment samples many months after an oil spill. This method, based on the procedure of Levy,[312] is described below.

Method

Sediment core samples were collected and sorted in polyethylene containers. An aliquot of beached oil obtained from rocks containing relatively large amounts of water and sand was dissolved in n-hexane and filtered through a Whatman No. 1 PS (phase separating) filter paper to remove solids and some of the water (approximately 58%). The rest of the water was removed by refluxing. A stock solution of the oil in n-hexane was prepared.

A stock solution of crude oil was prepared by dissolving 10 μl Arabian Light No. 1169 in spectroanalysed n-hexane (100 ml).

Standards containing approximately 5, 10, 20, 25, 40, and 50 mg oil per litre n-hexane were used to construct a standard curve.

Procedure

An oven-dried (60 °C overnight) sediment sample (10 g) was placed in a glass-stoppered 25 ml flask. The sediment was extracted with *n*-hexane (20 ml) by inversion, to dissolve the oil contained in the sand. The extract was filtered through a Whatman No. 42 filter paper. The absorption spectra of the extracted material (approximately 11 ml) was scanned between 210 nm and 350 nm and the absorbance at the crest of the peaks (approximately 256 nm and 228 nm) was measured, using a double beam scanning spectrophotometer, Beckman Model DN 1402 (slit width 5 mm, scanning rate 40 μm min^{-1}–50 cycles).

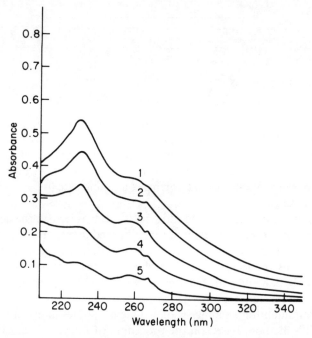

Figure 54 Ultraviolet absorption spectra of the Victoria Bay oil pollutant: (1) 43.4 mg l^{-1}; (2) 34.6 mg l^{-1}; (3) 25.9 mg l^{-1}; (4) 17.3 mg l^{-1}; (5) 8.7 mg l^{-1}

The ultraviolet absorption spectra of the standards are shown in Figures 54 and 55. The salient features of the spectra are the strong absorption peaks at 225 and 228 nm and the considerably weaker shoulder in the vicinity of 256 nm. Levy[312] showed that the ratio of the peak heights at 228 nm and 256 nm (*R* value) is constant for a particular oil but varies with different oils. This ratio was found to be independent of the concentration of oil over the range 8–78 mg l^{-1} (Tables 26 and 27). The peak height at either 228 nm or 256 nm can be used for quantitative analysis. The relationship between concentration of oil and peak height in the concentration range 8–78 mg l^{-1} is linear (Figure 56). Levy[312] found the *R*-value for crude and residual fuel oil to range from

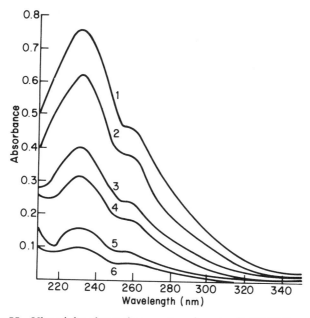

Figure 55 Ultraviolet absorption spectra of crude oil: (1) 77.7 mg l^{-1}; (2) 62.1 mg l^{-1}; (3) 38.8 mg l^{-1}; (4) 31.1 mg l^{-1}; (5) 15.5 mg l^{-1}; (6) 7.8 mg l^{-1}

Table 26 Ultraviolet absorbance characteristics of the Victoria Bay oil polluted at different concentrations

Concentrations (mg l^{-1})	Absorbance		Ratio A_{228}/A_{256}
	228 nm	256 nm	
8.7	0.105	0.071	1.47
17.3	0.218	0.151	1.44
25.9	0.347	0.239	1.45
34.6	0.443	0.306	1.44
43.3	0.541	0.373	1.45

The R value mean = 1.45; standard deviation = 0.01 and coefficient of variance (s.d./$m \times 100$) = 0.85% with $n=5$.

Table 27 Ultraviolet absorbance characteristics of a crude oil at different concentrations

Concentrations (mg l^{-1})	Absorbance		Ratio A_{228}/A_{256}
	228 nm	256 nm	
7.8	0.095	0.058	1.64
15.5	0.155	0.095	1.63
31.1	0.310	0.190	1.63
38.8	0.400	0.240	1.67
62.1	0.620	0.380	1.63
77.7	0.760	0.470	1.62

The R value mean = 1.64; standard deviation = 0.02 and coefficient of variance (s.d./$m \times 100$) = 1.07 with $n=6$.

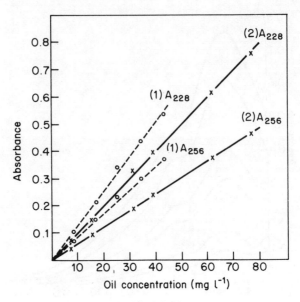

Figure 56 Calibration curve of (1) residual fuel oil concentration beween 8.7 mg l^{-1} and 43.3 mg l^{-1}; (2) crude oil concentration between 7.8 mg l^{-1} and 77.7 mg l^{-1}

Table 28 Ultraviolet absorbance characteristics and amount of extracted oil per 100 gram of sediment

Sample	R value	Calculated oil concentration mg oil per 100 g dry sediment	
		Victoria Bay standards	Crude oil standards
Vic 1/1/3	1.44	3.96	4.44
Vic 1/1/5*	1.58	6.37	7.10
Vic 1/1/10	1.45	1.65	1.88
Vic 2/2/3	1.41	1.83	2.08
Vic 2/2/5	1.50	1.20	1.38
Vic 2/2/10	1.40	0.91	1.06
Vic 2/3/3	1.47	2.61	2.94
Vic 2/3/5	1.50	2.54	2.86
Vic 2/3/10	1.44	1.92	2.18
Vic 3/1/3	1.44	3.38	3.87
Vic 3/1/5	1.41	11.04	12.64
Vic 3/1/10	1.44	1.08	1.20
Vic 3/2/3*	1.56	4.90	5.62
Vic 3/2/5	1.41	1.77	2.02
Vic 3/2/10	1.42	1.45	1.68
Vic 3/3/3	1.44	0.87	0.97
Vic 3/3/5	1.49	1.33	1.51
Vic 3/3/10	1.49	0.94	1.07
Arni 1	—	0	0

The R value mean = 1.45; standard deviation = 0.034 and coefficient of variance = 2.29% with n = 16; *omitted.

1.23 to 2.11. A sample of the spilled oil was very convenient for preparing curves. Alternatively, it is possible to compare oil concentrations within the R value range of a given class of oils. This is shown by the similarity of the calculated concentrations of oil obtained using the peaks for the Victoria Bay oil and the crude oil standards (Table 28).

The absorption spectra mainly reflect the composition of the aromatic compounds in a crude oil sample. The aromatic fractions account for 1.22–30% by mass of crude oil. Whenever environmental biodegradation has been studied, the aromatic hydrocarbons were found to be the most resistant. Since aromatic substances are rare in the marine environment, their detection is a good criterion of oil pollution in sediment. This was substantiated further by scanning the hexane extract of a sediment sample (10 g) from a pollution-free beach. This spectrum (Figure 57 Arni 1) has no peaks caused by the presence of aromatic compounds.

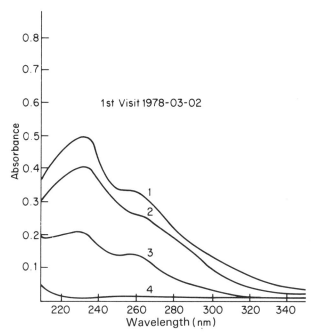

Figure 57 Ultraviolet absorption spectra of field sediment (1) Vic 1/1/3; (2) Vic 1/1/5, diluted 1 in 2; (3) Vic 1/1/10; (4) Arni 1

A recovery constant of 97% was obtained using 10 g of clean sediment samples spiked with 777 μg oil. The standard deviation was 0.014, and the coefficient of variance = 3.47% with $n = 10$.

The ultraviolet absorption spectra of extracts of oil-contaminated sediment samples and unpolluted beach (Arni 1) are shown in Figure 57–59 while the R value and concentrations of oil are given in Table 28. It can be seen that the

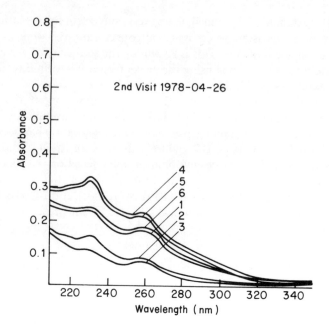

Figure 58 Ultraviolet absorption spectra of field sediment: (1) Vic 2/2/3; (2) Vic 2/2/10; (4) Vic 2/3/5; (6) Vic 2/3/10

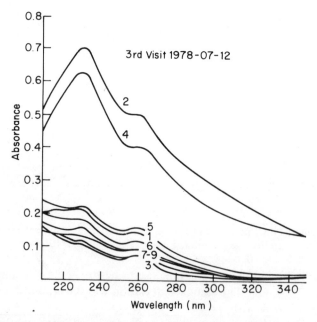

Figure 59 Ultraviolet absorption spectra of field sediment; (1) Vic 3/1/3; (2) Vic 3/1/5; (3) Vic 3/1/10; (4) Vic 3/2/3; (5) Vic 3/2/5; (6) Vic 3/2/10; (7) Vic 3/3/3/; (8) Vic 3/3/5; (9) Vic 3/3/10

R value for this particular oil was fairly constant. In fact, Levy[312] has shown the R value for a given oil to be constant for more than one year. This fact makes the method particularly useful in long-term pollution studies. Samples Vic 1/1/5 and Vic 3/2/3 are possible exceptions. Here the R values are outside the 95% confidence limit which indicates that it is very likely that pollution of the beach arose from an alternate source other than the suspected oil.

Hargrave and Phillips[313] have used fluorescence spectroscopy to evaluate concentrations of aromatic constituents in aquatic sediments. The oil concerned, a Venezuelan crude, contained about 35% by weight of aromatic constituents. Aromatic substances were extracted with n-hexane and fluorescence spectroscopy was used to produce a series of contour diagrams of fluorescence intensity at various excitation and emission wavelengths, in order to compare fluorescence spectral patterns of sample extracts and standard oils. Petroleum residues were determined and it was found that total oil concentrations ranged from 10 to 3000 μg g^{-1} wet sediment, with the highest concentrations occurring in sedimenting particles.

Standardized spiking experiments were performed by Hargrave and Phillips[313] with various amounts of Venezuela crude oil (approximately 35% aromatic compounds by weight) added to freshly collected beach sand and mud sediments in various concentrations. Recovery efficiencies of total oil added with n-hexane were always in excess of 85% (Table 29).

Several experiments were conducted to determine the optimum hexane:water ratio to be used for extraction. When no distilled water was present, wet sediment coagulated in hexane and fluorescence intensity in extracts was reduced in comparison with that achieved with water present. The addition of distilled water

Table 29 Comparison of fluorescence intensity in extracts of Bermuda beach sand by different extraction methods. Approximately 1.5 g wet sand added to either n-hexane (10 ml with 1 ml distilled water) or methylene chloride (2×10 ml extractions combined, evaporated under vacuum and residue taken up in 10 ml n-hexane). Coefficient of variation $V = S/\bar{X} \times 100$

Solvent	Treatment		No. samples	Fluorescence intensity* $\bar{X}$	S	Chart units V (%)
n-Hexane	5 min shaking		5	19	3.2	17.0
	15 min shaking		11	25	4.4	17.5
	60 min shaking		5	26	3.7	11.3
Methylene chloride	two 10 min shaking periods		7	22	3.1	14.2
n-Hexane	Soxhlet extraction (three	2 h	3	19	2.7	14.6
	successive aliquots of	3 h	3	2	0.2	17.6
	of the same sample)	17 h	3	0	0	0

*Maximum sensitivity 1000 chart units. 1 μg Guanipa crude oil in 10 ml n-hexane gives approximately 12.5 chart units of fluorescence with excitation at 310 nm, emission at 374 nm, slit widths 12 nm. Reprinted with permission from Hargrave and Phillips.[313] Copyright (1975) Applied Science Publishers Ltd.

Table 30 Comparison of fluorescence intensity in samples of Bermuda beach sand extracted with various volumes and proportions of distilled water and n-hexane. 1 g wet sand shaken 15 min on a Burrell wrist-action shaker

No. samples	Volume (ml)		Ratio water/ hexane	Fluorescence intensity*		Chart units V (%)
	hexane	water		$\bar{X}$	S	
3	1	20	0.05	20	1.3	6.5
8	5	20	0.25	19	4.3	23.3
4	1	10	0.10	28	1.1	4.0
4	2	10	0.20	26	0.9	3.3
4	5	10	0.50	15	4.5	30.6

*As described in footnote to Table 29
Reprinted with permission from Hargrave and Phillips.[313] Copyright (1975) Applied Science Publishers Ltd.

produced a slurry of sediment and hexane and the fluorescence intensity in extracts was increased (Table 30). Ratios of volume of water:hexane between 0.1 and 0.2 resulted in maximum extraction of fluorescing compounds and minimized variance between replicate samples.

Shaking with a wrist-action shaker for periods in excess of 15 min did not significantly increase the amount of fluorescing material removed from Bermuda sand samples (Table 29). Soxhlet extraction of sand with three successive 200 ml aliquots of n-hexane demonstrated that all fluorescing material could be removed during a 3 h period (Table 29). Total concentration, however, was only 80% of that observed in extracts derived by shaking, perhaps implying loss of volatile fluorescing compounds during extraction.

The most consistent results were obtained with sample sizes ranging from 0.1 to 2.0 g. Fluorescence per unit weight in samples <0.1 g was highly variable. Sample size up to 2 g did not affect relative recovery efficiencies which ranged from 90 to 100% of the maximum extracted (Figure 60).

Method

The following extraction procedure was adopted. After draining on n-hexane rinsed Whatman No. 1 filter paper over vacuum to remove excess water, wet sediment was placed in a 50 ml Pyrex centrifuge tube which had been soaked in soap and water for 12 h, rinsed with tap water, redistilled methylene chloride, and twice with redistilled hexane immediately before use. The amount of sediment used (0.25–2 g) depended on the particle size.

Distilled water (1 ml) and 10 ml redistilled n-hexane were added to each sediment sample to form a slurry. Centrifuge tubes were capped with hexane-rinsed foil-lined caps which were renewed for each sample. Tubes were shaken (maximum amplitude) on a wrist-action shaker (15 min) and then centrifuged (1500 rpm) for 10 min. Extracts could be left at this stage for several days without changes in fluorescence intensity occurring.

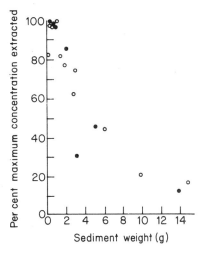

Figure 60 The effect of sample size on estimated oil concentration in mud (closed circles) from Bedford Basin sediment raps and in sand (open circles) from Eastern Passage. Each point represents a single extraction of wet sediment. Maximum concentrations in 1 g samples of mud (1920 $\mu g\,g^{-1}$) and sand (110 $\mu g\,g^{-1}$) estimated by comparison with Guanipa crude oil. Reprinted with permission from Hargrave and Phillips[313] Copyright (1975) Applied Science Publishers Ltd

Extracts were transferred to quartz fluorescence spectrophotometer cells (1 cm path length) using hexane-rinsed glass disposable pipettes. Emission spectra (280–520 nm, 20 nm intervals) were scanned over a range of excitation wavelengths (240–440 nm, 20 nm intervals) with a recording Perkin Elmer MP F-2A fluorescence spectrophotometer (all wavelength values are instrument settings and deviation from absolute values may be as great as 4 nm). Slit widths for both excitation and emission monochromators were always set at 12 nm and sensitivity varied from 1 to 3. Those areas of the spectral diagram where excitation and emission wavelengths approach each other were not considered since only the Raman band of the solvent could be recorded. The interval of scanning was chosen to provide sufficient detail for contour plotting. No decay in fluorescence of samples occurred during the interval (30 min) of exposure to the range of excitation wavelengths.

An off-line plotting routine, using a PDP 8 (Digital Equipment Corp.) computer to drive a Calcomp 563 Plotter, was used to generate contour diagrams relating fluorescence intensity to excitation and emission wavelengths. Fluorescence intensity of redistilled hexane blanks (excitation 310 nm, emission 374 nm, sensitivity setting 3) usually equalled 5 chart units and thus contour diagrams were only generated for sample extracts with fluorescence intensity at those wavelengths in excess of 50 chart units. Contour intervals, specified in the plotting programme, were chosen on the basis of the range of fluorescence intensity observed.

Special contour diagrams prepared from several crude and refined oils were compared with those generated by scanning sediment extracts. A match was assumed to exist when fluorescence peaks coincided. The degree of similarity between standard oil and sample spectra was quantified by plotting contours of difference in fluorescence intensity at all excitation:emission wavelengths after recalculating all fluorescence values relative to a maximum of 1000. Zero contours in these plots indicated spectral regions of no difference in fluorescence patterns. If excitation:emission wavelengths giving maximum fluorescence occurred within the region of minimal difference in relative fluorescence intensity between the spectra, these wavelengths could be used as fixed settings for calibration purposes. Linear calibrations were achieved by measuring the fluorescence intensity of a dilution series (0–30 μg oil/10 ml n-hexane). Fluorescence intensity in sample extracts, measured at similar wavelength settings, could then be expressed as an equivalent concentration of the standard oil. Sample extracts with high fluorescence intensity were diluted with n-hexane until concentrations fell within the calibration range.

Hargrave and Phillips[313] point out that application of the fluorescence method for estimating oil in water, sediments, and organisms requires that fluorescence be related to a standard. Wavelengths for measurement were chosen by scanning excitation and emission spectra and identifying peaks. An oil which contains similar excitation and emission maxima is then used as a standard to produce a calibration. Thus, total oil concentrations are not measured, only inferred, and the use of such calibration assumes that all fluorescence in samples is attributable to the presence of fluorescing compounds equivalent to those in the petroleum standard.

Several sources of error arise from the assumption that fluorescence intensity in samples is quantitatively related to a standard oil. Fluorescence analysis detects only aromatic compounds which comprise from 10 to 60% by weight of various oils, and this proportion changes during weathering and degradation. If oil is present in survey samples it is usually of unknown origin and composition and there may be significant differences between excitation:emission spectra of samples and some arbitrarily chosen standard. In addition, the presence of any heterocyclic compound which fluoresces at wavelengths similar to those of the aromatic compounds present in petroleum will interfere with the assumed quantitative relation to a standard oil. A comparison of fluorescence spectra for various aromatic molecules indicates a considerable degree of overlap, and separation of mixtures by chromatography is required before characterization by fluorescence spectral pattern is possible.

Parker and Barnes[314] and Lloyd[315-317] have demonstrated that fluorescence spectral analysis, using combinations of excitation and emission wavelengths, can be employed to characterize mixtures of aromatic compounds. Freegarde et al.[318] suggest that complete variations in fluorescence spectral patterns can be represented if emission spectra are scanned over a range of different excitation wavelengths and contours of equal fluorescence plotted. Such contour diagrams topographically indicate the combination of excitation and emission wavelengths which results in maximum fluorescence and permits comparison of fluorescence patterns produced by the aromatic substances present in different samples.

Fluorescence spectroscopy has been adapted as an alternative analytical method for estimating oil in various environmental samples. The method was originally used to detect oil in surveys of oil in sea water (Zitko and Carson;[319] Michalik and Gordon;[320] Levy;[321,322] Levy and Walton;[323] and Keizer and Gordon[324]) and in organisms and sediments (Zitko and Carson;[319] Scarratt and Zitko[325]).

Gas chromatography has found some applications in the determination of simple aromatics in water. Mel'kanovitskaya[326] has described a method for determining C_6–C_8 aromatics in subterranean waters. In this method the sample (25–50 ml) is adjusted to pH 8–9 and extracted for 3 min with 0.5 or 1.0 ml of nitrobenzene; the extract is washed with 0.3 ml of 5% hydrochloric acid or 5% sodium hydroxide solution and with 0.3 ml of water adjusted to pH 7. If necessary, oxygen-containing compounds are removed by washing 0.3 ml of the extract with 0.3 ml of 85% phosphoric acid and then with water. For the removal of complex esters, 0.3 ml of the extract is shaken with 3 drops of saturated methanolic hydroxylammonium chloride and 2 drops of saturated methanolic potassium hydroxide for several minutes and is then washed with 0.3 ml of 5% hydrochloric acid, 0.6 ml of 65% phosphoric acid and 0.3 ml of water. The purified extract is subjected to gas chromatography at 85 °C on a column (1 m × 4 mm) packed with 15% of polyoxyethylene glycol 2000 on Celite 545 (60–80 mesh) and operated with nitrogen (10 ml min^{-1}) as carrier gas, decane as internal standard and flame ionization detection.

Mirzayanov and Bugrov[327] have described a procedure for determining ethylbenzene and styrene in effluents in which the sample (50 g) is mixed with 0.1 g of cyclohexanol (internal standard) and extracted with carbon disulphide (25 ml). A portion (1–10 μl) of the extract is analysed by gas chromatography on a column (2 m × 3 mm) of Celite 545 supporting 5% of nitrile silicone rubber and operated at 135 °C, with nitrogen (33 ml min^{-1}) as carrier gas and a flame ionization detector. Concentrations down to 0.001% of ethylbenzene and styrene can be determined.

Wasik and Tsang[328,329] have described a method for the determination of traces of arene contaminants using isotope dilution gas chromatography. They used perdeuterated benzene as the isotope source and analysed solutions containing 10–20 mg l^{-1} of benzene and toluene. This method is most effective when the isotope can be completely separated from the parent compound and other contaminant, otherwise the isotope ratio must be determined by mass spectrometry.

Gas chromatography has also been used to distinguish between fossil fuels added to sediments through oil pollution and those hydrocarbons present in low concentrations as natural biogenic products (Blumer and Sass;[330] Farrington and Quinn[331]).

Dudova and Diterikhas[332] analysed down to 0.2 mg l^{-1} aromatics in underground water by a photometric procedure. The sample (100 ml) is shaken vigorously for 3 min with 10 ml of benzene-free hexane; the extract is filtered through paper, 3 ml of nitrating mixture added (10 g of dry NH_4NO_3 in 100 ml of conc. H_2SO_4), with vigorous stirring. The lower layer is poured into a test tube and heated on a boiling-water bath for 30 min then neutralized with conc. aqueous ammonia and extracted with 10 ml of ethyl ether. To determine benzene 1.0 ml of the extract is diluted with acetone to 5 ml and 0.5 ml of 20% aqueous sodium hydroxide added, and after 30 min is measured photometrically (green filter). To determine toluene and other aromatic hydrocarbons ethanol instead of acetone is used, only one drop of 5% sodium hydroxide solution is added and the colour measured without delay.

Continuous monitoring for oils

Although several methods have been developed for continuously monitoring oil or organic meterials in water by techniques such as ultraviolet spectroscopy[333-335] and infrared analysis[336,337] only a very limited number of published works deal with the detection of surface slicks. This is probably due to the inherent difficulties involved in their design and economic utilization.

An oil detector and warning device has been developed[338,339] mainly for waterworks intakes, but it is also claimed to have applications on inland waters generally. The apparatus automatically detects, reports, and records oil films down to $1 \mu\text{m}$ thickness. The whole device floats and therefore tolerates fluctuations in water levels. It operates on flowing waters between 0.1 and 1 m s^{-1} and differentiates between significant and insignificant pollutions. It is claimed to require little maintenance. The detector consists of a rotating plastic disc with fitted electrodes on its periphery. The depth of immersion is adjustable. Oil is removed from the water during rotation of the disc, which causes a decrease in conductance across the electrodes, and finally by a cleaning device before reimmersion.

Reflection measurements were used by Goolsby[340] to monitor water surfaces continuously for oil films at critical points on water streams or surfaces. Calculations[341] suggested that reflection of unpolarized light from an oil surface (refractive index $n = 1.40$) should occur with about 50% greater intensity than from a water surface ($n = 1.33$) for angles of incidence in the 5–30° range. Experiments with simple laboratory apparatus, i.e. an incandescent lamp, collimating lens, and a photodetector, gave a 100% increase in reflection when a drop of waste oil was added to a water surface. A prototype oil film monitor detected and recorded oil films on a refinery effluent stream flowing at 1 ft s^{-1}. Salinity, turbidity, and changes in water flow of two to three times had little effect on its behaviour. Operation on a tidal estuary is said to be possible. A baffle was required to eliminate surface turbulence, which is a critical factor, and to avoid fouling. The fouling of glass 'windows' on the device was a problem and in the prototype the response to oil was reduced by 50% in 6–8 weeks, although on inland waters weekly cleaning would probably be considered acceptable.

Mattson[342] has carried out investigations on the application of infrared spectroradiometry to the problem of remote detection and identification of oil slicks, without physical contact with the oil or water. The spectroradiometer incorporates a circular variable interference filter as a monochromator, and a small computer for instrumental control and data reduction. In the preliminary experiments, films of styrene and oleic acid were studied, and spectra between 8.5 and 12.5 μm obtained. Radiation was reflected from the oil film, gathered by a Cassegrainian collector, and focused on the circular variable filter wheel of the spectroradiometer, which in these particular experiments allowed scanning between 8.5 and 12.5 μm. Thin films of about 10 μ thickness tended to produce weakly characteristic styrene reflection spectra, but films of about 300 μm produced very definite styrene reflection spectra.

The desire to identify the oil, in addition to mere detection, necessitates a complex and expensive device, unsuitable for general use on inland waters. However, the principle of detection only, based on scanning the reflection over a constant limited range, say 3.3–3.6 μm would be a more practical proposition.

Fust et al.[343] have described an instrumental approach to the analysis of oil in water.

POLYAROMATICS

Many polyaromatic hydrocarbons (PAHs) in trace quantities have been shown to be directly carcinogenic to mammals. They are adsorbed on to particulate material which may be present in water samples and are also water-soluble to some extent, so that their occurrence in the environment has caused widespread concern. At least one hundred compounds of this type have been detected and characterized in environmental samples. The basic molecular structure consists of benzene rings either fused together or bridged by methylene side chains. Alkyl substituents also occur.

These compounds can be produced by biochemical degradation of other organic compounds under suitable conditions. They may occur in the environment from the combustion of materials such as wood or leaves. Other sources of aromatic material from which PAHs may be derived include crude oil which can contain 20% by weight of dicyclic and higher PAHs and high grade petrol, the aromatic content of which is over 50%. Unsaturated fatty acids, terpenoids, and steroids may also be potential PAH precursors. Some of the PAHs thus formed are relatively stable to further biodegradation. The significance of these compounds in the environment has been studied but analytical difficulties preclude the examination of large numbers of samples necessary in a detailed study.

Four techniques have been extensively used for the determination of polyaromatic hydrocarbons in water. These are gas chromatography or gas chromatography combined with mass spectrometry, fluorescence spectroscopy, high performance liquid chromatography and thin-layer chromatography. The gas-liquid chromatographic–mass spectrometric approach has the greater potential for obtaining a complete analysis of volatile environmental PAHs. Good separation is achieved by gas chromatography while mass spectrometry allows the best available identification of the separated compounds. This technique does, however, require relatively large samples, 10–100 ng of each component in fact being required in the injected sample. It is also expensive and unavailable in many laboratories. The World Health Organization recommends the two-dimensional thin-layer chromatographic method of Borneff and Kunte.[344] This allows the determination of nanogram quantities, but the number of compounds which are analysed is limited. Six compounds are determined. The procedure is lengthy and repeated handling of the standards increases working hazards.

The data obtained using liquid chromatographic techniques[345-352] indicates that a less effective separation of the more volatile PAHs occurs when comparisons are made with the data obtained using gas chromatography–mass spectrometric techniques although it might be pointed out that in the form of high performance liquid chromatography continual improvements in performance are being made in this technique.

Other techniques which have been applied to the determination of PAHs include X-ray, excited optical luminescence[353] and gel permeation chromatography,[354] isotope dilution gas chromatography[356] and collection on open pore polyurethane columns.[357]

Gas chromatography

Searl et al.[355] have described a method for determining PAHs including fluoranthrene, pyrene, and benzo(a) pyrene in coke oven effluents by the combined use of gas chromatography and ultraviolet absorption spectrometry. The sample is collected on a silver membrane filter, which is subsequently extracted with cyclohexane in a Soxhlet extractor. After addition of the internal standard, 1,3,5-triphenylbenzene, the extract is evaporated to small volume and subjected to gas chromatography on a column (10 ft × 1/8 in o.d.) of 2% of SE-30 on acid-washed and dimethylsilylated Chromosorb G (80–100 mesh). The column temperature is programmed from 175 to 275 °C at 4° per minute. The effluent is split so that 15% goes to a flame ionization detector and the rest is trapped in fractions corresponding to the recorded peaks for ultraviolet absorption determination of the compounds in the fractions.

Natusch and Tomkins[358] described a method for the isolation of PAHs by solvent extraction with dimethylsulphoxide, then back-extracted into pentane by addition of water to the dimethylsulphoxide. Partition coefficients for a variety of aliphatic and aromatic hydrocarbons and their derivatives were determined, to establish the types of compound which can be separated, and to examine the interactions in the PAH–dimethylsulphoxide system. The composition of the fractions extracted into dimethylsulphoxide was determined by gas chromatography.

A gas chromatographic method using a short packed column has been described by Frýcka.[359] This method is capable of separating critical pairs of PAHs, e.g. benzo(a)pyrene and benzo(e)pyrene in the course of a few minutes.

Dunn and Stich[360] have applied gas chromatography to the determination of PAHs (anthracene, phenanthrene, fluoranthrene, pyrene, chrysene, benz(a)anthracene, triphenylene) in solvent extracts of mussels and creosoted wood. Chromatography was carried out on dual columns, $1/8$ in × 10 ft packed with 2% OV-7 on Chromosorb W HP 80–100 mesh. Helium flow was set for an average linear gas velocity of 7 cm s^{-1} and detection was by flame ionization.

Chatot et al.[361] coupled thin-layer chromatography with gas chromatography to determine PAHs. The hydrocarbons were separated by two-dimensional thin-layer chromatography on prewashed alumina–cellulose acetate (2:1), with pentane as solvent in the first direction and ethanol–toluene–water (17:4:4) as solvent in the second direction. Zones fluorescent under ultraviolet radiation were extracted with benzene, the extracts were evaporated, and a solution of each residue in benzene (20 μl) was injected on to a stainless steel column (2 m × $1/8$ in) packed with 4.5% of SE-52 on Chromosorb G (DMCS). The carrier gas was nitrogen (60 ml min^{-1}), the column was temperature programmed from 100 to 290 °C at 3° per minute, and detection was by flame ionization.

Harrison et al.[362] studied the factors governing the extraction and gas chromatographic analysis of PAHs in water. Factors such as initial concentration, presence of suspended solids, and prolonged storage of the samples affected considerably extraction efficiencies. It is recommended that water samples should be collected directly into the extraction vessel and that analysis should be carried out as soon as possible after extraction.

Hellmann[365] and Acheson et al.[366] have also reported on the gas chromatography of PAHs.

Various workers[363,364,367-370] have studied the combination of a fluorescence detector with gas chromatography to the determination of PAHs.

Caddy and Meek[363] combined gas chromatography with a fluorescence detection technique for the determination down to 10 mg of PAHs in water. The water sample was adjusted to pH 2.5 then stirred with celite. The celite was then removed and extracted with cyclohexane. The cyclohexane was evaporated to a small bulk and this injected on to the gas chromatographic column (5% OV-1 or 5% Dexsil on 80–100 mesh DCMS Chromosorb W) and temperature programmed from 70 to 350 °C. After the separation the column-eluate passed to the onlet splitter so that 90% of the eluted material passed to the interface while the remaining 10% passed to the flame ionization detector. From the onlet splitter the column eluate passed to the interface where the eluated material was entrained with cyclohexane. This solution was pumped to the flow cell of a fluorimeter where measurements of fluorescence were continuously made. The fluorescence of the solution of the column eluate was monitored at wavelengths appropriate to some PAHs and the fluorimeter response was displayed on a chart recorder. In the fluorimeter, the samples were excited by an ultraviolet radiation band of 280–330 nm and the generated fluorescence was measured within the range 350–450 nm.

Linear concentration–peak area calibration curves were obtained using this technique for several hydrocarbons in the 0–3 μg range. Caddy and Meek[363] recommended this procedure for the determination of PAHs in environmental samples such as oils and sewage works sludges and for the determination of relatively non-biodegradable PAHs to fingerprint oils in conjunction with the flame ionization detector trace. They applied the technique, in this respect, to central heating fuel oil, lubricating oil, diesel oil, and heating oil. In activated sludge samples they found evidence for the presence of methyl pyrene and benzo(a)pyrene.

Various other workers[371–374] have investigated the application of combinations of gas chromatography with a fluorescence detector to the determination of nanogram quantities of PAHs.

Basu and Saxona[375] investigated the isolation of benzo(a)pyrene, fluoranthene, benzo(j)fluoranthene, benzo(k)fluoranthene, indeno(1,2,3-cd)pyrene, and benzo(ghi)perylene from water. A clean-up procedure is described which is capable of removing impurities introduced from water and foam plugs to the extent necessary for their interference-free detection. This isolation step is followed by gas chromatography using a flame ionization detector operated in a differential mode. The matched columns consisted of 6 ft × 1/8 in stainless steel packed with 3% Dexsil-300 on Chromosorb W (AW, DMCS treated, 100–120 mesh). The chromatograph was programmed in the following manner: initial temp. 200 °C, initial delay 2 min; programme rate 4 °C min^{-1}; final temp. 290 °C; final delay 8 min.

The stainless steel injection port liner was kept at 250 °C and the detector block at 300 °C. To avoid ghost peaks, the use of aluminium foil backed septa was found necessary. For injection of samples, a solvent flush technique was used to prevent injection error. The conditions used failed to separate the isomers benzo(j)fluoranthene and benzo(k)fluoranthene and therefore combined values are presented. The quantification of these PAHs was based on the response obtained with 1:1 mixture of the two compounds.

The chromatogram of the uncleaned foam eluate shown in Figure 61 points out that a clean-up procedure was required prior to quantification. The gas chromatogram of the foam eluate after purification by solvent partitioning is shown in Figure 61. Several impurity peaks remain in the PAH region, suggesting the necessity of further clean-up. These impurity peaks were eliminated by chromatography on a short Florisil column. The recovery of PAH from the entire clean-up procedure obtained on PAH-spiked foam eluate is presented in Table 31. The data show quantitative recovery for each PAH.

Acheson *et al.*[376] have studied the factors expected to affect the efficiency of extraction of PAHs from environmental water samples. Such factors include the initial concentration of PAH, the presence of suspended solids, and prolonged storage of the sample prior to analysis. These workers examined three

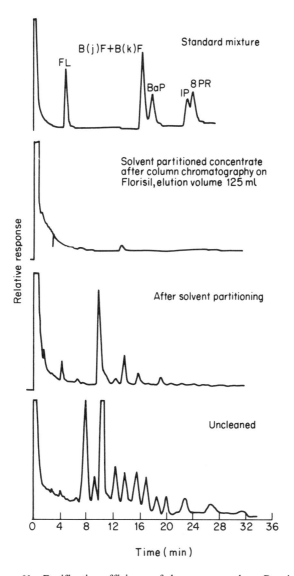

Figure 61 Purification efficiency of clean-up procedure. Reprinted with permission from Basu and Saxona[375]. *Environmental Science and Technology* **12**, 791. Copyright (1978) American Chemical Society

extraction procedures. The first involved conventional extraction in a separating funnel, previously reported by Wedgwood and Cooper[377-379] for the extraction of PAH from water. Second, a continuous solvent extraction procedure was used. The third method involved the use of an Ultra-Turrax (an ultrasonic mixer/homogenizer) as described by Borneff and Kunte.[380]

Table 31 Recovery of overall clean-up method determined by gas–liquid chromatography

Compound	Amount of standard added (μg)	Amount of standard recovered (μg)	Per cent recovery
FL	10.0	8.97	89.7
B(j+k)F	20.0	20.2	101.0
BaP	10.0	10.0	100.0
IP	10.0	8.93	89.3
BPR	10.0	9.15	91.5

FL, fluoranthene; B(j+k)F, benzo(j+k)fluoranthene; BaP, benzo(a)pyrene; IP, indeno(1,2,3-cd)pyrene; BPR, benzo(ghi)perylene.
Reprinted with permission from Basu and Saxona[375] *Environmental Science and Technology* **12**, 791. Copyright (1978) American Chemical Society.

In preliminary experiments, use of a separating funnel was found to give a low extraction efficiency compared to the other methods, and continuous extraction proved to be of comparable efficiency to the Ultra-Turrax, but was far more time-consuming. Consequently the Ultra-Turrax method was chosen for further intensive investigation. Dichloromethane was the preferred extraction solvent because of its good solvent properties for PAH, its low boiling point, and consequent ease of removal by distillation, and the infrequency of persistent emulsion formation when compared to some other solvents.

In the adopted procedure, highly purified water was dosed with PAH and extracted with distilled dichloromethane using the Ultra-Turrax. Subsequent careful evaporation of the solvent, after separation of the dichloromethane layer, and addition of octacosane as an internal standard was followed by analysis by gas chromatography on a Hewlett-Packard 7620 g.c. using balanced dual columns (3.5 m × 0.5 cm o.d. stainless steel; 3% OV-7 on 60, 80 mesh AW DCMS Gas Chromb Q) and nitrogen carrier (35 ml min^{-1} fitted with dual FID. A temperature of 260 °C was maintained for 8 min and then increased to 300 at 8 °C min^{-1}. Octacosane was used as an internal standard. A blank extraction involving no added PAH was used to determine background organic materials in the water and extraction solvent, which were in general found to be extremely low in concentration. Figures 62 and 63 show the variation in extraction efficiency with initial concentrations of pyrene and benzo(ghi)perylene in water. At higher concentrations efficiencies in the region of 80% comparable to those reported by Borneff and Kunte[381] are found. At lower initial concentrations, however, efficiencies may drop below 40%.

Acheson *et al.*[376] studied the effect of suspended solids, in the form of purified Fullers earth on the extraction efficiency of these two PAHs from water. Both the variation in extraction efficiency with the level of suspended solids and the variation with PAH concentration at a fixed level of suspended solids were investigated. The resultant efficiencies for extraction of pyrene and benzo(ghi)perylene are shown in Figures 62–64. The reduced efficiency with

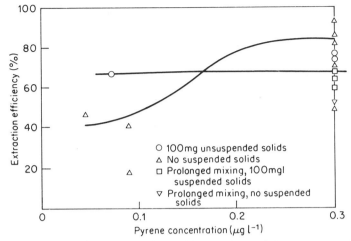

Figure 62 Variation of extraction efficiency with the initial concentration of pyrene showing the effect of suspended solids and prolonged mixing. Reprinted with permission from Acheson *et al.*[376] Copyright (1976) Pergamon Press Ltd

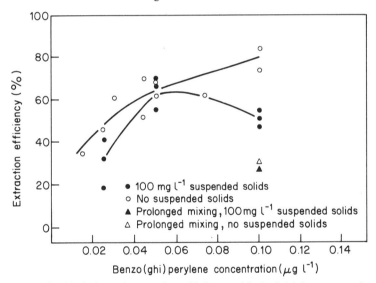

Figure 63 Variation of extraction efficiency with the initial concentration of benzo(ghi)perylene showing the effects of suspended solids and prolonged mixing. Reprinted with permission from Acheson *et al.*[376] Copyright (1976) Pergamon Press Ltd

increasing levels of suspended solids for both PAH studied appears to indicate adsorption on the solids, with a consequent increased ability to remain in the aqueous suspension. The variations in extraction efficiency with solute concentration at a fixed level of suspended solids are not readily explained: in particular the higher extractions obtained for lower concentrations of pyrene

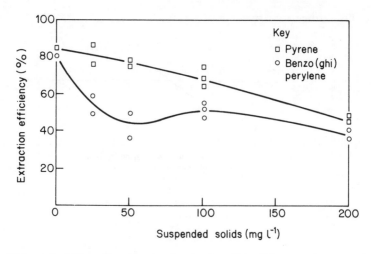

Figure 64 Effect of various levels of suspended solids upon the extraction efficiency for pyrene and benzo(ghi)perylene at initial concentrations of 0.3 and 0.1 μg l^{-1} respectively. Reprinted with permission from Acheson et al.[376] Copyright (1976) Pergamon Press Ltd

with 100 mg l^{-1} of suspended solids than without added solids would not be predicted.

Solvent extraction of the water sample for an extended period of 6 hours was shown to produce a lower extraction efficiency of PAHs than was obtained with a shorter extraction time, but, in this case, the extraction efficiency was not further reduced in the presence of suspended solids. Extraction efficiencies were the same with or without solids after 6 h mixing. These results may be explained either by adsorption of PAH on to the glass of the mixing (and extraction) vessel, or degradation (chemical or biological) within the vessel.

Two conclusions were drawn from these results and applied to the analysis of field samples. First, it is essential to sample water directly into the extraction vessel, and second, the extraction should be carried out as soon after sample collection as possible.

Continuous dichloromethane extraction for 36 and 72 h was compared with the above procedures. The results (Table 32) showed that extractions from pure water were very similar to those achieved using the Ultra-Turrax, but in the presence of suspended solids the efficiency of extraction was reduced substantially. Again, no clear explanation of the effect of suspended solids can be put forward.

Solvent extracts of environmental water samples, unlike extracts of the highly purified water contain substantial quantities of organic compounds other than PAH. These impurities were found not to affect adversely the analysis by thin-layer chromatography and fluorescence, but made gas chromatographic analysis of PAH extracts quite impossible. Hence, separation of extraneous organic compounds by the procedure of Hoffmann and Wynder[382] was employed by

Table 32 Comparison of extraction efficiencies (%) using continuous extraction and the Ultra-Turrax

Suspended solids (ppm)	Ultra-Turrax Pyr*	B(ghi)P†	Continuous extraction Time (h)	Pyr*	B(ghi)P†
0	84	82	36	83	55
100	68	53	36	27	16
100	68	53	72	34	15

*Initial concentration 0.3 µg l^{-1}.
†Initial concentration 0.1 µg l^{-1}.

Acheson et al.[376] and was found to produce a solution sufficiently purified to be readily amenable to gas chromatographic analysis. The efficiency of this procedure, which involves consecutive solvent–solvent partitions of PAH from aqueous methanol into cyclohexane and hence into nitromethane, was examined by the use of standard mixtures of PAH and was found to lie between less than 10% and 83% depending on the compound and the initial concentration (Table 33). At higher initial concentration the gas chromatographic analysis failed to reveal perylene in significant quantities, although an additional peak, eluted between octacosane and benzo(k)fluoranthene was apparent. This phenomenon is not readily explained, as both in the examination of this procedure at a lower initial perylene concentration and in the analysis of field samples using this purification procedure, perylene was detected in quantities comparable to those found by thin-layer chromatographic examination of an unpurified extract.

Thin-layer chromatography was carried out using the procedure of Borneff and Kunte[381] with slight modifications. The first development with iso-octane was omitted, and the plates were developed for a fixed distance of solvent front advance rather than a fixed time. After development, the plate was sprayed with a 2% w/v solution of tetrachlorophthalic anhydride in acetone: chlorobenzene

Table 33 Efficiencies of solvent partition purification procedures

Compound	Initial weight (µg)	Nitro-methane procedure	Recovery (%) DMSO modification	Initial weight (µg)	Nitro-methane procedure	DMSO modification
Fluoranthene	0.52	48	92	10.4	83	96
Pyrene	1.21	42	94	24.2	38	96
Benzo(a)anthracene + chrysene	1.54	50	90	30.8	64	92
Benzo(k)fluoranthene	0.60	54	91	12.0	59	94
Benzo(a)pyrene + Benzo(e)pyrene	1.01	36	95	20.2	33	99
Perylene	0.95	24	84	19.0	10	90
Indeno(1,2,3-ed)pyrene	0.38	49				100
Benzo(ghi)perylene	0.79	45	91	15.8	39	90
Coronene	0.50	44	93	10.0	82	110

Table 34 T.l.c. analysis of water samples ($\mu g\, l^{-1}$) (uncorrected for extraction efficiency)

Compound	Kew Bridge			Albert Bridge*	Tower Bridge			M4 Motorway		
	PAH measured in undosed sample	PAH added	PAH measured in dosed sample	PAH measured in undosed sample	PAH measured in undosed sample	PAH added	PAH measured in dosed sample	PAH measured in undosed sample	PAH added	PAH measured in dosed sample
Fluoranthene	0.11	0.05	0.16	0.15	0.27	0.11	0.44	1.10	0.11	1.10
Benzo(k)fluoranthene	0.06	0.06	0.08	0.03	0.09	0.13	0.25	0.32	0.13	0.50
Benzo(a)pyrene	0.10	0.07	0.16	0.12	0.26	0.13	0.52	0.65	0.13	0.98
Perylene	0.03	0.10	0.10	0.05	0.10	0.20	0.30	0.20	0.20	0.40
Indeno(1,2,3-ed)pyrene	0.04	0.04	0.08	0.08	0.16	0.08	0.24	0.32	0.08	0.48
Benzo(ghi)perylene	0.04	0.08	0.08	0.08	0.12	0.17	0.25	0.12	0.17	0.17

*This determination was not carried out in duplicate.

Table 35 G.l.c. analysis of water samples ($\mu g\,l^{-1}$) (uncorrected for extraction efficiency)

Compound	Kew Bridge			Albert Bridge*			Tower Bridge			M4 Motorway		
	PAH measured in undosed sample	PAH added	PAH measured in dosed sample	PAH measured in undosed sample	PAH added	PAH measured in dosed sample	PAH measured in undosed sample	PAH added	PAH measured in dosed sample	PAH measured in undosed sample	PAH added	PAH measured in dosed sample
Fluoranthene	0.18	0.05	0.20	0.02			0.18	0.11	0.21	0.49	0.11	0.64
Pyrene	0.26	0.13	0.29	0.05			0.23	0.25	0.52	0.43	0.25	0.86
Benzo(a)anthracene chrysene	0.14	0.16	†	0.27			0.53	0.32	0.52	0.39	0.32	0.84
Benzo(k)fluoranthene							0.43	0.13	0.68	0.49	0.13	1.06
Benzo(j)fluoranthene	0.24	0.06	0.52	0.15								
Benzo(b)fluoranthene												
Benzo(a)pyrene	0.21	0.11	0.36	†			0.13	0.21	0.33	0.57	0.21	1.00‡
Perylene	0.04	0.10	0.12	†			0.12	0.20	0.43	0.52	0.20	
Indeno(1,2,3-ed)pyrene	0.10	0.04	0.22	0.07			0.11	0.08	0.33	0.14	0.08	0.29
Benzo(ghi)perylene	0.04	0.08	0.08	0.04			0.03	0.17	0.18	0.07	0.17	0.21

*This determination was not carried out in duplicate.
†Not determined due to high background.
‡Peaks not resolved.

(10:1) and dried prior to measurement of fluorescence by irradiation with u.v. lamp ($\lambda = 360$ nm) and visual comparison with standard chromatograms.

The procedure of Hoffmann and Wynder[382] was not highly efficient and appeared to have two major drawbacks. The partition coefficient between cyclohexane and methanol for PAH was such that four time-consuming extractions were needed to allow a high recovery. Further, the high temperature necessary for evaporation of the nitromethane encouraged loss of PAH both by volatilization and thermal degradation. Haenni et al.[383] have described the use of dimethylsulphoxide as a solvent suitable for extraction of PAH from aliphatic solvents. It was incorporated in the procedure of Hoffmann and Wynder[382] in place of nitromethane, and the extracted PAH were isolated by subsequent dilution with water and back-extraction into cyclohexane which was then evaporated. This procedure had a much improved efficiency of 84–110% (Table 33).

Acheson et al.[376] compared the gas chromatographic and the thin-layer chromatographic methods on river water and motorway drainage samples. Samples were extracted with dichloroethane. Seven compounds were identified on the thin-layer plate (Table 34). Gas–liquid chromatographic analysis was also performed on the dichloromethane extracts, after purification by the procedure of Hoffmann and Wynder.[382] Results (Table 35) show a less accurate measurement of added PAH and indicated a generally lower precision than thin-layer chromatographic analysis, probably as a result of the difficulty of assessment of gas chromatographic peak areas in the presence of a variable background. In a small proportion of cases measurement of peak areas was rendered impossible by a high background, and no estimate of PAH concentration was made.

Both the gas chromatographic and the thin-layer chromatographic procedures have their disadvantages. The thin-layer method avoids extract purification and associated losses, but is of poor sensitivity for a number of compounds. Accuracy does, however, appear high for the compounds determined, as shown by analysis of PAH added to field samples. Gas–liquid chromatography has a more uniform sensitivity, but was unable to separate some isomeric compounds, and consequently some carcinogenic PAHs are determined only in combination with non-carcinogenic isomers. The gas chromatographic procedure is more susceptible to interference by background organic materials and appears to be of lower accuracy than thin-layer chromatography, although it clearly is able to give results of the correct order of magnitude.

Capillary gas chromatography

The use of glass capillary gas chromatography for determination of PAHs in cigarette smoke was first reported by Grimmer.[384] For several types of sample, the initial separation of PAHs from water is still accomplished by some workers by slightly modified versions of this method, viz. extraction into cyclohexane, washing the cyclohexane extract with dimethylformamide, and gas chromatographic determination of the individual components. One or two internal standards are used in quantitative determinations. Glass capillary columns with 80,000–120,000 theoretical plates are required.

Owing to their very low solubility in water PAHs occurring in industrial waste water are largely adsorbed on suspended solids. For example, Kadar et al.[385] showed that waste water from the aluminium industry has a PAH content of 10–150 μg l^{-1} after filtering through a Micropore filter.

The quantitative removal of organic pollutants from water is usually achieved by extraction into an organic solvent. This process is, however, time-consuming and often necessitates the use of large volumes of solvent. In order to circumvent these drawbacks, adsorption on a solid phase, such as Amberlite XAD, porous polyurethane resin and Tenax (a porous polymer based on 2,6-diphenylphenylene oxide) is often used.

Kadar et al.[385] extended this work by studying the efficiencies of Tenax for the removal of PAHs from standard water solutions. The method is applied to waste water samples from an aluminium plant. In this method the water samples were passed through the Tenax column at the rate of about 5 ml min^{-1}. Residual water was removed from the Tenax by passing nitrogen gas through the column. The Tenax material was then transferred to a Soxhlet apparatus. PAH and other organic compounds were extracted by reflux for 4 h with 35 ml of acetone. The Tenax can then be dried and reused. Preliminary experiments indicate that the extraction time can be reduced to 10–15 min by the use of ultrasonic extraction. Residual water is removed from the acetone extract by passing the solution through a small column of anhydrous sodium sulphate and washing with 5 ml of acetone. The extract is evaporated on a water bath at 50 °C in a stream of nitrogen to a volume of approximately 0.5 ml.

The residual solution was transferred to a thin-layer plate. The thin-layer plates were developed in a 4:1 v/v mixture of hexane and benzene. The time required for the development is 30–35 min. A single development of the plate is sufficient for separating PAH from other compounds such as paraffins, naphthenes, acids, and phenols. The spots were located visually under a u.v. lamp and the areas containing aromatic compounds were marked. The appropriate portions of the thin-layer were scraped into a glass flask. PAHs were extracted by vigorous shaking for 2 h with 5 ml of chloroform. Extraction of PAHs from the thin-layer material is a critical step in the procedure. Extraction into approximately 1 ml of chloroform by treatment in an ultrasonic bath for about 10 min was shown to be the best procedure. The internal standard was added to the chloroform in advance so that the samples can be analysed directly after centrifugation. After filtration or centrifugation, the solution may be used for spectrophotometric determination of total PAHs and gas chromatographic determination of individual PAH components.

The conditions for the gas chromatography were as follows:

Column: glass capillary, 25 m × 0.28 mm coated with SE-30, detection by flame ionization detector.

Carrier gas: helium at a flow rate of 2 ml min^{-1}.
Injector temperature: 300 °C.
Detector temperature: 300 °C.
Volume injected: 1 μl.
Column temperature: initially 100 °C, held for 4 min before programme starts.
Programming: 3° min^{-1}.
Final temperature: 260 °C.

Kadar *et al.*[385] found that the separation of PAHs from water on the Tenax column was the most critical step in the analytical procedure (Figure 65). Unsatisfactory results are often due to neglect of details at this step. The most important operating parameters are the height of the Tenax column and the flow rate of the water sample through the column. In this work, the height of the column was set at 10 cm without any appreciable decrease in PAH recovery. The optimum flow rate was 8–10 ml min^{-1}, for a Tenax column of 10 cm height and 13 mm diameter.

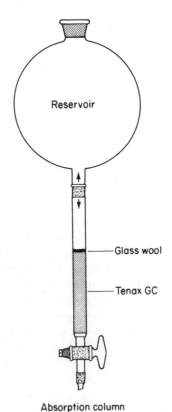

Figure 65 Sorption column with Tenax bed

Table 36 PAH yield from water samples

Hydrocarbon	Amount $\mu g\,l^{-1}$	Liquid–liquid extraction* $\mu g\,l^{-1}$	Kadar method[385] undiluted sample		Kadar method[385] 10 times diluted sample	
			$\mu g\,l^{-1}$	± s.d.†	$\mu g\,l^{-1}$	± s.d.
Anthracene	11.0	10.3	10.2	0.5	0.8	0.1
Pyrene	32.0	30.8	30.2	0.7	2.6	0.1
Chrysene	43.0	41.3	40.3	0.6	3.2	0.3
3-Methylcholanthrene	12.0	10.9	11.0	0.3	0.9	0.1
Total	98.0	93.3	91.7		7.5	
Total yield		95	94		77	

*Glass capillary column g.c. method.
†All standard deviations ($\mu g\,l^{-1}$) are based on 4 parallel results.

The PAH values obtained by gas chromatography are presented in Table 36. As indicated, excellent agreement is obtained at the 100 ng ml^{-1} level between the method and a much more time-consuming liquid–liquid extraction technique. Table 36 shows the overall recovery of PAHs from water at the 100- and the 10 ng ml^{-1} levels. At the 100 ng ml^{-1} level the recovery is 90–95%. At the 10 ng ml^{-1} level the recovery decreases to 70–90% and the spread of the analytical results increases. At or below the 10 ng ml^{-1} level, the amount of PAHs recovered may be increased by using a larger volume of water, for instance 5 or 10 litres.

In Table 37 and Figure 66 are shown results obtained for waste water sample from an aluminium plant. Parallel determinations were carried out on samples of 500 ml each. The standard deviation for the series is approximately 1 $\mu g\,l^{-1}$.

Unused Tenax may contain small amounts of low-molecular-weight components which can interfere with PAH components of the group chrysene, benzofluoranthene, benzo(e)pyrene, on a thin-layer chromatographic plate. It is therefore recommended that unused Tenax be extracted with acetone overnight. As shown in Figure 67 the acetone extract of unused Tenax gives rise to fluorescent spots on the thin-layer plate, with a retention factor of about 0.4. Since several PAH components have about this retention factor, these artefacts will interfere in the thin-layer procedure. Judging from their intensity, this interference will be small at the 100 ng ml^{-1} level, but may become dominant at lower concentrations. These artefacts will also interfere in the gas chromatographic analysis.

When analysing industrial waste water, it is advisable to protect the Tenax column against irreversible contamination. This can be achieved by using a short Celite 545 prefilter.

The thin-layer plate must be preconditioned in chloroform before use. This preconditioning is particularly important when the PAH level is very low, 10 ng ml^{-1} or less. One-dimensional thin-layer chromatography will give a good separation of semipolar PAH components from non-polar paraffins and

Table 37 Typical results for waste water from an aluminium plant

Substance	Sample 1 Run 1 ($\mu g\,l^{-1}$)	Sample 1 Run 2 ($\mu g\,l^{-1}$)	Sample 2 Run 1* ($\mu g\,l^{-1}$)	Sample 2 Run 2 ($\mu g\,l^{-1}$)
Biphenyl	3.2	3.6	<1	<1
Fluorene/fluorenone	6.3	8.4	2.9	3.2
Phenanthrene	16.9	23.1	14.2	14.0
Anthracene	2.8	2.8	1.1	1.2
Fluoranthene	20.8	18.9	10.8	12.4
Pyrene	15.3	12.7	5.6	6.0
Benzo(a)fluorene	3.2	3.4	1.6	1.5
Benzo(b)fluorene	2.8	3.0	1.3	1.3
Benzdiphenylensulphide	3.4	3.5	2.0	1.7
Benzo(a)anthracene	2.5	2.8	5.6	5.5
Chrysene/triphenylene	5.8	6.0	15.6	16.0
Benzo(b,k)fluoranthene	6.8	6.8	38.1	38.0
Benzo(e)pyrene	2.6	2.7	16.2	16.4
Benzo(a)pyrene	1.3	1.5	7.0	7.4
Dibenzoanthracene	3.4	4.3	8.2	8.0
Anthanthrene	<1	<1	3.2	3.2
Coronene	<1	<1	1.9	2.0
Dibenzopyrene	<1	<1	4.0	4.3

*The chromatogram for this run is shown in Figure 66.

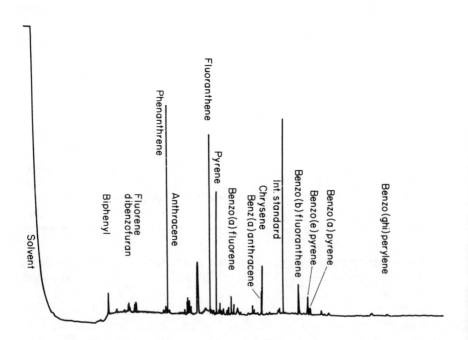

Figure 66 Typical chromatogram of industrial sample with standard added

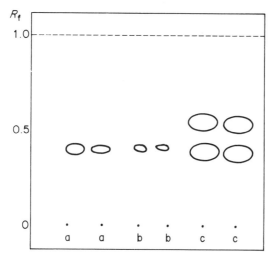

Figure 67 T.l.c. of unused Tenax extracted with acetone; a,b, two different Tenax batches (unused) extracted with acetone; c, acetone PAH solution, 25 μg of PAH per spot

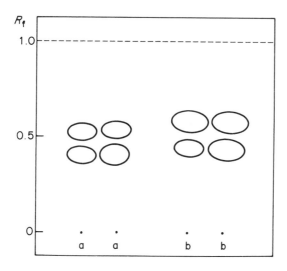

Figure 68 T.l.c. of PAH on Kieselgel 60 F254 (Merck); (a) sorbed on Tenax and extracted into acetone (initial concentration corresponds to 25 μg of PAH per spot); (b) initial acetone solution 25 μg of PAH per spot

naphthenes which will follow the solvent front, and from the more polar acids and phenols which will remain at the bottom of the plate. Figure 68 shows a thin-layer chromatogram of four PAH components. Under ultraviolet light, the chromatographic spots may serve as a visual indication of the amount of PAH. In Figure 68 it is noticeable that the total area of the chromatographic spots on the left-hand side (a) of the plate is less than the area to the right (b). Both sets of chromatographic spots correspond to the same initial amount of PAH (50 μg). The spots to the left, however, correspond to PAH components which have passed through the analytical procedure and thus have suffered a loss of about 10%. This loss is visible on the plate.

Giger and Schnaffner[386] described a glass capillary gas chromatographic method for the determination of PAHs in lake and river sediments. PAHs are isolated by a sequence of solvent extraction, gel filtration, and adsorption chromatography and individual concentrations determined by gas chromatography.

Bjorseth et al.[387] described a capillary gas chromatographic method for determining PAHs in sediments and mussels. Up to 34 PAHs were identified, some carcinogenic.

Saxena et al.[388] have also studied the application and limitations of high efficiency glass columns to the analysis of PAHs.

Gas chromatography–mass spectrometry

Benoit et al.[389] have investigated the use of macroreticular resins, particularly Amberlite XAD-2 resin, in the preconcentration of Ottawa tap water samples prior to the determination of 50 different PAHs by gas chromatography–mass spectrometry.

Water samples were prepared as follows: sampling cartridges, containing 15 g Amberlite XAD-2 macroreticular resin that had been previously cleaned by the method of McNeil et al.,[390] were rinsed with 250 ml acetone and washed with at least 1 l of purified water. The cartridges were attached to a potable water supply and the flow of water was controlled at ca. 70 ml min^{-1}. When 300 ml of water had been passed through the cartridge, the cartridge was disconnected from the tap and as much water as possible was removed from the cartridge by careful draining followed by the application of vacuum from a water aspirator. The XAD-2 resin was eluted with 300 ml of 15:85 v/v acetone hexane solution at a flow rate of ca. 5 ml min^{-1} (all solvents were of 'distilled in glass' quality and were redistilled in an all glass system). The organic layer was dried by passage through a drying column containing anhydrous sodium sulphate over a glass wool plug. Both the sodium sulphate and the glass wool plug were cleaned by successive washings with methylene chloride, acetone, and hexane prior to use. The dried solution was concentrated to a volume of ca. 3 ml, using a rotary evaporator, then quantitatively transferred with acetone to a graduated vial and was further concentrated, using a gentle stream of dry nitrogen gas, to a final volume of 1 ml.

To analyse the solvent extracts a 10 μl aliquot of the concentrated extract was injected into a Finnigan 4000 g.c.–m.s. instrument coupled to a 6110 data system. A 3% OV-17 provided the best separation of the detectable PAHs. A 1.8 m × 2 mm i.d. glass column, packed with 3% OV-17 of 80–100 mesh Chromosorb 750, was operated at an initial temperature of 100 °C for 1 min and was programmed to a final temperature of 225 °C at a rate of 3° min^{-1} and held at that temperature for the remainder of the analysis. The flow of helium carrier gas was set at 20 ml min^{-1} and the injection port temperature set at 200 °C. The glass jet separator and the ion source temperatures were set at 260 °C and 250 °C, respectively. Data acquisition was under the control of the Finnigan 6110 data system. The mass range, 35–400 amu, was scanned at a rate of 2.1 s per scan and the mass spectra (*ca.* 1000) stored on magnetic disc for subsequent analysis.

To test the effectiveness of their method of analysis for PAHs in drinking water Benoit *et al.*[389] prepared and analysed a control blank and a recovery study of 32 selected PAHs from XAD-2 resin was conducted.

None of the 55 compounds contained in the standard solution was detected in the concentrated extract from the control blank. This indicates that the XAD-2 resin is effective for the removal of these compounds from drinking water and that none of the reference compounds originates from the precleaned XAD-2 resin. However, when the amounts of PAH loaded and recovered are compared (Table 38), and average recovery of 0.84 is observed, with recoveries ranging from 0.57 to 1 of the loaded material. The weighed average recovery was 0.88 of loaded material. The fate of the unrecovered material was not established although, based on the results of the control blank, it is not likely that these materials were carried away by the effluent water.

Ottawa drinking water samples were analysed in order to obtain some indication of whether the results are representative of the general background level of anthropogenic contamination. Aliquots of the reference standard solution (50 PAHs and 5 O-PAHs) and the concentrated extracts from XAD-2 resin were analysed consecutively by g.c.–m.s. under identical operating conditions. As is evident from Figure 69, a representative gas chromatogram of the drinking water, XAD-2 resin extract concentrate, as reconstructed from the total ion current, contained a multitude of poorly defined peaks. Complete mass spectra, free of extraneous ions, could rarely be obtained from such data despite the background subtraction routine possible with the data system and, hence, individual components of the concentrate was achieved from mass chromatograms (Figure 70) which were reconstructed from selected ion currents rather than the total ion current. Mass chromatograms for selected ions that were characteristic of the compound of interest were obtained by searching the accumulated data for the ion of interest and recording the abundance of this ion as a function of retention time. As an example, the mass chromatograms of three ions — m/e 128, m/e 142, and m/e 154 are superimposed in Figure 70. The location of the peaks corresponding to the compounds of interest are indicated by asterisks in each chromatogram. For m/e 128 the asterisked peak corresponds to the molecular ion of naphthalene, for m/e 142 to the molecular ions of 2-methylnaphthalene and 1-methylnaphthalene, respectively, in order of increasing retention time and for m/e 154 to the molecular ions of biphenyl, 2-vinylnaphthalene and acenaphthalene, respectively, in order of increasing retention times. The retention times for each standard were established by

Table 38 The recovery of selected polycyclic hydrocarbons from Amberlite XAD-2 macroreticular resin

Compound	Amount loaded (ng)	Fraction recovered
Naphthalene	625	0.57
2-Methylnaphthalene	1200	0.88
1-Methylnaphthalene	625	0.71
2-Ethylnaphthalene 2,6-Dimethylnaphthalene	>2300	>0.86
Biphenyl	775	0.66
1,3-Dimethylnaphthalene	975	0.81
2,3-Dimethylnaphthalene 1,4-Dimethylnaphthalene	>2250	>0.82
4-Phenyltoluene	600	0.85
Diphenylmethane 3-Phenyltoluene	>2075	>0.76
Acenaphthene	625	0.60
Bibenzyl	975	0.65
1,1-Diphenylethylene	1625	0.98
cis-Stilbene	575	1.0
2,3,5-Trimethylnaphthalene	700	0.75
3,3'-Dimethylbiphenyl	1625	1.0
Fluorene	750	0.89
4,4'-Dimethylbiphenyl	700	0.90
trans-Stilbene 9,10-Dihydrophenanthrene	>1075	>0.84
Phenanthrene Anthracene	>1700	>1.0
Triphenylmethane	800	0.87
Fluoranthene	650	1.0
Pyrene	725	1.0
1,2-Benzfluorene 2,3-Benzfluorene	>1550	>0.99
Triphenylene Benz(a)anthracene Chrysene	>1650	>1.0
Average	1110	0.84

Reprinted with permission from Benoit et al.[389] Copyright (1979) American Chemical Society

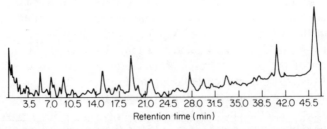

Figure 69 Reconstructed gas chromatogram (total ion current) from Ottawa drinking water Amberlite XAD-2 resin extract. Reprinted with permission from Benoit et al.[389] Copyright (1979) American Chemical Society

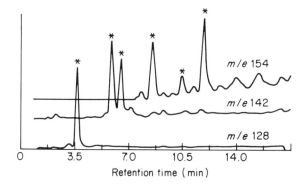

Figure 70 Reconstructed mass chromatogram for ions m/e 128, m/e 142 and m/e 154 from Ottawa drinking water Amberlite XAD-2 resin extract. Reprinted with permission from Benoit et al.[389] Copyright (1979) American Chemical Society

analysis of the reference standard solution and the data from the XAD-2 resin extracts were then searched for the ion of interest within the appropriate time region. In all instances the molecular ion and the next most abundant ion were selected as the characteristic ions which are listed in Tables 39 and 40.

A compound was considered identified if the two characteristics of the compound of interest were found to elute from the column within the retention time window (± 0.1 min) of the reference standard and to be in the relative abundance ratio ($\pm 20\%$) observed in the mass spectrum of the pure compound. For most, but not all, compounds screened unique identification was possible. In some instances however, coeluting isomers yielding similar mass spectra could not be resolved sufficiently to allow unequivocal identification. Such coeluting isomers are grouped together in Table 39 and are indicated by > beside the concentration values which is the sum of the contributions from all coeluting isomers. Furthermore is is emphasized that, because of the large number of compounds contained in the field sample extract, it was not possible to eliminate entirely from all the ion peaks of interest, contributions from possible interfering species. This was particularly true for methyl substituted PAHs for which numerous positional isomers may elute within a narrow time window. In many cases only a small number of the possible positional isomers were available commercially and could be included in the reference standard. Hence, unequivocal identification of positional isomers was often not possible.

Quantitative estimations of the detectable PAHs and O-PAHs in Ottawa drinking water were obtained by comparison of the areas of the two characteristic ion peaks (Tables 39 and 40) in the mass chromatograms of the reference standard and the field sample, respectively. The average of the concentrations for the two ions is presented in Table 39 (PAHs) and 40 (O-PAHs) for the two water samples analysed. No corrections were made for incomplete recovery. Of the 50 PAHs in the standard used by Benoit et al.[389] 38 are detected in at least one of the two drinking water samples tested. In sample 1 (February 1978)

Table 39 Polycyclic aromatic hydrocarbons detected in Ottawa drinking water sampled in January (2) and February (1) 1978

Compound	Ions monitored		Rel. ret. time[a]	Concentration (ng l^{-1})	
				1	2
Naphthalene	128	102	1.00	6.8	4.8
2-Methylnaphthalene	142	141	1.59	2.4	4.6
1-Methylnaphthalene	142	141	1.75	1.0	2.0
Azulene	128	102	1.90	n.d.	n.d.
2-Ethylnaphthalene	156	141	2.26	>0.70	2.1
2,6-Dimethylnaphthalene	156	141	2.32		
Biphenyl	154	153	2.30	0.70	1.1
1,3-Dimethylnaphthalene	156	141	2.51	1.9	1.1
2-Vinylnaphthalene	154	153	2.68	n.d.	n.d.
2,3-Dimethylnaphthalene	156	141	2.69	>0.68	14
1,4-Dimethylnaphthalene	156	141	2.69		
3-Phenyltoluene	168	167	2.74	0.20	1.5
Diphenylmethane	168	91	2.88	1.4	2.8
4-Phenyltoluene	168	167	2.94	0.20	3.7
Acenaphthylene	152	151	3.00	0.05	n.d.
Acenaphthene	154	153	3.25	0.20	1.8
Bibenzyl	182	91	3.41	1.9	1.5
1,1-Diphenylethylene	180	179	3.48	>7.4	n.d.
cis-Stilbene	180	179	3.59		
2,2-Diphenylpropane	196	181	3.62	n.d.	n.d.
2,3,5-Trimethylnaphthalene	170	155	3.71	0.65	5.2
3,3'-Dimethylbiphenyl	182	167	4.03	0.31	5.2
Fluorene	166	165	4.15	0.15	2.2
4,4'-Dimethylbiphenyl	182	167	4.18	0.57	7.0
4-Vinylbiphenyl	180	178	4.56	n.d.	n.d.
Diphenylacetylene	178	89	4.90	0.05	n.d.
9,10-Dihydroanthracene	180	179	5.03	0.66	n.d.
trans-Stilbene	180	179	5.20	>0.47	9.2
9,10-Dihydrophenanthrene	180	179	5.29		
10,11-Dihydro-5H-dibenzo(a,d)cycloheptane	194	179	6.03	0.40	n.d.
Phenanthrene	178	89	6.08	>0.52	2.2
Anthracene	178	89	6.14		
1-Phenylnaphthalene	204	203	6.77	n.d.	n.d.
1-Methylphenanthrene	192	191	7.06	n.d.	11
2-Methylanthracene	192	191	7.25	0.51	0.70
9-Methylanthracene	192	191	7.59	n.d.	0.70
9-Vinylanthracene	204	203	7.82	n.d.	n.d.
Triphenylmethane	244	167	8.04	n.d.	n.d.
Fluoranthene	202	101	8.41	0.55	1.9
Pyrene	202	101	8.90	0.53	1.7
9,10-Dimethylanthracene	206	191	8.99	0.19	n.d.
Triphenylethylene	256	178	9.25	0.08	n.d.
p-Terphenyl	230	115	9.44	n.d.	n.d.
1,2-Benzfluorene	216	108	9.64	n.d.	n.d.
2,3-Benzfluorene	216	108	9.73	n.d.	n.d.
Benzylbiphenyl	244	167	9.75	n.d.	n.d.

(continued)

Table 39 *(continued)*

Compound	Ions monitored		Rel. ret. time[a]	Concentration (ng l^{-1})	
				1	2
1,1′-Binaphthyl	254	126	10.95	n.d.	n.d.
Triphenylene	228	114	11.5		
Benz(a)anthracene	228	114	11.6	>8.1	3.3
Chrysene	228	114	11.8		

[a]Retention times are relative to the retention time of naphthalene (3.81 min).
Reprinted with permission from Benoit *et al.*[389] Copyright (1979) American Chemical Society.

Table 40 Oxygenated polycyclic aromatic hydrocarbons detected in Ottawa drinking water sampled in January (2) and February (1) 1978

Compound	Ions monitored		Rel. ret. time[a]	Concentration (ng l^{-1})	
				1	2
Xanthene	182	181	4.83	0.20	0.10
9-Fluorenone	180	152	5.93	0.90	1.5
Perinaphthenone	180	152	7.70	0.28	0.15
Anthrone	194	165	7.90	1.4	n.d.
Anthraquinone	208	180	8.11	2.4	1.8
Naphthalene	128	102	1.00		

[a]Retention times are relative to the retention time of naphthalene (3.81 min).
Reprinted with permission from Benoit *et al.*[389] Copyright (1979) American Chemical Society.

36 PAHs (Table 39) ranging in concentration from 0.05 to 8.1 ng l^{-1} and in sample 2 (January 1978) 30 PAHs (Table 39), ranging in concentration from 0.05 to 14 ng l^{-1}, were detected. Twenty-eight PAHs and four O-PAHs were detected in both samples analysed. The lower concentration of 0.05 ng l^{-1} represents the lower limit of detection of this method of analysis. There was an appreciable variation in the concentrations of most of the PAHs detected in the two drinking water samples; however, all compounds detected were found to be in the low ng l^{-1} range. This suggests that the observed concentrations of PAHs and O-PAHs are representative of the background level of contamination. The mean concentrations of PAH in samples 1 and 2 was 1.4 ng l^{-1} for a total weight of detected PAHs of 50.4 ng l^{-1} for a total weight of detected PAHs of 114 ng l^{-1}, respectively.

Of the five O-PAHs in the standard, all five, ranging in concentration from 0.20 to 2.4 ng l^{-1}, are detected in sample 1 and four, ranging in concentration from 0.10 to 1.8 ng l^{-1} are detected in sample 2 (Table 40). The mean concentrations and total weights of detected O-PAHs in the two samples were 1.0 and 5.2 ng l^{-1} for sample 1 and 0.91 and 3.7 ng l^{-1} for sample 2 respectively. It is noteworthy that for three of the oxygenated compounds (anthrone, anthraquinone, and 9-fluorenone) detected, the parent compound is also detected in the drinking water sample. Thus, the oxygenated species could possibly originate from the oxidation of the parent compound in the aqueous media.

Matsumoto and Hanya[391] used gas chromatography–mass spectrometry to identify and estimate ng l^{-1} amounts of squalane in river water samples and sediments taken in the Tokyo area. Water samples were extracted three times with ethyl acetate and then chromatographed on a silica gel column. The column volumes of hexane eluate were analysed using gas chromatography–mass spectrometry. Squalane was identified in all samples of river waters and sediments, as well as in night soil and sewage-works effluents and sludges, with concentrations of 0.46–1.7 μg l^{-1} in river waters and 0.86–15 μg g^{-1} of dry sediment.

Tan[392] devised a rapid simple sample preparation technique for analysing PAHs in sediments. PAHs are removed from the sediment by ultrasonic extraction and isolated by solvent partition and silica gel column chromatography. The sulphur removal step is combined into the ultrasonic extraction procedure. Identification of PAH is carried by gas chromatography alone and in conjunction with mass spectrometry. Quantitative determination is achieved by addition of known amounts of standard compounds using flame ionization and multiple ion detectors.

Lao et al.[393] described a computerized gas chromatographic–mass spectrometric analysis of PAHs in environmental samples such as coke oven emissions, coal tar, airborne particles, and wood preservative sludges. This method involves a three-step method: preliminary separation of polycyclic aromatic hydrocarbons by solvent and/or column chromatography, identification by gas chromatogram combined with a quadruple mass spectrometer and data processor, and measurement by computerized gas chromatography using internal standards. The efficiencies of different Dexsil gas chromatographic columns were also evaluated during the study.

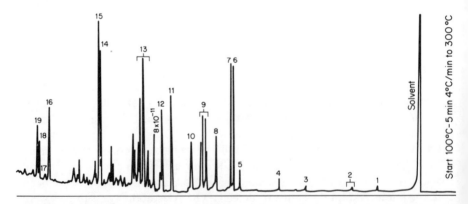

Figure 71 Capillary column gas chromatograph of anthracene oil. 1, Naphthalene; 2, Methylnaphthalenes; 3, Acetnaphthalene; 4, Fluorene; 5, Methylfluorene(?); 6, Phenanthrene; 7, Anthracene; 8, Carbazole; 9, Methylphenanthrenes + methylanthracenes; 10, Methylcarbazole; 11, Fluoranthene; 12, Pyrene; 13, Methylfluoranthenes + methylpyrenes + benzofluorenes; 14, Benz(a)anthracene; 15, Chrysene; 16, Benz(j)fluoranthene + benz(k)fluoranthene; 17, Benzofluoranthene(?); 18, Benzo(e)pyrene; 19, Benzo(a)pyrene

Table 41 G.c.–m.s. conditions

G.c. conditions	
Capillary column:	50 m × 1 mm i.d. OV-1 glass PLOT
Injection:	SGE Splitless 250 °C
Oven temperature:	50 °C for 2 min then programmed at 8 °C min^{-1} to 250 °C
Carrier gas:	Helium at 4 ml min^{-1} (make-up 16 ml min^{-1})
Detector:	AEI MS-30
Sample volume:	2 µl
M.s. conditions	
Ionization voltage:	24 eV
Trap current:	300 µA
Resolution:	1000
Source temperature:	250 °C
Separator:	SGE single jet at 240 °C
Sensitivity: 50	(2.5 kV) (Electron multiplier gain setting)
Mass range:	20–500 amu
Scan speed:	1 s decade^{-1}
Interscan delay:	1 s

Lao et al.[394] have also investigated methods for the estimation of PAHs in air samples.

Crane et al.[395] using a gas chromatographic method detected over 100 PAHs in anthracene oil used in the manufacture of water pipe lining materials (Figure 71). They determined the concentration of several of these compounds. They also applied gas chromatography–mass spectrometry to the identification of some of these compounds. The analytical conditions outlined above (Table 41) were applied to cyclohexane extracts of water extracts of the pipe lining materials.

Fluorescence spectroscopy

Early work on the determination of PAHs in water using fluorescence spectroscopy was carried out by Scholz and Altmann[396] who devised a method for the determination of down to 0.1 ng^{-1} of benzo(a)pyrene in water. In this method the sample is extracted with cyclohexane, and the concentrated extract is subjected to thin-layer chromatography on silica gel H impregnated with polyoxyethylene glycol 1000, with benzene–hexane (1:3) as solvent. The benzo(a)pyrene is located in 365 nm radiation by reference to standards, and is extracted from the adsorbent with cyclohexane. The solution is evaporated, and a solution of the residue in dioxan is subjected to fluorimetry at 429 nm (excitation at 365 nm). Down to 0.1 ng of benzo(a)pyrene per litre of sample could be determined, with a relative error in the range 1–10 ng l^{-1} of ± 15%.

Schwarz and Wasik[397] carried out fluorescence measurements of benzene, naphthalene, anthracene, pyrene, fluoranthene and benzo(e)pyrene in water. They reported fluorescence spectra, quantum yields, and concentration dependencies for these aromatic hydrocarbons in water to ascertain the applicability of measuring PAH in aqueous systems by spectrofluorimetry. The fluorescence quantum yields of benzene, naphthalene, anthracene, pyrene, fluoranthene, and benzo(e)pyrene in water are, respectively, $5.3 \pm 0.5 \times 10^{-3}$, 0.16 ± 0.02, 0.25, 0.69 ± 0.06, 0.20 ± 0.01, and $\sim 0.3 \pm 0.1$. With the exception of pyrene, oxygen quenching of the fluorescence concentration dependence was measured by photon counting the fluorescence intensity relative to the excitation light intensity. All the PAH fluorescences exhibited a linear dependency on the concentration. For a fluorescence signal-to-noise ratio of 1, the detection limits are as follows: naphthalene, $0.03\ \mu g\,l^{-1}$, anthracene, $0.03\ \mu g\,l^{-1}$, pyrene, $0.15\ \mu g\,l^{-1}$, fluoranthene, $0.17\ \mu g\,l^{-1}$, and benzo(e)pyrene, $0.10\ \mu g\,l^{-1}$.

The absorption spectra of the PAH and benzene in water are presented in Figures 72–75. In Figures 76 and 77, the excitation spectra of anthracene and benzo(e)pyrene are shown, and they match their absorption spectra in ethanol. A change of solvent from water to ethanol had no observable effect on the absorption spectra of benzene and the PAH.

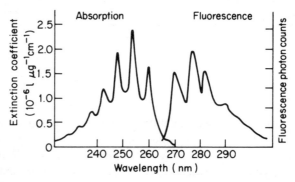

Figure 72 Absorption spectrum of benzene in water and a fluorescence spectrum of $3.2 \times 10^4\ \mu g\,l^{-1}$ of benzene in water. Reprinted with permission from Schwarz and Wasik.[397] Copyright (1976) American Chemical Society

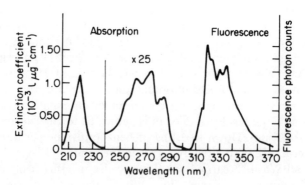

Figure 73 Absorption spectrum of naphthalene in water and a fluorescence spectrum of $3.5 \times 10^3\ \mu g\,l^{-1}$ of naphthalene in water. Reprinted with permission from Schwarz and Wasik.[397] Copyright (1976) American Chemical Society

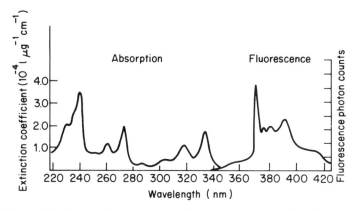

Figure 74 Absorption spectrum of pyrene in water and a fluorescence spectrum of 165 μg l^{-1} pyrene in water. Reprinted with permission from Schwarz and Wasik.[397] Copyright (1976) American Chemical Society

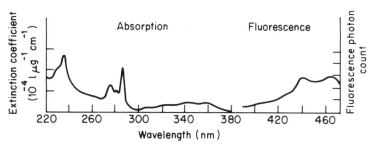

Figure 75 Absorption spectrum of fluoranthene in water and a fluorescence spectrum of 240 μg l^{-1} fluoranthene in water. Reprinted with permission from Schwarz and Wasik.[397] Copyright (1976) American Chemical Society

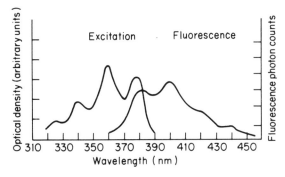

Figure 76 Excitation spectrum of anthracene in water in the near u.v. region and a fluorescence spectrum of 25 μg l^{-1} of anthracene in water. Reprinted with permission from Schwarz and Wasik.[397] Copyright (1976) American Chemical Society

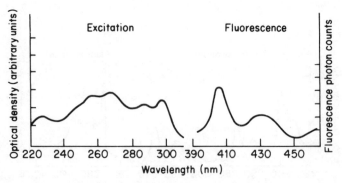

Figure 77 Excitation spectrum of benzo(e)pyrene in water in the 210–330 nm region and a fluorescence spectrum of $\sim 4\,\mu g\,l^{-1}$ benzo(e)pyrene in water. Reprinted with permission from Schwarz and Wasik.[397] Copyright (1976) American Chemical Society

The values of the PAH extinction coefficients in water as a function of wavelength are shown in Figures 72–75 and are presented in Table 42 along with those of anthracene and benzo(e)pyrene. To verify the applicability of these values to the PAH absorption spectra in water, the PAH solubilities were calculated from the optical densities of the saturated solutions and these extinction coefficients. The agreement between the calculated solubilities and the literature values is quite good for most of the values.

The PAH and benzene fluorescence spectra in water are also shown in Figures 72–77. They match their respective fluorescence spectra in ethanol. As in ethanol, the PAH exists as dispersed molecules in water and not as dimers and/or

Table 42 Extinction coefficients and saturated concentrations of PAH in water

PAH	Transition of interest	Extinction coefficient $l/\mu g\,cm$	Calcd satd H_2O concs $(\mu g\,l^{-1})$	Lit. values of satd concs $(\mu g\,l^{-1})$
Naphthalene	$^1A \to {}^1L_a$ 268 nm	3.9×10^{-5}	2.2×10^4	2.3×10^4 1.3×10^{4a}
Anthracene[1]	$^1A \to {}^1B_b$ 253 nm	1.1×10^{-3}	30	39,[b] 75[c]
Pyrene	$^1A \to {}^1B_b$ 240 nm	3.5×10^{-4}	171	165[c]
Benzo(e)pyrene	$^1A \to {}^1B_b$ 297 nm	2.5×10^{-1}	~ 4	3.5[c]
Fluoranthene	$^1A \to {}^1L_a$ 288 nm	1.6×10^{-4}	236	240[c]

[a]Determined by removing an aliquot of the saturated solution and dissolving it in ethanol to measure its o.d.; [b]determined by liquid chromatography; [c]determined by a nephelometric technique.
Reprinted with permission from Schwarz and Wasik.[397] Copyright (1976) American Chemical Society.

crystalline aggregates. In the latter case, red shifts of the fluorescence peaks and/or extra broad bands would appear in the water solvent.[398] These bands would be concentration dependent and would hamper identification of the fluorescences in a mixture of PAH in water.

The PAH in water can be identified by their fluorescence spectra. In Figures 73–77 the fluorescence spectra of the PAH are unique for each PAH. Even in spectral regions where there is overlap of the fluorescences such as those of anthracene, pyrene, and benzo(e)pyrene, the anthracene peaks are at 375 and 400 nm, those of pyrene at 370 and 395 nm and those of benzo(e)pyrene at 405 and 430 nm.

Schwarz and Wasik[397] also investigated in detail the fluorescence–concentration dependence for these substances and established an equation relating these parameters. A typical plot of fluorescence photon counts versus anthracene concentration is shown in Figure 78. There are several problems associated with the direct fluorimetric measurement of PAH in environmental water samples. Suspended matter can increase light scattering and reduce the sensitivity of the method. There is also the possibility of heavy metal ion quenching of the fluorescence intensity. This quenching would be similar to that of dissolved oxygen. These interferences can be removed by gas stripping the PAH and some water vapour out of the water sample. The PAH and water vapour would be collected in a liquid nitrogen trap and analysed by a fluorescence analysis. Another problem is the limited spectral resolution of the excitation and fluorescence measurements in identifying the PAH constituents in a sample mixture.

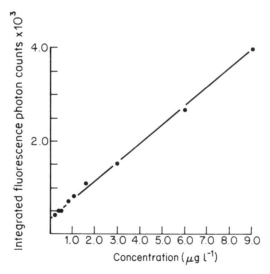

Figure 78 Plot of the fluorescence photon counts of anthracene in water as a function of concentration. Reprinted with permission from Schwarz and Wasik.[397] Copyright (1976) American Chemical Society

Recently, on the basis of the data of animal experiments to establish doses of benzo(a)pyrene which did not produce cancer, Shabad[399] calculated a maximum permissible concentration of benzo(a)pyrene in water of 0.3 ng l^{-1}.

Muel and Lacroix[400] were the first to utilize low temperature spectrofluorimetry to determine the benzo(a)pyrene content of drinking water samples by using the standard addition method, but this compound was undetectable in the small volume water samples examined. Jäger and Kassovitzova[401] determined the concentration of benzo(a)pyrene in drinking water down to concentration of 3 ng ml^{-1} with a relative error of 40% and found that the accuracy was greatly influenced by the presence of other organic compounds. Snow samples and soils were analysed for PAH by Gurov and Novikov[402] and Stepanova[403] et al. developed a procedure for the quantitative analysis of a mixture of benzo(a)pyrene and other PAHs in sewage and other industrial exhausts, based on preliminary thin-layer chromatographic separation and low temperature spectrofluorimetric quantification.

In a more recent and detailed investigation, Khesina and Petrova[404] employed this low temperature technique in the determination of benzo(a)pyrene and seven other PAHs in extracted waste water, after a preseparation by column chromatography. Other work concerned with the determination of benzo(a)pyrene in environmental waters by this method for public health purposes has been reviewed by Andelmann and Snodgrass[405] and recently by Andelmann and Suess.[406]

Further work on the fluorimetric determination of benzo(a)pyrene in water was conducted by Monarca et al.[410] These workers utilized low temperature spectrofluorimetry using the Shpol'skii effect.[411-414] This technique is capable of high selectivity and sensitivity in the determination of PAH compounds at 77 K and can permit this determination in mixtures without lengthy initial chromatographic separation procedures (Hood and Winefordner;[407] Lavalette et al.;[408] Gaevaya and Khesina[409]). The aim in this work was to develop a method sufficiently sensitive to determine benzo(a)pyrene at the 0.3 ng l^{-1} level as prescribed by Shabad.[399] The Environmental Protection Agency has been bound, under a Consent Decree, to set maximum concentration limits in effluent waters for a group of 'unambiguous priority pollutants'.[415] Included on this list are several polycyclic aromatic hydrocarbons. The World Health Organization has recommended that the total concentration of six specific PAH compounds, fluoranthene, benzo(d)fluoranthene, benzo(k)fluoranthene, benzo(a)pyrene, benzo(ghi)perylene, and indeno(1,2,3-ed)pyrene, not exceed 200 ng l^{-1} in domestic water.[416,417] The Federal Republic of Germany has legislated a maximum limit for PAHs of 250 ng carbon l^{-1} of drinking water. The basic assembly of the apparatus employed was similar to that previously described by Kirkbright and de Lima.[412]

A 1 m grating monochromator (Rank Hilger Ltd, Monospek 1000) with an aperture of f/8 and a reciprocal linear dispersion of 0.8 nm mm^{-1} at the exit slit was employed in conjunction with a 50 mm EMI 62565 photomultiplier. The signals were amplified using a microammeter (RAC, Model WV-84C) before being recorded directly on a potentiometric chart recorder. A 125 W mercury lamp was used as excitation source. The 375 and 325 nm wavelengths of excitation were isolated by means of interference filters. An adapted Aminco cold finger Dewar-flask sample cell was used. A coil of Nichrome wire was positioned round the transport quartz base of the Dewar. This wire was heated by passing a low a.c. current through it in order to minimize frosting and thus avoiding light scattering. Alternatively, dry nitrogen can be circulated or a vacuum may be maintained in the sample compartment to achieve the same result. A further source of light scattering, caused by bubbling of liquid nitrogen, may be eliminated by the addition of a small amount of liquid helium to the liquid nitrogen.

Silica tubes (3 mm i.d., 5 mm o.d.) were used as sample cells.

For the extraction procedure 5 l separating funnels were used in conjunction with a stirrer; the cyclohexane (AnalaR, BDH) was distilled twice at 30 °C under vacuum in a rotating evaporator. The recovered cyclohexane was dried with anhydrous sodium sulphate previously washed with cyclohexane.

n-Octane (AnalaR, BDH) was purified by percolating through activated silica gel (60–120 mesh) to remove traces of benzo(a)pyrene impurity[418] and cyclohexane (spectrosol for ultraviolet spectroscopy).

The widespread occurrence of the PAHs in the air of the laboratory and from other sources and the possible contamination of the samples from other interfering substances requires that all operations are undertaken with extremely clean glassware. Therefore it is recommended that glassware be cleaned with acetone and detergents and then carefully rinsed with distilled water prior to the extraction procedure. This cleaned apparatus should remain in contact with a solution of potassium permanganate for 12 h and prior to use be rinsed with distilled water. The graduated flasks were cleaned with detergents and then with acetone and finally with the octane-cyclohexane solution solvent.

Preliminary recovery experiments on synthetic solutions of benzo(a)pyrene in distilled water taken through the extraction and measuring procedure indicated a mean recovery of benzo(a)pyrene of 92.3% with a relative standard deviation of 21.3%. A second extraction of the 2.5 l water samples resulted in an additional 4% of benzo(a)pyrene, thus for routine analysis a single extraction was considered satisfactory.

An octane–cyclohexane solvent (9:1 v/v) was employed for the quantitative determination of benzo(a)pyrene using its quasilinear luminescence emission spectrum at 77 K. This was recorded using an excitation wavelength of 375 nm at at scanning speed of 2.5 nm mm^{-1} with a 0.1 mm spectrometer slit (0.08 nm spectral half band pass). The intensity of the luminescence was measured at the 403.0 nm maximum. As shown in Figure 79 good linearity was obtained in the range of concentration 10^{-8}–10^{-6} M between the signal intensity of 403.0 nm and the concentration of benzo(a)pyrene in the octane–cyclohexane solvent, and furthermore good precision can be obtained for measurements in the 1–25 ng ml^{-1} range which is required for analysis of real water samples (Figure 80).

Examination of a solution of benzo(a)pyrene, indeno(1,2,3-ed)pyrene and fluoranthene in the concentration ratio 1:1:20 showed that there is no evident quenching effect over the concentration range for benzo(a)pyrene from 10^{-8} to 10^{-7} M, whereas at 10^{-6} M the average depression of the benzo(a)pyrene luminescence response at 403.0 nm in this mixture was *ca.* 20% (Figure 79).

No interference from benzo(k)fluoranthene observed by other workers[419] in room temperature spectrophotofluorimetry was observed by Monarca *et al.*[410] at low temperature. Benzo(k)fluoranthene shows maximum luminescence emission at 403 nm when an excitation wavelength of 375 nm is employed; although this is close to that of benzo(a)pyrene at equal concentrations benzo(k)fluoranthene shows only *ca.* 10% of the benzo(a)pyrene response at 403 nm (Figure 80).

To analyse water samples Monarca *et al.*[410] extracted the samples (5 l) with cyclohexane and the extraction was diluted with 1 ml of octane–cyclohexane (9:1) solvent solution; excited at 375 (and 325) nm and the luminescence emission intensities obtained at 403 nm were compared with those obtained from a calibrating curve prepared for benzo(a)pyrene in octane–cyclohexane. The sample solutions were then diluted and the benzo(a)pyrene concentration in each was determined by the standard addition method.

Using this method Monarca *et al.*[410] found 0.5 ng l^{-1} benzo(a)pyrene in drinking water and between 32 and 38 ng l^{-1} in river waters. The limit of detection of the method was 0.5 ng l^{-1} or better using 2.5–5 l samples.

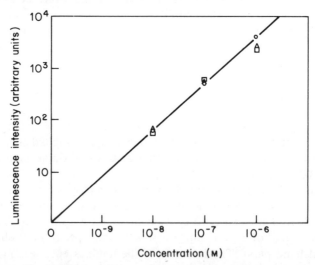

Figure 79 Luminescence intensity (arbitrary units) *vs.* concentration graphs for benzo(a)pyrene alone and in admixture: ○, Benzo(a)pyrene; □, Benzo(a)pyrene in admixture with Indeno(1,2,3-ed)pyrene and Fluoranthene (ratio 1:1:20); △, Benzo(a)pyrene in admixture with Indeno(1,2,3-ed)pyrene, Benzo(k)fluoranthene, and Fluoranthene (ratio 1:1:20). Reprinted with permission from Monarca *et al.*[410] Copyright (1979) Pergamon Press Ltd.

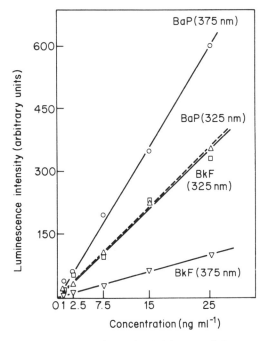

Figure 80 Luminescence intensity (arbitrary units) concentration graphs for benzo(a)pyrene and benzo(k)fluoranthene at different excitation wavelengths: ○, Benzo(a)pyrene (B(a)P) emission at 403.0 nm with 375 nm excitation; □, Benzo(k)fluoranthene (B(k)F) emission at 403.3 with 325 nm excitation; △, Benzo(a)pyrene (B(a)P) emission at 403.0 nm with 325 nm excitation; ▽, Benzo(k)fluoranthene (B(k)F) emission at 403.3 nm with 375 nm emission. Reprinted with permission from Monarca et al.[410] Copyright (1979) Pergamon Press Ltd.

Ogan et al.[420] have described a fluorimetric procedure capable of determining as little as 0.2 ng l^{-1} of 7 H benz(de)anthracene-7-one (benzanthrone), fluoranthene, perylene, benzo(a)pyrene and benzo(ghi)perylene in drinking water samples. The polycyclic aromatic hydrocarbons are determined by adsorption on to an extraction column, followed by desorption and separation in a single step using reversed phase liquid chromatography. The method is described below.

Apparatus: A Perkin-Elmer Model 601 liquid chromatography with a Perkin-Elmer Model 204-A or Model 650-10 LC spectrofluorimetric detector was used. An integrator was used to measure peak areas. The chromatography conditions are given in Table 43. The extraction column was mounted in place of the loop on a loop injection valve as shown in Figure 81. The sample was pumped either with a reciprocating pump or a syringe pump. High purity acetonitrile and methanol were used. Water was filtered, deionized, and purified by a carbon mixed bed system.

Stock solutions of the PAHs were made in tetrahydrofuran or preferably, chloroform at concentrations of approximately 500 μg ml^{-1}. These solutions

Table 43 Chromatography conditions

Liquid chromatography	Perkin-Elmer Model 601
Detector	Perkin-Elmer 204A spectrofluorimeter
	$\lambda_{ex} = 365$ nm
	$\lambda_{em} = 455$ nm (with 390 nm cutoff filter)
Mobile phase	38% Acetonitrile, 15% methanol in water
	1.0 ml min^{-1}
Temperature	65 °C
Analytical column	Perkin-Elmer HC-ODS (10 μm C_{18})
	0.26 × 25 cm
Extraction column	Perkin-Elmer ODS Sil-X-II Packing
	(40 μm C_{18})
	0.26 × 10 cm

Reproduced from the *Journal of Chromatographic Science* by permission of Preston Publications, Inc.

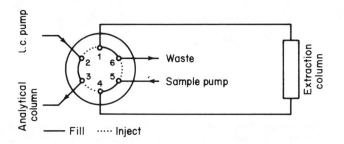

Figure 81 Connection of the extraction column to the loop injection valve. The solvent path is indicated for the valve in the Fill position (—) and the Inject position (····). Reproduced from the *Journal of Chromatographic Science* by permission of Preston Publications, Inc.

Table 44 Detection limit (direct injection)

Compound	Detection limit (pg)
7H-Benz(de)anthracene-7-one (BA-one)	125
Fluoranthene (Fth)	70
Perylene (Per)	60
Benzo(a)pyrene (B(a)Py)	50
Benzo(ghi)perylene (B(ghi)Per)	220

Reproduced from the *Journal of Chromatographic Science* by permission of Preston Publications, Inc.

were used to make a stock PAH mixture in methanol. Portions of this stock PAH mixture were then diluted with 20% methanol in water to give the test samples. Glassware used for the aqueous solutions was silanized with dimethyldichlorosilane. The concentrations of the PAH in the various samples are listed in Table 44.

Extraction column

The extraction column 0.26 × 10.0 cm, was packed with Perkin-Elmer ODS-Sil-X-II, a C_{18} bonded phase pellicular packing (40 μm diameter particle size) which provides high trapping capacity with low flow resistance. One end of the column tubing was fitted with a standard end fitting, and then the packing material was added in small amounts from the other end. After each addition, the packing was compacted by dropping the column from a height of about 5 cm several times. A standard end fitting completed the extraction column.

Procedure

Approximately 800 ml of sample is needed. This is taken in a Teflon container.

Methanol is added to the sample to a final concentration of 20% v/v. With the injection valve in Figure 81 in the Inject position, 400 ml of this sample is pumped through the sample pump in order to equilibrate the pump and accessory pump surfaces with the sample. The sample pump is stopped, the injection valve turned to Fill, and the sample pump started again. This directs sample through the extraction column (see Figure 81) and out to waste. A volume of 100 ml, determined by collecting the waste in a graduated cylinder, is pumped through the extraction column. The sample pump is stopped when 100 ml have been pumped and the valve turned to Inject. This directs the l.c. mobile phase through the extraction column in the opposite direction from which the sample was pumped, i.e. the column is backflushed. The compounds are desorbed by the mobile phase and carried to the analytical column.

While the chromatogram is developing, the injection valve is turned to Fill and the sample pump started again. As before, 100 ml are pumped through the extraction column and the pump stopped. When the first chromatogram is complete, the valve is turned to Inject, which initiates the analysis of this second 100 ml sample.

The chromatogram resulting from injection of 100 μl of a mixture of PAH standards is shown in Figure 82. The detection limits for each compound examined are listed in Table 44. The excitation and emission wavelengths chosen are a compromise among the optimum values for each compound; consequently, the detection limit for a particular compound can be made even smaller. Table 45 lists the detection limits for the optimized wavelength settings using the Perkin-Elmer Model 650-10 LC fluorescence detector.

Ogan et al.[420] found that adsorption of the PAHs on to glass and metal surfaces was a major problem throughout their study. They found that only methanol gave a significant improvement, hence all samples were made up to contain 20% methanol, including the solution used for direct injection. Also, the mobile phase was modified to include 15% methanol.

Although the methanol markedly improved the recovery efficiency, it was still much less than 100% due to adsorption on glass and metal surfaces.

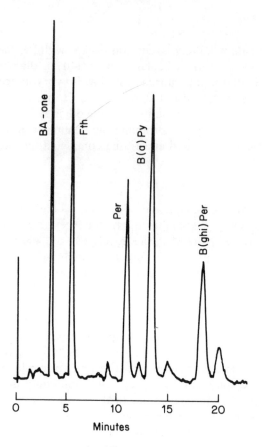

Figure 82 Chromatogram resulting from direct injection of the PAH mixture (for key see Table 44). Amounts injected: 7.7 ng BA-one, 3.5 ng Fth, 1.7 ng Per, 2.0 ng B(a)Py, and 5.6 ng B(ghi)Per. Reproduced from the *Journal of Chromatographic Science* by permission of Preston Publications, Inc.

Table 45 Detection limits (optimized conditions)*

Compound	λ_{ex} (nm)	λ_{em} (nm)	Detection limit (pg)
Fluoranthene (Fth)	360	460	3
Perylene (Per)	380	440	1
Benzo(a)pyrene (B(a)Py)	365	407	0.8
Benzo(ghi)perylene (B(ghi)Per)	365	410	6

*Perkin-Elmer Model 650-10 LC fluorescence detector.
Reproduced from the *Journal of Chromatographic Science* by permission of Preston Publications, Inc.

Ogan et al.[420] were able to obtain full recovery of the PAHs by fully saturating the adsorption sites on the metal pump, the system was first exposed to the sample overnight. The following morning, this sample was replaced with a fresh portion of the same sample batch. An hour later, this was replaced with a third portion of the same sample, which was used for the actual analysis. This equilibration method was used in the above experiments. In subsequent experiments, it was determined that equilibration of the pump and extraction

Table 46 Recovery efficiency

Compound	Recovery
7H-Benz(de)anthracene-7-one (BA-one)	97%
Fluoranthene (Fth)	101%
Perylene (Per)	98%
Benzo(a)pyrene (B(a)Py)	94%
Benzo(ghi)perylene (B(ghi)Per)	96%

Reproduced from the *Journal of Chromatographic Science* by permission of Preston Publications, Inc.

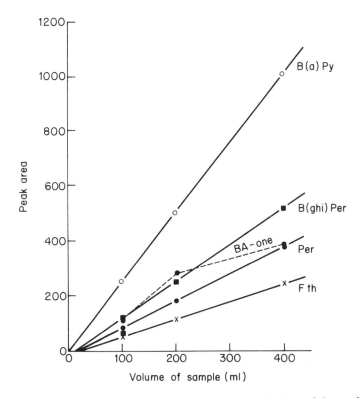

Figure 83 Peak area versus sample volume for 1:500 dilution of the stock PAH mixture (for key see Table 44). Reproduced from the *Journal of Chromatographic Science* by permission of Preston Publications, Inc.

column system was attained after 400 ml of sample had been pumped through the system.

The complete recoveries in Table 46 not only reflect the efficiency of the desorption step, but also the high efficiency of this packing material in extracting and retaining these compounds from very dilute solutions. These experiments demonstrate that this pellicular C_{18} bonded phase packing material is compatible with the microparticulate C_{18} bonded phase column for purposes of concentrating and analysing these PAHs in aqueous solution.

The conditions under which breakthrough begins to occur are an important characteristic of a trapping system. The linearity in Figure 83 for fluoranthene, perylene, benzo(a)pyrene, and benzo(ghi)perylene indicates that breakthrough is not a problem with these compounds for sample volumes at least as large as 400 ml, while there is evidently breakthrough for benzanthrene for a 400 ml sample. The linearity in Figure 84 between peak area and PAH concentration for all five compounds in a 100 ml sample indicates that there is no breakthrough due to saturation of column capacity.

Most of the work described by Ogan et al.[420] was concerned with the analysis of synthetic solutions although some results are reported for actual drinking water samples (Figure 85).

Dunn and Stick[421,422] have described a monitoring procedure for PAHs particularly benzo(a)pyrene in marine sediments and organisms in coastal waters. The procedures involve extraction and purification of hydrocarbon fractions from the sediments or organisms, and determination of compounds by thin-layer chromatography and fluorimetry, or gas chromatography.

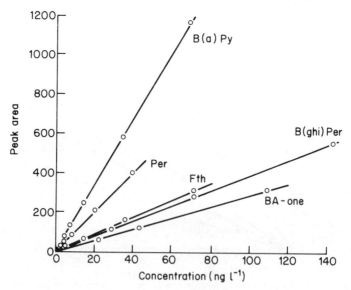

Figure 84 Peak area versus PAH concentration for 100 ml samples pumped through the extraction column (for key see Table 46). Reproduced from the *Journal of Chromatographic Science* by permission of Preston Publications, Inc.

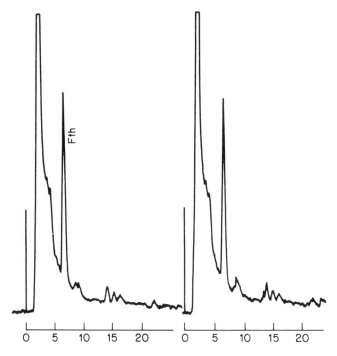

Figure 85 Chromatograms resulting from pumping two successive 200 ml portions of a drinking water sample (tap water plus methanol to 20% (v/v)) through the extraction column. The large peak was identified as fluoranthene (Fth) on the basis of retention time. Reproduced from the *Journal of Chromatographic Science* by permission of Preston Publications, Inc.

Method

Extraction procedure [422]

Solvents were practical or technical grade for economy and were redistilled before use in an all glass still with a Vigreaux reflux column. Solvents were stored in glass-stoppered containers and were protected from contact with rubber or plastic at all times.[423] Dimethylsulphoxide was spectral grade and was used without further purification, while hexadecane was practical grade and was purified before use by passage through a column of activated silica gel.[424] Sodium sulphate and potassium hydroxide were reagent grade and were used without further purification. Florisil (60–100 mesh) was from Matheson Coleman and Bell, batches of 250 g of the adsorbent were thoroughly washed by decantation with distilled water to remove fines and sodium sulphate.[425] The wet material was placed in a 600 ml fritted glass Buchner funnel, excess water was removed by suction, and the adsorbent washed in the funnel by stirring with two 300 ml portions of methanol. Excess methanol was removed by suction, and the adsorbent dried overnight *in vacuo* at 60 °C. The cleaned Florisil was activated at 250 °C for 18 h then cooled and partially reactivated by the addition of 2% water. Cellulose-acetate (20% acetylated) for thin-layer chromatography was from Macherey-Nagel and Co. Thin-layer plates were formed by coating a 0.5 mm layer of cellulose-acetate (25 g in 100 ml of ethanol) on glass plates 10 × 20 cm. Plates were cleaned before use by development in the running solvent.

Extraction

To avoid possible photodecomposition of PAH, all extraction and purification procedures were carried out under subdued yellow tungsten light. Twenty to forty grams of tissues were placed in a 300 ml flask and 150 ml of ethanol, 7 g of potassium hydroxide, two or three boiling chips, and an aliquot of radioactive benzo(a)pyrene (either 1000 dpm ^{14}C-benzo(a)pyrene, *ca.* 5 μg, or 25,000 dpm ^{3}H-benzo(a)pyrene, *ca.* 0.1 ng) were added. The tissue was digested by refluxing gently for 1.5 h with occasional swirling to prevent sticking to the bottom of the flask. The digest was added while hot to 150 ml of water in a 2 l separatory funnel, and the digestion flask rinsed out with an additional 50 ml of ethanol. The water–ethanol mix was extracted three times with 200 ml of iso-octane, and the iso-octane extracts were combined and washed with 4 × 200 ml warm (60 °C) water.

Wet sediment samples weighing 10–20 g were refluxed in 100 ml of ethanol with 5 g of potassium hydroxide, boiling chips and radioactive tracer. The contents of the flask were then poured into a 250 ml Erlenmeyer flask, and the sediment was allowed to settle by gravity for 5 min. The supernatant was decanted through a glass wool plug into 150 ml of water in a separatory funnel. The sediment was washed by decantation with two additional portions of ethanol (50 ml each) which were also added to the separatory funnel. The remainder of the processing was as for tissue samples.

Column chromatography on Florisil

A column of 30 g of Florisil covered with 60 g of sodium sulphate was prepared in a glass column (40 × 400 mm) with a coarse fritted glass disc. The column was prewashed with 100 ml of iso-octane, and the iso-octane extract of the sample then passed through. The column was allowed to drain and was then washed with two 100 ml portions of fresh iso-octane, allowing the column to drain briefly between each addition. Polycyclic aromatic hydrocarbons were eluted from the column with 3 × 100 ml benzene. The combined eluate was reduced to 5 ml by rotary evaporation, 50 ml of iso-octane were added and the volume again reduced to 5 ml to remove the benzene.

Dimethyl sulphoxide extraction

PAH were extracted from the 5 ml of iso-octane with 3 × 5 ml dimethyl sulphoxide. The dimethyl sulphoxide extracts were combined with 30 ml of water, and the PAH extracted into 2 × 10 ml iso-octane. The iso-octane extracts were combined, washed three times with 20 ml of water, and dried by passage through 10 g of sodium sulphate in a 15 ml coarse fritted glass Buchner funnel.

Thin-layer chromatography

The extract was reduced in volume to approximately 0.1 ml using rotary evaporation and then a stream of nitrogen and was applied as a narrow streak to the origin of a cellulose-acetate thin-layer plate. A standard of 10 ng benzo(a)pyrene was applied as a spot to one side of the plate, and the plate was developed with ethanol–toluene–water (17:4:4). The plate was positioned in the development tank with the bottom of the plate supported approximately 5 cm from the base of the 5 cm from the base of the tank. Sufficient developing solvent was present to wet the bottom of the plate. The plates were then partially air-dried, and the benzo(a)pyrene band was located and outlined under long wave ultraviolet light. The benzo(a)pyrene band, with an R_f of approximately 0.3 after 2 h development, was always the lowest fluorescent band on the plate.

The adsorbent at the position of the benzo(a)pyrene band was scraped off the plate while still damp and placed in a 15 ml fine fritted Buchner funnel. The benzo(a)pyrene was removed from the cellulose-acetate by washing with 4 × 4 ml hot (65 °C) methanol, using gentle suction. The methanol was added to 10 ml of a solution of 20% hexadecane in iso-octane, and the methanol and iso-octane were removed by rotary evaporation to leave the benzo(a)pyrene in 2 ml of hexadecane, ready for fluorimetry.

Fluorimetric measurement of benzo(a)pyrene

Benzo(a)pyrene was measured fluorimetrically in hexadecane using the baseline technique of Kunte.[426] Samples and standards of 10–200 ng benzo(a)pyrene ml^{-1} in hexadecane were excited at 365 nm in an Aminco-Bowman spectrophotofluorimeter, and the emission spectrum was recorded from 375 to 500 nm. An artificial baseline was drawn between minima in the fluorescence spectrum occurring at 418 and 448 nm, and the height of the peak at 430 nm above this baseline was measured. When necessary highly fluorescent samples were diluted with hexadecane to bring their fluorescence within the range of the standards used.

Figure 86 shows a flow chart of the procedures. The left portion of the chart shows the basic procedure for the isolation of PAH and the fluorimetric determination of benzo(a)pyrene. Measurement of the amount of radioactivity in the final fluorimetry sample permits the calculation of the percentage recovery of the compound through the procedures. The iso-octane eluate from the Florisil column and the iso-octane phase from the dimethylsulphoxide extraction are normally discarded when samples are processed for the fluorimetric determination of benzo(a)pyrene. These fractions, which are designated as fractions 1 and 2 in Figure 86, may be saved and analysed for paraffinic and

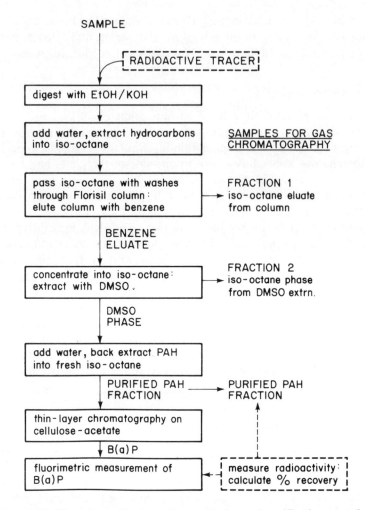

Figure 86 Flow chart of hydrocarbon extraction and purification procedures. Reproduced by permission of the *Journal of the Fisheries Research Board of Canada*

other hydrocarbons by gas chromatography. In addition, a portion of the purified PAH fraction may be diverted from thin-layer chromatography and analysed by gas chromatography.

Calculation of percentage recovery and amount of benzo(a)pyrene originally present in sample

After fluorimetry, the amount of radioactive benzo(a)pyrene internal standard in each sample is determined by scintillation counting. The recovery of benzo(a)pyrene is calculated by comparing the amount of radioactivity added at the beginning of the digestion procedure with the amount recovered in the fluorimetry sample. The amount of benzo(a)pyrene determined by fluorimetry was then corrected if necessary for the contribution of radioactive tracer (this correction is negligible if the higher specific activity ^{3}H-B(a)P is used), and the net amount of benzo(a)pyrene was then corrected for losses in the procedure. The amount of benzo(a)pyrene originally present in the sample was then expressed as μg benzo(a)pyrene kg^{-1} wet weight of tissue or dry weight of sediment.

The overall recovery of benzo(a)pyrene was generally 60–80% for tissue samples (mussels, clams, oysters, or fish) and 50–70% for sediment samples. Recoveries of benzo(a)pyrene from the alcoholic potassium hydroxide digest plus water into iso-octane were essentially quantitative for tissue samples and generally over 90% for sediment samples.

Recoveries from the Florisil column were generally over 90%, but fluctuated somewhat depending on the nature of the samples and the method of preparation of the Florisil. Losses occurred by tailing of benzo(a)pyrene on the column during elution with benzene, rather than by failure of benzo(a)pyrene to be retained by the column during elution with iso-octane. Fully activated Florisil was unsuitable as a chromatography substrate as it gave extensive tailing and poor recoveries, but the addition of 2% water to the activated material gave an adsorbent with consistently good chromatography properties. With all samples the violet fluorescing benzo(a)pyrene band on thin-layer plates was associated with a blue fluorescing band with a slightly greater R_f value. This band was only partially resolved from the benzo(a)pyrene band during normal chromatography but was completely resolved from the benzo(a)pyrene by using extended chromatography. The R_f value of this band, as well as the ultraviolet and fluorescence spectra of material isolated from this band, was identical to those of an authentic sample of benzo(b)fluoranthene.

Estimation of precision of benzo(a)pyrene determinations in mussels at the 10–20 ng kg^{-1} level ranged from an s.d. of 0.4 to 1.45. Shoreline mussel samples had a mean benzo(a)pyrene content of 0.55 mg kg^{-1} net weight with an s.d. of 0.11. Samples stored for 12 weeks at $-10\,°$C showed no significant change in benzo(a)pyrene content suggesting that this is an adequate method of sample storage. The analytical procedure takes 3–4 hours operator time per sample.

Dunn and Stich[421] applied this method to sediment samples in the vicinity of a sewage outfall showing elevated levels of benzo(a)pyrene. Mussels (*Mytilus edulis*) taken from the outer Vancouver harbour showed lower benzo(a)pyrene levels in the summer than in the winter, perhaps a result of seasonal discharges of sewage and storm drain water into the harbour. Elevated levels of benzo(a)pyrene in mussels growing near creosated timbers or pilings suggested that creosote may be a significant source of this substance in the marine environment. Direct evidence for this suggestion was obtained by comparison of gas chromatographic profiles of polycyclic aromatic hydrocarbons isolated from mussels and from creosoted wood.

Various other workers have investigated the fluorimetric determination of PAHs in sea water,[427] marine sediments[428] (benzo(a)pyrene and perylene), and aquatic fauna.[429] In the latter method hydrocarbons are separated from lipids by column chromatography on alumina, identifying and determining the oil fractions by fluorimetry using pyrene as a fluorescence standard. The detection limits of crude, Bunker C, and creosote oil were 100, 50, and 100 μg g^{-1} of lipid respectively. Fluorescence emission spectra of seal and fish oils could not be attributed to polynuclear aromatic hydrocarbons, but creosote oil was detected in shellfish samples in concentrations from 202 to 325 μg g^{-1} of lipid.

High performance liquid chromatography

An early reference to the use of high performance liquid chromatography in the analysis of PAHs is the work of Jentoft and Gouw (1968)[430] and Vaughan et al. (1973).[431] These workers used ultraviolet, visible and fluorescence detection and were able to detect 0.4 ng of anthracene and 15 ng of acenaphthalene.

Lewis[432] in 1975 gave comprehensive details of a method using counter-current extraction and high performance liquid chromatography with fluorimetric detection for estimating PAHs in water.

Sorrell et al.[433] have listed 15 PAHs commonly found in water (Table 47) and have described a high performance liquid chromatographic method for determining 13 trace compounds. Limits of detection are as low as 1 ng l^{-1}, well below the collective limit of 200 ng l^{-1} for six PAHs recommended by the World Health Organization.

In later work Sorrell and Reding[434] present an extension of this technique, for analysing 1–3 ng l^{-1} amounts of 15 polynuclear aromatic hydrocarbons in environmental water samples using high pressure liquid chromatography, preceded by a clean-up using alumina, with ultraviolet monitoring and fluorescence emission–excitation spectra for identification. The analytical procedure is outlined in Figure 87.

Method

A comparison of the h.p.l.c. chromatogram of an extract of tap water with that of an extract of spiked tap water eluted through alumina showed the effectiveness of the clean-up in removing u.v.-absorbing interferences. The alumina also fractionated 17 PAHs into three fractions. The two early-eluting fractions, which were discarded, may contain hydrocarbons, benzene, naphthalene, alkyl-substituted benzenes and naphthalenes, and other low-molecular-weight compounds.

The h.p.l.c. chromatograms of the three retained fractions from the standard PAHs taken through the analytical procedure are presented in Figures 88–90. The chromatography was performed isocratically and the column temperature was controlled. This permitted specific PAHs to be initially identified from their

Table 47 PAHs commonly found in water

Structure	IUPAC name	Mol. wt.	Relative carcinogenicity	Abbreviation
	Benzo(ghi)perylene	276	−	B(ghi)P
	Chrysene	228	−	Ch
	Fluoranthene	202	−	Fl
	Indeno(1,2,3-cd)pyrene	276	+	IP
	Phenanthrene	178	?	Ph
	Perylene	252	−	Per
	Pyrene	202	−	Pyr
	Anthracene	178	?	An
	Benzo(a)anthracene	228	+	B(a)A
	Benzo(b)fluoranthene	252	+ +	B(b)F
	Benzo(j)fluoranthene	252	+ +	B(j)F
	Benzo(k)fluoranthene	252	−	B(k)F
	Benzo(a)pyrene	252	+ + +	B(a)P
	Benzo(e)pyrene	252	+	B(e)P

+ + +, active; + +, moderate; +, weak; ?, unknown; −, inactive.

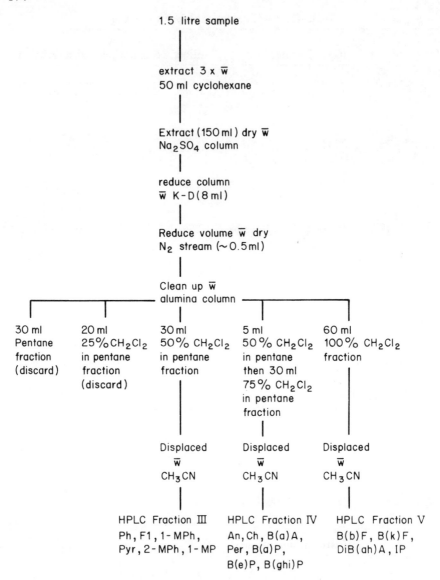

Figure 87 Analytical scheme for PAHs

retention times (which varied less than 2% over a 9 month period), and eliminated the need for column re-equilibration. Although some coeluting PAHs (e.g. anthracene and phenanthrene) had been placed into different fractions, it was clear that no single u.v. wavelength was capable of resolving all of the PAHs within a fraction. However, instead of further manipulating the chromatography, the sensitivity and resolution of the analysis were optimized by selecting the available u.v. wavelengths so as to minimize the response of

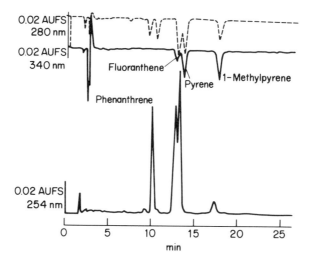

Figure 88 H.p.l.c. analysis of fraction III from alumina column (*ca.* 10 ng of standards). Reprinted with permission from Sorrell and Reding.[434] Copyright (1979) Elsevier Science Publishers B.V.

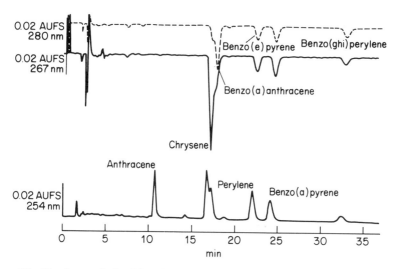

Figure 89 H.p.l.c. analysis of fraction IV from alumina column (*ca.* 10 ng of standards). Reprinted with permission from Sorrell and Reding.[434] Copyright (1979) Elsevier Science Publishers B.V.

the interferences near a relative absorption maximum of the PAH of interest (Table 48). Concentrations of the PAHs were determined from their u.v. responses. The sensitivity of the u.v. detectors, defined as a signal-to-noise ratio of 2, ranged from 0.25 to 1 ng l^{-1}.

All but two (1-methyl phenanthrene and 2-methyl phenanthrene) of the 17 PAHs could be sufficiently resolved from coeluting PAHs and other

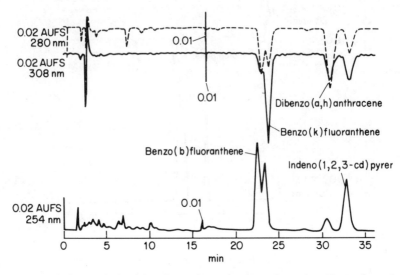

Figure 90 H.p.l.c. analysis of fraction V from alumina column (*ca.* 10 ng of standards). Reprinted with permission from Sorrell and Reding.[434] Copyright (1979) Elsevier Science Publishers B.V.

Table 48 Wavelengths selected to quantitate 15 PAHs

Compound	Abbreviation	Wavelength (nm)
Phenanthrene	Ph	254
Fluoranthene	Fl	340
Pyrene	Pyr	340
1-Methylpyrene	1-MP	340
Anthracene	An	254
Chrysene	Ch	267
Benzo(a)anthracene	B(a)A	280
Perylene	Per	254
Benzo(e)pyrene	B(e)P	280
Benzo(a)pyrene	B(a)P	254
Benzo(ghi)perylene	B(ghi)P	280
Benzo(b)fluoranthene	B(b)F	254
Benzo(k)fluoranthene	B(k)F	308
Dibenzo(a,h)anthracene	DiB(ah)A	280
Indeno(1,2,3-cd)pyrene	IP	254

Reprinted with permission from Sorrell and Reding.[434] Copyright (1979) Elsevier Science Publishers B.C.

interfering compounds by using the variable wavelength and the two fixed wavelength (254 and 280 nm) detectors. In addition, despite the alumina clean-up, some interferences with a 254 nm response coeluted with these PAHs and with 1-methyl phenthrene. Analysis by g.c.–m.s. showed that di-*n*-octyl adipate and phthalate esters were in this group of interferences. However, since neither the interferences nor 1-methyl phenanthrene or 2-methyl phenanthrene had a significant response at 340 nm, the concentrations of fluorene, pyrene, and 1-methyl phenanthrene could be measured (Figure 88). Since no selective wavelength could be found for 1-methyl phenanthrene and 2-methyl phenanthrene they were not quantified by Sorrell and Reding.[434] Table 48 shows optimum wavelengths for which 15 PAHs could be quantitated at concentrations of $1-3 \, \text{ng} \, \text{l}^{-1}$.

Fluorescence as a detection system, has the advantage of greater sensitivity and less susceptibility to interferences, since fewer compounds are fluorescent than are u.v. absorbent. Given the capability to scan both emission and excitation wavelengths, fluorescence provided more unique spectra for PAHs than m.s., and greatly augmented the preliminary identifications of PAHs via retention time.

Definitive fluorescence spectra of all the PAH standards (except indeno(1,2,3-cd)pyrene) in Table 48 were obtained by trapping the compounds in the flow cell and scanning the excitation and emission spectra. A spectrum of 6 ng of B(a)P detected in a raw water sample superimposed on that of 7 ng of a B(a)P standard (Figure 91) shows the type of confirmation of identity afforded by this detection technique. Even in the presence of coeluting fluorescent compounds, the spectral features were individually recorded by manipulating the emission and excitation wavelengths (the benzo(e)pyrene spectrum also appears to indicate the presence of a third compound). Indeno(1,2,3-cd)pyrene provided a very amorphous fluorescence spectrum which may be a characteristic of indeno(1,2,3-cd)pyrene or due to a consistently present fluorescent interference.

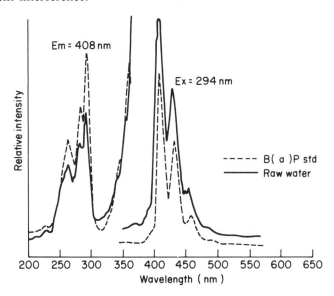

Figure 91 Comparison of Benzo(a)Pyrene B(a)P standard with fluorescence spectrum of untreated water sample. Reprinted with permission from Sorrell and Reding.[434] Copyright (1979) Elsevier Science Publishers B.V.

Since the same u.v. wavelengths were recorded for each compound within a fraction, various peak height ratios of these responses could be computed. It has been noted that these ratios can sometimes be used to identify compounds in samples[311] and they were calculated at 280 and 254 nm for the four PAHs which had responses apparently free of interferences at these wavelengths (Table 49). However, except for phenanthrene, there was a significant variation between the ratios for samples and those for standards which indicated that interferences were still present in the samples. Thus, the u.v. ratios were primarily used to determine the purity of the chromatographed compounds. If either these ratios or the fluorescence spectra indicated the presence of significant interferences, only an upper bound on the concentrations of the PAHs could be determined.

Table 49 Comparison of u.v. ratios* measured for PAHs in standards and samples

Compound for key see Table 48	n	Ratio standard	n	samples
Ph	12	0.17 ± 0.02	20	0.18 ± 0.01
B(a)P	18	0.73 ± 0.01	5	0.81 ± 0.03
B(ghi)P	18	1.42 ± 0.08	5	1.52 ± 0.22
IP	13	0.47 ± 0.02	3	0.40 ± 0.03

*Ratio of 280/254 nm peak heights.
n = number of analyses performed.
Reprinted with permission from Sorrell and Reding.[434] Copyright (1979) Elsevier Science Publishers B.V.

In earlier work using this procedure, Sorrell[435] had shown the recoveries of six PAHs from distilled water to range from 61 to 91% and to average $78 \pm 8\%$ (Table 50). Subsequent measurements of the recoveries of eleven PAHs from seven raw or finished water samples were over a 9-month period ranged from 53 to 116% and averaged $86 \pm 12\%$ (Table 50).

In a further study Sorrell and Reding[434] spiked similar concentrations of pyrene, fluoranthene, 1-methylphenanthrene, perylene, benzo(a)pyrene, and benzo(ghi)pyrene into four samples of unextracted tap water and the recoveries were determined by analysing the extracts directly (i.e. without alumina clean-up). Corresponding unspiked tap water samples were also extracted and analysed directly to determine the background concentration of the PAHs. After correcting for this background, the recoveries were calculated and ranged from 78 to 99%, averaging 88%. The average recovery (for all the PAHs) from these three series of experiments was about 85%.

While the recovery efficiencies allowed quantification of PAH concentrations at the time of extraction, the relevance of those values to the actual concentration at the time of sampling may be less certain. A preliminary experiment showed a complete loss of seven PAHs when spiked (about 27 ng l^{-1} each) into Cincinnati tap water and stored at 5 °C for 18 days. The identical water with no chlorine residual (excess sodium sulphite added) showed only a small loss

Table 50 Per cent recovery from various types of waters

Compounds	Distilled water*	Environmental waters[†‡]							Mean
		A	B	C	D	E	F	G	
Phenanthrene	78 ± 1	83	96	87	98	—	—	—	91 ± 7
1-Methylpyrene	—	—	—	98	85	92	93	72	88 ± 10
Chrysene	—	86	86	87	84	86	93	86	87 ± 3
Benzo(a)anthracene	—	81	65	89	84	92	85	76	82 ± 9
Perylene	79 ± 7	92	75	105	95	94	82	75	88 ± 11
Benzo(a)pyrene	80 ± 4	89	54	110	105	79	67	73	82 ± 20
Benzo(ghi)perylene	89 ± 3	90	83	116	111	79	62	82	89 ± 19
Benzo(b)fluoranthene	76 ± 6	89	87	91	96	92	89	91	91 ± 3
Benzo(k)fluoranthene	—	80	84	91	91	93	76	95	87 ± 7
Dibenzo(ah)anthracene	—	93	90	92	92	79	53	84	83 ± 14
Indeno(1,2,3-cd)pyrene	68 ± 7	94	89	86	95	80	55	89	84 ± 14

*The mean recovery of 3 samples spiked at 10 ng l^{-1}.
†Spiked at 20 ng l^{-1}.
‡A,B,D,G, treated waters; C,E, untreated waters; F, partially treated.
Reprinted with permission from Sorrell and Reding.[434] Copyright (1979) Elsevier Science Publishers B.V.

Table 51 Per cent recovery of PAHs from tap water*

Compound[†] for key see Table 48	Non-reduced[‡] 8 day	Sulphite reduced[§] 2 day	8 day
1-MP	0	79	66
Per	0	68	41
B(a)P	0	36	18
B(ghi)P	0	92	63

*Samples stored in dark at 5 °C (pH 7.5).
†Spiked at about 27 ng l^{-1} each.
‡Chlorine residual was about 1.5 mg l^{-1}.
§Chlorine residual was < 0.1 mg l^{-1}.
Reprinted with permission from Sorrell and Reding.[434] Copyright (1979) Elsevier Science Publishers B.V.

for five of the PAHs and a somewhat larger loss for benzo(a)pyrene and perylene. A similar 8 day study of the same tap water source looked at only the four PAHs which did not occur naturally. The results showed total and partial losses of the PAHs in the presence and absence of a chlorine residual, respectively. Again, benzo(a)pyrene and perylene showed greater losses than the other PAHs (Table 51).

To examine this apparently greater loss of benzo(a)pyrene and perylene, sodium sulphite (25 mg l^{-1}) was added to 12 bottles of Cincinnati tap water. Eight of these samples were also spiked with about 20 ng l^{-1} each of six PAHs, and all were stored for varying amounts of time at 5 °C. The unspiked water samples were extracted and analysed on the initial and fifth days to determine and subtract the background PAH concentrations. The spiked samples were also analysed after 0, 2, 5, and 8 days. The results showed a gradual but

Table 52 Stability* of PAHs in tap water (sulphite reduced)

Compound‡ for key see Table 48	Days of storage†			
	0	2	5	8
Fl	87	73	88	83
Pyr	93	79	87	86
1-MP	102	93	83	87
Per	89	59	50	47
B(a)P	89	42	27	24
B(ghi)P	102	84	88	88

*As measured by per cent recoveries.
†Stored in the dark at 5 °C.
‡Spiked at about 20 ng l^{-1}.
Reprinted with permission from Sorrell and Reding.[434] Copyright (1979) Elsevier Science Publishers B.V.

noticeably greater loss of benzo(a)pyrene and perylene versus that of the other PAHs (Table 52). From a log plot of the concentration (as peak height) of four of the PAHs versus time, it was clear that the benzo(a)pyrene and perylene concentrations decreased at unequal rates.

Sorrell and Reding[434] conducted a time storage experiment by analysing on different days, identical samples of tap water to which sodium sulphite and a pellet spiked with six PAHs had been added (Table 53). The results indicated the utility of the spiked pellet as a method control, although the loss of the unstable PAHs was less than determined earlier. However, the range of initial recoveries observed with Table 50 and without alumina clean-up (Tables 51–53) were comparable; thus, the observed decrease in recoveries of perylene and benzo(a)pyrene with time is probably due to a mechanism unrelated to the analytical procedure used.

Table 53 Time storage* of spiked pellet in tap water

Compound‡ for keys see Table 48	Days of storage†	
	0	2
Fl	69	73
Pyr	78	78
1-MP	92	89
Per	81	74
B(a)P	72	63
B(ghi)P	94	89

*As determined by per cent recoveries.
†Stored in dark at 5 °C.
‡Spiked at about 20 ng l^{-1}.
Reprinted with permission from Sorrell and Reding.[434] Copyright (1979) Elsevier Science Publishers B.V.

Hunt et al.[436] have shown that all six representative PAHs tested in the World Health Organization standards for drinking water, (i.e. fluoranthene, 3,4-benzfluoranthene, (benzo(b)fluoranthene), 11,12-benzfluoranthene (benzo(k)fluoranthene), 3,4-benzpyrene (benz(a)pyrene), 1,12-benzperylene (benzo(ghi)perylene), and indeno(1,2,3-cd)pyrene, can be separated successfully by high performance liquid chromatography, using as a packing material, phthalimidopropyltrichlorosilane bonded to microparticulate silica gel with a non-polar mobile phase of toluene–hexane. They used a non-polar phase of toluene–hexane (1:10) and achieved good separations as indicated in Figure 92.

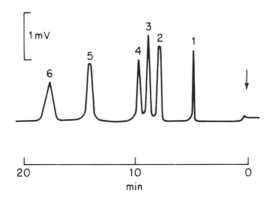

Figure 92 Separation of six PAH Standards on PPS. Key: 1, fluoranthene; 2, benzo(k)fluoranthene; 3, benzo(b)fluoranthene; 4, benzo(a)pyrene; 5, indeno(1,2,3-cd)pyrene; 6, benzo(ghi)perylene. Conditions: mobile phase, 10% toluene in hexane at a flow rate of 2 ml min^{-1}; detection, Amino Fluoromonitor at ⅓ full sensitivity. Reprinted with permission from Hient.[436] Copyright (1978) Pergamon Press Ltd

Das and Thomas[437] used fluorescence detection in high performance liquid chromatography to determine nine PAHs in occupational health samples including process waters. The nine compounds studied were, benzo(a)anthracene, benzo(k)fluoranthene, benzo(a)pyrene/fluoranthene, chrysene, benzo(k)fluorene, perylene, benzo(e)pyrene, dibenz(ah)anthracene, and benz(ghi)perylene.

The method involves the use of a deuterium light source and excitation wavelengths below 300 nm. Limits of detection in the 0.5–1.0 pg range were obtained for several PAH compounds of environmental or toxicological significance. Limits of detection close to subpicogram levels were obtained (e.g. benzo(a)anthracene, 0.6 pg; benzo(k)fluoranthene, 0.4 pg; benzo(a)pyrene, 1.1 pg). Precision studies gave a relative standard deviation from 0.32 to 2.66% (e.g. benzo(a)anthracene, 0.33%; benzo(k)fluoranthene, 0.70%; benzo(a)pyrene, 0.50%). The system allows the use of dilute solutions, thus eliminating the usual clean-up procedures associated with trace analysis.

Method

Reagents

Spectrograde cyclohexane and acetonitrile were used. Distilled, deionized water shown to be free of fluorescent impurities was used. Stock solutions of the standards were prepared by dissolving in cyclohexane at $\mu g\,ml^{-1}$ concentration.

Appropriate dilutions with cyclohexane followed by replacing the cyclohexane with 75% acetonitrile were made to obtain the concentrations of PAH of $pg\,\mu l^{-1}$ prior to h.p.l.c. analysis of the standards. The stock solutions were stored at 4 °C when not in use.

Apparatus

A Spectra-physics Model 3500B Liquid Chromatograph equipped with a variable wavelength fluorescence detector, Model FS 970 Spectrofluoro Monitor (Schoeffel Instrument Corp, Westwood NJ) and a variable wavelength u.v. detector, Model SF 770 Spectroflow Monitor (Schoeffel instrument Corp) connected in series was used with a 25 cm long 2.1 mm i.d. Dupont Zorbax ODS column (C_{18} reverse phase) of 5–7 μm particle size. The u.v. detector had a deuterium lamp for the wavelength range of 190–400 nm and had an 8-μl flow cell with 1 mm bore and a 10 mm light path between the quartz windows. The FS 970 Spectrofluoro Monitor had a low volume flow cell (5 μl) and featured a continuously selectable monochromatic excitation energy over the entire ultraviolet–visible spectrum by utilizing a highly stabilized deuterium lamp (190–400 nm) or tungsten-halogen lamp (350–600 nm). The fluorescence emission measurement system consisted of a very sensitive end-on photomultiplier tube and a solid state photometer with filter changer. The fluorometer was equipped with a set of six interchangeable emission cut-off filters, 370, 389, 418, 470, 550, and 580 nm (transmitting above stated wavelengths) to select emission spectra of interest, and with special ultraviolet–visible band type prefilters (Corning 7.54 and 7.51) for excitation.

H.p.l.c. in reverse phase mode was performed isocratically with 82% acetonitrile in water. The mobile phase was degassed before use by applying vacuum for a few minutes while agitating the solvent mixture. The flow rate was 0.3 ml min^{-1}, the column pressure was between 300 and 500 psi, the temperature was 25 °C maintained by a thermostated oven, and a chart speed of 0.2 in min^{-1}. The injection was made with a sample loop operated by a rotary valve using a 10 μl injection volume. The wavelengths used in u.v.

Figure 93 Liquid chromatogram of a mixture of nine polycyclic aromatic hydrocarbons with u.v. absorption detection. Chromatographic conditions: two 1 mm i.d. × 25 cm Zorbax ODS column, 82:18(v/v) CH_3CNH_2O, 25 °C, 300–500 psi, flow rate 0.3 ml min^{-1}, 10 μl injection of PAH dissolved in 75% CH_3CN. (a) u.v. 254 nm at 0.01 AUFS, (b) u.v. 280 nm at 0.01 AUFS. Reprinted with permission from Das and Thomas.[437] Copyright (1978) American Chemical Society

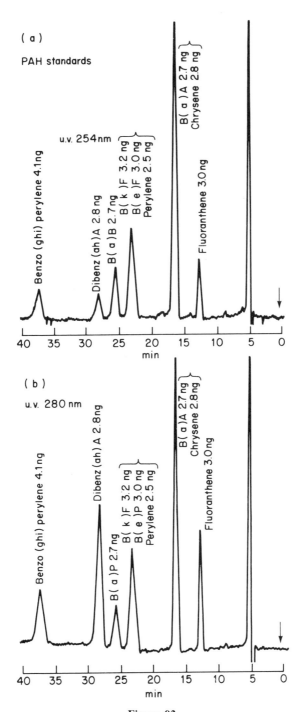

Figure 93

detection were 254 and 280 nm and the following three sets of excitation and emission wavelengths conditions were used in fluorescence detection with excitation prefilter Corning 7.54: (a) λ_{ex} 280 nm, $\lambda_{em} > 389$ nm, (b) λ_{ex} 250 nm, $\lambda_{em} > 370$ nm, and (c) λ_{ex} 240 nm, $\lambda_{em} > 470$ nm. The analyses of the individual PAH, synthetic mixture, and the samples were performed under identical conditions.

Peaks observed in the sample chromatographic profiles were identified by four methods. First, tentative identifications were made on the basis of the retention times and comparison with standards. Second, standards were added to the samples and the mixtures rechromatographed. A quantitative increase in the peak height was taken as further characterization. The third method of identification was by the use of a different fluorescence technique, i.e. analysing samples under three optimal fluorescence conditions (λ_{ex} 280/$\lambda_{em} > 389$, λ_{ex} 250/$\lambda_{em} > 370$, and λ_{ex} 240/$\lambda_{em} > 470$) and comparing with standards. Finally, identification of peaks was made by determining the peak height ratio at two different wavelengths (λ_{ex} 280/$\lambda_{em} > 389$ and λ_{ex} 250/$\lambda_{em} > 370$) and comparing with the ratios obtained for the standards. Peak height was used for quantitation of PAH in the samples.

Figures 93(a) and 93(b) show the chromatographic profiles of the PAH obtained with u.v. detection at 254 nm and 280 nm respectively. It is seen from the figures that the microparticulate column gives good separation of fluoranthene, benzo(a)pyrene, dibenz(ah)authracene, and benz(ghi)perylene, but fails to resolve either chrysene from B(a)authracene or benzo(k)fluoranthene, benzo(e)pyrene, and perylene from each other in the multicomponent mixture.

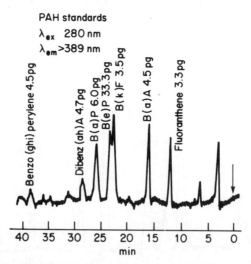

Figure 94 Liquid chromatogram of a mixture of polycyclic aromatic hydrocarbons showing sensitivity of fluorescence detection. Chromatographic conditions same as in Figure 93. Reprinted with permission from Das and Thomas.[437] Copyright (1978) American Chemical Society

Das and Thomas[437] discuss in particular difficult separations such as chrysene from benzo(a)anthracene and perylene from benzo(e)pyrene and show how these separations can be achieved.

Shown in Figure 94 is a fluorescence chromatogram of a synthetic PAH mixture containing 3.3 pg of fluoranthene, 4.5 pg benzo(a)anthracene, 3.5 pg benzo(k) fluoranthene, 33.3 pg benzo(e)pyrene, 6.0 pg benzo(a)pyrene, 4.7 pg dibenz(ah) anthracene, and 4.5 pg benz(ghi)perylene at λ_{ex} 280 nm and $\lambda_{em} >$ 389 nm (4.6 pg chrysene and 4.5 pg perylene in the mixture do not respond under the fluorescence conditions used). The detection limits (twice the signal-to-noise ratio) of the PAH are estimated from the chromatogram and are listed in

Table 54 Detection limits of standard PAH

Compound	Amount (pg)
Fluoranthene (λ_{ex} 280 nm/$\lambda_{em} >$ 389 nm)	0.5
Benzo(a)anthracene (λ_{ex} 280 nm/$\lambda_{em} >$ 389 nm)	0.6
Benzo(k)fluoranthene (λ_{ex} 280 nm/$\lambda_{em} >$ 389 nm)	0.4
Benzo(e)pyrene (λ_{ex} 280 nm/$\lambda_{em} >$ 389 nm)	5.1
Benzo(a)pyrene (λ_{ex} 280 nm/$\lambda_{em} >$ 389 nm)	1.1
Dibenz(ah)anthracene (λ_{ex} 280 nm/$\lambda_{em} >$ 389 nm)	2.3
Benz(ghi)perylene (λ_{ex} 280 nm/$\lambda_{em} >$ 389 nm)	3.0
Chrysene (λ_{ex} 250 nm/$\lambda_{em} >$ 370 nm)	2.3
Perylene (λ_{ex} 250 nm/$\lambda_{em} >$ 370 nm)	0.6

Reprinted with permission from Das and Thomas.[437] Copyright (1978) American Chemical Society.

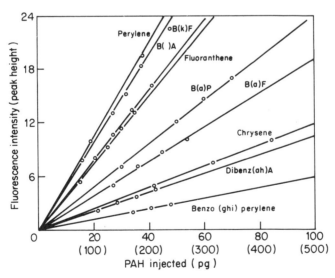

Figure 95 Calibration curves of polycyclic aromatic hydrocarbons showing linearity of h.p.l.c. fluorescence analysis. 0–100 pg concentration range for all PAH except B(a)P (0–500 pg). Reprinted with permission from Das and Thomas.[437] Copyright (1978) American Chemical Society

Table 54. Also shown in the table are the detection limits of chrysene and perylene determined individually at λ_{ex} 250 nm and $\lambda_{em} > 370$ nm.

These results show that the lower limits of detection for most of the PAH are close to subpicogram levels which are about three orders of magnitude lower than the minimum detectable concentration obtained by u.v. detection.

Calibration plots for the PAH in the synthetic mixture are shown in Figure 95. The h.p.l.c. fluorescence response was found to be linear over a wide concentration ranging from 1 to 100 pg of each PAH in the mixture (0–500 pg for benzo(e)pyrene. The correlation coefficients for these plots are between 0.999 and 1.000. For quantitative analysis, it is necessary to recheck these calibrations frequently.

Applications

Case studies to test the applicability of the described fluorescence detection technique were carried out by h.p.l.c. analysis of PAH in real life examples. Figure 96 shows the chromatograms of recycle waste water. The particulate sample which was run under three optimal fluorescence conditions was carried out by directly injecting the dilute effluent. The major PAH detected are fluoranthene, benzo(a)pyrene, benzo(k)fluoranthene and benzo(a)anthracene.

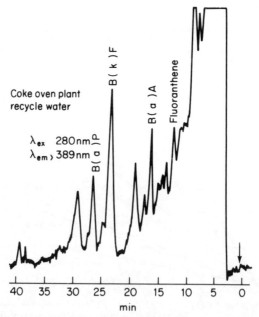

Figure 96 Analysis of coke oven plant dilute recycle waste water for polycyclic aromatic hydrocarbons by direct injection. Chromatographic conditions same as in Figure 93. Reprinted with permission from Das and Thomas.[437] Copyright (1978) American Chemical Society

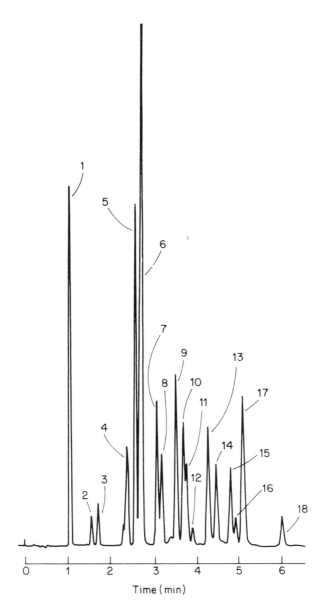

Figure 97 Analysis of polyaromatic hydrocarbons. Column: 125 × 4.6 mm i.d. C_{18} bonded-phase packing; 5 μm particles. Mobile phase: acetonitrile–water, linear gradient from 60 to 100% acetonitrile in 5 minutes, at 4 ml min^{-1}; inlet pressures: 3700 psig (25.5 MPa) initial, 2200 psig (15.2 MPa) final; ambient temperature; u.v. detector at 254 nm. Peaks: 1, benzene; 2, impurity; 3, naphthalene; 4, fluorene + acenaphthalene; 5, phenanthrene; 6, anthracene; 7, fluoranthene; 8, pyrene; 9, benzo(b)fluoranthene; 10, chrysene; 11, benzo(a)anthracene; 12, impurity; 13, benzo(e)pyrene + perylene; 14, benzo(a)pyrene; 15, dibenz(ah)anthracene; 16, dibenz(ac)anthracene; 17, benzo(ghi)perylene + indeno(1,2,3-cd)pyrene; 18, coronene. Copyright (1982) Perkin Elmer Ltd.

Oyler et al.[438] have determined aqueous chlorination reaction products in amounts down to 1 ng l^{-1} of PAHs using reversed phase high performance liquid chromatography–gas chromatography. The method involves filtration through a glass microfibre filter and concentration on a high performance liquid chromatography column. The PAH material is then eluted using an acetonitrile water gradient elution technique. The fractions are injected separately on to a gas chromatographic column equipped with a photoionization detector.

Workers at Perkin-Elmer[439] have studied the high speed separation of PAHs using a C_{18} bonded phase packings (5 μm particles) using both isocratic and gradient elution. Figure 97 shows the analysis of several PAH standards using the 5 μm bonded phase column with gradient elution from 60 to 100% acetonitrile in 5 minutes at a flow rate of 4 ml min^{-1}. Although not completely resolving all of the components, this method offers a rapid analysis which is adequate for many of these compounds. These high speed columns can also be used in a very high resolution mode. At the expense of analysis time, very high resolution can be attained by connecting two columns in series. An example of this is shown in Figure 98 where two 3 μm columns are coupled resulting in an efficiency of about 28,000 plates under isocratic conditions. These workers utilized these coupled columns, in this case, with gradient elution, to analyse the same PAH standards. The analysis time is increased to 26 minutes, but the extremely high resolution achieved is quite apparent. This very high resolution is extremely useful in analysing complex samples containing many components.

Schönmann and Kern[440] have used on-line enrichment for ppb analysis of PAHs in water by high performance liquid chromatography. These workers point out that concentration is a key problem in verifying water quality at part-per-billion or part-per-trillion levels. Classical techniques such as freeze drying, extraction and evaporation, steam distillation or preconcentration by adsorption have serious limitations in terms of contamination and sample loss. They used an on-line trace enrichment technique allowing direct high performance liquid chromatographic analysis of aqueous samples containing very low concentrations of PAHs.

This trace enrichment method is based on the affinity of non-polar pollutants for reverse phase chromatography supports. When aqueous samples are passed through a reverse phase column these compounds and any other non-polar organic compounds present in the sample are immobilized at the head of the column. When detectable quantities of pollutants have been accumulated on the column, they can be analysed by introducing a mobile phase of the desired eluant strength.

They used a Varian Model 5000 liquid chromatograph with three reservoir capability for on-line enrichment methods: one reservoir can be used to quantitatively transfer large sample volumes into the column and a binary gradient can then be introduced from the remaining reservoirs. An analytical method employing all three reservoirs was programmed to successfully analyse increasing sample volumes automatically (Figure 99). Sample volumes were varied from 50 ml to 120 ml and detector response (as measured by peak height)

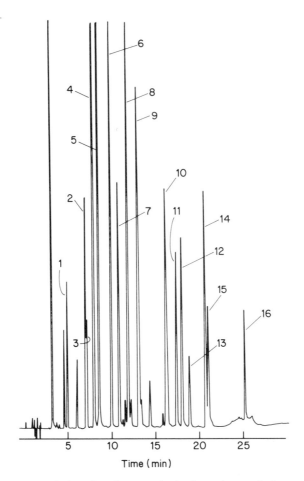

Figure 98 Analysis of polyaromatic hydrocarbons. Column: two 100 × 4.6 mm i.d. in series, C_{18} bonded-phase packing; 3 μm particles. Mobile phase: acetonitrile–water, linear gradient from 65 to 90% in 20 minutes, at 1.8 ml min^{-1}; inlet pressure: 4500 psig (31.0 mPa) initial, 2700 psig (18.6 MPa) final; ambient temperature; u.v. detector at 254 nm. Peaks: 1, naphthalene; 2, fluorene; 3, acenaphthalene; 4, phenanthrene; 4, anthracene; 6, fluoranthene; 7, pyrene; 8, benzo(b)fluoranthene; 9, chrysene + benzo(a)anthracene; 10, benzo(e)pyrene + perylene; 11, benzo(a)pyrene; 12, dibenz(ah)anthracene; 13, dibenz(ac)anthracene; 14, benzo(ghi)perylene; 15, indeno(1,2,3-cd)pyrene; 16, coronene. Copyright (1982) Perkin Elmer Ltd

was found to be linear when plotted against sample volume (Figure 100). Sample volumes of as much as 120 ml could be concentrated with no observable peak broadening.

The on-column enrichment technique does have some limitations which must be taken into consideration for its successful use. First, compounds which have lower affinity for reverse phase material may begin to move from the column

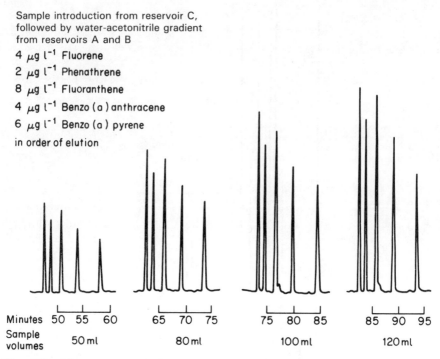

Figure 99 Direct preconcentration on column using different sample volumes (standards in acetonitrile). Reprinted with permission from Schönemann and Kern,[440] Copyright (1981) Varian AG

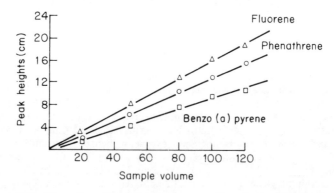

Figure 100 Linearity of peak heights vs. sample volume. Reprinted with permission from Schönemann and Kern,[440] Copyright (1981) Varian AG

head during the sample concentration stage, resulting in peak broadening or analyte breakthrough in cases of large sample volumes or compounds with very low reverse phase K' values. This situation limits the utility of on-column concentration for polar phenols, for example. The second limitation of the technique is the time required for introduction of the sample. The pressure limits of most reverse phase columns require that the sample be introduced at flow rates of a few millilitres per minute, leading to long analysis times for large sample volumes. This problem can be circumvented by concentrating the sample at high flow rate into a low-pressure-drop precolumn, followed by switching in the analytical column to in-line configuration before the introduction of the mobile phase.

The reverse phase enrichment technique is used to determine solvent impurities and is, in fact, employed in the quality control of h.p.l.c.-grade water. It is therefore essential to work with clean solvents to avoid interference from dissolved impurities. Even double-distilled water may contain trace-level organic impurities and it may be necessary to pass water over a reverse phase column prior to use in the preparation of standard solutions and mobile phases. The sample itself should be optically clear and free of any particulate material which might plug the column, hydraulic lines, or in-line filters. Waste water with high levels of contamination should be prefiltered to remove solid material. The on-column enrichment method finds its most useful application in the analysis of pollutants in the sub-ppm range in drinking water, ground water, lake and river water.

To demonstrate the use of the technique Schönemann[440] analysed a lake water sample for PAHs using fluorescence detection with selective filters (Figure 101). The compounds cannot be completely identified by retention times, so an aliquot of the sample was spiked with two compounds suspected to be present. A number of polynuclear aromatics were found in a concentration below 100 ng l^{-1}. The rather polar phenols were trapped as well and can be determined in the lower ppb range as shown in Figure 102.

The method is useful for quantitative analysis, provided that the peak heights are linear with concentration (Figure 103). This was the case for most of the

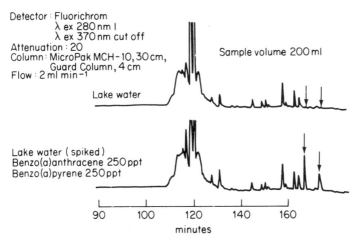

Figure 101 Lake water analysis. Reprinted with permission from Schönemann and Kern.[440] Copyright (1981) Varian AG

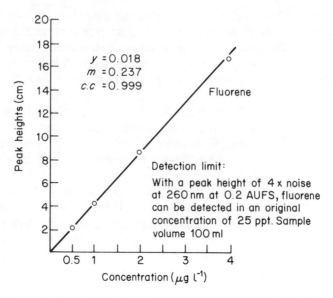

Figure 102 Linearity peak heights *vs.* concentration. Reprinted with permission from Schönemann and Kern.[440] Copyright (1981) Varian AG

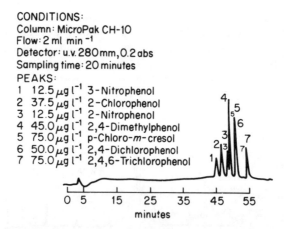

Figure 103 Phenols sample volume 40 ml. Reprinted with permission from Schönemann and Kern.[440] Copyright (1981) Varian AG

12 PAHs investigated (Figure 104). The recoveries were good (70–100%) for most PAHs considering the large enrichment factor of 10^4. Lower recovery values were obtained for some of the high molecular weight PAHs mainly due to their low solubility in water. Such limited solubility should be taken into account when standards are prepared (only aqueous solutions with a maximum of 10–20% acetonitrile can be trapped on reverse phase columns). In a natural

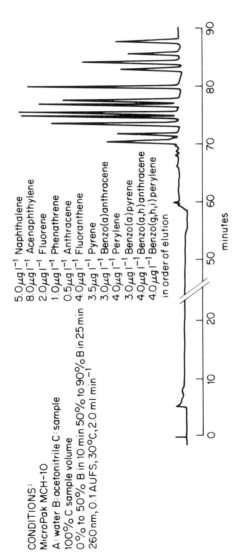

Figure 104 Polynuclear aromatics sample volume 100 ml. Reprinted with permission from Schönemann and Kern.[440] Copyright (1981) Varian AG

water sample, however, experience has shown that trace enrichment recovers most dissolved polynuclear alcohols and phenols.

Various other workers[441-450] have studied the application of high performance liquid chromatography to the determination of PAHs in water samples. Hagenmaier et al.[449] used a reverse phase high pressure liquid chromatography procedure for the determination of trace amounts of polycyclic aromatic hydrocarbons in water. Different column packing materials were tested, in conjunction with non-polar stationary phases of various polarities, for separation efficiency, detection limits, and long-term stability. The method was suitable for concentrations as low as $2 \, \text{ng} \, l^{-1}$ in a 1 litre sample. Compounds studied included fluoranthene, benzofluoranthene isomers, benzopyrene and perylene derivatives.

O'Donnell[451] screened Irish rivers for fluoranthene and benzopyrene using high performance liquid chromatography on a reverse phase C_{18} column. The solvents were methanol–water (1:1) and pure methanol. Detection was possible at concentrations of 0.018 ppm. Levels averaged less than one-sixth of the WHO and EEC recommended tolerance limits for both compounds.

Thin-layer chromatography

Early work on the application of thin-layer chromatography to the separation and determination of PAHs was carried out by several workers.[452-459]

Borneff and Kunte[452] extract a 10 litre sample of water with 600 ml pure benzene, the benzene is removed and the residue examined by thin-layer chromatography. Crane et al.[459] have modified this method. Gilchrist et al.[453] carried out preparative scale thin-layer chromatography followed by mass spectrometric examinations of the separated bands. Kay and Latham[454] carried out gel-permeation column chromatography (in a rather time-consuming method) followed by thin-layer chromatography of the eluted fractions. Nowacka-Barezyk et al.[456] separated benzo(k)fluoranthene from other PAHs originating in atmospheric samples by thin-layer chromatography on 10% acetylated cellulose.

Saxena et al.[460] used polyurethane foams to concentrate trace quantities of six representatives of polynuclear aromatic hydrocarbons (fluoranthene, benzo(k)fluoranthene, benzo(j)fluoranthene, benzo(a)pyrene, benzo(ghi)perylene, and indeno(1,2,3-cd)pyrene) prior to regular screening of these compounds in US raw and potable waters. Final purification and resolution of samples was by two-dimensional thin-layer chromatography, followed by fluorimetric analysis and quantification.

In this method the PAHs are collected by passing water through polyurethane foam plug. Water is heated to $62 \pm 2 \, °C$ prior to passage and flow rate is maintained at approximately $250 \, \text{ml} \, \text{min}^{-1}$ to obtain quantitative recoveries. The collection is followed by elution of foam plugs with organic solvent, purification by partitioning with solvents and column chromatography on Florisil, and analysis by two-dimensional thin-layer chromatography on cellulose-acetate–alumina plates followed by fluorimetry and gas–liquid chromatography using flame ionization detection. The latter method was less sensitive than thin-layer chromatography. Employing this method and a sample volume of 60 l, PAHs were detected in all the water supplies sampled. Although the sum of the six representative PAHs in drinking waters was small (0.9–15 ppt), the values found for raw waters were as high as 600 ppt.

In further work Saxena et al.[461] and Basu and Saxena[462] showed that the polyurethane foam plug method had an extraction efficiency for PAHs of at least 88% from treated waters and 72% from raw water.

Foam retention efficiencies of the six PAHs from spiked laboratory tap water at 25 ppb are shown in Table 55. The data confirm that polyurethane foam plugs under suitable conditions not only effectively concentrate benzo(a)pyrene but other PAHs as well. Foam plugs concentrated PAH almost quantitatively from finished water at lower concentrations also (Table 56). The high PAH retention

Table 55 Foam retention efficiencies for PAH from spiked tap water

Compound	Amt added to water (μg)	Per cent retention
Fluoranthene (FL)	100	100
Benzo(j)fluoranthene (BjF)	100	88[a]
Benzo(k)fluoranthene (BkF)	100	
Benzo(a)pyrene (BaP)	100	81
Indeno(1,2,3-cd)pyrene (IP)	100	89
Benzo(ghi)pyrene (B(ghi)P)	100	91

[a]Combined value given since the compounds could not be separated on the g.l.c. column.
Water source, laboratory tap water; water volume, 4 l; concn of each PAH, 25 ppb; detection method, g.l.c.–FID
Reprinted with permission from Saxena et al.[461] Copyright (1977) American Chemical Society.

Table 56 Foam retention efficiencies for PAH from treated water

Compound for key see Table 55	Concn in aqueous phase (ng l^{-1})	Amt retained by foam from 1 l of water (ng)	Per cent retention
Fl	278.6	260.4	93.5
BjF	48.3	47.4	98.1
BkF	51.7	50.6	97.9
BaP	36.4	33.6	92.3
IP	25.5	23.9	93.7
BPR	22.6	19.8	87.6

Water source, laboratory tap water; water volume, 60 l; concn of fluoranthene, 500 ppt; all others, 100 ppt; detection method, t.l.c.–fluorimetric.
Reprinted with permission from Saxena et al.[461] Copyright (1977) American Chemical Society.

efficiencies were also maintained with heavily polluted surface waters (Table 57). It is unclear why the retention values for three of the PAHs are well above 100%. A possible explanation for this may be the inability of cyclohexane extraction to quantitatively recovery these PAHs from the heavily polluted water; this will give rise to lower PAH concentration in the aqueous phase than actually present.

Table 57 Foam retention efficiencies for PAH from raw water *(for key see Table 55)*

Compound	Concn in aqueous phase (ng l^{-1})	Amt retained by foam from 1 l of water (ng)	Per cent retention
Fl	289.1	343.7	118.9
BjF	77.6	94.0	121.1
BkF	66.1	55.6	84.1
BaP	74.5	59.7	80.1
IP	85.2	61.2	71.8
BPR	23.9	28.3	118.4

Water source, Onondaga Lake water; volume, 30 l; Concn of fluoranthene, 500 ppt; all others, 100 ppt; detection method, t.l.c.–fluorimetric.
Reprinted with permission from Saxena et al.[461] Copyright (1977) American Chemical Society

Table 58 shows the detection limits of the six PAHs using the polyurethane foam preconcentration method with g.l.c.–FID or t.l.c.–fluorimetric analysis. The detection of fluoranthene and benzo(a)pyrene in t.l.c.–fluorimetry is restricted by the background levels of these compounds contributed from the foam plugs. Their detection limit is assumed to be twice the background fluorescence level. In the case of the g.l.c.–FID method, the detection limits for PAH are based on a minimum output response of five times the background noise level and a maximum volume of 5 µl from a total of 100 µl concentrate.

Table 58 Limit of detection of PAH with foam preconcentration coupled with t.l.c.–fluorimetric or g.l.c.–FID method *(for key see Table 55)*

	T.l.c.–fluorimetric detection		G.l.c.–FID detection	
Compound	Absolute limit (ng)	Limit in 60 l water (ng l^{-1})	Absolute limit (ng)	Limit in 60 l water (ng l^{-1})
Fl	140.0	2.3	13.6	4.5
BjF	7.5	0.1	10.1[a]	3.4[a]
BkF	5.0	0.1		
BaP	10.0	0.2	11.9	4.0
IP	10.0	0.2	14.7	4.9
BPR	20.0	0.3	14.9	5.0

[a]Combined values given since the compounds could not be separated on the g.l.c. column.
Reprinted with permission from Saxena et al.[461] Copyright (1977) American Chemical Society.

Details of this procedure are given below. During the concentration of PAH from water on foam plugs, several other contaminants were also concentrated and some of these got eluted during PAH elution. In addition several impurities originating in the foam were also leached during the elution process. The latter impurities could be partially eliminated only by precleaning of the plugs with cyclohexane and/or benzene by batch or Soxhlet extraction. The impurities interfered with the analysis of PAH; therefore, a clean-up procedure was devised. The levels of impurities and subsequently the extent of clean-up necessary are directly dependent upon the sample volume and number of foam plugs employed for concentration. A sample volume of 60 l was adequate for detection of PAH. Sixty litres of unspiked finished water was passed through six precleaned[462] foam columns each containing two plugs, in three successive steps maintaining the water temperature at $62 \pm 2\,°C$ and flow rate at 250 ± 10 ml min^{-1}. Each column was eluted with 30 ml acetone and 125 ml cyclohexane. The combined extract was concentrated and subjected to clean-up. At no time was the PAH mixture allowed to proceed to complete dryness since this has been shown to result in loss of PAH.[463]

Additional clean-up involving a short Florisil column was found to be necessary for further separation from impurities. Chromatographic grade Florisil (60–100 mesh) was washed with methanol and 1:1 hexane–benzene and activated for at least 4 h at 130 °C. The Florisil was cooled to room temperature, and 8 g was transferred to a glass column (1.5 × 30 cm) with benzene by slurry method and washed with an additional 100 ml of benzene prior to passing sample concentrate through it. The flow chart illustrated in Figure 105 depicts the complete clean-up procedure. Thin-layer chromatography was performed using aluminium oxide acetylated cellulose plates which were prepared as described by Borneff and Kunte[452] except that 2% (w/v) $CaSO_4 \cdot 2H_2O$ (200 mesh) was added to the slurry to increase binding of the layer to the surface. The plates were developed in two dimensions; n-hexane–benzene (4:1 v/v) and methanol–ether–water (4:4:1 v/v).

The emission and excitation spectra used for identification of the suspected PAH spots were run directly on the plates at room temperature with a thin-film scanner attached to a spectrophotofluorimeter. This procedure eliminated the losses usually encountered during removal of PAH spots for fluorescence measurement in solvents and the interferences arising due to solvent interactions.

For quantification of the spots, the excitation wavelength was fixed at 365 nm, and the fluorescence intensities were measured at the following wavelengths (nm): Fl, 458; BjF, 427; BkF, 428; BaP, 427; and BPR, 416 (for key see Table 55).

Navra'til et al.[464] have also investigated the use of high capacity open pore polyurethane columns for the collection and preconcentration of PAHs from water.

Weil et al.[465] described a semiquantitative test for the detection of PAHs in drinking water. A total of 395 West German drinking water samples were analysed and only 22 were found to contain more than half the permitted limit of 250 ng l^{-1} (expressed as organic carbon) for the sum of six individual PAH compounds. A method was devised for distinguishing the heavily contaminated

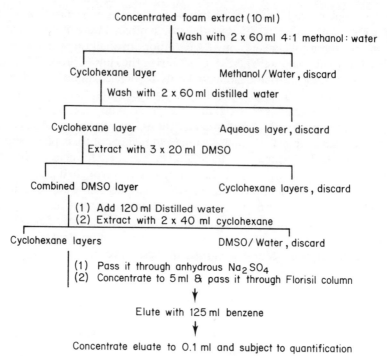

Figure 105 Flow chart of clean-up method. Reprinted with permission from Saxena et al.[461] Copyright (1977) American Chemical Society

samples (more than 125 ng l^{-1}) from the remainder, which involves high performance thin-layer chromatography on prepared RP-18 plates of a cyclohexane extract. The chromatogram is developed by acetonitrile–methylene chloride–water (9:1:1), and assessed by visual observation under a u.v. lamp. The method was capable of distinguishing samples containing more than 50 ng PAH compounds per litre.

Kunte[466] investigated the interference effects of 44 PAHs on a thin-layer chromatographic method for determining fluoranthene, benzo(b)fluoranthene, benzo(k)fluoranthene, benzo(a)pyrene, benzo(ghi)perylene and indeno(1,2,3-cd) pyrene according to the West German drinking water regulations. Overlapping occurred but these overlaps did not interfere with the determination of these six compounds, nor was quantitative determination of the six compounds influenced by the addition of a soil extract or a mixture of 18 PAHs occurring in the environment. Figure 106 illustrates the separations achieved by these workers.

Payne[477] carried out a field investigation of benzopyrene hydroxylate induction as a monitor for marine petroleum pollution. Isaaq et al.[467] isolated stable mutagenic ultraviolet photodecomposition products of benzo(a)pyrene by thin-layer chromatography.

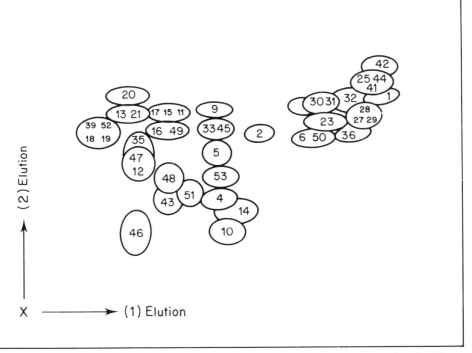

Figure 106 Two dimensional thin-layer chromatogram of 44 PAHs. Reprinted with permission from Kunte.[466] Copyright (1980) Springer-Verlag

Spot no.	Substance
1	+ Anthracene
2	+ Benz(a)anthracene
3	+ Benzo(ghi)fluoranthene
4	+ + Benzo(b)fluoranthene
	(= Benz(e)acephenanthrylene)
5	+ + Benzo(k)fluoranthene
6	+ Benzo(a)fluorene
7	+ Benzo(b)fluorene
8	Benzo(c)fluorene
9	+ Benzo(e)pyrene
10	+ + Benzo(a)pyrene
11	+ + Benzo(ghi)perylene
12	Benzo(b)chrysene
13	+ Coronene
14	+ Chrysene
15	Benzo(b)triphenylene
	(= Dibenz(a,c)anthracene)
16	+ Dibenz(a,h)anthracene
17	+ Dibenz(a,j)anthracene
18	Dibenzo(b,k)perylene
19	Dibenzo(b,n)perylene
20	Dibenzo(def,p)chrysene
	(= Dibenzo(a,l)pyrene)
21	Naphtho(1,2,3,4-def)chrysene
	(= Dibenzo(a,e)pyrene)
22	Dibenzo(fg,op)naphtacene
	(= Dibenzo(e,l)pyrene)
23	+ + Fluoranthene
24	Fluorene
25	+ 2-Methylanthracene

Spot no.	Substance
26	2-Methylfluorene
27	+ 2-Methylphenanthrene
28	+ 3-Methylphenanthrene
29	4,5-Methylenphenanthrene
30	4-Methylpyrene
31	1-Methylpyrene
32	2-Methylpyrene
33	+ Perylene
34	+ Phenanthrene
35	Picene (= Benzo(a)chrysene)
36	+ Pyrene
37	Naphtacene
38	+ Triphenylene
39	Dibenzo(h,rst)pentaphene
41	9-Methylanthracene
42	9-Phenylanthracene
43	+ Anthanthrene
	(= Dibenzo(def,mno)chrys
44	9,10-Dimethylanthracene
45	3-Methylcholanthrene
46	Dibenzo(b,def)chrysene
	(= Dibenzo(a,h)pyrene)
47	Benzo(rest)pentaphene
	(= Dibenzo(a,i)pyrene)
48	+ + Indeno(1,2,3-cd)pyrene
49	Pentaphene
50	7,12-Dimethylbenz(a)anthracene
51	Dibenzo(ghi,k)fluoranthene
52	Tribenzo(b,j,l)fluoranthene
53	Benzo(j)fluoranthene

Harrison et al.[468] examined the effects of water chlorination upon levels of some PAHs in water including a study of the effects of pH, temperature, contact time, and chlorinating agent. Thieleman[469] investigated by thin-layer chromatography the action of chlorine dioxide on acetnaphthalene, anthracene, carbazole, chrysene, phenanthrene, and pyrene.

GREASES

The traditional and widely accepted method for determining grease in waste water is based on organic solvent extraction, solvent evaporation, and weighing.[470,472] The method is time-consuming, both in the analyst's time and in duration. A rather large sample of waste water is required for the traditional method, and it is difficult to manipulate the samples to assure a homogeneous aliquot. Simplification of the methodology is fraught with difficulties because of the extreme complexity of waste water in composition and variability.

Maxcy[471] has described a simpler volumetric procedure, outlined below, for the determination of grease in waste water.

Method

Stock samples of approximately 4 l were acidified to pH 1.0 by the addition of sulphuric acid and stored at 2 °C. The waste water was blended in a Waring blender so that 100 ml aliquots could be taken as needed.

100 ml of a representative sample are pipetted into a 1000 ml separatory funnel, 100 ml of ethanol are added and the mixture is shaken for 30 s. 100 ml of trichlorotrifluoroethane are added and the mixture shaken for 2 min. Combinations of ethers were also tried. The trichlorotrifluoroethane is drained into a 400 ml beaker and 100 ml of trichlorotrifluoroethane are added to the separatory funnel, shaken for 2 min, and the trichlorotrifluoroethane drained into the beaker.

Glass beads are added to promote gentle boiling. The volume of trichlorotrifluoroethane in the beaker is reduced to 5–10 ml by evaporation on a hot plate. The trichlorotrifluoroethane containing the grease is quantitatively transferred to a Babcock skim milk bottle by rinsing the beaker with two separate 5 ml portions of trichlorotrifluoroethane. A small glass bead is added to the Babcock bottle and by using a water bath heating from 40° to 100 °C the trichlorotrifluoroethane is evaporated to apparent dryness. The Babcock bottle is exposed to approximately 135 °C and 56 cm (22 in) vacuum for 10 min. The bottle is cooled for convenient handling, 10 ml of 0.12 M trisodium phosphate added, then the bottle is warmed to approximately 90 °C. The bottle is swirled to suspend the grease, then cooled to room temperature. If the suspended grease exists as free-floating globules in a clear solution of trisodium phosphate, distilled water is added to the bottle with frequent swirling to bring the solution into the capillary. If the grease appears to be dispersed, then with frequent swirling 20 ml of 1% sulphuric acid are added, followed by distilled water to bring the mixture into the capillary. The mixture is centrifuged for 5–10 min in a Babcock centrifuge or equivalent. The time of sample centrifugation is dependent on how readily the greasy matter first agglomerates at the base of the bottle neck and then moves into the capillary. A vortex mixer may prove useful in moving grease into the capillary. The bottle is then warmed to approximately 90 °C and centrifuged for 5 min. Agglomerated grease remaining at the base of the neck should move into the capillary at this time, forming a measurable greasy column. Greasy overflow from the capillary is prevented through elimination of trapped air bubbles and solution level adjustment for fluid expansion during heating. The quantity of grease is read in hundredths on the graduated capillary, then multiplied by 0.18 because the calibration of the Babcock skim milk test bottle is based on an 18 g sample.

Typical results (Table 59) indicate a generally good response of the volumetric test to alteration in grease content. Three of the four samples showed precise agreement with the theoretical while one sample showed 114% recovery. The latter figure was not surprising because the volumetric test is only calibrated to the nearest 0.01% and the comparative tests were based on analysis of two components for the mixture.

Table 59 Results for waste water contamination with milk or synthetic waste water

Waste water (%)	Synthetic waste water (%)	Milk (ml)	Calculated volumetric reading (in hundredths)	Observed volumetric reading (in hundredths)	Recovery (%)
50	50	0	14	16	114
75	25	0	10	10	100
100	0	0.9	7	7	100
100	0	1.8	9	9	100

Reprinted with permission from Maxcy.[471] Copyright (1976) Water Pollution Control Federation.

The volumetric test was used to determine the grease content of natural waste water and a dilute solution of a synthetic waste water. A stock synthetic waste water was prepared by making a paste of 10.1 g cooking fat and 11.7 g wheat flour, mixing it in a Waring blender to 200 ml, and then transferring it to a 1 l flask. Warm water and 5 g of soap were added to make 1 l. Samples of natural waste water were then altered by the addition of synthetic waste water, then extracted for the volumetric test. The calculated and observed results with the volumetric test were compared (Table 60). There was quantitative recovery of the cooking fat, but only an average of 75% of the milk fat was recovered. The latter result was probably caused by the relatively stable emulsion of extremely small fat globules covered with adsorbed material in homogenized milk. Incomplete contact of fat and solvent might be expected.

Cook et al.[472] have described improved methods for the determination of oil and grease in water. This method utilizes the semiautomatic oil monitoring

Table 60 Effectiveness of the volumetric test in recovering components of waste water

Extraction solvent	Component added	Theoretically expected recovery	Observed recovery	Recovery (%)
Trichlorotrifluoroethane	cooking fat	0.0455	0.045	99
Ethers*	cooking fat	0.0549	0.054	98
Trichlorotrifluoroethane	milk	0.0234	0.0162	69
Ethers	milk	0.0234	0.0180	77
Trichlorotrifluoroethane	milk	0.0468	0.0378	81
Esters	milk	0.0468	0.0360	77

*An equal mixture of ethyl ether and petroleum ether.
Reprinted with permission from Maxcy.[471] Copyright (1976) Water Pollution Control Federation.

instrument, the Horiba OCMA-200. These workers compared results obtained by this procedure with these obtained by a US Environmental Protection Agency reference method.[473] The Horiba instrument is a very convenient device to use since sample handling and contamination possibilities are almost nil. An acidified, 10 ml oil-in-water sample is injected via a syringe into an extraction chamber with an equal amount of solvent, either Freon or carbon tetrachloride. A vibrating plunger is activated for a given amount of time, causing intimate mixing of the sample and solvent. The plunger is deactivated, and the solvent and oil mixture is allowed to separate. A valve is then opened, permitting the solvent mixture to flow through a filter into the infrared chamber where the amount of 3.5 micron light attenuation by the oil in solvent is measured.

This wavelength was chosen because it represents a major absorption region of most hydrocarbons which comprise the oil or grease. With proper calibration, the instrument then reads directly in parts per million of oil. Analysis time can range from 50 seconds to 30 minutes, depending on the solvent and the required extraction time. Freon was chosen as the solvent with an extraction time of 30 minutes, which is adequate for all known oils. Because the Horiba instrument has a full scale measurement capability of 20 mg l^{-1}, mixtures of higher oil content are measured by dilution with oil-free water.

The Horiba device was calibrated using both carbon tetrachloride and Freon as solvents. Both solvents yielded excellent calibration curves as shown in Figure 107. If Freon is to be used, two modifications must be performed on the instrument to compensate for the lower boiling point of the solvent. First, the internal solvent filter heater must be disconnected. Second, all oil analysis must be conducted with the instrument cover removed to provide better convective cooling, thereby avoiding zero and span drift. The installation of an auxiliary fan or internal ducting changes could possibly maintain internal temperature values compatible with Freon, thus eliminating the drift problem.

Cook et al.[472] showed that adjustment of the sample to pH 2 enhanced the extraction efficiency of the instrument and also that a 30 min Freon extraction time was adequate.

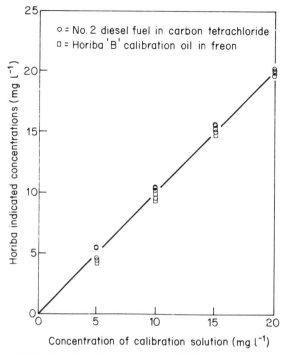

Figure 107 Horiba Oil Analyzer Calibration Curve. (Data generated using the standard vendor calibration procedure.)

Cook et al.[472] also investigated factors which affected the accuracy of the reference method referred to earlier.[347] The reference method currently requires three sample extractions with 30 ml Freon each. However, as shown in Table 61 larger amounts of solvent are required to recover at least 80–90% of the soluble oil. The use of 30 ml solvent yields rather erratic and inconsistent results, while the use of 150 ml solvent per extraction (for a given extraction time) gives values closer to the actual concentrations. The data are shown in the table only to indicate the importance of the use of adequate quantities of solvent.

Table 61 Oil analysis results utilizing 30 ml and 150 ml quantities of Freon for the same extraction time

Total extraction (min)	Calibration solution oil concentration (mg l^{-1})	30 ml solvent results (mg l^{-1})	150 ml extraction results (mg l^{-1})
12	10	5.8	9.6
12	20	9.0	16.4
40	40	33.2	37.2
30	80	18.4	62.4

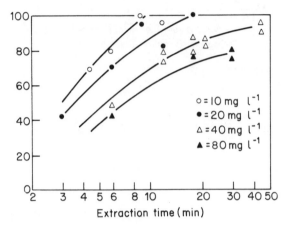

Figure 108 Effects of extraction time on soluble oil measurements utilizing the modified method

In the reference method, total extraction time is 6 minutes which, as shown in Figure 108, is inadequate for recovery of the oil. As shown, the optimum extraction time is a function of the oil concentration and, for 80 mg l^{-1}, it must be as high as 60 minutes for extraction of approximately 80% of the oil. Mechanical stirring was utilized for the long extraction times rather than the recommended manual shaking method. Instead of simply predrying a flask at 103 °C and storing it in a desiccator prior to taring as indicated in the procedure, it was found by Cook et al.[472] that more stable tare weight values are obtained by predrying the flask at 200 °C, storing it in a desiccator, and weighing it in a desiccating balance containing silica gel. The tare weight is determined when the balance/flask system reaches an equilibrium value for at least 1 hour. This assures the analyst that the tare weight is relatively free from moisture adsorption errors.

The solvent evaporation step of the reference method uses a 70 °C water bath for evaporation and a 15 minute exposure to an 80 °C water bath for drying. Air is drawn through the flask with a vacuum pump for 1 minute after the drying step. The flask is then stored in a desiccator for 30 minutes and weighed. More precise results are obtained by using a well controlled flask heater and an all-glass single-stage distilling apparatus. Solvent evaporation is then accomplished by adjustment of the heat input to the system for light boiling while the vapour temperature is monitored. Evaporation is discontinued when the vapour temperature exceeds 48 °C (the boiling point of Freon TF) or when the flask is visibly dry. The flask is then cooled by flushing with ultra-pure argon.

Final weight values are obtained by placing the flask in the desiccating balance and monitoring the weight until equilibrium is reached. Figure 109 shows that adsorbed moisture weights are significant and that desiccating times must be long enough to compensate for this potentially significant error. Obviously, specified desiccating times cannot be used, and balance/flask equilibrium should be the criteria for valid measurements.

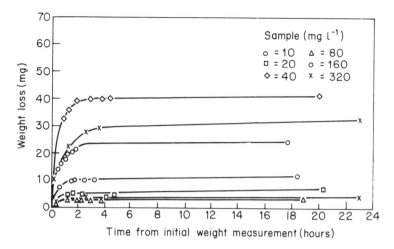

Figure 109 Weight loss *vs.* time curves for distilling flasks used with various oil samples after solvent evaporation

Cook et al.[472] concluded that, with some modifications, the precision and accuracy of the reference method for oil and grease analysis can be greatly improved. Critical factors are: quantities of extracting solvents used, extraction time, and moisture adsorption on the gravimetric flask. The use of 150 ml of Freon rather than 30 ml, 45 minutes extraction time, and careful desiccating procedures during weighing are recommended.

The semiautomatic Horiba oil analyser can be correlated with the modified reference method, if care is taken to use the proper extraction time and to prevent instrument overheating (which results in excessive zero and span drift) when Freon is used as the solvent. The sample must be properly acidified just prior to its injection into the analyser. With these precautions, reasonable correlation can be shown for both soluble and insoluble oils and greases.

Cleverley[478] has carried out a comparison of various lubricating greases by infrared spectroscopy in the 4000–650 cm^{-1} region. This work would be useful in classifying greases originating in extracts of water samples.

Various workers[474–476] have discussed the volumetric determination of greases in sewage, sludge, and industrial wastes.

REFERENCES

1. Brunnock, J. V., Duckworth, D. F. and Stephens, G. G. *J. Inst. Petrol.*, **54**, 310 (1968).
2. Duewer, D. L., Kowalski, B. R. and Schatski, T. F. *Anal. Chem.*, **47**, 1573 (1975).
3. Kawahara, F. K. and Ballinger, D. G. *Ind. Eng. Chem., Prod. Res. Dev.*, **9**, 553 (1970).
4. Lynch, P. F. and Brown, Ch. W. *Environ. Sci. Technol.*, **7**, 1123 (1973).
5. Mattson, J. S., Mattson, C. S., Spencer, M. J. and Starks, S. A. *Anal. Chem.*, **49**, 297 (1977).

6. Erhardt, M. and Blumer, M. *Environ. Pollut.*, 179 (1972).
7. Clark, H. A. and Jurs, P. C. *Anal. Chem.*, **47**, 374 (1975).
8. Rasmussen, D. V. *Anal. Chem.*, **48**, 1562 (1976).
9. Association of British Chemical Manufacturers and Society for Analytical Chemistry. *Recommended Methods for the Analysis of Trade Effluents*, Cambridge, Heffer, 124 pp. (1958).
10. Ruebelt, C. *Helgoländer wiss. Meeresunters.*, **16**, 306 (1967).
11. Beynon, L. R. *Methods for the Analysis of Oil in Water and Soil*. The Hague, Strichting CONCAWE, 40 pp. (1968).
12. Kawahara, F. K. *Laboratory Guide for the Identification of Petroleum Products*. 1014 Broadway, Cincinnati, Ohio, U.S. Dept of the Interior, Federal Water Pollution Central Administration, Analytical Quality Central Laboratory, 41 pp. (1969).
13. Lively, L. *Engng Bull. Ext. Ser. Purdue Univ.*, **118**, 657 (1965).
14. Davies, A. W. *Bull. Cent. belge Étud. Docum. Eaux.*, **330**, 252 (1971).
15. Caruso, S. C. *Devs. appl. Spectrosc.*, **6**, 323 (1967).
16. Jeltes, R. and Veldink, R. *J. Chromat.*, **27**, 242 (1967).
17. Jeltes, R. *Wat. Res.*, **3**, 931 (1969).
18. Peake, E. and Hodgson, G. W. *J. Am. Oil Chem. Soc.*, **43**, 215 (1966).
19. Peake, E. and Hodgson, G. W. *J. Am. Oil Chem. Soc.*, **44**, 696 (1967).
20. Adams, I. M. *Process Biochem.*, **2** (5), 33 (1967).
21. Institute of Petroleum: Standardization Committee. *J. Inst. Petrol.*, **56**, 107 (1970).
22. Maehler, C. Z. and Greenberg, A. E. *J. sanit. Engng Div. Am. Soc. civ. Engrs.*, **94**, 969 (1968).
23. Rook, J. J. H_2O, **4**, 385 (1971).
24. McAuliffe, C. *Chem. Geol.*, **4**, 225 (1969).
25. Goma, G. and Durand, G. *Wat. Res.*, **5**, 545 (1971).
26. Trent River Authority. Private communication. Trent River Authority, Meadow Lane, Nottingham.
27. Cole, R. D. *J. Inst. Petrol.*, **54**, 288 (1968).
28. Ramsdale, S. J. and Wilkinson, R. E. *J. Inst. Petrol.*, **54**, 326 (1968).
29. Johnson, W. and Kawahara, F. K. *Proc. 11th Conf. Gt. Lakes Res.*, **550** (1968).
30. Ellerker, R. *Wat. Pollut. Control*, **67**, 542 (1968).
31. Dewitt Johnson, W. and Fuller, F. D. *Proc. 13th Conf. Gt. Lakes Res.*, **128** (1970).
32. Blumer, M. *Mar. Biol.*, **5**, 195 (1970).
33. Le Pera, M. E. *Identification and Characterization of Petroleum Fuels using Temperature-programmed Gas–liquid Chromatography*. Springfield, Va., National Technical Information Service, Rep. no. AD 646-382 (1966).
34. Brunnock, J. V. *J. Inst. Petrol.*, **54**, 310 (1968).
35. Duckworth, D. F. *Aspects of Petroleum Pollutant Analysis*. *'Water Pollution in Oil'* ed. by P. Hepple, London, Institute of Petroleum, p. 165 (1971).
36. Blumer, M. *Mar. Biol.*, **5**, 196 (1970).
37. Sanders, W. N. and Maynard, J. B. *Anal. Chem.*, **40**, 527 (1968).
38. Gouw, T. H. *Anal. Chem.*, **42**, 1394 (1970).
39. Cole, R. D. *Nature, Lond.*, **233**, 546 (1971).
40. Lysyj, I. and Newton, P. R. *Anal. Chem.*, **44**, 2385 (1972).
41. Lysyj, I. and Newton, P. R. *Anal. Abstr.*, **22**, 459 (1972).
42. Kawahara, F. K. *J. Chromat. Sci.*, **10**, 629 (1972).
43. Jeltes, R. and den Tonkelaar, W. A. M. *Water (H_2O)*, **5**, 288 (1972).
44. Ahmadijian, M. and Brown, C. W. *Environ. Sci. Technol.*, **7**, 452 (1973).
45. McMullen, A. I., Monk, J. F. and Stuart, M. J. *International Laboratory*, **60–61** Jan–Feb, 5447 (1975).
46. Jeltes, R., Bunghardt, E., Thijsse, T. R. and den Tonkelaar, W. A. M. *Chromatographia*, **10**, 430 (1977).

47. Cole, R. D. *J. Inst. Petrol.*, **54**, 288 (1968).
48. Vos, L. L., Bridie, A. L. and Herzbery, S. H_2O, **10**, 277 (1977).
49. Garra, M. E. and Muth, J. *Environ. Sci. Technol.*, **8**, 249 (1974).
50. Freegarde, M., Hatchard, C. G. and Parker, C. A. *Lab. Pract.*, **20**, 35 (1971).
51. Erhardt, M. and Blumer, M. *Environ. Pollut.*, **3**, 179 (1972).
52. Zafiron, O. C., Myers, J. and Freestone, F. *Marine Pollution Bulletin*, **4**, 87 (1973).
53. Boylan, D. B. and Tripp, B. W. *Nature (Lond.)*, **230**, 44 (1971).
54. Bulten, J. N., Morris, B. F. and Sass, J. *J. Bermuda Biological Research, Special Publication No. 10*, 350 pp. (1973).
55. Zafiron, O. C. and Oliver, C. *Anal. Chem.*, **45**, 952 (1973).
56. Hertz, H. S., May, W. E., Chesler, S. N. and Gump, B. H. *Environ. Sci. Technol.*, **10**, 900 (1976).
57. Rasmussen, D. V. *Anal. Chem.*, **48**, 1562 (1976).
58. Institute of Petroleum. London, *J. Inst. Petrol.*, **56**, 107 (1970).
59. Wilson, C. A., Ferreto, E. P. and Coleman, H. J. *Am. Chem. Soc., Div. Chem., Pepr.*, **20**, 613 (1975).
60. Garza, M. E. and Muth, J. *Environ. Sci. Technol.*, **8**, 249 (1974).
61. Adlard, E. R., Creaser, L. F. and Matthews, P. H. *Anal. Chem.*, **44**, 297 (1972).
62. Blumer, M. and Saas, J. *Science*, **176**, 1120 (1972).
63. Ramsdale, S. J. and Wilkinson, R. E. *J. Inst. Petrol.*, **54**, 326 (1968).
64. Adlard, E. R., Creaser, L. F. and Matthews, P. H. D. *Anal. Chem.*, **44**, 64 (1972).
65. Brunnock, J. V., Duckworth, D. F. and Stephens, G. G. *J. Inst. Petrol.*, **54**, 310 (1968).
66. McKay, T. R. *Proc. Ninth Int. Symp. on Gas Chromatography, Montreux, Switzerland*, pp. 33–38 October (1972).
67. Adlard, E. R. and Matthews, P. H. D. *Nutur. phys. Sci.*, **233**, 83 (1971).
68. Millard, J. P. and Arvesen, J. C. *Appl. Optics*, **11**, 102 (1972).
69. Garra, M. E. and Muth, J. *Sci. Technol.*, **8**, 249 (1974).
70. Albaigés, J. and Albrecht, P. *Int. J. Environ. Anal. Chem.*, **6**, 171 (1979).
71. Brunnock, J. V., Duckworth, D. F. and Stephens, G. G. *J. Inst. Petrol.*, **54**, 310 (1968).
72. Ramsdale, S. J. and Wilkinson, R. E. *J. Inst. Petrol.*, **54**, 326 (1968).
73. Erhardt, M. and Blumer, M. *Environ. Pollut.*, 179 (1972).
74. Adlard, E. R., Creaser, L. F. and Matthews, P. H. D. *Anal. Chem.*, **44**, 64 (1972).
75. Bailey, N. J. L., Jobson, A. M. and Rogers, M. A. *Chem. Geol.*, **11**, 203 (1973).
76. Henderson, W., Wollrab, V. and Eglington, G. *Advances in Organic Geochemistry, 1968*. Eds. Schenk, P. A. and Havenaar, I., p. 181, Pergamon, Oxford (1969).
77. Gallegos, E. J. *Anal. Chem.*, **43**, 1151 (1971).
78. Rubinstein, I., Strausz, O. P., Spyckerelle, C., Crawford, R. J. and Westlake, D. W. S. *Geochim. Cosmochim. Acta*, **41**, 1341 (1977).
79. Albaigés, J., Borbon, J. and Salgre, P. *Tetrahedron Lett.*, 595 (1978).
80. Rubinstein, I. and Albrecht, P. *J. Chem. Soc. Chem. Commun.*, 957 (1975).
81. Van Dorsselaar, A., Albrecht, P. and Ourisson, G. *Bull. Soc. Chim. France.*, 165 (1977).
82. Kimble, B. J., Maxwell, J. R., Philp, R. P., Eglington, G., Albrecht, P., Ensminger, A., Arpino, P. and Curisson, G. *Geochim. Cosmochim. Acta*, **38**, 1165 (1974).
83. Rubinstein, I., Sieskind, O. and Albrecht, P. *J. Chem. Soc., Perkin Trans.*, **1**, 1833 (1975).
84. Mulheirn, L. J. and Ryback, G. *Advances in Organic Geochemistry, 1975*. Eds. Campos, R. and Goni, J., p. 173, Enadimsa, Madrid (1977).
85. Seifert, W. K. and Moldowan, J. M. *Geochem. Cosmochim. Acta*, **42**, 77 (1978).
86. Reed, W. E. *Geochim. Cosmochim. Acta*, **41**, 237 (1977).
87. Van Dorsselaar, A., Schmitter, J. M., Albrecht, P., Claret, J. and Connan, J. Paper presented at the *8th International Meeting on Organic Geochemistry, Moscow* (1977).

88. Pym, J. G., Ray, J. E., Smith, G. W. and Whitehead, E. V. *Anal. Chem.*, **47**, 1627 (1975).
89. Ensminger, A., Thesis. Univ. Louis Pasteur. Strasbourg. 149 pages (1977).
90. Hertz, H. S., May, W. E., Chesler, S. N. and Gump, B. H. *Environ. Sci. Technol.*, **10**, 900 (1978).
91. May, W. E., Chesler, S. N., Cram, S. P., Gump, B. H., Hertz, H. S., Enagonio, D. P. and Dyszel, S. M. *J. Chromatogr. Sci.*, **13**, 535 (1975).
92. Chesler, S. N., Gump, B. H., Hertz, H. S., May, W. E., Dyszel, S. M. and Enagonio, D. P. *NBS Technical Note No. 889*, 73 pp., Washington, DC (1976).
93. Smith, D. H. *Anal. Chem.*, **44**, 536 (1972).
94. Walker, J. D., Colwell, R. R., Hamming, M. C. and Ford, H. T. *Environ. Pollut.*, **9**, 231 (1975).
95. Brown, R. A. and Huffman, H. L. *Science*, **101**, 847 (1976).
96. Walker, J. D., Colwell, R. R., Hamming, M. C. and Ford, H. I. *Bull. Environ. Contam. Toxicol.*, **13**, 245 (1975).
97. Triems, K. *Chem. Tech. Berl.*, **20**, 596 (1968).
98. Triems, K. and Heinze, O. *Anal. Abstr.*, **16**, 2556 (1969).
99. Farrington, J. W., Teal, J. M., Quinn, J. G., Wade, T. and Burns, K. *Bull. Environ. Contam. Toxicol.*, **10**, 129 (1973).
100. Millson, M. F. *J. Chromat.*, **50**, 155 (1970).
101. Done, J. N. and Reid, W. K. *Separ. Sci.*, **5**, 825 (1970).
102. Schuldiner, J. A. *Anal. Chem.*, **23**, 1676 (1951).
103. Herd, M. *Analyst (London)*, **78**, 383 (1953).
104. Crump, G. B. *Nature*, **193**, 674 (1962).
105. Krieger, H. *Gas-u WassFach*, **104**, 695 (1963).
106. Lambert, G. *Bull. Cent. Belge Etud. Docum. Eaux*, **20**, 271 (1966).
107. Cross, B. T. and Hart, M. *Medmenham, Water Research Association Internal Laboratory Report IRL 62.*
108. Matthews, P. J. *Water Pollution Control*, **67**, 588 (1968).
109. Matthews, P. J. *J. appl. Chem (London)*, **20**, 87 (1970).
110. Ruebelt, C. *Helgoländer Wiss. Meeresunters*, **16**, 306 (1967).
111. Ruebelt, C. *Haustech. Essen. Vortragsveroeff.*, **65**, 231 (1970).
112. Killer, F. C. A. and Amos, R. *J. Inst. Petrol.*, **52**, 315 (1966).
113. Coates, J. P. *J. Inst. Petrol.*, **57**, 209 (1971).
114. Sauer, W. A. and Fitzgerald, G. E. *Environ. Sci. Technol.*, **10**, 893 (1976).
115. Wasik, S. P. *J. Chromatogr. Sci.*, **12**, 845 (1974).
116. Frankenfield, J. W. and Schulz, W. *Identification of Weathered Oil Films Found in the Marine Environment*, DOT Contract No. CG-23, 035-A. Exxon Government Research Lab. Linden, NJ 1974 (available through the National Technical Information Service, Springfield, Va, 22151, Accession No, ADA-015883).
117. Brown, L. R., Pabst, G. S. and Light, M. *Marine Pollution Bulletin*, **9**, 81 (1978).
118. Whittle, P. J. *Analyst*, **102**, 976 (1977).
119. Fudge, R. and Nicholas, P. V. *T. Ass. Analysts*, **8**, 31 (1970).
120. Rosen, O. Severn River Authority Private communication.
121. Voege, F. A. and Vanesche, H. L. *J. Am. Wat. Wks-Assn.*, **56**, 1351 (1964).
122. Cole, R. D. *J. Inst. Petrol.*, **54**, 288 (1968).
123. Machler, C. Z. and Greenberg, A. E. *Sanit. J. Engng. Div. Am. Soc. Civ. Engrs*, **94**, 969 (1968).
124. Kawahara, F. K. *Laboratory Guide for the Identification of Petroleum Products*. 1014, Broadway, Cincinnati, Ohio, USA. Dept. of the Interior Federal Water Pollution Control Admin. Division of Water Quality Research. Analytical Quality Control. Laboratory, 41 pp. (1969).
125. Johnson, W. and Kawahara, F. K. *Proc. 11th Conf. Gt Lks Res.* 550 (1968).
126. Kawahara, F. K. *Environ. Sci. Technol.*, **3**, 150 (1969).

127. Kawahara, F. K. and Ballinger, D. *Ind. Engng Chem. Prod. Res. Rev.*, **9**, 553 (1970).
128. Mattson, J. S. and Mark, H. B. *Anal. Chem.*, **42**, 234 (1970).
129. Mattson, J. S. and Mark, H. B. *Environ. Sci. Technol.*, **3**, 161 (1969).
130. Kawahara, F. K. and Ballinger, D. G. *Ind. Engng Chem. Prod. Res. Dev.*, **9**, 553 (1970).
131. Kawahara, F. K. *J. Chromatogr. Sci.*, **10**, 629 (1972).
132. Mattson, J. S., Mark, H. B., Kolpack, R. L. and Schutt, C. E. *Anal. Chem.*, **42**, 234 (1970).
133. Mark, H. B., Yu, T. C., Mattson, J. S. and Kolpack, R. L. *Environ. Sci. Technol.*, **6**, 833 (1972).
134. Baier, R. E. *J. Geophys. Res.*, **77**, 5062 (1972).
135. Mattson, J. S. *Anal. Chem.*, **43**, 1872 (1972).
136. Powell, D. L., Darland, E. J. and Williams, T. R. *Anal. Lett.*, **4**, 479 (1971).
137. Pierre, L. J. *Appl. Optics*, **12**, 2035 (1973).
138. Ahmadjian, M. and Brown, C. W. *Anal. Chem.*, **48**, 1257 (1976).
139. Lynch, P. F. and Brown, C. W. *Environ. Sci. Technol.*, **7**, 1123 (1973).
140. Kawahara, F. K. *J. Environ. Sci. Technol.*, **3**, 150 (1969).
141. Mattson, J. S. *Anal. Chem.*, **43**, 1872 (1971).
142. Bogatie, C. F. *Tappi*, **57**, 103 (1974).
143. Kawahara, F. K. *Environ. Sci. Technol.*, **3**, 150 (1969).
144. Kawahara, F. K. and Ballinger, D. C. *Ind. Engng Chem. Prod. Res. Dev.*, **4**, 553 (1970).
145. Dyer, J. R. *Applications of Absorption Spectroscopy of Organic Compounds*, 1st Edition, Prentice Hall, Englewood Cliffs, NJ (1965).
146. Whittle, P. J., McCrum, W. A. and Horne, M. W. *Analyst (London)*, **105**, 679 (1980).
147. Simard, R. G., Hasegawa, I., Bandaruk, W. and Headington, C. E. *Anal. Chem.*, **23**, 1384 (1951).
148. Strichting CONCAWE, The Hague, *Report 9/72* (1972).
149. US EPA *Manual of Methods for Chemical Analysis of Water and Wastes*, p. 232, Storet No. 00560 (1974).
150. American Petroleum Institute. *Manual on Disposal of Refinery Waster*, Vol. IV, Method 733-58 (1958).
151. Coles, G. P., Dill, R. M. and Shull, D. L. *Am. Chem. Soc., Div. Pet. Chem. Prep.*, **20**, 641 (1975).
152. Greenfield, M. *Environ. Sci. Technol.*, **7**, 636 (1973).
153. Levy, E. M. *Water Research*, **6**, 57 (1972).
154. Freegarde, M. *Lab. Pract.*, **20**, 35 (1971).
155. Thruston, A. D. and Knight, R. W. *Environ. Sci. Technol.*, **5**, 64 (1971).
156. Lloyd, J. B. F. *J. Forens. Sci. Soc.*, **11**, 83 (1971).
157. Lloyd, J. B. F. *J. Forens. Sci. Soc.*, **11**, 153 (1971).
158. Lloyd, J. B. F. *J. Forens. Sci. Soc.*, **11**, 235 (1971).
159. Parker, C. A. and Barnes, W. J. *Analyst (London)*, **85**, 3 (1960).
160. Wakeham, S. G. *Environ. Sci. Technol.*, **11**, 272 (1977).
161. US Environmental Protection Agency Industrial Environment Research Laboratory, Cincinnati, Ohio, p. 256 (32428) (1979).
162. Bryan, D. E. and Guinn, V. P. Springfield, Virginia National Information Service, USA. *EC Report No GA 9889*, p. 134 (1970).
163. Guinn, V. P. and Bellanca, S. C. *Washington USA PO National Bureau of Standards Special Publication, No 312*, p. 93 (1967).
164. Bryan, D. E., Guinn, V. P., Hackleman, R. P. and Lukens, H. R. *Report Atomic Energy Commission U.S. GA-9889*, p. 134 (1970).
165. Lukens, H. R., Bryan, D., Hiatt, N. A. and Schlesinger, H. L. *Report Atomic Energy Commission U.S. GULF-RT-A-10684*, p. 134 (1971).

166. Lukens, H. R. *Report Atomic Energy Commission US GULF-RT-A-10973* p. 74 (1972).
167. Flaherty, J. P. and Eldridge, H. B. *Appl. Spectrosc.*, **24**, 534 (1970).
168. Scott, K. A. and Bean, B. M. *U.S. Patent No 3574550*, Washington Patent Office (1967).
169. Brunnock, J. V. *J. Inst. Petrol.*, **54**, 310 (1968).
170. Louis, R. *Chemie. Ingr-Tech.*, **40**, 538 (1968).
171. Kubo, H., Bernthal, R. and Wildeman, R. *Anal. Chem.*, **50**, 899 (1978).
172. Stuart, R. A. and Branch, R. D. *Instrument News* (Perkin-Elmer Corporation, Norwalk, Connecticut), **21**, No. 2 (1970).
173. Levy, E. M. *Water Research*, **6**, 57 (1972).
174. Eldridge, H. B. and Flaherty, J. P. *Ind. Engng Chem. Prod. Res. Dev.*, **9**, 422 (1970).
175. Budnikov, G. K. and Medyantseva, E. P. *Zhur. Analit. Khim.*, **28**, 301 (1973).
176. Lieberman, M. *Report U.S. Environment Protection Agency. EPA-RZ-73-102* 184 pp. (1973).
177. Association of British Chemical Manufacturers and Society for Analytical Chemistry *Recommended Methods for the Analysis of Trade Effluent*, Cambridge, Heffer, p. 124 (1958).
178. Milner, A. Private communication.
179. Sherratt, J. G. *Analyst (London)*, **81**, 518 (1956).
180. Rogen, J. *Chemical and Physicochemical Analysis of Water*, Paris, Dunod, p. 700 (1971).
181. American Society for Testing and Materials Standards with Related Materials, Part 23. *Water and Atmospheric Analysis*, Philadelphia, The Society, **XVIII**, 1036 pp. (1969).
182. American Public Health Association *Standard Methods for Examination of Water and Wastewater*, 13th edition, New York, The Association XXXV, 874 pp. (1971).
183. Taras, M. J. and Blum, K. A. *J. Water Pollution Control*, **40**, 404 (1968).
184. Lively, L. *Engng Bull. Ext. Serv. Purdue University*, **118**, 657 (1965).
185. Jeltes, R. and Veldink, R. *J. Chromat.*, **27**, 242 (1967).
186. McAucliffe, C. *J. Phys. Chem.*, **70**, 1267 (1966).
187. Benyon, L. R. *Methods for the Analysis of Oil in Water and Soil*. The Hague, Strichting CONCAWE 40 pp. (1968).
188. Jeltes, R. *Wat. Res.*, **3**, 931 (1969).
189. Jeltes, R. *Anal. Abstr.*, **15**, 3633 (1968).
190. Goma, G. and Durand, G. *Wat. Res.*, **5**, 545 (1971).
191. Lur'e, Yu. Yu., Panova, V. A. and Nickolaeva, Z. V. *Gidrokhim. Water*, **55**, 108 (1971). Ref. *Zhurkhim.*, 199D (1971) (16) Abstr. No. 16G258.
192. Jeltes, R. and Van Tonkelaar, W. A. M. *Wat. Res.*, **6**, 271 (1972).
193. Bridie, A. L., Bos, J. and Herzberg, S. *J. St. Petrol.*, **59**, 263 (1973).
194. Desbaumes, E. and Imhoff, C. *Wat. Res.*, **6**, 885 (1972).
195. Morgan, N. L. *Bull. Environ. Contam. Toxicol.*, **14**, 309 (1975).
196. Department of the Environment, and National Water Council, HM Stationery Office, London. *Methods for the Examination of Waters and Associated Materials*. 12 pp. (R22BENV) (1979).
197. May, W. E., Chesler, S. N., Cram, S. P., Gump, B. H., Hertz, H. S., Enagonio, D. P. and Dryzel, S. M. *J. Chromatogr. Sci.*, **13**, 535 (1975).
198. Khazal, W. J., Vejrosta, J. and Novák, J. *J. Chromat.*, **157**, 125 (1978).
199. Drozd, J. and Novák, J. *J. Chromat.*, **152**, 55 (1978).
200. Drozd, J. and Novák, J. *J. Chromat.*, **136**, 37 (1977).
201. Novák, J. *Quantitative Analysis by Gas Chromatography*. Marcel Dekker, New York, p. 107 (1975).
202. Drozd, J., Novák, J. and Rijks, J. *J. Chromat.*, **158**, 471 (1978).

203. Franken, J. J., Rutten, G. A. F. M. and Rijks, J. A. *J. Chromat.*, **126**, 117 (1976).
204. Blomberg, L. *J. Chromat.*, **115**, 365 (1975).
205. Schomburg, G. and Husmann, H. *Chromatographia*, **8**, 517 (1975).
206. Grob, K. and Grob, G. *J. Chromat.*, **125**, 471 (1976).
207. Grob, K., Grob, G. and Grob, K. Jr. *Chromatographia*, **10**, 181 (1977).
208. Bouche, J. and Verzele, M. *J. Gas Chromat.*, **6**, 501 (1968).
209. Rijks, J. A. Thesis, Eindhoven University of Technology, p. 20 (1973).
210. McAucliffe, C. *J. Chem. Tech.*, **1**, 46 (1971).
211. Kaiser, L. E. and Oliver, B. G. *Anal. Chem.*, **48**, 2207 (1976).
212. McAucliffe, C. *J. Chem. Geol.*, **4**, 225 (1969).
213. Swinnerton, J. W. and Linnenbom, V. *J. J. Gas Chromat.*, **5**, 570 (1967).
214. Novák, J., Zlutický, J., Kubelka, V. and Mostecký, J. *J. Chromat.*, **76**, 45 (1973).
215. Grob, K. *J. Chromat.*, **84**, 255 (1973).
216. Grob, K. and Zúrcher, F. *J. Chromat.*, **117**, 285 (1976).
217. Grob, K. and Grob, G. *J. Chromat.*, **90**, 303 (1974).
218. Grob, K., Grob, G. and Grob, K. Jr. *J. Chromat.*, **106**, 299 (1975).
219. Desbaumes, E. and Imhoff, C. *Wat. Res.*, **6**, 885 (1972).
220. Novák, J., Zlutický, J., Kubelka, V. and Mostecky, J. *J. Chromat.*, **76**, 45 (1973).
221. Polak, J. and Lu, B. *Anal. Chim. Acta*, **63**, 231 (1974).
222. Kaiser, R. *J. Chromat. Sci.*, **9**, 227 (1971).
223. Colenutt, B. A. and Thorburn, S. *Int. J. Environ. Stud.*, **15**, 25 (1980).
224. Wasik, S. P. *J. Chromat. Sci.*, **12**, 845 (1974).
225. Jeltes, R. R. and Veldink, R. J. *Chromatography*, **27**, 242 (1967).
226. Jeltes, R. *Wat. Res.*, **3**, 931 (1969).
227. Jeltes, R. *Water Pollution by Oil.* Ed. P. Hepple, London Institute of Petroleum, p. 43 (1971).
228. Simard, R. C., Hasecawa, W., Bondaruk, W. and Headington, C. E. *Anal. Chem.*, **23**, 1384 (1951).
229. Ruebelt, C. *Helgoländer Wiss. Meeresunters*, **16**, 306 (1967).
230. Fastabend, W. *Chemie-Ingr-Tech.*, **37**, 728 (1965).
231. Ruebelt, C. *Gas-u-Wass Fach.*, **108**, 893 (1967).
232. Ruebelt, C. *Z. Anal. Chem.*, **221**, 299 (1966).
233. Osipov, V. M. *Khimiya Tekhnol. Topl. Masel.*, **16**, 52 (1971).
234. Golubeva, M. T. *Lab. Delo.*, **11**, 665 (1966).
235. Lindgreen, C. G. *J. Am. Wat. Wks. Ass.*, **49**, 55 (1957).
236. Ludzack, F. J. and Whitfield, C. E. *Anal. Chem.*, **28**, 157 (1956).
237. Benyon, L. R. *Methods for the Analysis of Oil in Water and Soil.* The Hague, Strichting CONCAWE 40 pp. (1968).
238. Hughes, D. R., Belcher, R. S. and O'Brien, E. J. *Bull. Environ. Contam. Toxicol.*, **10**, 170 (1973).
239. Suzuki, R., Yamaguchi, N. and Matsumoto, N. *Japan Analyst*, **23**, 1296 (1974).
240. Rosen, A. A. and Middleton, F. M. *Anal. Chem.*, **27**, 709 (1955).
241. Webber, L. A. and Burks, C. E. *Anal. Chem.*, **24**, 1086 (1952).
242. Santo, T., Hagiwara, K. and Ozawa, Y. *Japan Analyst*, **21**, 1235 (1972).
243. Brown, R. A., Searl, T. D., Elliot, J. D., Phillips, B. G., Brandon, D. E. and Horoghan, H. *Proc. of 1973 Conference on Prevention and Control of Oil Spills.* Washington DC p. 505 (1973).
244. Rodier, J. *Chemical and Physicochemical Analysis of Water.* Paris, Dunod, 700 pp. (1971).
245. Boisselet, L. *World Petroleum Congress. Proc. 5th New York* pp. 165–175 (1959).
246. Hellmann, H. *Deutsch. gewaesserk. Mitt.*, **13**, 19 (1969).
247. Hellmann, H. and Zehle, H. *Z. Anal. Chem.*, **265**, 245 (1973).
248. Gruenfeld, M. *Environ. Sci. Technol.*, **7**, 636 (1973).
249. Mallevialle, J. *Wat. Res.*, **8**, 1071 (1974).

250. Martin, P. and Geyer, D. *Korrespondez Abwasser*, **21**, 202 (1974).
251. Geyer, D., Martin, P. and Adrian, P. *Gas-u-Wasser (ach Wasser, Abwasser)*, **119**, 72 (1978).
252. Götz, R. *Facum Städte Hygiene*, **29**, 10 (1978).
253. Ahmed, S. M., Beasley, M. D., Etromson, A. C. and Hites, R. A. *Anal. Chem.*, **46**, 1858 (1974).
254. Schatzberg, P. and Jackson, D. F. *U.S. Coast Guard Report No 734209.9*. Washington DC, November (1972).
255. Uchiyama, M. *Wat. Res.*, **12**, 299 (1978).
256. Carsin, J. L. *Revue Internationale d'Oceanographic Medicale*, **48**, 77 (1977).
257. Hellmann, H. *Z. Anal. Chem.*, **244**, 44 (1969).
258. Mark, H. B. *Environ. Sci. Technol.*, **6**, 833 (1972).
259. Geobgen, H. G. *Haustech. Essen, Vortragsveroeff*, **231**, 55 (1970).
260. Benyon, L. R. *Methods for the Analysis of Oil and Soil*. The Hague, Strichting, CONCAWE (1968).
261. Edgar, D. *J. Chromat.*, **43**, 271 (1969).
262. Berthold, I. *Z. Anal. Chem.*, **240**, 320 (1968).
263. Koppe, P. and Muhle, L. A. *Vom Wasser*, **35**, 42 (1968).
264. Semenov, A. D., Stradomanskaya, A. G. and Zurbina, L. F. *Gidrokhim Mater*, **53**, 51 (1971). Ref. *Zhur Khim* 19GD (1971).
265. Goretti, G., Marsella, I., Petronio, B. M. *Rio ital. Sostanze grasse*, **51**, 66 (1974).
266. Hunter, O. *Environ. Sci. Technol.*, **9**, 241 (1975).
267. Sinel'nikov, V. E. *Gidrokhim Mater*, **50**, 127 (1969). Ref. *Zhur. Khim.* 19GD (1969) (24) Abstr No 24 G 264.
268. Bauer, K. and Driescher, H. *Fortschr Wasserchem ihrer Grenzgeb*, **10**, 31 (1968).
269. Parker, C. A. and Barnes, W. J. *Analyst (London)*, **85**, 3 (1960).
270. Dadashev, Kh. K. and Agamirova, S. I. *Vop. Issled. neft, i nefteprod. razrabotki protseeor pererabotki neft.; Obsled. Zav. Ustanovok*, No. 1, p. 81 (1957).
271. Rychkova, V. I. *Elekt. Sta., Mosk.*, **40**, 76 (1969).
272. Alekseeva, V. and Gol'dina, Ts. A. *Zav. Lab.*, **16**, 35 (1950).
273. Yudilevich, M. M. Luminescence Methods for the Determination of the Content of Mineral Oils in Water (in Russian) Moscow: *Gosudarst. Energet. Izdatel* (1959).
274. Yudilevich, M. M. Luminescence method and apparatus for analysis of water-oil emulsions: in Russian *Metody Lyumineestsentn. Analiza* (Minsk: Akad. Nauk Belorum S.S.R.) Sbornik, 87 (1960).
275. Yudilevich, M. M. In Russian, *Peredoye Metody Khim. Teknol. i Kontrolya Proizv.*, 280 (1964).
276. Yudilevich, M. M. In Russian, *Vodopodgotovka, Vod. Rezhim Khim Kontrol. Parosilovykh Ustanovkakh*, S. B. Statei, No. 2, 173 (1966).
277. Kyrge, Kh. In Russian, *Morsk. Flot.*, **19**, 31 (1959).
278. Leger, A. *Fr. Pat. 1*, 560,844 (Paris, Patent Office).
279. Leonchenkova, E. T. In Russian, *Obogashch. Rud*, **5**, 24 (1960).
280. Shklyar, I. V. In Russian, *Trudy vses. nauchno-issled, geologora zved. Inst.*, No. 155, 341 (1960).
281. Pochkin, Yu. N. and Massino, O. A. In Russian, *Kazan. ed. Zh.*, **3**, 81 (1968).
282. Freegarde, M., Hatchard, C. G. and Parker, C. A. *Lab. Pract.*, **20**, 35 (1971).
283. Danyl, F. and Nietsch, B. *Mikrochemie mikrochem. Acta*, **39**, 333 (1952).
284. Nietsch, B. *Angew. Chem.*, **66**, 571 (1954).
285. Nietsch, B. *Gass. Wass. Warme*, **10**, 66 (1956).
286. Nietsch, B. *Mikrochim. Acta*, 171 (1956).
287. Leoy, E. M. *Wat. Res.*, **5**, 723 (1971).
288. Bauer, K. and Driescher, H. *Fortschr. Wasserchem. ihrer Grenzgeb*, **31** (10) (1968).
289. Sinel'nikov, V. E. *Kazan med Zhim.*, **3**, 83 (1968).
290. Wade, T. L. and Quinn, J. G. *Marine Pollution Bulletin*, **6**, 54 (1975).

291. Golden, J. *Techniques et Sciences Municipales*, **71**, 17 (1976).
292. Liu, D. L. *Water and Sewage Works*, **125**, 40 (1978).
293. Zsolnay, A. and Kiel, W. *J. Chromat.*, **90**, 79 (1974).
294. Witmer, F. E. and Goilan, A. *Environ. Sci. Technol.*, **7**, 945 (1973).
295. Lee, E. G. H. and Walden, C. C. *Wat. Res.*, **4**, 641 (1970).
296. Freegarde, M., Hatchard, O. G. and Parker, C. A. *Lab. Pract.*, **20**, 35 (1971).
297. Imanvilov, L. A. *Transp. Khranenie Nefteprod. Uglevovdorodn Syr'ya*, **4**, 7 (1968).
298. Saltzman, R. S. *Analysis Instrum.*, **6**, 79 (1968).
299. Obnorlenskii, P. A. *Khimiya Tekhnol. Topl. Masel*, **15**, 54 (1970).
300. Nadzhafova, K. N. In Russian, *Trudy vses. nauchno-issled. Inst. Vodosnabzh., Kanaliz., Gidrotekhn. Sooruzhenii Inzh. Gidrogeol.*, No. 23, 107 (1970).
301. Minasyan, K. V. In Russian, *Prom. Arm.*, **11**, 50 (1969).
302. Osipov, V. M. and Belova, T. D. In Russian, *Khimiya Tekhnol. Topl. Masel*, **13**, 56 (1968).
303. Skotnikova, L. A. In Russian, *Izv. Vyssh. ucheb. Zaved., Neft. Gaz.*, **12**, 111 (1969).
304. Polinskaya, R. E. In Russian, *Proekt. Issled. Rab., Neftedob. Prom., 'Giprovostokneft'* No. 3, 106 (1967).
305. Lurje, Yu. Yu. In Russian, *Nauch. Soobshch. Vses. Nauchn-Issled. Inst. Vodosnabzh., Kanaliz., Gidrotekhn. Sooruzhenii, i Inzh. Gidrogeol., Ochistka Prom. Stochn. Vod., Moscow* 34 (1963).
306. Harva, O. and Somersalo, A. *Acta chem. fenn.*, **31**, 384 (1958).
307. Przybylski, Z. *Chemia analit.*, **8**, 601 (1963).
308. Reisus, K. *Fortschr. Wasserchem. ihrer Grenzgeb.*, **10**, 43 (1968).
309. Levy, E. M. *Wat. Res.*, **5**, 723 (1971).
310. Moneva, M. and Angelieva, R. In Russian, *Khig. Zdraveopazvane*, **11**, 286 (1968).
311. Hennig, H. F. O. *Marine Pollution Bulletin*, **10**, 234 (1979).
312. Levy, E. M. *Wat. Res.*, **6**, 57 (1972).
313. Hargrave, B. T. and Phillips, G. A. *Environ. Pollut.*, **8**, 193 (1975).
314. Parker, C. A. and Barnes, W. J. *Analyst (London)*, **85**, 3 (1960).
315. Lloyd, J. B. F. *J. Forens. Sci. Soc.*, **11**, 83 (1971).
316. Lloyd, J. B. F. *J. Forens. Sci. Soc.*, **11**, 153 (1971).
317. Lloyd, J. B. F. *J. Forens. Sci. Soc.*, **11**, 235 (1971).
318. Freegarde, M., Hatchard, C. G. and Parker, C. A. *Lab. Pract.*, **20**, 35 (1971).
319. Zitko, V. and Carson, W. V. *Tech. Report Fish. Res. Bd. Can.*, No. 217, 29 pp. (1970).
320. Michalik, P. A. and Gordon, D. C. *Tech. Report Fish. Res. Bd. Can.* No. 284, 26 pp. (1971).
321. Levy, E. M. *Wat. Res.*, **5**, 723 (1971).
322. Levy, E. M. *Water Air Soil Pollution*, **1**, 144 (1972).
323. Levy, E. M. and Walton, A. *J. Fish. Res. Bd. Can.*, **30**, 261 (1973).
324. Keizer, P. D. and Gordon, D. C. *J. Fish. Res. Bd. Can.*, **30**, 1039 (1973).
325. Scarratt, D. J. and Zitko, V. *J. Fish. Res. Bd. Can.*, **29**, 1347 (1972).
326. Mel'kanovitskaya, C. G. *Guirokhim. Mater*, **53**, 153 (1972). Ref. *Zhur. Khim.* 19GD (1972) (12) Abstract No 12G265.
327. Mirzayanov, V. S. and Bugrov, Yu. F. *Zavrod Lab.*, **38**, 656 (1972).
328. Wasik, S. P. and Tsang, W. *Anal. Chem.*, **42**, 1649 (1970).
329. Wasik, S. P. and Tsang, W. *Anal. Abstr.*, **21**, 1854 (1971).
330. Blumer, M. and Sass, J. *Marine Pollution Bulletin*, **3**, 92 (1972).
331. Farrington, J. W. and Quinn, J. G. *Estuary and Coast Mar. Sci.*, **1**, 71 (1973).
332. Dudova, M. Ya. and Diterikhas, O. D. *Gidrokhim. Mater*, **50**, 115 (1969). Ref. *Zhur. Khim.* 19GD (1969) (23) Abstract No. 23G246.
333. Saltzman, R. S. *Analysis Instrum.*, **6**, 79 (1968).
334. Braemer, H. C. *Wat. Sewage Wks.*, **113**, 275 (1966).
335. Braemer, H. C. *Proc. Am. Chem. Soc., Div. Wat. Waste Chem.*, 1966.

336. Gregory, G. R. E. C. and Palmer, P. L. *Apparatus for Detecting Oil in Water*. Brit. Pat. 1, 221, 066 (London Patent Office).
337. Muskewity, G. *Apparatus for the Continuous Determination of Oils in Liquid and Gaseous Media*. Ger. Pat. 1, 810, 604 (Berlin Patent Office).
338. Grabbe, F. *Industrieabwasser*, 45 (1968).
339. Emschergenossen-schaft, Essen. *Eff. Wat. Treat. J.*, **10**, 667 (1970).
340. Goolsby, A. D. *Environ. Sci. Technol.*, **5**, 356 (1971).
341. Vasicek, A. *Optics of Thin Films*. Amsterdam, North-Holland Pub. Co., pp. 26–41 (1960).
342. Mattson, J. S. *Environ. Sci. Technol.*, **5**, 415 (1971).
343. Fust, H. W., Kreider, R. E. and Gardiner, K. W. Seventeenth Annual ISA Analysis Instrumentation Symposium, Houston, Texas, USA. April (1971). *Analysis Instrumentation*, **9**, Publ. Instrument Society of America, Pittsburgh. May (1971). Paper 45 Automated Instrumental Approach for oil in water.
344. Borneff, J. and Kunte, H. *Arch. Hyg. Bakt.*, **152**, 202 (1968).
345. Jentoft, R. E. and Gouw, T. H. *Anal. Chem.*, **40**, 1787 (1968).
346. McKay, J. F. and Latham, D. R. *Anal. Chem.*, **45**, 1050 (1973).
347. Vaughan, C. G., Wheals, B. B. and Whitehouse, H. J. *J. Chromat.*, **78**, 203 (1973).
348. Melchiorri, C., Chiacchiarini, L., Grilla, A. and D'Area, S. U. *N. Ann. Ig. Mier.*, **24**, 279 (1973).
349. Monarcha, S., Conti, R., Scassellati Sforzolini, G. and Savino, A. *Iq. Mod.*, **69**, 331 (1976).
350. Borneff, J. In *Fate of Pollutants in the Air and Water Environments*, Ed. Suffet, I. H. Chichester, J. Wiley, Vol. 2, 393 (1977).
351. Outkiewicz, T., Ryborz, S. and Maslowski, J. *Environ. Prot. Engng*, **4**, 263 (1978).
352. Jentoft, R. E. and Gouw, T. H. *Anal. Chem.*, **40**, 1787 (1968).
353. Woo, C. S., D'Silva, A. P. and Fassell, D. A. *Anal. Chem.*, **52**, 159 (1980).
354. McKay, J. F. and Latham, D. R. *Anal. Chem.*, **45**, 1050 (1973).
355. Searl, T. D., King, W. H. and Brown, R. A. *Anal. Chem.*, **42**, 954 (1970).
356. Wasik, J. P. and Tsang, W. *Anal. Chem.*, **42**, 1649 (1970).
357. Navralil, J. D., Sievers, R. E. and Walton, W. F. *Anal. Chem.*, **49**, 2260 (1977).
358. Natusch, D. F. S. and Tomkins, B. A. *Anal. Chem.*, **50**, 1429 (1978).
359. Fryćka, J. *J. Chromat.*, **65**, 432 (1972).
360. Dunn, B. P. and Stich, H. F. *J. Fish. Res. Bd Can.*, **33**, 2040 (1976).
361. Chatot, G., Jequier, W., Jay, M., Fontages, R. and Obaton, P. *J. Chromat.*, **45**, 415 (1969).
362. Harrison, R. M., Perry, R. and Wellings, R. A. *Wat. Res.*, **10**, 207 (1976).
363. Caddy, D. E. and Meek, D. M. *Technical Reprint TR36*. Water Research Centre, Stevenage Laboratory, Elden Way, Stevenage, Herts. (1976).
364. Caddy, D. E. and Meek, D. M. *Proc. Anal. Div. Chem. Soc.*, **13**, 45 (1976).
365. Hellmann, H. Z. *Anal. Chem.*, **275**, 109 (1975).
366. Acheson, M. A., Harrison, R. M., Perry, R. and Wellings, R. *Wat. Res.*, **10**, 207 (1976).
367. Bowman, M. C. and Beroza, M. *Anal. Chem.*, **40**, 535 (1968).
368. Burchfield, H. P., Wheeler, R. J. and Bernos, J. B. *Anal. Chem.*, **43**, 1976 (1971).
369. Freed, D. J. and Faulkner, L. R. *Anal. Chem.*, **44**, 1194 (1971).
370. Robinson, J. W. and Goodbread, J. P. *Anal. Chim. Acta*, **66**, 239 (1973).
371. Bowman, M. C. and Benoza, M. *Anal. Chem.*, **40**, 535 (1968).
372. Burchfield, H. P., Wheeler, R. J. and Bernos, J. B. *Anal. Chem.*, **43**, 1976 (1971).
373. Freed, D. J. and Faulkner, L. R. *Anal. Chem.*, **44**, 1194 (1971).
374. Robinson, J. W. and Goodbread, J. P. *Anal. Chim. Acta*, **66**, 239 (1973).
375. Basu, D. K. and Saxona, J. *Environ. Sci. Technol.*, **12**, 791 (1978).
376. Acheson, M. A., Harrison, R. M., Perry, R. and Wellings, R. A. *Wat. Res.*, **10**, 207 (1976).

377. Wedgwood, P. and Cooper, R. L. *Analyst (London)*, **78**, 170 (1953).
378. Wedgwood, P. and Cooper, R. L. *Analyst (London)*, **79**, 163 (1954).
379. Wedgwood, P. and Cooper, R. L. *Analyst (London)*, **81**, 42 (1956).
380. Borneff, J. and Kunte, H. *Arch. Hyg. Bakt.*, **148**, 585 (1964).
381. Borneff, J. and Kunte, H. *Arch. Hyg. Bakt.*, **153**, 220 (1969).
382. Hoffmann, D. and Wynder, E. L. *Anal. Chem.*, **32**, 295 (1960).
383. Haenni, E. O., Havard, J. W. and Joe, F. L. *J. Ass. Off. Agric. Chem.*, **45**, 67 (1962).
384. Grimmer, G. *Erdöl und Kohle*, **25**, 339 (1972).
385. Kadar, R., Nagy, K. and Fremstad, D. *Talanta*, **27**, 227 (1980).
386. Giger, W. and Schnaffner, C. *Anal. Chem.*, **50**, 243 (1978).
387. Bjorseth, A., Knutsen, J. and Skei, J. *Science of the Total Environment*, **13**, 71 (1979).
388. Saxena, J., Basu, D. K. and Kozuchowski, J. *Method Development and Monitoring of Polynuclear Aromatic Hydrocarbons in Selected US Water*, Health Effects Research Laboratory, TR-77-563, No. 24, Cincinnati, Ohio (1977).
389. Benoit, F. M., Label, G. L. and Williams, D. T. *Int. J. Anal. Chem.*, **6**, 277 (1979).
390. McNeil, E. E., Olson, R., Miles, W. F. and Rajabalee, R. J. M. *J. Chromat.*, **132**, 277 (1977).
391. Matsumoto, G. and Hanya, T. *J. Chromat.*, **194**, 199 (1980).
392. Tan, Y. L. *J. Chromat.*, **176**, 319 (1979).
393. Lao, R. C., Thomas, R. S. and Monkman, J. L. *J. Chromat.*, **112**, 681 (1975).
394. Lao, R. C., Thomas, R. S., Oja, J. and Dubois, L. *Anal. Chem.*, **45**, 908 (1973).
395. Crane, R. I., Fielding, M., Gibson, T. M. and Steel, C. M. Water Research Centre, Medmenham, UK. Technical Report No TR158 *A Survey of PAH Levels in British Waters*. January (1981).
396. Scholz, L. and Altmann, H. J. *Z. Anal. Chem.*, **240**, 81 (1968).
397. Schwarz, F. P. and Wasik, S. P. *Anal. Chem.*, **48**, 524 (1976).
398. Forster, T. and Kasper, K. *Z. Electrochem.*, **50**, 976 (1955).
399. Shabad, L. M. *Hyg. Sanit.*, **36**, 155 (1971).
400. Muel, B. and Lacroix, G. *Bull. Soc. Chim.* 2139 (1960).
401. Jäger, J. and Kassovitzova, B. *Chem. List.*, **62**, 216 (1968).
402. Gurov, F. I. and Novikov, Yu. V. *Hyg. Sanit.*, **36**, 409 (1971).
403. Stepanova, M. I., Il'ina, R. I. and Shaposhnikov, Yu. K. *J. Anal. Chem., USSR*, **27**, 1075 (1972).
404. Khesina, A. Ya. and Petrova, T. B. *J. Spectrosc. USSR*, **18**, 622 (1973).
405. Andelmann, J. B. and Snodgrass, J. E. *CRC Crit. Rev. Environ. Contr.*, **4**, 69 (1974).
406. Andelmann, J. B. and Suess, M. J. *Bull. Wld Hlth Org.*, **43**, 479 (1970).
407. Hood, L. V. S. and Winefordner, J. D. *Anal. Chim. Acta*, **42**, 199 (1968).
408. Lavalette, D., Muel, B., Habert-Habert, M., René, L. and Latarjet, R. *J. Chem. Phys.*, **65**, 2141 (1968).
409. Gaevaya, T. Ya and Khesina, A. Ya. *J. Anal. Chem. USSR*, **29**, 1913 (1974).
410. Monarca, S., Causey, B. S. and Kirkbright, G. F. *Wat. Res.*, **13**, 503 (1979).
411. Nurmukhametov, R. N. *Russ. Chem. Rev.*, **53**, 180 (1969).
412. Kirkbright, G. F. and de Lima, C. G. *Analyst (London)*, **99**, 338 (1974).
413. Causey, B. S., Kirkbright, G. F. and de Lima, G. C. *Analyst (London)*, **101**, 367 (1976).
414. Farooq, R. and Kirkbright, G. F. *Environ. Sci. Technol.*, **10**, 1018 (1976).
415. *Sampling and Analysis Procedures for Screening of Industrial Effluents for Priority Pollutants*. US Environmental Protection Agency, Environmental Monitoring and Support Laboratory, Cincinnati, Ohio, April 1977.
416. *International Standards for Drinking Water*, third edition, World Health Organization, Geneva, p. 37 (1971).

417. Cathrone, B. and Fielding, M. *Proc. Anal. Div. Chem. Soc.*, **15**, 155 (1978).
418. Fedonin, V. F., Tolikina, N. F., Belyatskaya, O. N. and Gul, V. E. *Zhur. Anal. Khim.*, **20**, 1022 (1965).
419. International Union of Pure and Applied Chemistry IUPAC. *Pure Appl. Chem.*, **40**, 36 (1974).
420. Ogan, K., Katz, E. and Slavin, W. *J. Chromatogr. Sci.*, **16**, 517 (1978).
421. Dunn, B. P. and Stich, H. F. *J. Fish. Res. Bd Can.*, **33**, 2040 (1976).
422. Dunn, B. P. *Environ. Sci. Technol.*, **10**, 1018 (1976).
423. Kordan, H. A. *Science*, **149**, 1382 (1965).
424. Howard, J. W., Teage, R. T., White, R. H. and Fry, B. E. *J. Ass. Off. Anal. Chem.*, **49**, 595 (1966).
425. Mills, P. A. *J. Ass. Off. Anal. Chem.*, **51**, 29 (1968).
426. Kunte, H. *Arch. Hyg. Bakt.*, **151**, 193 (1967).
427. Hiltabrand, R. T. *Marine Pollution Bulletin*, **9**, 19 (1978).
428. Maher, W. A., Bagg, J. and Smith, J. D. *Int. J. Environ. Anal. Chem.*, **7**, 1 (1979).
429. Zitro, V. *Bull. Environ. Contam. Toxicol.*, **14**, 621 (1975).
430. Jentoft, R. E. and Gouw, T. H. *Anal. Chem.*, **40**, 1787 (1968).
431. Vaughan, C. G., Wheats, B. B. and Whitehouse, M. J. *J. Chromat.*, **78**, 203 (1973).
432. Lewis, W. M. *Water Treat. Exam.*, **24**, 243 (1975).
433. Sorrell, R. K., Dressman, R. C. and McFarren, E. F. Environment Protection Agency, Cincinnati, Ohio. *High Pressure Liquid Chromatography for the Measurement of PAHs in Water*. Report No 29344 p. 39 (1977). Paper presented at Water Quality Technology Conference, Kansas City, Missouri 5/6 Dec 1977.
434. Sorrell, R. K. and Reding, R. *J. Chromat.*, **185**, 655 (1979).
435. Sorrell, R. K., Dressman, R. C. and McFarren, E. F. *AWWA-Water Quality Technology Conference, Kansas City, Mo., December 5–7 1977*, American Water Works Association, Denver, Colo, p. 3A-3 (1978).
436. Hunt, D. C., Wild, P. J. and Crosby, N. T. *Wat. Res.*, **12**, 643 (1978).
437. Das, B. S. and Thomas, G. H. *Anal. Chem.*, **50**, 967 (1978).
438. Oyler, A. R., Bodenner, D. L., Welch, K. J., Hukhoxen, R. K., Carlson, R. M., Kopperman, H. L. and Caple, R. *Anal. Chem.*, **50**, 837 (1978).
439. Perkin Elmer Ltd. *The Analytical Bulletin* No. 1, p. 4. (2/2 1982).
440. Schönmann, M. and Kern, H. *Varian Instrument Applications*, **15**, 6 (1981).
441. Lewis, W. M. *Water Treat. Exam.*, **24**, 243 (1975).
442. Crathorne, B. and Fielding, M. *Pract. Anal. Div. Chem. Soc.*, **11**, 155 (1978).
443. Crane, R. I., Crathorne, B. and Fielding, N. In *Hydrocarbons and Halogenated Hydrocarbons in Aquatic Environments (Environmental Science Research Series, Vol 16)*. Ed. Asghan, B. K. Plenum, New York (1976).
444. Dong, M., Locke, D. C. and Ferrand, E. *Anal. Chem.*, **48**, 368 (1976).
445. Krstulovic, A. M., Rosie, D. M. and Brown, P. R. *Anal. Chem.*, **48**, 1383 (1976).
446. Lankmayr, E. P. and Muller, K. *J. Chromat.*, **170**, 139 (1979).
447. Nielsen, T. *J. Chromat.*, **170**, 147 (1979).
448. Crosby, N. T., Hunt, D. C., Philip, L. A. and Patel, I. *Analyst (London)*, **106**, 135 (1981).
449. Hagenmaier, H., Feirabend, R. and Jager, W. *Zeitschrift für Wasser und Abwasser Forschung*, **10**, 99 (1971).
450. Black, J. J., Dymerski, P. P. and Zapisek, W. F. *Bull. Environ. Contam. Toxicol.*, **22**, 278 (1979).
451. O'Donnell. *Ir. J. Environ. Sci.*, **1**, 77 (1980).
452. Borneff, J. and Kunte, H. *Arch. Hyg. Bakt.*, **153**, 220 (1969).
453. Gilchrist, C. A., Lynes, A., Steel, G. and Whitham, B. T. *Analyst (London)*, **97**, 880 (1972).
454. Kay, J. F. and Latham, D. R. *Anal. Chem.*, **45**, 1050 (1973).
455. Stepanova, M. I., Il'ina, R. I. and Shaposhnikov, *Zh. analit. Khim.*, **27**, 1201 (1972).

456. Nowacka-Barezyk, K., Adamiak-Ziemba, J. and Janina, A. Z. *Chemia Anglit.*, **18**, 223 (1973).
457. Hellman, H. Z. *Anal. Chem.*, **275**, 109 (1975).
458. Crathorne, B. and Fielding, M. *Proc. Anal. Div. Chem. Soc. London*, **11**, 155 (1978).
459. Crane, R. I., Crathorne, B. and Fielding, M. *Determination of Polycyclic Aromatic Hydrocarbons in Water*. Water Research Centre LR 4407 (1980).
460. Saxena, J., Basu, D. K. and Kozuchowski, J. US National Technical Information Service, Springfield, Va. *Report No. PB276635* p. 94 (1977).
461. Saxena, J., Kozuchowski, J. and Basu, D. K. *Environ. Sci. Technol.*, **11**, 682 (1977).
462. Basu, D. K. and Saxena, J. *Environ. Sci. Technol.*, **12**, 791 (1978).
463. Fox, A. and Staley, A. M. *Anal. Chem.*, **48**, 992 (1976).
464. Navra'til, J. D., Sievers, R. E. and Walton, H. F. *Anal. Chem.*, **49**, 2260 (1977).
465. Weil, L., Grimmer, G., Hellmann, H., de Jong, B., Kunte, H., Sanneborn, M. and Stroker, I. *Zeitschrift für Wasser und Abwasser Forsching*, **13**, 108 (1980).
466. Kunte, H. *Fresenius Zeitschrift für Analytische Chemis*, **301**, 287 (1980).
467. Issaq, H. J., Andrews, A. W., Janini, G. M. and Barr, E. W. *J. Liquid Chromat.*, **2**, 319 (1979).
468. Harrison, R. M., Perry, R. and Wellings, R. A. *Environ. Sci. Technol.*, **10**, 1151 (1976).
469. Thieleman, H. *Mikrochim. Acta*, **5**, 575 (1972).
470. *Standard Methods for the Examination of Water and Wastewater*, 13th Edn, Amer. Pub. Health Assn., Washington, DC (1971).
471. Maxcy, R. B. *J. Wat. Pollut. Contr. Fedn*, **48**, 2809 (1976).
472. Cook, P. P., Duvall, P. M. and Bourke, R. C. *Water and Sewage Works*. Reference No. 1978 R71-R72 and R74-R78.
473. US Environmental Protection Agency. *Methods for Chemical Analysis of Water and Wastes*, EPA-625-/6-74-003, STORET No. 00556, 1974.
474. Pomeroy, R. and Wakeman, C. M. *Ind. Engng Chem. Anal. Ed.*, **13**, 795 (1941).
475. Gilcreas, F. W. *Sew. and Ind. Wastes*, **25**, 1379 (1958).
476. Ullman, W. W. and Sanderson, W. W. *Sew. and Ind. Wastes*, **31**, 8 (1959).
477. Payne, J. F. *Science*, **191**, 945 (1976).
478. Cleverley, B. J. *Forens. Sci. Soc.*, **8**, 69 (1968).

Chapter 2

Detergents

ANIONIC DETERGENTS

La Noce[1] in 1969 reviewed methods then available for determining anionic and non-ionic surface active agents in water. He reviewed methods for the determination of anionic surfactants by titration with a cationic reagent colorimentrically by complexing with methylene blue or methyl green or by infrared spectroscopy. Since then, work has been done on the development of methods based on atomic absorption spectroscopy as discussed below.

Atomic absorption spectroscopy

Le Bihan and Courtot Coupez,[2,3] analysed fresh water and sea water as follows. To 1 litre of filtered sea water was added, with shaking, 10 ml of hydrochloric acid and 10 ml of 0.023 M copper 1,10-phenanthroline sulphate. After 5 min 43 ml of isobutyl methyl ketone were added, shaken vigorously for 1 min, allowed to stand for 5 min, and, after separating the phases, re-extracted with 25 ml of the ketone. The copper was determined by atomic absorption spectrometry using the 324.7 nm line. A blank was prepared from sea water containing about 1% of the amount of detergent to be determined. Calibration graphs must be prepared for each anionic detergent. The method is applicable to fresh water if sodium chloride is added to prevent emulsion formation, and is applicable to cationic detergents by a different method. The following species (wt per litre) do not interfere: CO^{II} and Ni^{II} (10 mg), HSO_4^- (3 g), $H_2PO_4^-$ (0.2 g), B_R^- (10 mg), and I^- (2 mg), Fe^{3+} (30 mg), CrO_4 (1 mg), SCN^- (20 mg) and NO_3^- (1 mg) produce a small constant increase in the atomic absorption, and hydrogen sulphide must be eliminated by oxidation. Non-ionic detergents do not interfere. Le Bihan and Courtot Coupez[4] used the same complex and flameless atomic absorption spectroscopy to determine anionic detergents. Crisp et al.[5] were the first to use *bis*(ethylenediamine) Cu(II) ion for the determination of anionic detergents. They determine the concentration of detergents by flame atomic absorption spectroscopy or by a colorimetric method. The colorimetric method was more sensitive with a limit of detection of $0.03\,\mu g\,l^{-1}$ (as linear alkyl sulphonic acid) compared to $0.06\,\mu g^{-1}$ for atomic absorption spectroscopy. Their method is applicable to fresh and sea water. Crisp et al.[6] determined anionic detergents in fresh estuarine and sea water, at the ppb level. The detergent anions in a 750 ml water sample are extracted with chloroform as an ion association compound with the *bis*(ethylenediamine) copper(II) cation and determined by atomic absorption spectrometry using a graphite furnace atomizer. The limit of detection (as linear alkyl sulphonic acids) is $2\,\mu g\,l^{-1}$. Gagnon[7] has described a rapid and sensitive atomic absorption spectrometric method developed from the work of Crisp et al.[5] for the determination of anionic detergents at the ppb level in natural waters. The method described below is based on determination by atomic absorption spectrometry using the *bis*(ethylenediamine) copper(II) ion. The method is suitable for detergent concentrations up to $50\,\mu g\,l^{-1}$, but it can be extended up to $15\,mg\,l^{-1}$. The limit of detection is $0.3\,\mu g\,l^{-1}$.

Method

An atomic absorption spectrophotometer equipped with a deuterium arc background corrector and graphic furnace was used. A hollow cathode copper lamp was used and the lamp current was about 10 mA. The wavelength was 324.7 nm for copper determination. Argon was used as the purging gas and gas flow was set at interrupted mode for 3 s during atomization. The drying temperature was set at 100 °C for 12 s, the charring temperature was 700 °C for 12 s and the atomization temperature was 2700 °C for 4 s.

Reagents

Ethylendiamine copper(II) reagent. 62.3 g of copper sulphate pentahydrate and 49.6 g of ammonium sulphate are dissolved in water, 45.1 g (50 ml) of 1,2-diaminoethane are added and diluted to 1 l with water. The reagent is stable for at least a month.

Standard reference anionic detergent solution. Sodium dodecyl sulphate USP grade, was purified by recrystallization twice from ethanol. A $20\,mg\,l^{-1}$ solution was prepared daily.

Procedure

100 ml of water sample are placed into a separatory funnel and the pH adjusted to 6–9. A white precipitate is formed when the pH is not adjusted. To this is added 5 ml of ethylenediamine–Cu(II) reagent and 5 ml of chloroform. The ratio of sample to reagent to chloroform is maintained at 20:1:1. The solution is shaken for 1 min and allowed to stand until the phases separate, then the chloroform layer is run into a small glass stoppered tube taking care to minimize evaporation of the chloroform. A small portion (10 µl) of the chloroform is then injected directly into the graphite furnace.

This procedure is suitable for a concentration range of 0–50 μg l^{-1} of detergents (as sodium dodecyl sulphate). Its applicability can be extended up to 15 mg ml^{-1} by establishing a standard curve in the adequate concentration range or by dilution of the chloroform layer. The calibration curve was linear and the calculated sensitivity was 0.3 μg l^{-1} for sea water and fresh water. The blank value for chloroform generally ranged from 0.030 to 0.080 absorbance unit which can be reduced by washing the chloroform with 0.1 M nitric acid. Figure 110 represents the uncorrected and corrected absorbance values of different detergent concentrations in u.v. oxidized sea water.

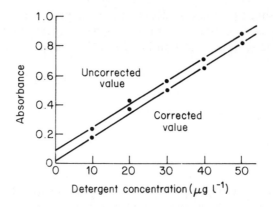

Figure 110 Absorbance values for detergent concentration (μg l^{-1}) before and after correction of the blank. Reprinted with permission from Gagnon[7]. Copyright (1979) Pergamon Press Ltd

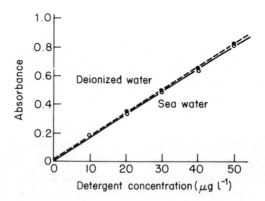

Figure 111 Comparison of absorbance values obtained for detergent concentration in deionized water and in ultraviolet oxidized sea water. Reprinted with permission from Gagnon[7]. Copyright (1979) Pergamon Press Ltd

A comparison of the absorbance values of deionized water and sea water is shown in Figure 111. There is no interference due to the ionic species in sea water. The sea water used had a salinity of 30 ppt and contained 740 μg l^{-1} of NO_3^-, 90 μg l^{-1} of PO_4^{3-}, 24 μg l^{-1} NH_4^+, 87 μg l^{-1} of NO_2^- and 1.8 μg l^{-1} of SiO_2. It was first filtered through 0.45 micron filter and u.v. oxidized for 3 h to break down the organic matter and, thereby render organic bound copper non-extractable by chloroform.

The recovery of different concentrations of detergents added to sea water was used to evaluate the accuracy of the method. Six duplicate determinations were carried out at different levels (Table 62). The recovery is 80% at 1 μg l^{-1} but reaches 90% at 10 μg l^{-1}. The recovery is 97% or better at higher concentrations.

Table 62 Recovery of sodium dodeylsulphate (SDS) in sea water

SDS added (μg l^{-1})	Number of samples	Mean SDS found (μg l^{-1})	Mean recovery
1.0	10	0.8	80
10.0	6	9.2	92
20.0	6	18.4	92
30.0	6	29.1	97
40.0	6	38.8	97
50.0	6	49.5	99

Reprinted with permission from Gagnon.[7] Copyright (1979) Pergamon Press Ltd.

At detergent concentration of the mg l^{-1} (ppm) level environmental copper concentrations produce no problem. But at the μg l^{-1} (ppb) level interference by natural organic chelators can occur. A copper concentration between 1.3 and 2.9 μg l^{-1}, an important fraction (25%) of the dissolved copper in sea water, is extractable in chloroform. This fraction can vary considerably (10–50%) depending on the sample. Filtration and subsequent ultraviolet irradiation of sea water reduce these values to 3–4%. A concentration of 0.3–0.7 μg l^{-1} of chloroform-extractable copper can cause an error of 2–4% (if 25% of the dissolved copper is extractable) at a detergent concentration of 10 μg l^{-1}. At a detergent concentration of the order of 1 μg l^{-1}, however, the error can be as important as 10–20%. This level can fluctuate considerably depending on the water sample. It is important to measure the chloroform-extractable copper in the sample to evaluate the possible error and, therefore to get a better accuracy of the detergent concentrations.

Spectrophotometric methods

Earlier spectrophotometric methods for the determination of anionic detergents were based on the use of methylene blue, Rhodamine B, Ferroin,[8,11] azure A,[10] and methylene green.[9]

Stroehl and Kurzak[8] studied the absorption spectra and stability of complexes formed between anionic detergents and methylene blue and methyl green. The absorption maxima for the methylene blue and methyl green complexes of seven anionic surfactants were found to lie between 657 and 671 nm and between 613 and 628 nm, respectively. These workers suggested that in routine determinations the measurements should be made at 660 and 620 nm, rather than at 650 and 615 nm, respectively. The rate of change of extinction with time for the same series of complexes was also measured. Differences due to the anions were insignificant, but fading of the colour is much more rapid in methylene blue than in methyl green complexes.

Taylor and Fryer[11] determined anionic detergents in sewage and sewage effluents, using iron(II) chelates (ferroin). To a 5 ml sample containing less than 0.5 μmole of anionic detergent at pH from 4 to 10 is added 0.5 ml of 2 M sodium acetate—2 M acetic acid (pH 5) and 1 ml of 0.895 mM ferroin (for wholly liquid samples) or 1 ml of 8.95 mM ferroin (for samples containing suspended solids). The sample is extracted with chloroform (2×2 ml), diluted to 5 ml with chloroform, and the extinction measured at 512 nm.

Taylor and Williams[12] studied the selective determination of anionic surfactants using ionic association compounds of these compounds with iron(II) chelates. The effects of solution variables including ionic strength, pH and buffer concentration, reagent excesses, solution volume, type of solvent, and the selection of reagents and conditions of extraction and separation are examined, and a method is proposed for the selective determination of homologous surfactants. A method is also proposed for the determination of surfactants of various chain lengths.

Taylor and Waters[13] studied the extraction constants of association compounds of anionic surfactants with iron(II) chelates of the ferroin type. The extraction constants of these chelates distributed between water and chloroform were correlated with the structure of the surfactant and the ligand. Taylor *et al.*[14] also studied the use of iron(II) chelates of 1,10-phenanthaline and its derivitives, bipyridyl and terpyridyl in the estimation of anionic surfactants in chloroform extracts of water samples. The extent of extraction increases with increasing chain length of surfactants and is characterized by a breakthrough point (i.e. the chain length above which the surfactant is extracted and below which it is not). This point depends on the choice of iron(II) chelate and extracting solvent. A selective procedure for determining surfactants in simulated sewage liquor and river water based on experimental control of the 'breakthrough point' and involving spectrophotometric determination of the extracted complexes, is described. The effect of foreign ions is also discussed by the workers.

Wudzinska and Ponikowska[15] developed automated methods for the determination of anionics in surface water and waste water based on the use of methylene blue. The method involves colorimetric determination of the concentration of the products of reaction of the anionic surfactants with methylene blue. These complexes are extracted into chloroform and measurements are made at 650 nm in a flow through cell. Samples of waste water should be diluted with 10% aqueous sodium sulphate to ensure separation of the chloroform extract. The method can be used for determining surfactants in concentrations equivalent to 0.05–1 mg of sodium hexacosylbenzene sulphonate per litre. For samples containing 1, 0.5, and 0.1 mg of sodium hexacosylbenzene sulphonate per litre the respective standard deviations were 3.5, 2.5, and 2.5 μg l^{-1}. Kazarac et al.[16] have also discussed methylene blue based methods for determining anionic detergents in sea water samples. The technique is based on solvent extraction and preconcentration of detergents using Wickbold's apparatus.[17] A method using azure A instead of methylene blue has been proposed by Den Tonkelaar and Bergshoeff.[18] Workers at the Water Research Centre [21] have described a methylene blue based autoanalysis method for determining 0–1 μg l^{-1} anionic detergents in water and sewage effluents. This method was based on the work of Longwell and Maniece[19] and subsequently modified by Abbot[9] and by Södergren.[20] The Water Research Centre report describes the method in detail and discusses its precision and accuracy.

Wang[22] and Rodier[23] have also used methylene blue for the determination of anionic detergents. The British Standards method on anionic detergents is based on the use of methylene blue.[24]

Began et al.[25] have described a spectrophotometric method for the determination of anionics based on benzene extraction of a detergent–Rhodamine B complex. Various other dyes have been used for the estimation on anionic detergents in fresh water. Thus Wang and Langley[26] have compared azure A with methylene blue as a reagent for the rapid determination of anionic detergents of the linear alkyl sulphonate type and recommended their method as an alternative to the lengthy standard methylene blue procedure described by the American Public Health Association[27] and ASTM.[28] In the same method they estimate cationic detergents with methyl orange.

Method

Reagents

Stock linear sulphonate (LAS) solution: an amount of the reference material, equal to 1.000 g of LAS on a 100% active basis, is weighed, dissolved in distilled water and diluted to 1 l to obtain a concentration of 1.00 ml = 1.00 mg of LAS. The solution should be stored in a refrigerator to minimize biodegradation.

Standard linear alkylate sulphonate (LAS) solution: 50 ml of stock LAS solution is diluted to 1 l with distilled water.

Chloroform, anhydrous: glass wool.

Methylene blue reagent: 625 mg of methylene blue (Eastman No. P573 or equivalent) are dissolved in 400 ml of distilled water. Then 10 ml of concentrated sulphuric acid are added gradually to the 400 ml mixture and shaken until dissolution is complete. The solution is diluted to 500 ml.

Strong oxidizing reagent for cleaning glassware: about 20 g of technical grade potassium dichromate are dissolved in 40 ml of hot water in a 1000 ml volumetric flask, cooled under tap water and when cold, 350 ml of concentrated sulphuric acid are added, with constant swirling. The reagent is then kept in a glass stoppered bottle.

Apparatus

Graduate cylinders, 50 ml; separatory funnel, 250 ml; filtering funnels 65 mm; and spectrophotometer or filter photometer.

Procedure

Calibration curves preparation. (1) Preparation of anionic surfactant's calibration curve

Prepare a series of separatory funnels with 0, 0.50, 1.00, 1.50, 2.00, 2.50, and 3.00 ml of the standard LAS solution. Add sufficient water to make the total volume 50 ml in each funnel. Then, treat each standard as described below. Plot a calibration curve of μg LAS *vs*, absorbance (or % transmittance) or mg l^{-1} LAS *vs*. absorbance. The following test is carried out to establish whether the sample contains anionic detergents or cationic detergents.

An aliquot of the water sample is pipetted into a separatory funnel and diluted to 50 ml with distilled water. To this is added 1 ml of azure A reagent, 5 ml of buffer solution, and 25 ml of chloroform. The separatory funnel is stoppered and shaken vigorously for 30 seconds. The sample is allowed to stand undisturbed for 5 min after shaking. The chloroform will settle as a lower layer. If the chloroform layer is blue in colour, anionic surfactants are present in the water sample. If the chloroform layer is colourless, the sample does not contain anionic surfactant.

An aliquot of the water sample is pipetted into a separatory funnel and diluted to 50 ml with distilled water. To this is added 1 ml of azure A reagent, 5 ml buffer solution, and 25 ml of chloroform to each separatory funnel. The funnel is stoppered and shaken vigorously for 30 seconds. Each sample is allowed to stand undisturbed for 5 min after shaking. The chloroform will settle as a lower layer.

A small plug of glass wool is wedged in the stem of a filtering funnel. The filtering funnel is placed above a clean, dry test cell (1 cm light path), the chloroform layer is filtered through the glass wool to remove the water and the treated chloroform collected in the cell. The absorbance of the chloroform solution at 623 nm is determined against a blank of chloroform. The equivalent LAS anionic content from the calibration curve of anionic surfactant is determined.

Note that the methylene blue reagent described above can be used to replace azure A reagent. However, the optimum wavelength for maximum absorbance is 652 nm when the methylene blue reagent is used. The azure A dye reacts with anionic surfactants to form a chloroform-soluble blue-coloured complex in the presence of chloroform (Figure 112). The intensity of blue colour in the chloroform layer is proportional to the concentration of the 'azure A–anionic surfactant complex'. The colour intensity of the azure A surfactant complex can then be measured by making colorimetric or spectrophometric readings of the chloroform solution at the optimum wavelength 623 nm. The absorbance curve of the azure A–LAS complex is illustrated in Figure 113.

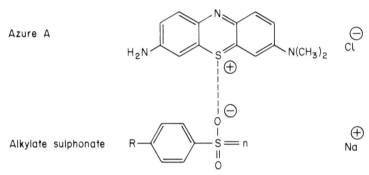

Figure 112 Blue coloured complex of anionic surfactant with azure A. Reprinted with permission from Wang and Langley.[26] Copyright (1977) Springer Verlag, N.Y.

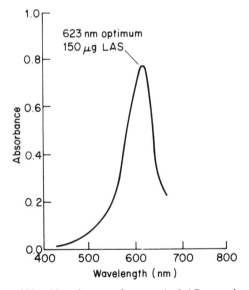

Figure 113 Absorbance of azure A–LAS complex. Reprinted with permission from Wang and Langley.[26] Copyright (1977) Springer Verlag, N.Y.

Janeva and Borissova Pangarova[29] pointed out that in the spectrophotometric determination of anionic detergents there are many difficulties, because of strong interference of other anionics (chloride, nitrate, bicarbonate, phosphate) which are present in much higher concentrations than the surfactants. These interferences can be decreased by using suitable dyes added in small quantities. They investigated 19 cationic dyes of the Remacryl, Maxilon, Astrazon, Basacryl, and Diorlin types as analytical reagents. Remacrylblau B and Remacrylrot 2BL were found to be the most suitable. They proposed routine procedures for the extractional spectrophotometric determination of anionic surfactants in waters by use of these two cationic dyes.

In acid medium anionic surfactants form blue (with Remacrylblau B) and red (with Remacrylrot 2BL) complexes which can be extracted with chloroform under conditions in which the dye remains in the water phase. The absorption spectra of the chloroform phases are shown in Figure 114. The absorption

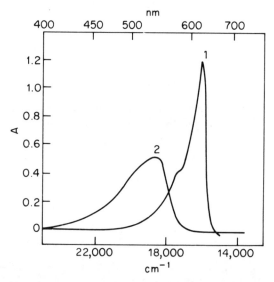

Figure 114 Absorption spectra of the chloroform extracts of the complexes of lauryl sulphate with: (1) Remacryblau B, and (2) Remacrylrot 2BL, 30 μg of LS, 1 cm cell, blank — chloroform

maximum of the complex of lauryl sulphate with Remacryblau B is at 623 nm and that of the complex with Remacrylrot 2BL at 538 nm. It is found that 0.25 ml of the Remacrylrot B solution (0.50 ml of Remacrylrot 2BL solution) is enough for complete bonding of 40 μg of lauryl sulphate when 0.04% solutions of the dyes are used. It was found that only 5 min are necessary for extraction of the complexes, and a single extraction gives about 95% recovery. The reaction in the water phase occurs immediately and a delay of 20 min before extraction is without effect. The absorbance of the organic phase can be measured

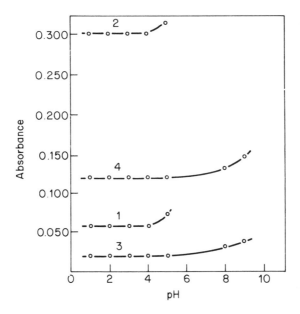

Figure 115 Absorbance of the organic phase as a function of pH 1 cm cell, 2 µg of LS (curves 1 and 3) and 10 µg of LS (curves 2 and 4); 1 and 2, Remacrylblau B, 623 nm; 3 and 4, Remacrylrot 2BL, 538 nm

immediately after the extraction and remains constant for an hour. There is a large pH range over which the absorbance remains constant (pH 1–4 for Remacrylblau B, pH 1–5 for Remacrylrot 2BL, Figure 115). In alkaline medium Remacrylblau B is extracted along with its lauryl sulphate complex. This effect is negligible with Remacrylrot 2BL and the determination can be carried out with it in weakly alkaline medium. This is an important advantage of Remacrylrot 2BL in spite of the lower sensitivity, because waters can be analysed without preliminary treatment.

Janeva and Borissova-Pangarova[29] investigated the effect of several ions at two concentrations of lauryl sulphate. These solutions contained calcium 156 mg l^{-1}, magnesium 55 mg l^{-1}, sodium 283 mg l^{-1}, potassium 32 mg l^{-1} chlorine 32 mg l^{-1} nitrate 100 mg l^{-1}, bicarbonate 750 mg l^{-1} and sulphate 210 mg l^{-1}. They found that the degree of interference depends on the quantity of dye used. Thus using 0.25 ml of Remacrylblau B or 0.5 ml of Remacrylrot 2BL (0.04% solutions) for 50 ml of water the absorbance of the model solution is the same as that for 1 µg of lauryl sulphate. The absorbance is only half this when the same volumes of 0.01% solutions are used, enough for the determination of 10 µg of lauryl sulphate. This is suitable for drinking waters. In the surface waters the concentration of the surfactants is higher and the interference when working with 0.04% solutions of the dyes can be neglected.

Method

Determination of anionic surfactants in surface waters

The sample, containing not more than 30 µg of surfactants is placed in a separating funnel and diluted to 50 ml with distilled water. To this is added 5 ml of 0.1 N sulphuric acid, 0.25 ml of 0.04% Remacrylblau B solution (or 0.5 ml of Remacrylrot 2BL solution) and 10 ml doubly distilled chloroform. The mixture is shaken for 10 min and the organic phase transferred first into a test tube and then into the curvette. The absorbance is measured at 623 nm (for Remacrylblau B) or 538 nm (for Remacrylrot 2BL) against a blank obtained by treating 50 ml of distilled water in the same way of the sample. The calibration curve should cover the range 0–30 µg of lauryl sulphate in 50 ml.

Determination of anionic surfactants in drinking waters

The procedure is the same as for the determination of anionic surfactants in surface waters, but 0.01% solutions of the dyes must be used and the calibration curve covers the range 0.10 µg of lauryl sulphate in 50 ml.

The lowest quantity of anionic surfactants (expressed as lauryl sulphate) which could be determined with the use of a 1 cm cell and an absorbance of 0.020 is 0.014 mg l^{-1} (with Remacrylblau B) and 0.030 mg l^{-1} (with Remacrylrot 2BL). Standard deviations are listed in Table 63.

Table 63 Statistical characteristics

Lauryl sulphate	Standard deviation (mg l^{-1})	
	Remacrylblau B	Remacrylrot 2BL
0.04	0.002	0.003
0.20	0.008	0.005
0.60	0.013	0.016

The relative standard deviations (%) for the four dyes calculated from the 12 results for three quantities of lauryl sulphate added to the model solution are as follows:

	Lauryl sulphate (6 µg)	Lauryl sulphate (15 µg)	Lauryl sulphate (30 µg)
Methylene blue	7.5	4.2	3.1
Azure	5.3	2.9	2.3
Remacrylblau B	4.7	2.2	2.2
Remacrylrot 2BL	4.3	3.3	1.3

It is seen in Table 64 that the azure A method gives rather higher results than the methylene blue method, perhaps because of the higher interferences.

Table 64

Model solutions (μg of LS added)	Methylene blue	LS found, mg l^{-1} azure A	Remacrylrot B	Remacrylrot 2BL
6	6.9	7.3	7.0	6.5
15	16.1	16.6	16.2	15.4
30	30.8	31.8	31.1	30.4
Natural waters				
1	0.010	0.020	0.016	0.012
2	0.124	0.160	0.134	0.130
3	0.152	0.176	0.160	0.150
4	0.260	0.344	0.256	0.268
5	0.340	0.356	0.348	0.348
6	0.660	0.700	0.690	0.660
7	0.680	0.780	0.700	0.680
8	0.160	1.400	1.270	1.200
9	0.540	1.620	1.540	1.540
10	3.000	3.640	3.600	3.100

Bhat et al.[30] used complexation with the *bis*(ethylenediamine) copper(II) cation as the basis of a method for estimating anionic surfactants in fresh estuarine and sea water samples. The complex is extracted into chloroform and copper measured spectrophotometrically in the extract using 1,2(pyridyl azo)-2-napthol. Earlier methods published by these workers described spectrophotometric or flame atomic absorption and graphic furnace atomic absorption based methods for evaluation of the *bis*(ethylenediamine) copper(II) anionic complex having the ability to determine anionics at levels below 100 μg l^{-1}. Bhat et al.[30] using the same extraction system were able to improve the detection limit of the method to 5 μg l^{-1} (as linear alkyl sulphonic acid) in fresh estuarine and sea water samples.

Method

Standard reference anionic surfactant solution: a solution containing 5.68% (w/w) active linear alkyl sulphonic acids (LAS) of mean molecular weight 316. This solution was used to prepare a stock standard solution containing 1000 mg LAS l^{-1} which was diluted further as required.

Copper-ethylenediamine reagent: 62.3 g of copper sulphate pentahydrate and 49.6 g of ammonium sulphate are dissolved in water. To this is added 45.1 g (50 ml) of 1,2-diaminoethane (ethylenediamine) and the solution is diluted to 1 l with water. The reagent is stable for at least a month.

1-(2-pyridylazo)-2-naphthol(PAN) reagent: 0.375 g of PAN are dissolved in absolute ethanol, 22.0 g (31.4 ml) of diethylamine are added and the solution diluted to 1 l with absolute ethanol. This reagent is stable for at least 2 months.

Procedure

A suitable volume of the water sample is placed in a 250 ml separating funnel and the pH adjusted to 5–9; 10.0 ml of copper-ethylenediamine reagent and 10.0 ml of chloroform are added, the funnel shaken for 1 min and allowed to stand until the phases separate. About 7 ml of the organic phase are run into a dry 15 ml graduated centrifuge tube, taking care that no droplets of the aqueous phase pass into the tube. The top of the tube is covered with aluminium foil and the tube is centrifuged at 2500 rpm for 5 min. A 5.0 ml aliquot of the clarified extract is pipetted into a small glass-stoppered flask, 1.0 ml of PAN reagent are added and the flask shaken to mix well. The absorbance of this solution is measured at 560 nm with a 5:1 chloroform–ethanol solution in the reference cell. The colour developed by the PAN reagent is stable for at least 24 h. A blank determination is performed with 200 ml of distilled water. The blank absorbance should not be more than 0.025. Calculate the surfactant concentration in the sample by comparison with standards.

Calibration

In the range 0–250 μg the amount of anionic surfactant present (y μg as LAS) could be calculated from the measured absorbance (x), obtained with 1 cm cells and corrected for the blank, by means of the equation $y = 186.4x$. The precision of the proposed method was assessed by carrying out repeated determinations of standard linear sulphuric acid solutions. The results are shown in Table 65.

Table 65 Precision of proposed method

Linear alkyl sulphuric acid taken[a]	$s(\mu g)$	s_r[b] (%)	LAS taken[a] (μg)	$s(\mu g)$	s_r(%)
2.5	0.3	12	50.0	1.2	2
10.0	0.3	3	250.0	2.0	1
25.0	0.5	2			

[a] 12 determinations were carried out at each level. Between 84 and 97% recovery of linear alkyl sulphonate was obtained from sea water samples.
[b] s_r% = relative standard deviation.
Reprinted with permission from Bhat et al.[30]. Copyright (1980) Elsevier Science Publishers B.V.

The limit of detection taken as the amount of surfactant which gave an absorbance equal to twice the standard deviation of a set of at least 10 absorbance readings at or near blank level, was found to be 1 μg, as linear sulphuric acid (corresponding to a concentration of 5 µg linear sulphonic acid l^{-1} for a sample volume of 200 ml). Between 84 and 97% recovery of linear alkyl sulphonate was obtained from sea water samples. Kalenichenko et al.[31] used a spectrophotometric method for estimating alkyl benzene sulphates in their biodegradation studies of these compounds.

Infrared spectroscopy

Ogden,[32] Frazee,[33] Mächlar,[34] and Ihara[35] have reported infrared methods for determining anionic detergents in water. Oba et al.[36] have described an infrared method for the microanalysis of anionic surfactants in waste waters and sewage.

The surfactants are extracted with chloroform as methylene blue complexes; and sulphate-type surfactants (fatty alcohol sulphate and fatty alcohol ethoxysulphate) are then removed by hydrolysis; residual sulphonate-type surfactants are released from the methylene blue complexes by ion-exchange and then converted to sulphonyl chloride derivatives for infrared spectroscopy. Sulphate-type surfactants are calculated from the differences in methylene blue active substances before and after hydrolysis. The types of detergents that can be determined by the Oba[36] method include linear alkylbenzene sulphate (LAS), branched alkylbenzene sulphonate (ABS), alpha-olefin sulphonate (AOS), fatty alcohol sulphate (AS), and fatty alcohol ethoxysulphate (AES). In this method anionic detergents such as LAS, ABS, AOS, AS, and AES contained in sewage or river waters were extracted by chloroform as methylene blue complexes. After evaporation of chloroform the residue obtained was hydrolysed to decompose sulphate-type surfactant methylene blue complexes. The amount of sulphate-type surfactants was calculated from the loss of methylene blue active substances (MBAS) level before and after hydrolysis.

After hydrolysis sulphonate-type surfactant-methylene blue complexes were first hydrogenated to saturate the AOS; second, ion-exchanged to remove methylene blue and then converted to sulphonyl chloride derivatives for infrared analysis.

Method

The flow sheet of the analytical procedure is shown in Figure 116. Estimation of sulphate-type surfactants. The MBAS level of the sample water is determined by Abbot's[9] method. The chloroform is evaporated after the determination of MBAS level and the residual methylene blue active substances are dissolved in 2 ml of ethanol. 45 ml of deionized water and 15 ml of concentrated hydrochloric acid are added to the residual methylene blue active substances solution and then refluxed for 1 h on a hot plate. After cooling, decomposed products are extracted twice with 50 ml of petroleum ether. The water layer is neutralized with sodium hydroxide solution and ethanol and small amounts of contaminating petroleum ether are removed by distillation. The methylene blue active substances level of the water portion is determined by Abbot's method.[9]

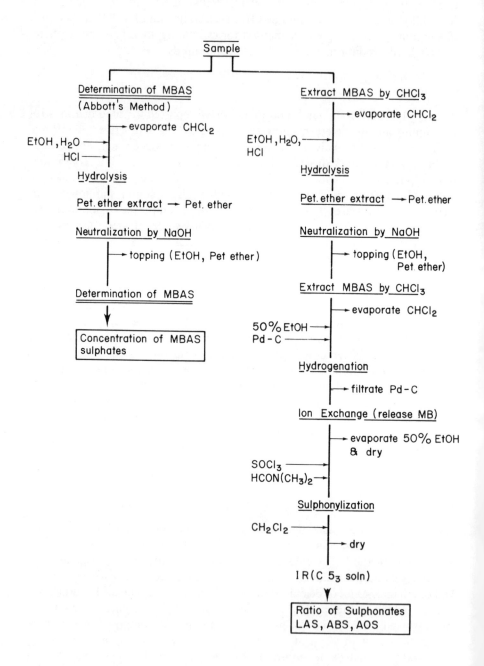

Figure 116 Flow sheet of infrared analysis. Reprinted with permission from Oba[36]. Copyright (1976) Pergamon Press Ltd.

Estimation of sulphonate type surfactants

Two separating funnels A and B (each 300 ml) containing 100 ml of alkaline sodium borate solution (95.35 g of sodium borate and 20 g of sodium hydroxide in 1 l solution) and 10 ml of methylene blue solution (2.1 g l^{-1}) are prepared and the previous extraction repeated five times with 100 ml of chloroform. Two separating vessels A and B (each 20 l polythene vessels equipped with a cock at the bottom of the vessel for separating) are prepared and the appropriate amount of sample water (containing more than 20 mg of methylene blue active substances) and the extracted solution in funnel A added to the solution in funnel B and 30 ml of 10 N sulphuric acid added to the vessel B. To vessel A is added 2 l of chloroform for methylene blue active substances extraction and the vessel shaken for 2 min. After shaking the chloroform layer is separated into vessel B. The same operation is repeated twice using 1 l of chloroform each time. Vessel B is shaken twice a second for 2 min and the chloroform layer separated. The chloroform layer is evaporated and the residue dissolved in 10 ml of ethanol. 45 ml of deionized water and 15 ml of concentrated hydrochloric acid are added to the ethanol solution and boiled gently for 1 h on a hot plate with refluxing. After cooling, the decomposed products (fatty alcohols etc.) resulting from hydrolysis are extracted with three portions of 100 ml of petroleum ether. The water layer is neutralized with sodium hydroxide solution and ethanol removed by distillation. The sulphonate-type surfactant methylene blue complexes are extracted three times with 100 ml of chloroform. The combined chloroform extracts are evaporated and the residual methylene blue active substances dissolved in 100 ml of 50% ethanol. Hydrogen is passed into the 50% solution of MBAS in the presence of 100 mg of palladium–carbon catalyst (Pd: 5%) for 2 h on a water bath at 50 °C. The catalyst is removed by filtration with a glass silter (G 4), the filtrate is passed through 19 ml of H type Dowex 50w X8 (50.100 mesh) column to remove sulphonate-type surfactants from methylene blue and the eluate collected in a beaker. An additional 100 ml of fresh 50% ethanol is passed through the column for washing, the solvent is evaporated in the eluate and the residue dried under reduced pressure. The sulphonyl chloride derivatives are prepared from the sulphonates by refluxing the mixture of the residue, 5 ml of thionyl chloride, and 0.5 ml of *N,N*-dimethylformamide in a small round-bottom flask with an attached condenser at 80–85°C on an oil bath for 45 min. Then nitrogen is passed into the mixture during the reaction, 5 ml of dichloromethane is added and the reaction mixture filtered through a glass filter (G 4). The filtrate is concentrated with a rotary evaporator to remove methylene chloride and excess thionyl chloride, the concentrated mixtures are dissolved in carbon tetrachloride and the insoluble complexes of thionyl chloride–formamide removed by filtration with filter paper. The carbon tetrachloride is evaporated after drying the solution over anhydrous sodium sulphate. The residue (sulphonyl chloride derivatives of sulphonate-type surfactants) is dried under reduced pressure. The sulphonyl chloride derivatives are dissolved in an appropriate amount of carbon disulphide to give 20–80% of transmittance in the infrared spectra. The spectrum of this solution is measured in two frequency ranges from 1000 to 700 cm^{-1} and from 700 to 500 cm^{-1} using a 0.5 mm KBr cell. Referring to calibration curves obtained with known amount of each sulphonate type surfactant, the ratio of each component of the sulphonates is calculated by solving three simultaneous equations according to Lambert Beer's Law at 640, 618, and 524 cm^{-1} for LAS, ABS, and AOS respectively.

Calculation of surfactant contents

Amount of sulphonate-type surfactants:

$$X = M_2 \frac{\text{av Mw}_1}{444.6}$$

Amount of sulphate-type surfactants:

$$Y = (M_1 - M_2) \frac{\text{av Mw}_2}{444.6}$$

$$\text{Per linear alkylbenzene sulphonate} = \frac{X}{X+Y}(L) \cdot 100$$

$$\text{Branched alkylbenzene sulphonate} = \frac{X}{X+Y}(A) \cdot 100$$

$$\text{Alpha-olefin sulphonate} = \frac{X}{X+Y}(O) \cdot 100$$

$$\text{Sulphates} = \frac{X}{X+Y} \cdot 100$$

where M_1 = amount of MBAS (methylene blue active substance) determined as sodium di-2-ethyl hexylsulphosuccinate; M_2 = amount of MBAS after hydrolysis; av Mw$_1$ = average mol. wt. of sulphonate-type surfactant contained in the sample calculated from the infrared measurement; av Mw$_2$ = average mol. wt. of sulphate-type surfactants; L = % of linear alkylbenzene sulphonate in sulphonates (obtained by infrared measurement) A = % of ABS in sulphonates (obtained by infrared measurements): O = % of alpha-olefin sulphonate in sulphonates (obtained by infrared measurement); 444.6 = mol.wt. of sodium di-2-ethyl hexyl sulphosuccinate.

Figure 117 shows the infrared spectra of a mixture sulphonyl chloride derivatives of alpha-olefin sulphonate, linear alkyl sulphonate and branched alkyl sulphonate and an unknown sample recovered from municipal raw sewage. Both infrared spectra are very similar in their pattern.

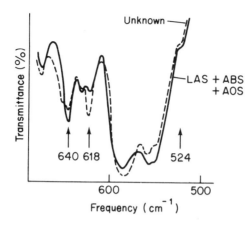

Figure 117 Infrared spectra a mixture of LAS, ABS and AOS sulphonyl chloride derivatives and an unknown sample. Reprinted with permission from Oba.[36] Copyright (1976) Pergamon Press Ltd.

When the method was applied to analysis of anionic surfactants in sewage, especially in effluent from a secondary sewage treatment plant, sometimes alpha-olefin sulphonate was detected at a higher content than expected and in this case the absorption peak at 810 cm^{-1} was also observed. On the other hand, the absorption peak at 810 cm^{-1} cannot be observed in infrared spectra of known mixtures of linear alkyl sulphonate, branched alkyl sulphonate, alpha-olefin sulphonate, fatty alcohol sulphate, or fatty alcohol ethoxysulphate.

Though the materials having the absorption peak at 810 cm^{-1} are not identified, some biodegradation intermediates of surfactants extractable as methylene blue active substances can be considered as one of the interferences. The presence of these biodegradation intermediates was not confirmed but methylene blue activity and infrared spectra of butyl-, hexyl-, octyl-, and decylbenzene sulphonate and sodium sulphophenyl undecanoate were examined by Oba et al.[36] who found that alkylbenzene sulphonates having six or more carbon atoms in their alkyl chain were extracted as methylene blue active substances quantitatively but sodium sulphophenyl undecanoate was not extracted quantitatively. The infrared absorption peak at 810 cm^{-1} was observed in infrared spectra of sulphonyl chloride derivatives of hexyl- and octylbenzene sulfonate. This absorption peak was not observed in the cases of alkylbenzene sulphonates having ten or more carbon atoms in thier alkyl chain. Furthermore, in infrared spectra of sulphonyl chloride derivatives of hexyl- and octylbenzene sulphonate, the interfering peak appeared at 524 cm^{-1} which was a key band for alpha-olefin sulphonates. At the same time, the absorption peak at 640 cm^{-1} specific for linear alkyl sulphonate was weakened. Therefore, if these materials exist, the amount of linear alkyl sulphonate could be underestimated and the amount of alpha-olefin sulphonate could be overestimated. It is necessary to make corrections. In the case of analysis of unknown samples which show an absorption peak at 810 cm^{-1} the analytical results should be corrected providing that hexyl and octylbenzene sulphonates are interfering materials. For convenience, this correction was made with 1:1 mixtures of hexyl- and octylbenzene sulphonate because no difference in molecular extinction coefficients between these two alkylbenzene sulphonates exists. Three kinds of calibration curves with a mixture (1:1) of hexyl-and octylbenzene sulphonate were made at 810 cm^{-1} and at 524 and 610 cm^{-1}. The correction for unknown samples was carried out as follows. If the absorption peak at 810 cm^{-1} appeared, the absorption intensity was converted into the concentration of the mixtures of the alkylbenzene sulphonates by the calibration curve at 810 cm^{-1}. First the concentration was changed into the corresponding absorbance by the calibration curve at 524 cm^{-1} which was converted into the apparently surplus concentration of alpha-olefin sulphonate by the calibration curve made on alpha-olefin sulphonate at 524 cm^{-1}. The net alpha-olefin sulphonate concentration was obtained by subtracting the apparent surplus alpha-olefin sulphonate concentration from the apparent alpha-olefin sulphonate concentration. Second, the concentration was changed into the corresponding absorbance by the calibration curve at 640 cm^{-1} which was then converted into linear alkyl sulphonate concentration by the calibration curve prepared with linear alkyl sulphonate at 640 cm^{-1}.

Ambe and Hanya[38] have combined the Longwell Maniece[39] and Abbott[9] methods using methylene blue with the infrared spectroscopic method of Sallee[40] to devise a method for the determination of alkylbenzene sulphonates. Methylene blue alkylbenzene sulphonate complexes give absorption peaks at 890 and 1010 cm^{-1}, the ratio of the heights being proportional to the ratio of the amount of sulphonate to the total amount of methylene blue sensitive substances in the complex. A graph of peak-height ratio *vs.* weight ratio is prepared by use of a standard substance. The filtered sample is shaken (50 ml) with 0.1 N sulphuric acid (1 ml), 0.025% methylene blue solution (1 ml) and 1,2-dichloroethane (20 ml) for 1 min. After washing the separated organic layer twice with 20 ml of 0.0013% solution of methylene blue in 0.004 N sulphuric acid containing also 0.022% of silver sulphate, its extinction is measured at 655 nm to give the total amount of substances active towards methylene blue. The organic layer is evaporated to dryness prior to pelleting with potassium bromide and infrared spectroscopy.

This method has been applied to bottom sediments and muds.[41] The mud sample is centrifuged to separate the water, dried at room temperature, ground, and sieved. This residue is extracted for 1 h at 80 °C with methanol–benzene (1:1), the extraction is repeated twice and the combined extracts are evaporated and the residue dissolved in water. Alkylbenzenesulphonates are then determined by infrared spectroscopy as described above.

Rand *et al.*[42] and Sallee *et al.*[40] and Nagai *et al.*[37] have also investigated infrared spectroscopic methods for the determination of alkylbenzene sulphonates.

Titration methods

Tsuji et al.[103] carried out microdeterminations of anionic and non-ionic detergents in water by a potentiometric method involving the use of cholinesterase. This method utilizes the phenomena that anionics inhibit the cholinesterase–butyrylthiocholine–enzyme system. A constant current is applied across two platinum electrodes immersed in a solution of butyrylthiocholine, iodine, and the containing sample. Cholinesterase is then added and changes in electrode potential due to the formation of thiol are plotted against them to determine the concentration of anionic detergent.

Wang et al.[43,108] discuss indirect two-phase titration methods for the determination of anioic detergents Wang et al.[44,45] also describe a direct titration procedure for anionics involving titration with 1,5-dimethyl-1,5-diazoundecamethylene polymethobromide, Wang et al.[45] have also described a two-phase titration procedure for the determination of linear alkyl sulphonate and branched chain alkylbenzene sulphonate in water. The method involves initial treatment of water sample with a known amount of a quaternary ammonium salt, adjusting the pH value and adding chloroform. The two-phase mixture is then titrated with standard sodium tetraphenylboron reagent. The procedure can be used for quantitative determination of anionic surfactants in both fresh and saline water. These workers have also published a method for the analysis of linear alkylsulphonate based on an expected stoichiometric reaction between the anionic surfactant and an added known cationic reagent (cetyldimethylbenzyl ammonium chloride) followed by back-titration of the excess cationic reagent with a standard solution of sodium tetraphenylboron.

Bock and Reisinger[47] and Wickbold[48] have described a method for the determination of sodium dodecylbenzene sulphonate based on precipitation with p-toluidine and titration with sodium hydroxide.

Other techniques

Zvonaric et al.[49] have described a polarographic method for the determination of the surfactant activity of sea water. Taylor and Waters[50] describe a radiometric estimation of traces (down to 0.0005 ppm) of anionic detergents in ground and potable waters. This method involves mixing the sample (20 ml) with 0.5 ml of ^{59}Fe-labelled ferroin solution (89.5 μm containing 0.1 μCi of ^{59}Fe per ml) mixed with 0.5 ml of sodium acetate–acetic acid buffer solution of pH 4. To this solution is added 2 ml of sample (diluted if necessary) containing less than 10 μg of anionic surfactant and 5 ml of chloroform saturated with water and the mixture shaken for 1 min, centrifuged for 5 min and a measured volume of the chloroform phase transferred to a special glass vessel for counting in a well-type NaI(Tl) crystal. A blank determination is carried out similarly. Use of an internal standard avoids interference from iron and from up to 1 ppm of fluoride.

Hart et al.[51] have described an indirect polarographic method for determining linear alkylbenzene sulphonates in sewage and tap water. The method is reliable for concentrations of 0.5 μg l^{-1} or more. In this procedure the sample was nitrated with fuming nitric acid, then adjusted to pH 12 and polarographed. Differential pulse polarograms of different types of sewage are illustrated in Figure 118. All of the sewage samples gave a reduction peak that occurred at the same potential as for the nitro derivative of 4-phenyldodecane sulphonate, i.e. -0.74 V.

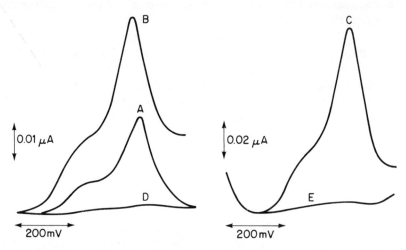

Figure 118 Differential pulse polargrams of nitrated sewage residues and blanks of nitration mixture. A, surfactant-free sewage; B, normal sewage; C, dosed sewage; D, blank; and E, blank. Supporting electrolyte Br buffer (pH 12) and initial potential -0.3 V. Reprinted with permission from Hart et al.[51] Copyright (1979) Royal Society of Chemistry

Table 66 Linear alkylbenzene sulphonate content of sewage determined by spectrophotometric and polarographic procedures

Day of sampling	Source of sample		Concentration (μg ml^{-1})	
			*Spectrophotometric method (methylene blue)	Polarographic method
1	Sewage stream	A	2.9	0.6
		B	0.8	1.9
		C	42.0	2.9
2		A	16.8	0.3
		B	67.0	1.4
		C	69.3	22.3
3		A	4.3	0.2
		B	12.0	1.4
		C	77.5	2.9

*Sewage stream A, detergent-free sewage; B, normal sewage; C, dosed sewage.
Reprinted with permission from Hart et al.[51] Copyright (1979) Royal Society of Chemistry.

The polarographic method was found to be suitable for the determination of linear alkylbenzene sulphonate with reasonable specificity. This is indicated by the results shown in Table 66 where, in general, the methylene blue spectrophotometric method gave higher results, probably because the methylene blue procedure gives a total anionic surfactant content, the so-called MBAS value (methylene blue active substances), whereas only surfactant species or other organic compounds with a benzene ring would be likely to interfere in the polargraphic method. Even then the E_p values of the derivatives of these species would not necessarily coincide with those of linear alkylbenzene sulphonates' nitro derivatives.

Samples of sewage were spiked with known amounts of 4-phenyldodecane sulphonate and subjected to polarographic linear alkylbenzene sulphonate

Table 67 Recoveries of linear alkylbenzene sulphonates from sewage streams A, B, and C where A is detergent free, B is normal and C is dosed with surfactant

Source of sample		Amount added (μg)	Amount found (μg)*	Recovery
Sewage steam	A	25	21.3	85.2
	A	25	20.5	82.0
	A	25	22.2	88.8
	B	100	95.2	95.2
	B	100	104.1	104.1
	B	100	91.5	91.5
	C	150	141.3	94.2
	C	150	146.7	97.8
	C	150	149.0	99.3

*Average of three determinations.
Reprinted with permission from Hart et al.[51] Copyright (1979) Royal Society of Chemistry.

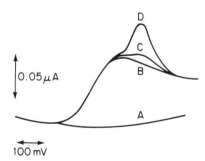

Figure 119 Differential pulse polarograms of tap water residues after nitration. A, blank; B, tap water; C, tap water + 11.2 μg of LAS; and D, tap water + 22.4 μg of LAS. Polarography was performed in Br buffer (pH 12) using an initial potential of −0.3 V. Reprinted with permission from Hart et al.[51] Copyright (1979) Royal Society of Chemistry

analysis and the recovery of added standard was calculated. These results are presented in Table 67.

Figure 119 shows the differential pulse polarograms of some of the nitrated residues of tap water. Tap water gave a broad peak which masked the reduction peak derived from less than 11.2 µg of linear alkylbenzene sulphonate. Recoveries of linear alkylbenzene sulphonate were calculated for the tap water samples spiked with 22.4, 33.6, and 44.8 µg of 4-phenyldodecane sulphonate per 50 ml. A baseline was constructed to the shoulders on each peak, and the peak heights measured to this. Recoveries were found to be 80, 88, and 91% in order of increasing linear alkylbenzene concentration.

Gas chromatography

Combined gas chromatography–mass spectometry has been used[52] to determine trace amounts of the individual components of alkylbenzene sulphonates as their methylsulphonate derivatives. Gas chromatographic analysis was performed using a gas chromatograph equipped with a flame ionization detector. A silanized glass column (2 m × 3 mm i.d.) was packed with 1.5% silicone OV-1 on Chromosorb W AW DMCS (80–100 mesh). Nitrogen was used as carrier gas with a flow rate of 40 ml min.$^{-1}$ The chromatogram was recorded as the total ion current monitor (TICM) at 20 eV. The molecular separator and ion source were maintained at 300° and 330 °C, respectively. Mass spectra were taken at 70 eV with an accelerator voltage of 3.5 kV.

In the work-up procedure a chloroform solution of methylene blue is shaken with a suitable volume (200–300 ml) of a sample solution. After shaking for 1 min, the chloroform layer is transferred to a 100 ml round-bottomed flask. If an emulsion is formed, it is removed by centrifugation. The extraction is repeated three times using 10 ml chloroform each time. The combined chloroform containing methylene blue and methylene blue alkylbenzene sulphonate complexes are evaporated to dryness under reduced pressure, then dissolved in a small amount of ethanol and passed through a column of cation-exchange resin (Dowex 50W × X8, 50–100 mesh, 50 × 10 mm i.d.; flow rate 0.3 ml min^{-1}) for removal of methylene blue. The ethanol eluate (*ca.* 10 ml) containing alkylbenzene sulphonate is then evaporated and the residue dissolved in *ca.* 20 ml of water. The aqueous solution is washed with chloroform (3 × *ca.* 5 ml), concentrated, transferred to a 1 ml ampoule (preignited 3 h at 500 °C) and then evaporated. After addition of *ca.* 10 mg phosphorus pentachloride the ampoule is sealed and maintained at 110 °C for 10 min on a hot plate. The sulphonyl chloride derivatives produced are extracted with *n*-hexane and the extract is transferred to another 1 ml ampoule and evaporated. After 0.5 ml methanol are added to the ampoule, which is sealed and maintained at 70 °C for 20 min, the methanol solution containing the methyl sulphonate derivatives produced is evaporated and the residue dissolved in *n*-hexane. The hexane solution is applied to the silica gel column (30 × 4 mm i.d.). The column is washed with three times its volume of *n*-hexane and then with one volume of

n-hexane–benzene (1:1). The eluate, comprising eight times the column volumes of n-hexane–benzene (1:1) is pooled and evaporated to a definite volume.

Alkylbenzene sulphonate was analysed as its methyl sulphonate derivatives by gas chromatography and gas chromatography–mass spectrometry. The concentration of alkylbenzene is determined by measuring the peak areas of the gas chromatogram and/or the mass fragmentogram and comparing these with linear dodecylbenzene sulphonate.

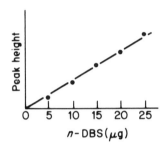

Figure 120 Calibration curve of dodecylbenzene sulphonate. Reprinted with permission from Hon-Hami and Hanya.[52] Copyright (1978) Elsevier Science Publishers B.V.

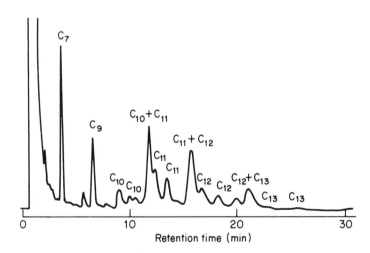

Figure 121 Gas chromatogram of ABS as methyl esters in river water. Column temperature, 230 °C. Reprinted with permission from Hon-Nami and Hanyo.[52] Copyright (1978) Elsevier Science Publishers B.V.

Figure 120 shows a calibration obtained by running 5–25 µg of dodecylbenzene sulphonate through the whole procedure. A linear relationship was obtained between the amount of dodecylbenzene sulphonate and the peak area on the chromatogram. The reproducibility was satisfactory. The dodecylbenzene sulphonate analysed was a mixture of $2\text{-}C_{12}$ (24.5%), $3\text{-}C_{12}$ (19.3%), and $4\text{-},5\text{-},6\text{-}C_{12}$ (56.2%). Using this method more than 1 µg of alkylbenzene sulphonate was determined accurately. No evidence was found by Hon-Nami and Hanya[52] for thermal decomposition of alkylbenzene sulphonates during gas chromatography.

Figure 121 shows a gas chromatogram of alkylbenzene sulphonate methyl esters in a river water sample. The pattern of the gas chromatogram was analogous to that of the linear alkylbenzene sulphonate standard. The assignment of the peaks was performed on the basis of the retention times and mass spectrum, and the individual components of alkylbenzene sulphonate were determined by mass fragmentography. For overlapped peaks on the gas chromatogram, more than two mass spectra were recorded. The result of the GLC–MS analysis is shown in Table 68. The total amounts of alkylbenzene sulphonate determined by mass fragmentography were in good accord with those determined by gas chromatography. Desulphonation gas chromatography has been applied[53–56] to the analyses of partially degraded linear alkylbenzene sulphonate mixtures.

Table 68 Distribution of alkylbenzenes sulphonate in river waters

Isomer	River 1	River 2
$2\text{-}C_{10}$	4.8	4.4
$3\text{-}C_{10}$	3.7	6.2
$4,5\text{-}C_{10}$	7.2	16.2
$2\text{-}C_{11}$	12.2	3.9
$3\text{-}C_{11}$	8.3	8.2
$4,5\text{-},6\text{-}C_{11}$	34.7	42.7
$2\text{-}C_{12}$	4.2	ND
$3\text{-}C_{12}$	3.4	1.2
$4\text{-},5\text{-}6\text{-}C_{12}$	13.7	17.1
$2\text{-}C_{13}$	0.8	ND
$3\text{-}C_{13}$	0.8	ND
$4\text{-},5\text{-},6\text{-},7\text{-}C_{13}$	6.2	ND

ND, not determined.
Reprinted with permission from Hon-Nami and Hanya.[52] Copyright (1978) Elsevier Science Publishers B.V.

High performance liquid chromatography

Taylor and Nickless[57] have described a paired-ion high performance liquid chromatographic technique for the separation of mixtures of linear alkylbenzene sulphonates and p-sulphophenylcarboxylate salts

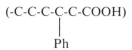

in river waters. Partially biodegraded linear alkylbenzene sulphonate was analysed by the same method. Structural information on metastable intermediates formed was provided by stopped-flow ultraviolet spectra, comparison of retention behaviour with that of standards, and analysis of collected fractions.

Samples (1.5 l) were concentrated for analysis by acidification with sulphuric acid to pH 2, followed by passage through a column containing 20 ml of XAD-4 resin at a flow rate of 7 ml min^{-1}. Compounds retained by the resin were eluted wtih 3×25 ml portions of methanol and the combined eluates evaporated to dryness then up to 2 μl.

The liquid chromatograph consisted of two high-pressure, reciprocating piston pumps, a septum injector, and a variable wavelength, double beam u.v. detector. Reversed-phase packing material mean particle size 5 μm was prepared by the reaction of LiChrosorb Si60 silica with octadecyltrichlorosilane. The mobile phase was a solution of cetyltrimethylammonium sulphite in methanol—distilled water.

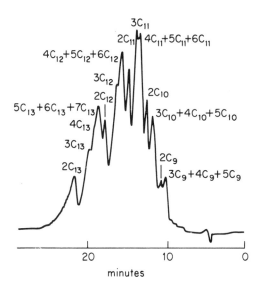

Figure 122 Paired-ion h.p.l.c. of ungraded LAS. Column: 250×4.6 mm, bonded C_{18} silica, $d_p = 5$ μm. Mobile phase: $13.7 \cdot 10^{-3}$ M $(CTMA^+)_2SO_4^{2-}$ in 87.5% methanol, 12.5% water, pH 5.4. Flow rate: 0.8 ml min^{-1}; u.v. detection at 224 nm, 0.5 AUFS. Reprinted with permission from Taylor and Nickless.[57] Copyright (1979) Elsevier Science Publishers B.V.

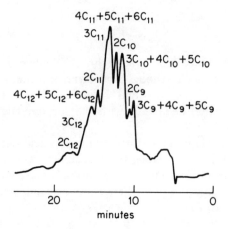

Figure 123 Paired-ion h.p.l.c. of partially biodegraded LAS. Conditions as in Figure 122. Reprinted with permission from Taylor and Nickless.[57] Copyright (1979) Elsevier Science Publishers B.V.

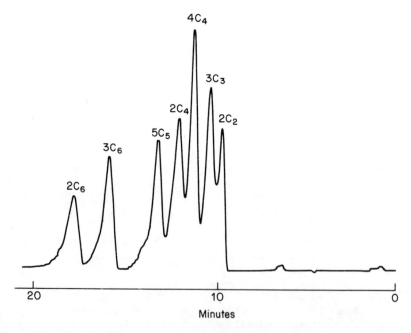

Figure 124 Paired-ion h.p.l.c. of a mixture of SPC salts. Column: 250×4.6 mm, bonded C_{18} silica, $d_p = 5$ μm. Mobile phase: $5.5 \cdot 10^{-3}$ M(CTMA$^+$)$_2$SO$_4^{2-}$ in 75% methanol, 25% water, pH 2.3. Flow rate: 0.8 ml min^{-1}; u.v. detection at 222 nm, 0.5 AUFS. Reprinted with permission from Taylor and Nickless.[57] Copyright (1979) Elsevier Science Publishers B.V.

Figure 122 shows the trace obtained from the analysis of undegraded linear alkylbenzene sulphonate mixture (sodium dodecylbenzene sulphonate). Peak identifications were made by coinjection with pure undegraded linear alkylbenzene sulphonate compounds and confirmed by analysis of the undergraded linear alkylbenzene sulphonate mixture by desulphonation followed by gas chromatography of the resultant alkylhenzenes on 15 m OV-1 support coated open-tubular column. Figure 123 shows the corresponding trace obtained for a partially biodegraded linear alkylbenzene sulphonate.

Under the conditions employed in Figures 122 and 123 p-sulphophenyl carboxylate salts were eluted close to the solvent front. In order to achieve separation of a mixture of p-sulphophenyl carboxylate standards the methanol content of the mobile phase was decreased by 87.5 to 75%. Figure 124 depicts the separation employing a mobile phase at pH 2.3. A 5.5×10^{-3} M solution of trimethylcetylammonium hydroxide was adjusted to pH 6.2, 4.6, and 2.2 with sulphuric acid. Each solution was employed as the mobile phase for the chromatography of a mixture of p-sulphophenyl carboxylate salts. The effect

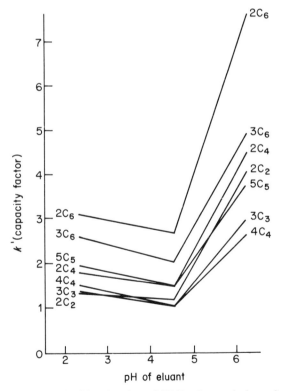

Figure 125 Paired-ion h.p.l.c. of SPC salts, variation of pH of mobile phase. Column: 250×4.6 nm, bonded C_{18} silica, $d_p = 5$ μm. Mobile phase: $5.5 \cdot 10^{-3}$ M(CTMA$^+$)$_2$SO$_4^{2-}$ in 75% methanol, 25% water. Reprinted with permission from Taylor and Nickless.[57] Copyright (1979) Elsevier Science Publishers B.V.

of pH on retention behaviour is illustrated in Figure 125. At pH 6.2, carboxylate groups are predominantly in the ionized form $R-CO_2^-$ which makes ion pairs with cetyltrimethylammonium ion. The decreased retention times observed at pH 4.6 are due to an increase in the concentration of the $R-CO_2H$ form, which do not make ion pairs with cetyltrimethylammonium ion.

The figures in Table 69 demonstrate that aqueous solutions of p-sulphophenyl carboxylate salts can be concentrated on XAD-4 resin. A solution of 3 ppm p-sulphophenyl carboxylate in distilled water was acidified with sulphuric acid to pH 2 and passed through a column of XAD-4 (20 ml). Retention by the resin was monitored by measurement of the ultraviolet absorbance (at 222 nm) of the eluate. Compounds adsorbed on the column were eluted with 3×25 ml methanol. The combined eluates were evaporated to dryness, dissolved in 2 ml water, and analysed by paired-ion high performance liquid chromatography. Retention by the resin of $2C_4$ p-sulphophenyl carboxylate and hence its overall recovery was greater at a column flow rate of 7 ml min^{-1} than at 20 ml min^{-1} p-Sulphophenyl carboxylate salts in neutral solution were poorly retained by XAD-4.

Table 69 Concentration of aqueous p-sulphophenyl carboxylate (SPC) solutions on XAD-4 resin. Experiment 1, 3 ppm $2C_2$, $2C_3$, $2C_4$, $2C_6$, SPC, flow rate 20 ml min^{-1}; experiment 2, 2 ppm $2C_4$, $3C_4$, $4C_4$, SPC, flow rate 7 ml min^{-1}

Parameter	Volume of solution passed (ml)	Experiment 1		Experiment 2	
Retention by resin (%)	100		87.9		
	500		61.8		
	950		54.2		94.6
Overall SPC recovery (%)		$2C_2$	21.0		
		$2C_3$	45.1		
		$2C_4$	72.5	$2C_4$	90.3
				$3C_4$	93.2
				$4C_4$	92.5

Reprinted with permission from Taylor and Nickless.[57] Copyright (1979) Elsevier Science Publishers B.V.

Riley and Taylor[58] have discussed the analytical concentration of dissolved organic materials including surfactants from sea water using Amberlite XAD-1 resin. Wang et al.[43-46] discussed carbon absorption for the concentration of linear alkylate sulphonates. Wickbold[17] has discussed the concentration and separation of anionic detergents from surface water on a gas–water interface. In this method a 1 litre sample in a glass tower (500 mm × 60 mm) is treated with concentrated hydrochloric acid (10 ml) and an upper layer of ethyl acetate (10 ml) is added. Nitrogen (50–60 l h^{-1}) is bubbled up through a glass frit at the base of the tower, and the surfactant is carried by the bubbles into the organic layer, which is replaced after 5 min. The combined organic layers are evaporated, the residue is dissolved in water (100 ml) and a little methanol, and the surfactant (100–200 μg) is determined by the method of Longwell and Maniece.[59]

Ultraviolet spectroscopy has been used for the determination of alkylbenzene sulphonates.[60, 61]

CATIONIC DETERGENTS

Wang[62] has described a simple titration method applicable to the analysis of cationic surfactants. Methyl orange and azure A were used as primary dye and secondary dye respectively. The method is free from interference by inorganic salts even at high concentrations as in sea water samples (Wang and Pek)[63]. This method, however, cannot accurately measure the ionic surfactant at low concentration ranges, such as below 1 mg l^{-1}. Subsequently Wang and Langley[26] devised a more sensitive colorimetric method for analysing cationic surfactants of the quaternary ammonium type in fresh water. Methyl orange will react with ionic surfactant to form a chloroform-soluble coloured complex in the presence of chloroform. The colour intensity of the chloroform layer is proportional to the concentration of the 'dye–ionic surfactant complex', and can then be measured by making spectrophotometric readings of the chloroform solution at the optimum wavelength.

Yellow coloured complex of methyl orange with cationic surfactant in chloroform

Method

Reagents

Stock cetyldimethylbenzylammonium chloride (CDBAC) solution: an amount of the reference material equal to 1.000 g of CDBAC on a 100% active basis is weighed and dissolved in distilled water and diluted to 1 litre; 1 ml = 1 mg of CDBAC.

Standard cetyldimethylbenzylammonium chloride (CDBAC) solution: 50.00 ml of stock CDBAC solution are diluted to 1 litre with distilled water; $1.00 \text{ ml} = 50.0 \, \mu\text{g}$ of CDBAC.

Methyl orange solution: 0.10 g of methyl orange are dissolved in a small amount of distilled water and diluted to 100 ml so that the concentration is 0.1% by weight.

Buffer solution: 250 ml of 0.5 M citric acid and 250 ml of 0.2 M disodium hydrogen orthophosphate are mixed together.

Preparation of cationic surfactant's calibration curve

A series of separatory funnels are prepared with 0, 0.50, 1.00, 1.50, 2.00, 2.50, 3.00, 3.50, and 4.00 ml of the standard CDBAC solution. Sufficient water is added to make the total volume 50 ml in each funnel. Each standard is then treated as described below. A calibration curve of μg of CDBAC vs. absorbance (or % transmittance), or mg l^{-1} CDBAC vs. absorbance, is plotted.

Procedure

Cationic surfactant analysis: An aliquot of the water sample is pipetted into a separatory funnel and diluted to 50 ml with distilled water. To the separatory funnel is added 1 ml of methyl orange reagent, 5 ml of buffer solution, and 25 ml of chloroform. The separatory funnel is stoppered and shaken vigorously for 30 seconds. The sample is allowed to stand undisturbed for 20 min after shaking. The chloroform will settle as a lower layer. A small plug of glass wool is wedged in the stem of a filtering funnel. The filtering funnel is placed above a clean, dry test cell (5 cm light path) and the chloroform layer is filtered through the glass wool to remove the water. The treated chloroform is collected in the cell. The absorbance of the chloroform solution is determined at 415 nm against a

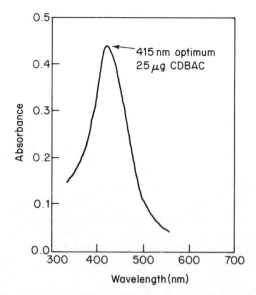

Figure 126 Absorbance of methyl orange–CDBAC complex. Reprinted with permission from Wang and Langley.[26] Copyright (1977) Springer Verlag, N.Y.

chloroform blank. The equivalent CDBAC content is determined from the calibration of cationic surfactant. The absorbance curve of the methyl orange–cationic surfactant complex is illustrated in Figure 126.

Waters and Kupfer[64] have described a titrimetric method for the determination of cationic detergents in biodegradation test liquids. The cationic surfactant is concentrated by evaporation of the sample and separated from the residue by solvent extraction as an ion-association compound; the extract is treated under non-aqueous conditions in Bio-Rad AG 1-X2 resin in the chloride form to remove interfering anionic components and the isolated cationic surfactant is determined colorimetrically as its disulphine blue ion-association compound which is extractable into chloroform. The method is not suitable for analysis of natural waters and waste waters which contain high levels of natural substances that react with disulphine blue.

Wang et al.[44] have described a direct titrimetric procedure for cationics employing poly(vinylsulphuric acid) potassium as titrant. A continuous solvent extraction method has been described for the spectrophotometric determination of cationic surfactants.[65]

Le Bihan and Courtot Coupez[2] analysed fresh water and sea water for cationic detergents by a method based on atomic absorption spectrometry of the copper–detergent complex (see section on anionic detergents).

The British Standard method[24] for cationic detergents is based on the addition to the sample of an excess of a standard anionic detergent and determination of excess anionic by a spectrophotometric methylene blue method.

NON-IONIC DETERGENTS

Various methods involving the use of solvent extraction and ion-exchange resins etc, have been employed to extract non-ionic detergents from water prior to analysis. Major methods of analysis used to date include methods based on atomic absorption spectrophotometry of detergent complexes, spectrophotometry of complexes, and liquid chromatography.

Atomic absorption methods

Courtot Coupez and le Bihan[66, 67] determined non-ionic detergents in sea and fresh water samples at concentrations down to 0.002 ppm by benzene extraction of the tetrathiocyanatocobaltate (II) $((NH_4)_2 [Co(SCN)_4])^-$ detergent ion pair followed by atomic absorption spectrophotometric determination of cobalt.

Courtot Coupez and Le Bihan[66] determined the optimum pH (7.4) for extraction of non-ionic surfactants with the above complex–benzene system. Cobalt in the extract is estimated by atomic absorption spectrometry after evaporation to dryness and dissolution of the residue in methyl isobutyl ketone. The method is applicable to surfactant concentrations in the range 0.02–0.5 mg l^{-1} and is not seriously affected by the presence of anionic surfactants.

The sample (500 ml) is filtered and sodium chloride (150 g) and the cobalt reagent[68] (75 ml) added. The pH is adjusted to 7.4 with concentrated aqueous ammonia and set aside for 15 minutes. The sample is extracted with benzene (30 ml) and centrifuged at 2000 rpm for 3 min. An aliquot (25 ml) is evaporated in a rotary evaporator and the residue dissolved in isobutyl methyl ketone (10 ml), aspirated into an air–acetylene flame, and the absorption measured at 240.7 nm with a lamp current of 10 mA, a slit width of 100 μm and a burner height of 4 mm. The detergent concentration is determined over the range 0–0.4 ppm by reference to a calibration graph. The average recovery is 60%. Interference of anionic detergents at concentrations of less than 10 ppm is prevented by extraction from an ammonium chloride medium.

Arpino et al.[69] performed microdeterminations of non-ionic ethoxylated surfactants in surface waters by precipitation with tungstophosphoric acid as described by Burttschell[70] and by the precipitation of tetrathiocyanatocobaltate (II) detergent ion pair, followed by determination of cobalt in the precipitate by atomic absorption spectrometry as described by Courtot Coupez and le Bihan[66]. Both methods were found by Arpino et al.[69] to give erratic results. Greff[71] has also described a widely used method for non-ionic detergents based on the benzene extraction of the ammonium tetrathiocyanatocobaltate (II) couplex followed by atomic absorption spectrophotometry at 320 nm. The method is simple, rapid, well suited to the routine analysis of large numbers of samples, and applicable to surfactant concentrations in the range 0.1–20 mg l^{-1}. With this extraction system, it is necessary to saturate the water samples with sodium chloride to improve extraction efficiency. A more serious disadvantage is interference by anionic surfactants.

In more recent work Crisp et al.[72] have described a method for the determination of non-ionic detergents concentrations between 0.05 and 2 mg l^{-1} in fresh, estuarine, and sea water based on solvent extraction of the detergent–potassium tetrathiocyanatozincate (II) complex followed by determination of extracted zinc by atomic absorption spectrometry. A method is described below for the determination of non-ionic surfactants in the concentration range 0.05–2 mg l^{-1}. Surfactant molecules are extracted into 1,2-dichlorobenzene as a neutral adduct with potassium tetrathiocyantozincate (II) and the determination is completed by atomic absoption spectrometry. With a 150 ml water sample the limit of detection is 0.03 mg l^{-1} (as Triton X-100). The method is relatively free from interference by anionic surfactants; the presence of up to 5 mg l^{-1} of anionic surfactant introduces an error of no more than 0.07 mg l^{-1} (as Triton X-100) in the apparent non-ionic surfactant concentration.

Method

Standard reference non-ionic surfactant solution: the reference non-ionic surfactant was Triton X-100 (gas chromatography grade). A stock standard solution of Triton X-100 was prepared, containing 1500 mg l^{-1}.

Zinc-thiocyanate reagent: 116 g of zinc sulphate heptahydrate, 312 g of potassium thiocyanate and 40 g of potassium acetate are dissolved in hot water and diluted to 2 l with water. The reagent is extracted with three 50 ml volumes of 1,2-dichlorobenzene before use.

1,2-Dichlorobenzene: the commercial solvent is passed through a 20 cm column of activated alumina before use and the used solvent recycled by washing it with water, drying over anhydrous calcium chloride, and passing through the alumina column.

Procedure

A 150 ml water sample, containing not more than 2 mg of non-ionic surfactant per litre, is placed in a 500 ml separating funnel, fitted with a Teflon stop-cock. The pH of the water sample is adjusted to 6–8. To the funnel is added 50.0 ml of zinc thiocyanate reagent and 20.0 ml of 1,2-dichlorobenzene, the funnel is shaken for 5 min, and allowed to stand until the phases separate. About 13 ml of the organic phase is run into a dry 15 ml graduated centrifuge tube, taking care that no droplets of the aqueous phase pass into the tube. The top of the tube is covered with aluminium foil and centrifuged at room temperature and 2500 rpm for 30 min. A 10.0 ml aliquot of the clarified extract is pipetted into a 25 ml volumetric flask and 10.0 ml of 0.1 M hydrochloric acid added. The flask is stoppered and shaken for 2 min. The phases are allowed to separate and the aqueous (upper) layer is aspirated directly from the flask into the atomic absorption spectrometer. The following atomic absorption spectrometer working conditions are used: oxidizing air–acetylene flame, 5 mA lamp current, 213.9 nm wavelength and 0.2 nm spectral band pass.

A blank determination is performed with 150 ml of distilled water. The blank absorbance should be small (0.000–0.001). The surfactant concentration in the sample (as mg Triton X-100^{-1}) is calculated by comparison with standards run simultaneously.

The precision of the method varies between 0.01 and 0.04 (S^2 mg l^{-1}) at concentrations of Triton X-100 between 0.1 and 2 mg l^{-1}, where S is the standard deviation.

Figure 127 shows the effect of polyethoxylate chain length. The response to the method of a range of *n*-nonylphenol ethoxylates containing up to 100 ethoxy units per molecule was studied. Response is greatest between 8 and approximately 40 ethoxy units. The effects of various ions on the recovery of Triton X-100 (1.00 mg l^{-1}) are shown in Table 70. Only sulphide, iron(III), aluminium(III), and chromium(III) ions interfere at concentrations likely to be found in contaminated waters. There is, however, no interference from sulphide at up to 100 mg l^{-1} if the water is treated with 1 ml of 30% hydrogen peroxide, shaken, and allowed to stand for 10 min before addition of the zinc thiocyanate reagent. Interference from up to 100 mg l^{-1} concentrations of iron(III), aluminium(III) or chromium(III) is suppressed by addition of 1 g of EDTA (disodium salt dihydrate) prior to the addition of the zinc thiocyanate reagent;

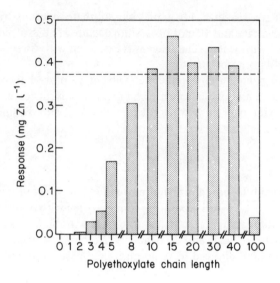

Figure 127 Effect of polyethoxylate chain length. Response of $1.00\,\text{mg}\,l^{-1}$ solutions of n-nonylphenol ethoxylates, containing 2–100 ethoxy units per molecule, is expressed as the concentration of zinc in the acid back-extract. The dashed line indicates the zinc concentration obtained from Triton X-100 ($1.00\,\text{mg}\,l^{-1}$). Reprinted with permission from Crisp et al.[72] Copyright (1979) Elsevier Science Publishers B.V.

Table 70 Allowable concentrations of foreign ions[a]

Concentration	Ion
5 M	Cl^-, Na^+
0.5 M	NO_3^-, SO_4^{2-}
1000 mg l^{-1}	F^-, Br^-, I^-, NO_2^-, SCN^-, CH_3, CO_2^-, K^+, Mg^{2+}
100 mg l^{-1}	$P_3O_{10}^{5-}$, NH_4^+, Ca^{2+}, Ni^{2+}, Co^{2+}, Mn^{2+}, Zn^{2+}
10 mg l^{-1}	Cu^{2+}
0.1 mg l^{-1}	S^{2+}, Fe^{3+}, Al^{3+}, Cr^{3+}

[a] 95–105% recovery of Triton X-100 ($1,00\,\text{mg}\,l^{-1}$) was obtained in the presence of the stated condition.
Reprinted with permission from Crisp et al.[72] Copyright (1979) Elsevier Science Publishers B.V.

the EDTA is unnecessary, however, if the metals are present mainly as a suspension of the hydroxides.

The performance of this method in the presence of anionic surfactants is of special importance, since most natural or waste water samples which contain non-ionic surfactants also contain anionic surfactants. Recovery data from Triton X-100 from water and sea water containing the anionic surfactants (linear alkylbenzene sulphonates) are given in Table 71. The presence of up to 5 mg l^{-1} linear alkylbenzene sulphonate increases the apparent concentration of non-ionic surfactant by a maximum of 0.07 mg l^{-1} (as Triton X-100). Soaps, such as sodium stearate, do not interfere with the recovery of Triton X-100 (1.00 mg l^{-1}) when present at the same concentration (i.e., 1.00 mg l^{-1}). Cationic surfactants, however, form extractable ion-association compounds with the tetrathiocyanatozincate ion and interfere with the method.

Table 71 Recovery of Triton X-100 from water and sea water in the presence of linear alkylbenzene sulphonates

Linear alkylbenzene sulphonates	Triton X-100 added (mg l^{-1})	Mean Triton X-100 found[a] (mg l^{-1})	
		in water	in sea water
5	0	0.04	0.03
5	0.10	0.15	0.15
5	1.00	1.04	1.04
5	2.00	2.07	2.00
0.5	0	0.03	0.03
0.5	0.10	0.13	0.15
0.1	0	0.03	0.03
0.1	0.10	0.13	0.15

[a]Mean of duplicate determinations.
Reprinted with permission from Crisp et al.[72] Copyright (1979) Elsevier Science Publishers B.V.

Spectrophotometric methods

Crabb and Persinger[73] have determined the molecular extinction coefficients of the cobalt thiocyanate complexes of nonylphenol–ethylene oxide adducts used in the spectrophotometric determination of the latter. These workers show compounds containing fewer than 3 ethylene oxide units not form a colour, and also that, for compounds of low molecular weight the apparent molecular extinction coefficient does not vary rectilinearly with chain length.

Belen'kii et al.[74] have applied the ammonium cobaltothiocyanate spectrophotometric method to the determination of non-ionic surfactants of the oxythylated alkylphenol type in coloured effluents. Activated charcoal removes from the solution both the dyes and the surfactants but the surfactants can then be separately desorbed by treatment of the charcoal with benzene–chloroform–light petroleum–ethanol (4:2:2:1) and determined by the ammonium cobaltothiocyanate method.

Brown and Hayes[75] and Grieff[68] have described spectrophotometric methods for non-ionic surfactants based on the ammonium cobaltothiocyanate.

Baleux[76] has described a spectrophotometric method for non-ionic polyoxyethylene surfactants in effluents using an iodide–iodine reagent. Although the method is especially applicable to waste effluent it is less satisfactory for the direct determination of the surfactants in natural waters.

Favretto and coworkers[77–80] in a series of papers have described spectrophotometric methods for non-ionic surfactants based on the formation of a sodium picrate surfactant adduct. In the original method[77] a sodium picrate–surfactant adduct is extracted from a strongly alkaline sodium nitrate solution into 1,2-dichloroethane and the extract absorbance is measured at 378 nm. Anionic surfactants interfere at concentrations greater than 0.2 mg l^{-1}. Allowance for this effect may be made by the determination of anionic surfactants (as methylene blue active substances) and reference to interference graphs.

Favretto et al.[81–83] have discussed in detail the use of 1,2-dichloroethane as an extraction solvent for the monodispense surfactants RO(CH$_2$CH$_2$O)nH (where R is p-tert-nonylphenyl and $3 \leqslant n \leqslant 11$). The method is based on the coordination reaction between the polyether chain and the potassium cation;[82] the cation complex is extracted into 1,2-dichloroethane as an ion pair with picrate which is a sensitive anionic chromophore for spectrophotometric determination at 378 nm.[86]

In a modification of their original method[77] devised to overcome interference by anionics and cationics and improve sensitivity Favretto et al.[78] consider particularly the analysis of polyoxyethylene n-dodeceyl ethers, the composition of which had been evaluated by gas-liquid chromatography.

The Favretto et al.[78] picric acid spectrophotometric method

Reagents

1,2-Dichloroethane: extra-pure grade freshly distilled.

Potassium nitrage–potassium hydroxide solution: 25.38 g of potassium nitrate are dissolved in 0.0125 M potassium hydroxide solution and diluted to 100 ml with water. This solution is 2.50 M in potassium nitrate.

Sodium nitrate–sodium hydroxide solution: in an 100 ml calibrated flask, 56.60 g of sodium nitrate are dissolved in 0.010 M sodium hydroxide solution and diluted to volume with water. This solution is 6.66 M in sodium nitrate.

Picric acid solution, 0.020 M: 0.46 g of picric acid (dried with phosphorus peutoxide) are dissolved in water in a 100 ml calibrated flask.

Sodium hydroxide solution, 0.10 M: prepared from Analar grade reagent.

Sulphuric acid, 0.10 M: prepared from Analar grade reagent.

Procedure

A 50.0 volume of the 1 mg l^{-1} aqueous solution of surfactant was extracted three times with 10.0 ml of freshly distilled 1,2-dichloroethane at 20 °C for 5 min min in a 100 ml separating funnel with a PTFE stockcock. All organic layers were transferred dropwise into a 100 ml flask and concentrated under vacuum to about 2–3 ml, which were then transferred quantitatively into a 50 ml calibrated flask with a few millilitres of 1,2-dichloroethane. The solvent was evaporated by means of a stream of nitrogen at 45 °C. The residue, first dissolved in 0.2 ml of 95% ethanol and then diluted with 20 ml of water, was processed as described under spectrophotometric calibration graphs. A reagent blank was also prepared and carried through the procedure.

Spectrophotometric calibration graphs

Calibration graphs were obtained at 20 °C by means of the following procedure. In a 50 ml calibrated flask, 20.0 ml of potassium nitrate–potassium hydroxide solution were added to an aliquot of standard solution of surfactant. The aliquots were chosen so as to give $0.1-1.0$ mg l^{-1} of surfactant in the final 50 ml of aqueous phase. After mixing and allowing to stand for 1 h, 5.00 ml of the aqueous solution of picric acid were added and the volume was made up to the mark with water. The mixed solution (pH 10.2) was allowed to stand for 15 min and then transferred into a separating funnel with a PTFE stopcock. A 5.00 ml volume of 1,2-dichloroethane was added and the mixture was shaken for 5 min. The organic layer was transferred dropwise into a conical centrifuge tube fitted with a polythylene stopcock and centrifuged at 1000 g for 5 min. The absorbance of the clear organic extract was measured at 378 nm in a 1 cm or 2 cm cell against a blank, prepared by extracting the reagents only. The absorbance of the reagent blank against the solvent is constant.

Calibration graphs were also constructed by using sodium as the co-ordinating cation. In this instance, 25.00 ml of an aqueous solution 6.66 M in sodium nitrate and 0.010 M in sodium hydroxide were added to an aliquot of standard solution of surfactant and processed as described above.

The following materials were employed as calibration standards: monodisperse polyoxyethylene n-alkyl ethers $C_{12}E_8$ and $C_{16}E_4$ (C_mE_n, where m is the carbon number of the alkyl chain and n is the degree of polymerization of the ethoxy group) were purified by repeated column chromatography in order to obtain a gas-liquid chromatographic purity of at least 99.9%. All surfactants were previously dried at 50 °C under vacuum (0.5 mmHg) for 3 h. Aqueous stock solutions (20 mg l^{-1}) were prepared weekly from surfactants previously dried with phosphorus pentoxide

When the absorbance of the 5 ml of organic extract is plotted against the surfactant concentration (c_a) present in 50 ml of aqueous phase, straight lines passing through the origin are obtained for both the monodisperse $C_{12}E_6$ ($A = abc_a$) and the polydisperse $C_{12}E_{\bar{n}}$ ($A = \bar{a}bc_a$) at least up to $c_a = 1.0$ mg l^{-1}. Table 72 summarizes the values of the slope ($\bar{a}$) of the calibration graphs obtained in the range $0.1 \leqslant c_a \leqslant 1$ mg l^{-1} at 20 °C. This slope has a direct working value, representing the absorbance ($b = 1$ cm) of the 5 ml of extract from 50.0 ml of aqueous phase containing 1.00 mg l^{-1} of polydisperse surfactant. In this table the a value of $C_{12}E_6$ is also reported. The slope obtained with K$^+$ is systematically higher than that obtained with Na$^+$, because polyoxyethylene complexes with K$^+$ are more stable than those with Na$^+$.

Figure 128 shows the variation of $\bar{a}$ as a function of $\bar{n}$. The standard deviation is represented by the size of the circles. The slope of both series increases with $\bar{n}$, but a maximum is observed at $\bar{n} \approx 10$ with Na$^+$, whereas with K$^+$ the curve

Table 72 Slopes of the calibration graphs for $C_{12}E_{\bar{n}}$ surfactants with K^+ and Na^+ as coordinating cations at 20 °C

Surfactant	K^+ (1.005 M)		Na^+ (3.335 M)	
	$\bar{a}$	$\pm s(N)$	$\bar{a}$	$\pm s(N)$
$C_{12}E_{\overline{3.8}}$	0.170	±0.005 (9)	0.146	±0.004 (9)
$C_{12}E_{\overline{7.1}}$	0.227	±0.003 (6)	0.202	±0.003 (7)
$C_{12}E_{\overline{8.0}}$	0.232	±0.002 (7)	0.214	±0.002 (6)
$C_{12}E_{\overline{9}}$	0.239	±0.001 (7)	0.217	±0.002 (6)
$C_{12}E_{\overline{10}}$	0.244	±0.002 (6)	0.218	±0.001 (6)
$C_{12}E_{\overline{11}}$	0.247	±0.001 (6)	0.217	±0.002 (6)
$C_{12}E_{\overline{12}}$	0.250	±0.002 (6)	0.216	±0.002 (7)
$C_{12}E_{\overline{15}}$	0.252	±0.002 (5)	0.216	±0.002 (6)
$C_{12}E_{\overline{17}}$	0.254	±0.001 (5)	0.214	±0.001 (7)
$C_{12}E_{\overline{23}}$	0.256	±0.002 (6)	0.213	±0.002 (6)
$C_{12}E_{\overline{6}}$	0.268	±0.001 (8)	0.230	±0.002 (7)

Reprinted with permission from Favretto *et al.*[78] Copyright (1978) Royal Society of Chemistry.

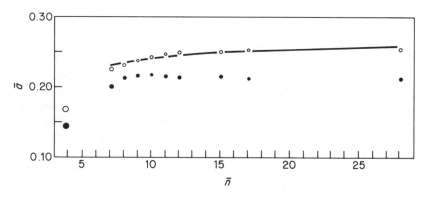

Figure 128 Graph of the slope of the calibration graphs, $\bar{a}$ values obtained with extracts from aqueous phases containing potassium ions at a concentration of 1.005 M and picric acid at a concentration of 2.0×10^{-3} M, pH 10.2; in the range $7.1 \leqslant \bar{n} \leqslant 28$ the value are empirically interpolated by the hyperbola $\bar{a} = 0.211\, \bar{n}/(1 + 0.774\, \bar{n})$, with a root mean square deviation of the observed from the calculated points of ±0.003. Closed circles, $\bar{a}$ values obtained with extracts from aqueous phases containing sodium ions at a concentration of 3.335 M and picric acid at a concentration of 2.0×10^{-3} M. Reprinted with permission from Favretto *et al.*[78] Copyright (1978) Royal Society of Chemistry

seems to tend towards an asymptote. At least in the range $7.1 \leqslant \bar{n} \leqslant 28$ the observed results can be empirically interpolated by an hyperbola through the origin ($\bar{a} = p\bar{n}/(1 + q\bar{n})$). A statistical *t*-test shows that, at least at the 0.95 probability level, the calculated asymptote for $\bar{n} \rightarrow \infty$ ($p/q = 0.273 \pm 0.004$) does not differ significantly from the value observed with $C_{16}E_6$ ($a = 0.268 \pm 0.001$).

To conclude, the replacement of NA^+ with K^+ improves the sensitivity (by 10–20%) in the spectrophotometric determination of $C_{12}E_{\bar{n}}$ surfactants. At least in the range of $\bar{n}$ values considered, this replacement also eliminates the maximum in the graph of A versus $\bar{n}$, and makes the relationship approach a simple hyperbola; if $\bar{n}$, is evaluated in some way, an absolute determination of the surfactant concentration is possible.

However, as the evaluation of $\bar{n}$ in polluted waters is not possible, and often at trace levels no information can be obtained on the hydrophobic group of polyoxyethylene non-ionic surfactants, it is preferable to express the concentration of non-ionics as 'potassium picrate active substances' (PPAS) by analogy with 'cobaltothiocyanate active substances' (CTAS) and 'methylene blue active substances' (MBAS) (the last term refers to anionic surfactants). The concentration of PPAS is usually referred to a standard monodisperse surfactant, in the same way as MBAS are expressed as sodium dodecylsulphate. Favretto et al.[78] suggest the use of $C_{12}E_6$ as a standard for the polyoxyethylene n-alkyl ethers for the following reasons:

(1) its preparation with the degree of purity necessary for a standard substance is easy;
(2) its response to potassium picrate reagent is high;
(3) it is stable for several years when stored in a desiccator; and
(4) it is sufficiently water-soluble for stock solutions with concentrations up to 25 mg l^{-1} to be prepared.

No difference in extraction efficiency was observed with various surfactants (see Table 73). Studies with $C_{12}E_{\bar{9}}$ showed that in the pH range 5.2–12.8 the efficiency of the extraction of the surfactant (at a concentration of 1.0 mg l^{-1}) from water by means of the proposed procedure is 98–99%.

Table 73 Recoveries of $C_{12}E_{\bar{n}}$

$\bar{n}$	pH	$r \pm s$ (%)
$\overline{3.8}$	7.18	98.9 ± 0.3
	5.18	98.2 ± 0.8
	7.18	98.3 ± 0.2
$\bar{9}$	11.91	98.3 ± 0.6
	12.77	99.1 ± 0.6
$\overline{17}$	7.18	98.4 ± 0.7

$r \pm s$ (%) is the mean recovery (at 20 °C) ± the root mean square deviation from four analyses in extraction with 1,2-dichloroethane from a 1.0 mg l^{-1} aqueous solution of surfactant.
Reprinted with permission from Favretto et al.[78]
Copyright (1978) Royal Society of Chemistry.

Synthetic anionic surfactants interfere in the spectrophotometric determination of non-ionic surfactants because the colourless surfactant anion competes with the yellow picrate in establishing equilibrium in the two-phase extraction system.[84] Systematic negative errors introduced by anionic surfactants are illustrated in Figure 129. No interference is observed up to 1.0 mg l^{-1} of sodium dodecylsulphate but the absorbance of the organic extract decreases

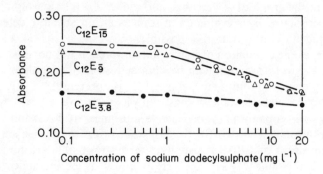

Figure 129 Absorbance, A, at 20 °C ($b = 1$ cm, $\lambda = 378$ nm) of extracts in 5 ml of 1,2-dichloroethane obtained from 50.0 ml of an aqueous phase containing 1.00 mg l^{-1} of $C_{12}E_{\bar{n}}$ and various concentrations of sodium dodecylsulphate. Reprinted with permission from Favretto et al.[78] Copyright (1978) Royal Society Chemistry

linearly with $\log_{10}c$ (where c mg l^{-1} is the concentration of anionic surfactant) up to at least 20 mg l^{-1}. Soap (as sodium stearate) does not interfere at concentrations up to at least 10 mg l^{-1}.

The interference of anionic surfactants is eliminated if the non-ionic surfactant is previously extracted with 1,2-dichloroethane. At least in the pH range 5.2–12.8 and up to 10 mg l^{-1} of anionic surfactants in the aqueous phase this solvent does not extract the latter; the test for methylene blue active substances gives a negative response on the extraction residue redissolved in water. However, when the concentration of the anionic surfactant is greater than 10 mg l^{-1} the recovery of the non-ionic surfactant decreases markedly. In this instance an appropriate dilution before the extraction is recommended. Cationic surfactants (e.g. dodecyltrimethylammonium chloride) give a very strong positive interference as they form an extractable salt with picrate. This interference is reduced about 15-fold on extraction with 1,2-dichloroethane.

Favretto et al.[78] also discuss the application of the picrate spectrophotometric method to the determination of poloxyethylene non-ionic detergents in sea water. They point out that sea water is a complex matrix, in which polyoxyethylene non-ionic surfactants cannot be determined directly with potassium picrate reagent (at pH 10.2 a precipitate is formed). In sea water polluted with urban liquid wastes, anionic surfactants are also normally present at concentrations one to five times greater than that of non-ionic surfactants. Therefore, the extraction with 1,2-dichloroethane must precede the spectrophotometric determination. As sea water contains trace amounts of solvent-extractable organic substances, which react with potassium picrate reagent, they are removed by further extraction of the organic phase with an acidic and then with an aqueous alkaline solution.

The following procedure was adopted by Favretto et al.[78] for samples of sea water containing 0.5–0.02 mg l^{-1} of non-ionic surfactants. A 250 g aliquot of sea water (filtered through a 0.8 μm Millipore filter) was extracted three times for 5 min with 25.0 ml of 1,2-dichloroethane in a 500 ml separating funnel with a PTFE stockcock. The organic layers were collected in a 100 ml separating funnel and extracted for 5 min with a 2.5 ml of 0.10 M sulphuric acid. The organic layer was collected in another 100 ml separating funnel and extracted with 2.5 ml of 0.10 M sodium hydroxide solution. The organic phase was concentrated under vacuum, transferred quantitatively into a 50 ml calibrated flask and evaporated to dryness at 40 °C under a stream of nitrogen. The residue was redissolved in 0.2 ml of 95% ethanol diluted with 20 ml of distilled water and treated with the potassium picrate reagents as described under spectrophotometric calibration graphs (see earlier).

The extract of another 250 g aliquot of sea water was evaporated to dryness as described above in a 50 ml calibrated flask, the residue was redissolved in 0.2 ml of 95% ethanol, 20.0 ml of an aqueous solution 2.50 M in potassium nitrate and 0.0125 M in potassium hydroxide were added and the volume was made up with water. This aqueous phase was extracted with 5.00 ml of 1,2-dichloroethane, the organic layer was centrifuged and the absorbance was read at 378 nm under the same conditions as the sample. This absorbance (matrix blank) added to that of the reagent blank, gives the absorbance of the total blank that is to be subtracted from the sample absorbance.

A mean value of $93 \pm 1\%$ was obtained in recovery experiments on $C_{12}E_{\bar{9}}$ (at an aqueous concentration of 0.10 mg l^{-1}) extracted from synthetic sea water by means of the above procedure. Therefore, a multiplication factor of 1.07 was adopted in correcting for the extraction losses. Washing with acidic and alkaline solutions does not change the overall extraction

Table 74 'Potassium picrate active substances' (PPAS) in the sea water from Trieste harbour

Date	MBAS	PPAS
September 29th, 1977	0.13	0.065
		0.070 (a)
		0.085 (uw)
October 5th, 1977	0.04	0.039
		0.044 (a)
		0.043 (uw)
October 7th, 1977	0.35	0.187
		0.200 (a)
		0.216 (uw)
October 24th, 1977	0.07	0.075 ± 0.003*
November 2nd, 1977	0.07	0.069
		0.071 (a)
		0.096 (uw)

PPAS are expressed as mg kg^{-1} of $C_{12}E_6$; unwashed extracts (uw) and alkaline washed extracts (a). *Mean ± root mean square deviation from four analyses. MBAS, methylene blue active substances.

Reprinted with permission from Favretto et al.[78] Copyright (1978) Royal Society of Chemistry.

efficiency appreciably, and it also remains unchanged on extraction of the non-ionic surfactants from synthetic sea water or from distilled water.

Table 74 gives the results of some determinations of potassium picrate active substances in waters sampled in Trieste harbour and also the effect of acidic and alkaline washing on the potassium picrate active substances value. In these samples, the matrix blanks expressed as a percentage of the potassium picrate active substances are always less than 2%. Replicate analyses gave a precision of about 4% expressed as the coefficient of variation.

Favretto et al.[78] applied gas–liquid chromatography to an evaluation of the polydispersity of polyoxythylene non-ionic surfactants. A 500 ml volume of water was extracted three times with 50 ml of 1,2-dichloroethane. The combined organic layers were extracted with 5.0 ml of 0.10 M sulphuric acid and then with 5.0 ml of 0.10 M sodium hydroxide solution. The purified extract was concentrated under vacuum, transferred into a conical glass min-vial and evaporated to a small volume (0.1 ml) by means of a stream of nitrogen. An aliquot of this solution was injected into the gas chromatographic column.

Chromatographic peaks attributable to the polydisperse surfactant $C_m E_{\bar{n}}$ were first recognized by using an internal standard a similar $C_m \pm E_{\bar{n}}$ surfactant (the comparison of the elution temperature of the peaks is a doubtful criterion as elution temperatures vary with the age of the column) mixed (approximately $1 + 1$ by mass) with a portion of residue solution. The values of n of the peaks pertaining to the distribution were then determined by introducing a monodisperse $C_m E_n$ as an internal standard in another aliquot of the residue solution $(1 + 10)$.

The gas–liquid chromatographic evaluation of $\bar{n}$ can obviously be performed on a single distribution $C_m E_{\bar{n}}$ but not on mixtures of distributions. Commercial surfactants always consist of mixed distribution. Only when the procedure is applicable, $\bar{n} = \Sigma x_n n$ is calculated from the observed distribution of the molar fraction (x_n) for various values of n. It should be checked that $\Sigma x_n = 1$.

Figure 130 shows as an example the chromatograms of the substances extracted with dichloroethane from the sample (see Table 74). Purification of the extract improves the trace of the chromatogram by eliminating or reducing peaks of extraneous substances. The chromatogram of the purified extract clearly shows the peaks of the polydisperse non-ionic surfactants. In this particular sample, their nature was tentatively identified as $C_{16}E_{\bar{n}}$ by comparison first with $C_{12}E_{\bar{8}}$ and then with $C_{16}E_4$. This monodisperse compound was also used for assigning the value of n to the $C_{16}E_n$ peaks.

With the chromatographic system adopted (non-polar stationary phase and short, low resolution column in order to elute polymers of low volatility), superimposition of peaks often occurs: for instance, the elution temperature of $C_{16}E_6$ is very close to that of $C_{12}E_4$. Therefore an unambiguous identification of the polymers in the distribution is not always possible, particularly in polyoxyethylene non-ionic surfactants of unknown origin.

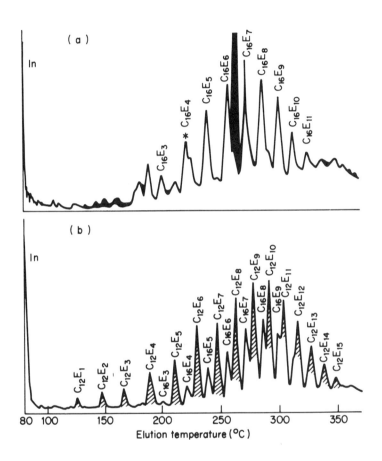

Figure 130 Gas chromatograms of polyoxyethylene non-ionic surfactants extracted from sea water sampled on October 7th 1977, near a sewage outlet in the harbour of Trieste. (a) Extract obtained with acidic and basic washings (the peaks of substances removed with washings are darkened); an asterisk marks the $C_{12}E_4$ peak identified by addition of the monodisperse compound as an internal standard. (b) Extract obtained with acidic and basic washings and added (about $1+1$ m/m) to the standard $C_{12}\overline{E_{8.0}}$. The shaded peaks are those of the standard. A Pye-Unicam 104 double-column chromatograph with flame ionization detectors with Pyrex glass columns (40 cm × 1.7 mm i.d.) packed with 80–100 mesh Gas-Chrom Q coated with 3% m/m GE SE-30, c.c. grade was used. Fractionation was carried out with a linear temperature programmer programmed from 80 to 370 °C at the rate of 10 °C min^{-1} and with injector and detector temperatures of 400 °C. The flow rates of the gases were nitrogen 45, hydrogen 45, and air 300 ml min^{-1}. The effective peak number was 0.7 for C_{22}/C_{23} n-alkanes. Reprinted with permission from Favretto et al.[78] Copyright (1978) Royal Society of Chemistry

Stancher and Tunis[85] have investigated the reaction mechanism in the determination of non-ionic surfactants in waters as potassium picrate active substances. Favretto et al.[80] have also determined polyoxyethylene alkyl ether non-ionic surfactants in waters at trace levels as potassium active substances. Various other workers[86–89] have investigated solvent extraction–spectrophotometric methods for the determination of polyoxyethylene non-ionic surfactants in water. Dozanska[90] has described a spectrophotometric method involving reaction with cobalt thiocyanate or tungstophosphoric acid for the determination of non-ionic detergents in sewage. Hey and Jenkins[91] have also reported on spectrophotometric methods utilizing cobalt thiocyanate.

The barium chloride–phosphomolybdic and spectrophotometric method for non-ionic surfactants is described below.[92, 93] The method involves the formation of a non-stoichiometric complex between the polyethoxylated compound, barium chloride and phosphomolybdic acid. The complex precipitates quantitatively from aqueous acidic solution, is isolated by centrifugation, redissolved in 2-methoxyethanol and the absorbance measured at 310 nm.

Procedure

To a 50 ml centrifuge tube are added 20.0 ml of the aqueous solution containing the polyethoxylated compound (not more than 1 mg), 1.0 ml of 20% (v/v) hydrochloric acid, 1.0 ml of 20% (w/v) barium chloride solution and 1.0 ml of 10% (w/v) phosphomolybdic acid solution. The tube is well shaken after each addition. When flocculation does not occur in the first minute (the slightest opalescence can be considered as a positive reaction), the solution should be stirred vigorously using a vortex mixer for a further minute. If no change occurs, the solution can be warmed slightly (to not more than 40 °C) and allowed to stand. The centrifuge tube is filled close to the top with 0.5 M hydrochloric acid and mixed by inversion. The tube should then be centrifuged for 5 min at 1500 g to compact the precipitate. The supernatant is then decanted off by gently inverting the tube. The tube is again filled with 0.5 M hydrochloric acid, mixed, centrifuged, and decanted as before. The precipitate in the bottom of the tube is dissolved by the addition of a mixture of 5 ml of 2-methoxyethanol and 1 ml of concentrated hydrochloric acid. The dissolution of the precipitate can be greatly speeded up by vigorous vortex mixing, although some warming may also be needed. The solution is transferred into a 25 ml graduated flask, made up to the mark with pure 2-methoxyethanol and the absorbance measured at 310 nm against a blank of the same solvent composition. Normally a check standard (20 ppm) of the polyethoxylated compound under investigation should be determined at the same time with each batch of analyses.

Straight-line calibration plots were obtained for secondary alcohol ethoxylates, alkylphenylethoxylates, and polyethylene glycols in the range 1–20 ppm, with good agreement between duplicates. Two important points should be noted. First, the slope of the calibration graph (that is, the sensitivity) is not the same for all different chemical types. Second, the detection limit is determined by the lowest concentration of polyoxyethylated compound that will precipitate as the complex from aqueous solution.

Liquid chromatography

Huber et al.[94] have described a column chromatographic method for the rapid separation and determination of non-ionic detergents of the polyoxyethylene glycol monoalkylphenyl type and applied it to the analysis of water samples. Such adducts ranging in chain length from 1 to 20 ethylene oxide units were separated on 23 cm columns of PEG-400 on Spherosil with three mobile phases, viz, 2,2,4-trimethylpentane, 2,2,4-trimethylpentane–CCl_4(2:1), and 2,2,4-trimethylpentane–CCl_4 (1:2) representing the extremes in polarity for separating the oligomers in a reasonable time. Use of the first mobile phase permitted separation of the oligomers produced by the reaction between one mole of phenol and from 3 to 7 moles of ethylene oxide, and of the second mobile phase those produced similarly with from 7 to 14 moles of ethylene oxide (but with no separation of the shorter oligomers). With use of the third mobile phase there was no separation of oligomers, but for each sample a peak of Poisson distribution was obtained that represented the oligomer distribution of the sample. Detection was by ultraviolet absorption; the relative precision was 0.4% and the limit of detection was 0.2 μg.

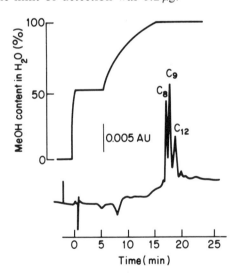

Figure 131 Typical liquid chromatogram of the separation of POE octylphenyl, nonylphenyl, and dodecylphenyl ethers by gradient elution with a holding process. C-8, C-9 and C-12 correspond to POE octylphenyl, nonylphenyl, and dodecylphenyl ethers respectively. Flow rate in elution: 1 ml min^{-1}. Reprinted with permission from Otsuki and Shiraishi.[95] Copyright (1979) American Chemical Society

Otsuki and Shiraishi[95] used reversed phase absorption liquid chromatography and field desorption mass spectrometry to determine polyoxyethylene alkylphenyl ether non-ionic surfactants in water. Figure 131 shows a typical liquid chromatogram of the separation of polyoxyethylene octylphenyl, nonylphenyl and dodecylphenyl ethers obtained by gradient elution with a holding process. After holding the mobile phase composition of 50:50 water–methanol, polyoxyethylene alkylphenyl ethers were eluted by a further initiation of the gradient in the order octylphenyl, nonylphenyl, and dodecylphenyl ethers, indicating that the retention time was related to the number of alkyl carbon atoms. Although the efficiency of adsorption was nearly constant in the range from 1 to 4 ml min^{-1}, sensitivity increased with decrease in flow rate of gradient elution. Thus 1 ml min^{-1} was used in gradient elution. Holding time in gradient elution showed no effect on retention time for 20 min in the mobile phase composition of 50:50 water–methanol. Calibration graphs of polyoxyethylene, octylphenyl, and nonylphenyl ethers with a full scale range of 0.05 absorbance unit were linear between 0 and 20 μg. Since polyoxyethylene dodecylphenyl ethers used as a standard contained a considerable content of impurities, as shown below by the field desorption mass spectrum, the accurate determination of polyoxyethylene dodecylphenyl ethers was impossible.

The results of recovery tests from fortified, filtered tap water are given below. At the 1 mg l^{-1} level, the recovery was more than 96% using 10 ml samples and at the 50 μm l^{-1} level, it was more than 71% using 200 ml samples.

	Recoveries	
	1.25 mg l^{-1}	0.05 mg l^{-1}
Concentration in fortified sample		
Polyoxyethylene octylphenyl ethers	100%, 98%	78%, 94%
Polyoxyethylene nonylphenyl ethers	96%, 97%	71%, 75%

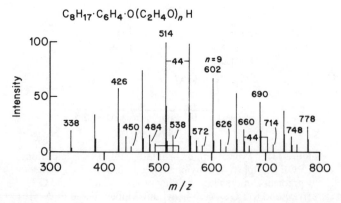

Figure 132 FD mass spectrum of POE octylphenyl ethers. Reprinted with permission from Otsuki and Shiraishi.[95] Copyright (1979) American Chemical Society

Figure 132 shows the field desorption mass spectrum of polyoxyethylene octylphenyl ethers. Each peak is due to each molecular ion with a different degree of polymerization of ethylene oxide because any fragment in the range of less than 100 m/z was not observed. The difference between two peaks corresponds to the difference of molecular weight of ethylene oxide, 44 m/z. The peak at 602 m/z is equivalent to polyoxyethylene ether with the degree of polymerization of 9 of ethylene oxide. The other small peaks would be due to impurities having different alkyl group or dialkyl groups. The peaks at 484, 528, 572, 616, 660, 704, and 748 m/z may be due to polyoxyethylene nonylphenyl ethers because of the difference of 14 m/z by methylene group. It is possible that the other peaks at 450, 494, 538, 582, 626, 670, 714, and 758 m/z are due to polyoxyethylene dioctylphenyl ethers. Figure 133 shows the field desorption mass spectrum of polyoxyethylene nonylphenyl ethers. The peak at 528 m/z corresponds to polyoxyethylene nonylphenyl ether with the degree of polymerization of 7 of ethylene oxide. The other small peaks at 366, 410, 454, 498, 542, 586, and 630 m/z may be due to polyoxyethylene decanylphenyl ethers

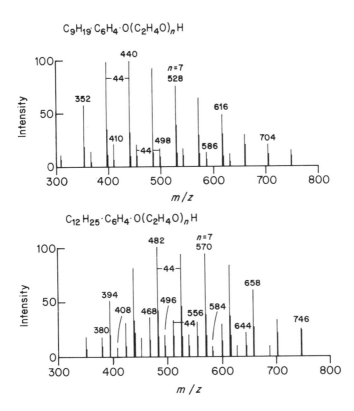

Figure 133 FD mass spectrum of polyoxyethylene nonylphenyl ethers. Reprinted with permission from Otsuki and Shiraishi.[95] Copyright (1979) American Chemical Society

as an impurity because of the difference of 14 m/z by methylene group. The field desorption spectrum of polyoxyethylene dodecylphenyl ethers suggests that this reagent contains a considerably high content of impurities such as polyoxyethylene undecylphenyl and tridecylphenyl ethers, and when the reagent is used as a standard material, purification is necessary. A peak at 570 m/z corresponds to polyoxyethylene dodecylphenyl ethers with the degree of polymerization of 7 of ethylene oxide.

From the field description mass spectra of standard samples, a table for identification of polyoxyethylene alkylphenyl ethers and determination of the degree of polymerization of ethylene oxide was constructed as shown in Table 75; n is the number of alkyl carbon atoms and m is the degree of polymerization of ethylene oxide. When the field desorption mass spectrum having the peak pattern with the difference of 44 m/z was obtained such as the peaks at 484, 528, 572, 616, and 660 m/z. Table 75 would tell that those peaks are due to polyoxyethylene nonylphenyl ethers with the degree of polymerization of 6–10 of ethylene oxide. Table 76 also shows the identification of polyoxyethylene dialkylphenyl ethers and determination of the degree of polymerization of ethylene oxide based on calculations of the molecular weight.

Table 75 Table for identification of poly(oxyethylene) alkylphenyl ethers from field desorption mass spectra

	$C_nH_{2n+1} \cdot C_6H_4 \cdot (C_2H_4O)_mH$												
n	4	5	6	7	8	9	10	11	12	13	14	15	16
6	354	398	442	486	530	574	618	662	706	750	794	838	882
7	368	412	456	500	544	588	632	676	720	764	808	852	896
8	383	426	470	514	558	602	646	690	734	778	822	866	910
9	396	440	484	528	572	616	660	704	748	792	836	880	924
10	410	454	498	542	586	630	674	718	762	806	850	894	938
11	424	468	512	556	600	644	688	732	776	820	864	908	952
12	438	482	526	570	614	658	702	746	790	834	878	922	966
13	452	496	540	584	628	672	716	760	804	846	892	936	980

Reprinted with permission from Otsuki and Shiraishi.[95] Copyright (1979) American Chemical Society.

Table 76 Table for identification of poly(oxyethylene) dialkylphenyl ethers from field desorption mass spectra

n	3	4	5	6	7	8	9	10	11	12	13	14	15
6	394	438	526	526	570	614	658	702	746	790	834	878	922
7	422	466	510	554	598	642	686	730	774	818	862	906	950
8	450	494	538	582	626	670	714	758	802	846	890	934	978
9	478	522	566	610	654	698	742	786	830	874	918	962	1006
10	506	550	594	638	682	726	770	814	858	902	946	990	1024
11	534	578	622	666	710	754	798	842	886	930	974	1018	1062
12	562	606	650	694	738	782	826	870	914	958	1002	1046	1090
13	590	634	678	722	766	810	854	898	942	986	1030	1074	1118

Reprinted with permission from Otsuki and Shiraishi.[95] Copyright (1979) American Chemical Society.

Figure 134 shows the field desorption mass spectrum of Triton X-100 as an example of identification of polyoxyethylene alkylphenyl ethers. From Table 75 it is seen that the peaks at 426, 470, 514, 558, 602, and 646 m/z are due to polyoxyethylene ethers with the degree of polymerization of 3–12 of ethylene oxide. The other peaks at 538, 582, 626, 670, 714, 758, and 802 m/z indicate the presence of polyoxyethylene dioctylphenyl ethers with the degrees of polymerization of at least 5–11 of ethylene oxide. This result demonstrates that the main component in Triton x-100 is polyoxyethylene octylphenyl ethers with the degree of polymerization of 3–12 of ethylene oxide and it also contains polyoxyethylene dioctylphenyl ethers with a degree of polymerization of at least 5–11 of ethylene oxide.

Figure 135 shows the field desorption mass spectrum of the concentrate of the fraction containing polyoxyethylene octylphenyl and nonylphenyl ethers separated by gradient elution with a 10 min holding process after adsorption of standard mixture sample. It is apparent that polyoxyethylene octylphenyl and nonylphenyl ethers in water can be recovered by reversed phase adsorption chromatography.

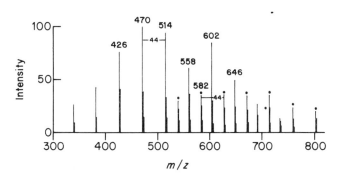

Figure 134 FD mass spectrum of Triton X-100. Reprinted with permission from Otsuki and Shiraishi.[95] Copyright (1979) American Chemical Society

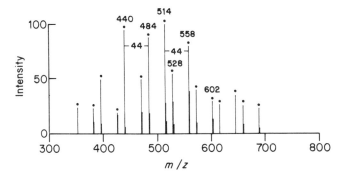

Figure 135 FD mass spectrum of the fraction containing POE octylphenyl and nonylphenyl ethers separated by gradient elution. Reprinted with permission from Otsuki and Shiraishi.[95] Copyright (1979) American Chemical Society

Otsuki and Shiraishi[95] conclude the mixtures of polyoxyethylene alkylphenyl ethers can be separated by the affinity between the chain length of the alkyl group in the polyoxyethlene alkylphenyl ethers and the octadecyl group of the bonded stationary phase, and the solubility in the solvent at the mobile phase composition during gradient elution. This reversed phase adsorption chromatography involves injection of the sample either by direct injection (maximum 2 ml) or by pumping a certain volume through the column. The choice depends on whether the sample contains high levels of polyoxyethylene alkylphenyl ethers. In case of the sample containing more than 1 mg l^{-1} direct injection should be used. In the application of the method to environmental water samples, the reason for selecting a hyperbolic programming and for introducing a holding process during gradient elution is to elute most organic compounds adsorbed from water sample in the early stage of the gradient to avoid overlapping of the peaks of the polyoxyethylene alkylphenyl ethers eluted in the later stage. Most organic compounds adsorbed from filtered natural water samples and having an absorbance near 280 nm were eluted during a holding process of about 15 min when using a mobile phase composition of 50:50 water–methanol. Possible interfering compounds in environmental samples may be dialkylphthalate esters having more than eight carbon atoms in an alkyl group and an absorbance near 280 nm such as di-2-ethylhexylphthalate but the peak can be confirmed by field desorption mass spectrometry. The other possible compounds are alkylbenzene sulphonates, but they are not adsorbed on the C_{18} column because the sodium sulphonate group has strong hydrophilic property.

Cassidy and Niro[96] have applied high speed liquid chromatography combined with infrared spectroscopy to the analysis of polyoxyethylene surfactants and their decomposition products in industrial process waters. Molecular sieve chromatography combined with infrared spectroscopy gives a selective method for the analysis of trace concentrations of these surfactants. These workers found that liquid-solid chromatography and reversed phase chromatography are useful for the characterization and analysis of free fatty acids.

Nickless and Jones[97] and Musty and Nickless[98,99] evaluated Amberlite XAD-4 resin as an extractant for down to 1 ppm of polyethylated secondary alcohol ethoxylates, $(R(OCH_2CH_2)_nOH))$ surfactants and their degradation products from water samples. This resin was found to be an effective adsorbent for extraction of polyethoxylated compounds from water except for polyethylene glycols of molecular weight less than 300. Flow rates of 100 ml min^{-1} were possible using 5 g of resin, and interfering compounds can be removed by a rigorous purification procedure. Adsorption efficiencies of 80–100% at 10 ppb were possible for non-ionic detergents using distilled water solutions. The main purpose of the work of Nickless and Jones[97] was the investigation of secondary alcohol ethoxylate as it proceeds through the water system. Associated with this is polyethylene glycol, a likely biodegradation product. Alkylphenol ethoxylate was also considered but only as a possible interferent that should be differentiated in order to allow fuller characterization of secondary alcohol ethoxylate residues.

Nickles and Jones[97] used in their study fairly pure secondary alcohol ethoxylate standards (R(OCH$_2$CH$_2$)$_n$OH) where n is 3, 5, 7, 9, and 12, also alkylphenol ethoxylate standards (R–C$_6$H$_4$–(OCH$_2$CH$_2$)$_n$OH) where n is 5, 7, 8, and 9, and polyethylene glycols (H-(OCH$_2$CH$_2$)$_n$OH) where n is 2, 3, 9, and 22. The barium chloride–phosphomolybdic acid spectrophotometric method[92,93] and also thin-layer chromatography[87] were used to determine polyoxyethylated compounds in column effluents. Nickless and Jones[97] tested the efficiences of three column materials for adsorption of secondary alcohol ethoxylate from solutions and found that Amberlite XAD-4 resin combined a high adsorption on the column with a high subsequent desorption of the ethoxylate from the column with ethanol (Table 77). Subsequently acetone was found to be a superior desorption solvent.

Table 77 Adsorption efficiencies of tested adsorbents

Adsorbent	SAE 9EO unadsorbed (%)	SAE 9EO desorbed by ethanol (%)*
Polyurethane foam	54	26
Activated carbon (coconut charcoal)	<0.5	<0.5
XAD-4	<0.5	115

*<0.5% corresponds to the limit of detection (ppm) for SAE 9EO.
Reprinted with permission from Nickless and Jones.[97] Copyright (1978) Elsevier Science Publishers B.V.

The XAD-4 results were very encouraging but the measured desorption was well over 100% (Table 77) and was an indication of interfering compounds still present in the resin from its manufacture. These interfering compounds were polyanionic in nature and Nickless and Jones[97] devised a procedure involving successive washing with acetone–hexane (1:1), acetone, and ethanol of XAD-4 resin to remove impurities before its use in the analyses of samples.

Results obtained by this procedure indicated a fairly rapid drop in an adsorption efficiency of the resin for polyethylene glycols between 9EO and 3EO. It is not clear if the drop in efficiency is roughly linear with a shortening in chain length, but it appears that the resin might be limited to the study of polyethylene glycols in water for chain length greater than 7EO.

The ability of XAD-4 resin to adsorb with high efficiency polyethoxylated compounds at very low concentrations is very important, if it is to be used for the examination of all types of water systems, including sea water. When the concentration is very low, then this necessitates the processing of large volumes of water in order to obtain sufficient material for characterization, using methods such as infrared, ultraviolet, and n.m.r. spectroscopy, as well as liquid chromatography. The performance of the resin in this respect was evaluated by using solutions of a mixture of three model compounds, secondary alcohol ethoxylate 9EO, alkylphenol ethoxylate 9EO, and polyethylene glycol 9EO. Recoveries varied from 82% at the 0.01 ppm level to 98% at the 1 ppm level.

Jones and Nickless[100] continued their study of Amberlite XAD-4 resins by examining polyethoxylated materials before and after passage through a sewage works. Samples from the inlet and outlet of the sewage works and from the adjacent river were subjected to a three-stage isolation procedure and the final extracts were separated into a non-ionic detergent and a polyethylene glycol. The non-ionic detergent concentration was 100 times lower (8 ng ml^{-1}) in the river than in the sewage effluent. Thin-layer chromatography and ultraviolet, infrared, and n.m.r. spectroscopy were used to identify, in the non-ionic detergent component, alkylphenol ethoxylates (the most persistent), secondary alcohol ethoxylate, and primary alcohol ethoxylate.

Three stage isolation procedure

Stage 1

The water sample was first passed down a column of XAD-4 resin to extract surface active materials on to the resin. The organic adsorbed from the water sample by the XAD-4 resin were eluted off with four solvent systems to give two fractions:

(1) 2 ml of methanol–water (1:1) followed by 20 ml of methanol;
(2) 50 ml of acetone, followed by 20 ml of acetone–n-hexane (1:1).

Each fraction was collected and evaporated to dryness in a stream of filtered air on a hot water bath. Fraction 2 contained most of the polyethoxylated material, but significant amounts were also present in fraction 1. The residue from fraction 1 was treated with 10 ml of acetone, decanted into a small tube, centrifuged, and the clear acetone layer poured into the beaker containing fraction 2. The combined extracts were again evaporated to dryness.

Stage 2: liquid-solid chromatography

The polyethoxylated material contained in fraction 2 from the XAD-4 elution will still be a relatively minor part of the residue. The indications were that most of the unwanted organic compounds extracted from the river water would be medium to non-polar in character. There would therefore be overlap with the medium polar oligomers of secondary alcohol ethoxylate and alkylphenol ethoxylate (i.e. those with only 1, 2, or 3 ethylene oxide units per molecule). Silica gel was found to be the best chromatographic adsorbent and the correct choice of eluting solvent strength is important for efficient separation. a (7:3) ethyl acetate–benzene extract of fraction 2 was poured down a column of silica gel. The ethyl acetate–benzene insoluble residue contained highly polar polyethoxylated material and was dissolved and loaded on to the column when more polar solvents were used later in the eluting procedure. A typical elution procedure is shown in Table 78. Most of the unwanted organic compounds came through in fractions 1 and 2 and were discarded. Fractions 3 and 4 (less than 1% of total) were combined and kept for thin-layer chromatographic examination. Fractions 5 and 6 were combined and usually contained greater than 98% of the total polyethoxylated material present in the original extract, together with some highly polar compounds imparting a faint yellow colour to the solution.

Table 78 Typical elution procedure for an XAD-4 extract on silica gel

Fraction no.	Eluting solvent and type of residue obtained
1	25 ml of ethyl acetate–benzene (7:3). A yellow band eluted off, containing most of the non-polar to medium polar compounds in the residue.
2	A further 25 ml of ethyl acetate–benzene (7:3). More yellow coloured material eluted.
3	30 ml of pure ethyl acetate. Further amounts of yellow coloured residue containing a trace amount of polyethoxylated material.
4	20 ml of ethyl acetate–acetone (1:1). Less yellow coloured residue eluted and further trace amounts of polyethoxylated material.
5	70 ml of pure acetone. Large amounts of polyethoxylated material, no yellow colour.
6	30 ml of acetone–water (3:2). More polyethoxylated material plus some yellow-brown residue.

Reprinted with permission from Jones and Nickless[100]. Copyright (1978) Elsevier Science Publishers B.V.

Stage 3

The combined fractions 5 and 6 obtained from liquid-solid chromatography needed further separation for two reasons:

(1) the fractions still contained significant amounts of other organic compounds which were found to be mainly acidic in character; and

(2) meaningful results could only be obtained from spectroscopic examination if the polyethoxylated material is divided into the non-ionic detergent components and the polyethylene glycol component. Jones and Nickless[100] solved both of these problems in one operation using liquid chromatography based on a procedure by Nadeau and Waszaciak.[101] With this procedure using water as a stationary phase (on Celite) and chloroform–benzene as the mobile phase, very good separations of polyethylene glycol from secondary alcohol ethoxylate and alkylphenol ethoxylate can be obtained. By simply changing the stationary phase to dilute sodium hydroxide solution, unwanted acidic compounds can be removed with the polyethylene glycol fraction. The separation scheme is shown in Figure 136.

Having separated the mixture into a secondary alcohol ethoxylate plus alkylphenol ethoxylate and polyethylene glycol components, Jones and

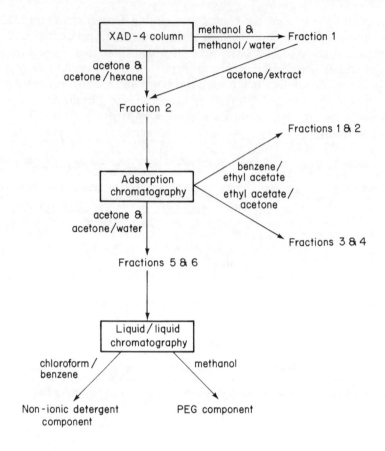

Figure 136 Complete separation scheme for polyethoxylated material. Reprinted with permission of Jones and Nickless.[100] Copyright (1978) Elsevier Science Publishers B.V.

Nickless examined these fractions using thin-layer chromatography[87] and ultraviolet, infrared, and n.m.r. spectroscopy.

In the thin-layer chromatographic separations ethyl acetate–acetic acid–water, 4:3:3 was used for quantitative information, where a compact spot is obtained for non-ionic detergents (secondary alcohol ethoxylate and alkylphenol ethoxylate combined), and an elongated spot of lower R_f value for polyethylene glycol. Ethyl acetate–acetic acid–water, 70:15:15 was used for information on the molecular weight distribution of secondary alcohol ethoxylate and alkylphenol ethoxylate. Alkylphenol ethoxylate gives a 'string' of well resolved spots and secondary alcohol ethoxylate a long unresolved streak. These patterns will be superimposed if both are present in the residue.

Ultraviolet spectroscopy was used to evaluate alkylphenol ethoxylates as only this type of compound gives a peak at 277 nm with a characteristic shoulder at 285 nm (Figure 137). Chloroform was the solvent used and after measurement the sample can be reconcentrated in a stream of air back to 0.5 ml.

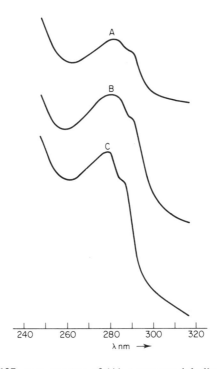

Figure 137 u.v. spectra of (A) a commercial alkylphenol ethoxylate, APE 9EO, (B) sewage effluent extract; and (C) Roberts river extract. Reprinted with permission of Jones and Nickless.[100] Copyright (1978) Elsevier Science Publishers B.V.

Table 79 Concentrations of non-ionic detergent and polyethylene glycol found in water samples

Sample identification volume processed	Weight of acetone-soluble XAD-4 extract (mg)	Weight of non-ionic detergent and PEG calculated from t.l.c. (mg)		Approximate concentration of non-ionic detergent and PEG in the water sample (ppm)	
		Non-ionic†	PEG	Non-ionic†	PEG
1 Sewage influent (14 l)	88.0	10.0	0.25	0.7	0.02
2 Sewage influent after storing for 3 days (3.5 l)	29.6	1.5	0.8	0.4	0.2
3 Sewage effluent (30 l)	17.7	2.0	0.6	0.07	0.02
4 River water downstream of sewage outlet (46 l)	33.5	0.4	—*	0.008	—

*Present, not measured.
†Standard alcohol ethoxylates plus alkylphenol ethoxylates.
Reprinted with permission from Jones and Nickless.[100] Copyright (1978) Elsevier Science Publishers B.V.

The results obtained by applying these procedures to sewage works and river water samples are shown in Table 79.

Infrared spectra of secondary alcohol ethoxylates and alkylphenol ethoxylates isolated from samples are shown in Figure 138. The general appearance of both types is very similar with the broad strong peak at 1100–1120 cm^{-1} characteristic of a polyethoxylate grouping. The only clearly recognizable difference between them is the very sharp aromatic peak present in the alkylphenol ethoxylate spectrum at 1500–1505 cm^{-1}. n.m.r. spectroscopy was able to distinguish between secondary alcohol ethoxide and primary alcohol ethoxylate.

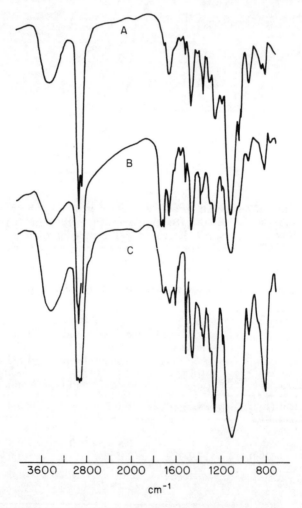

Figure 138 i.r. spectra of (A) sewage influent extract; (B) sewage effluent extract; and (C) Roberts river water extract. Reprinted with permission of Jones and Nickless.[100] Copyright (1978) Elsevier Science Publishers B.V.

Thin-layer chromatography

The method used by Patterson et al.[87] in biodegradation studies is described below.

A measured amount of the solution containing the polyethoxylated compound (usually in acetone or chloroform) is spotted on to the starting line of a silica gel thin-layer chromatographic plate. The plate is then developed using one of the two solvents described below depending on the information required. After development the plate is dried at 100 °C cooled, and sprayed with a modified Dragendorff reagent. The compounds show up as deep pink spots or streaks against a pale yellow background. The sensitivity is reasonably high and detection limits for nearly all of the model compounds are between 0.5 and 5 μg.

Solvent A

This solvent is composed of 40 parts of ethyl acetate, 30 parts of water and 30 parts of acetic acid. Secondary alcohol ethoxylate and alkylphenol ethoxylates produce compact spots, all with approximately the same R_f value of 0.8. Polyethylene glycols on the other hand, appear as short streaks of much lower R_f value, the value decreasing with the increasing number of ethoxide units. Quantitative results are obtained by comparison of the colour intensity of the unknown with a range of standards.

Solvent B

This solvent is composed of 70 parts of ethyl acetate, 15 parts of acetic acid and 16 parts of water. The solution is not polar enough to affect movement of polyethylene glycols which remain near the starting line. The situation is different for the secondary alcohol ethoxylates and alkylphenol ethoxylates which are partially or completely resolved in terms of the polyethoxide chain lengths, the shorter the polyethoxide chain the higher the R_f value. Alkylphenol ethoxides containing different numbers of ethylene oxide units show a well defined series of spots, which may spread over the whole length of the plate, if the chain length distribution is broad. Secondary alcohol ethoxylates with different numbers of ethylene oxide units on the other hand do not resolve into spots, but each produces a streak which varies in intensity along the length.

Miscellaneous methods

Various attempts have been made to develop titration methods for non-ionic surfactants. Wickbold[102] has described a method in which the surfactants are extracted from aqueous samples (mixed with sodium bicarbonate) by the passage of nitrogen (50–60 l h^{-1}) saturated with ethyl acetate through the sample, covered by a layer of ethyl acetate for 5 min. After evaporation of the extract and dissolution of the residue in 5 ml of methanol and 40 ml of water, and adjustment of the pH to 4–6, 30 ml of precipitation reagent (a 1:2 mixture of 29% aqueous barium chloride and a solution containing 1.7 g of basic bismuth(III)nitrate, 65 g of potassium iodide, and 220 ml of acetic acid per litre) is added. The precipitate formed is dissolved in 30 ml of hot ammonium tartrate solution (12.4 g of tartaric acid and 18 ml of 25% aq. ammonia per litre) and after adjustment of the pH to 4.6, is titrated potentiometrically (platinum and calomel electrodes) with 0.5 mN-pyrrolidine-1-carbodithioate. Tsuji et al.[103] have described a potentiometric titration method for non-ionic surfactants utilizing cholinesterase. This method utilizes the phenomenon that non-ionics weaken the inhibitory effect of the cholinesterase butyryl-thiocholine system.

La Noce[1] has described methods for determining non-ionic surfactants gravimetrically, turbidimetrically, colorimetrically by precipitation as a metal complex and determination of the metal (Mo, W, Bi, or Co), by thin-layer chromatography, or by infrared spectroscopy.

McCracken and Datyner[104] have determined non-ionics in aqueous solution by a method based on foam entrainment. The method involves measuring the weight of foam plus liquid entrained by passage of a regulated air flow during a given time. This weight can be correlated with the initial concentration of detergent, e.g. with measurement at 25 °C, the relationship is almost rectilinear for 0.015 to 0.06% of detergent.

Wickbold[17] has described a method for separating non-ionics from surface water at a gas–water interface in which a 4 litre sample is treated with sodium bicarbonate (10 g) in a glass tower (830 mm × 99 mm) and ethyl acetate (200 ml) is used for the collection of the surfactant from a stream of nitrogen ($90–100 \, l \, h^{-1}$). After evaporation of the combined organic phases, the surfactant is determined by any suitable procedure.

Continuous automated extraction techniques for removal of low concentrations of non-ionics from water have been developed.[105, 106] In this system the sample is mixed with solvent and passed through a vertical coil partly packed with glass ballotini. Separation of the phases begins in the upper (empty) part of this coil and is completed by means of a double-weir separator, liquid control being by means of phase-sensing electrodes and solenoid valves. The system has been used for extracting 5–25 μg amounts of non-ionic detergent from effluents containing concentrations of up to 0.1 ppm, with typical extraction efficiencies of 95–100%.

Linhart[107] has described a method for separating detergents into cationic, anionic, and non-ionic types on a mixed bed of a cation exchange resin (e.g. Lewatit S1020, H_+ form) and an anion-exchange resin (Lewatit M5020, OH^- form). He then determined the detergents by a polarographic procedure. This method is based on the damping of the polarographic maximum of oxygen by the surfactants and on the fact that in mM potassium chloride it is not the molarity but the percentage by weight of the surfactants that is proportional to this damping. This proportionality (which is rectilinear over the range $0–100 \, mg \, l^{-1}$) applies equally to anionic, cationic, and non-ionic surfactants containing up to 90 ethoxy units and to polyoxyethylene glycols containing >400 ethoxy units.

The British standard method[24] for non-ionic detergents is based on either colorimetry by extraction with chloroform and measurement of the colour formed with ammonium cobaltothiocyanate or by thin-layer chromatography after extraction with chloroform with location of the spots with a modified Burger reagent and comparison with standards.

REFERENCES

1. La Noce, I. T. *Riv. ital. Sostanze grasse*, **46**, 673 (1969).
2. Le Bihan, A. and Courtot Coupez, J. *Bull. Soc. Chim. France* **1**, 406 (1970).
3. Courtot Coupez, J. and Le Bihan, A. *Analyst Lett.*, **2**, 211 (1969).
4. Le Bihan J. and Courtot Coupez, J. *Analyst Lett.*, **10**, 759 (1977).
5. Crisp, P. T., Eckert, J. M. and Gibson, N. A., *Anal. Chem. Acta*, **78**, 391 (1975).
6. Crisp, P. T., Eckert, J. M., Gibson, N. A., Kirkbright, G. I. and West, T. S. *Anal. Chim. Acta*, **87**, 97 (1976).
7. Gagnon, M. J. *Wat. Res.*, **13**, 53 (1979).
8. Stroehl, G. W. and Kukzak, D. Z. *Z. Anal. Chim.*, **242**, 88 (1968).
9. Abbot, D. C. *Analyst (Lond.)*, **87**, 286 (1962).
10. Tonkelar, C. and Bergshoeff, G. *Wat. Res.*, **3**, 31 (1969),
11. Taylor, C. G. and Fryer, B. *Analyst (Lond.)*, **94**, 1106 (1969).
12. Taylor, C. G. and Williams, J. *Anal. Chim. Acta,* **69**, 373 (1974).
13. Taylor, C. G. and Waters, J. *Anal. Chim. Acta,* **69**, 363 (1974).
14. Taylor, C. G., Waters, J. and Williams, P. V. *Anal. Chim. Acta,* **69**, 373 (1974).
15. Wudzinska, I. and Ponikowska, A., *Chemia. analit.*, **16**, 375 (1971).
16. Kazarac, Z., Cosovic, B. and Branica, M. *Mar. Sci. Commun.*, **1**, 147 (1975).
17. Wickbold, R. *Tenside Deterg.*, **8**, 61 (1971).
18. Den Tonkelaar, W. A. M. and Bergshoeff, G. *Wat. Res.*, **3**, 31 (1969).
19. Longwell, J. and Maniece, W. D. *Analyst (Lond.)*, **80**, 167 (1955).
20. Södergren, A. *Analyst (Lond.)*, **91**, 113 (1966).
21. Petts, K. W., and Parkes, D. Water Research Centre Report TR 75 *Determination of Anonic Surfactants within a Concentration Range 0–1.0 mg l^{-1} in Sewage Effluents and Waters by Autoanalysis.* May 1978.
22. Wang, L. K. *J.Am. Water Wks Ass.*, **67**, 19 (1975).
23. Rodier, J. *L'Analyse Chimique et Physico-Chimique de l'Eau.*, 326 (1971).
24. British Standards Institution *BS 2690 Part II*, (1971).
25. Begin, C., Tudoriu, E. and Rusu, V. *Chemia. analit.*, **2**, 46 (1972).
26. Wang, L. K. and Langley, D. F. *Arch. Environ. Contam. Toxicol.*, **5**, 447 (1977).
27. American Public Health Association, *Standard Methods for the Examination of Water and Wastewater*, 13th edition, pp. 339–346 (1971).
28. ASTM Standards, Part 23, *Water Atmospheric Analysis*, American Society for Testing and Materials, Philadelphia, Pa., 619 (1972).
29. Janeva, B. and Borissova-Pangarova, R. *Talanta*, **25**, 279 (1978).
30. Bhat, S. R., Eckert, J. M. and Gibson, N. A. *Anal. Chim. Acta*, **116**, 191 (1980).
31. Kalenichenko, K. D., Mal'tev, V. N. and Gamayunov, N. S. *Hydrobiological Journal,* **15**, 73 (1979).
32. Ogden, C. P., Webster, H. L. and Halliday, J. *Analyst (Lond.)*, **86**, 22 (1961).
33. Frazee, C. Z. and Crider, R. O. *J. Am. Oil Chem. Soc.*, **41**, 334 (1964).
34. Machlar, C. Z., Cripps, J. M. and Borkley, A. F. *J. Water Pollut. Control Fedn,* **39**, R92 (1967).
35. Ihara, I., Ozaki, M. Yukagaka Ishiguro, *Riv. ital. Sostanze grasse*, **19**, 1043 (1970).

36. Oba, K., Miura, K., Sekiguchi, H., Yagi, R. and Mori, A. *Wat. Res.,* **10**, 149 (1976).
37. Nagai, T., Hashimoto, S., Mori, A. and Yamane, I. *Kagyo Kakaku Zasshi,* **73**, 1968 (1970).
38. Ambe, Y. and Hanya, T. *Japan Analyst,* **21**, 252 (1972).
39. Longwell, N. and Maniece, O. *Anal. Abstr.,* **2**, 2244 (1955).
40. Sallee, O. *Anal. Chem.,* **28**, 1822 (1956).
41. Ambe, Y. *Environ. Sci. Technol.,* **7**, 542 (1973).
42. Rand, M. C., Greenberg, A. E. and Taras, M. J. *Standard Methods for the Examination of Water and Wastewater,* American Public Health Association/American Water Works Association/Water Pollution Control Federation, 14th Ed., 603 (1975).
43. Wang, L. K., Kao, S. I., Wang, M. H., Kao, J. K. and Loskhin, A. L. *Ind. Engng Chem. Prod. Res. Devel.,* **17**, 186 (1978).
44. Wang, L. K., Wang, M. H. and Kao, J. T. *Water, Air and Soil Pollution,* **9**, 337 (1978),
45. Wang, L. K., Yang, J. Y., Ross, R. G. and Wang, M. H. *Wat. Res. Bull.,* **11**, 267 (1975).
46. Wang, L. K., Yang, J. Y. and Wang, M. H. *Proc. of 28th Industrial Waste Conference USA,* 76–82 (1973).
47. Bock, R. and Reisinger, H. *Z. Anal. Chem.,* **241**, 10 (1968).
48. Wickbold, R. *Fette Seifen Anstrichmittell,* **57**, 164 (1955).
49. Zvonaric, T., Zutic, V. and Branica, M. *Thalassia Jugoslavica,* **9**, 65 (1973).
50. Taylor, C. G. and Waters, J. *Analyst (Lond.),* **97**, 533 (1972).
51. Hart, J. P., Franklin Smythe, W. and Birch, B. J. *Analyst (Lond.),* **104**, 853 (1979).
52. Hon-Nami, H. and Hanya, T. *J. Chromat.,* **161**, 205 (1978).
53. Swisher, R. D. *J. Water Pollut. Control. Fedn,* **35**, 877 (1963).
54. Huddleston, R. L. and Allred, R. C. *Dev. Ind. Microbiol.* **4**, 24 (1963).
55. Swisher, R. D. *J. Water Pollut. Control. Fedn,* **35**, 1557 (1963).
56. Leidner, H., Gloor, R. and Wuhrmann, O. *Tenside Deterg.,* **13**, 122 (1976).
57. Taylor, P. W. and Nickless, G. *J. Chromat.,* **178**, 259 (1979).
58. Riley, J. P. and Taylor, D. *Anal. Chim. Acta,* **46**, 307 (1969).
59. Longwell, N. and Maniece, O. *Anal. Abstr.,* **2**, 2244 (1955).
60. Uchiyama, M. *Wat. Res.,* **11**, 205 (1977).
61. Uchiyama, M. *Wat. Res.,* **13**, 847 (1979).
62. Wang, L. K. *Evaluation of Improved Two-phase Titration Methods and a Field Test Kit for Analysing Ionic Surfactants in Water and Wastewater.* Calspan Corporation, Buffalo, New York, Technical Report No. ND-5296-M-3, 66 pp. (1973).
63. Wang, L. K. and Pek, S. L. *Ind. Engng Chem.,* **14**, 308, (1975).
64. Waters, J. and Kupfer, W. *Anal. Chim. Acta,* **85**, 241 (1976).
65. Kawase, J. and Yamanaka, M. *Analyst (Lond.),* **104**, 750 (1979).
66. Courtot Coupez, J. and Le Bihan, A. *Anal. Lett.,* **2**, 567 (1969).
67. Courtot Coupez, J. and Le Bihan, A. *Riv. ital. Soztanze grasse,* **20**, 672 (1971).
68. Grieff, O. *Anal. Abstr.,* **13**, 3650 (1966).
69. Arpino, A., Calatroni, C. and Jacini, G. *Riv. ital. Sostanze grasse,* **51**, 140 (1974).
70. Burttschell, O. *Anal. Abstr.* **14**, 6490 (1967).
71. Greff, R. A., Setzkorn, E. A. and Leslie, W. D. *J. Am. Oil Chem. Soc.,* **42**, 180 (1965).
72. Crisp, P. T., Eckert, J. M. and Gibson, W. A. *Anal. Chim. Acta,* **104**, 93 (1979).
73. Crabb, N. T. and Persinger, H. E. *J. Am. Oil Chem. Soc.,* **45**, 611 (1968).
74. Belen'kii, S. M., Bel'fon, N. Y., Glazyrina, L. P. and Nekitina, N. A. *Zav. Lab.,* **34**, 1441 (1968).
75. Brown, N. and Hayes, O. *Anal. Abstr.,* **3**, 1432 (1956).

76. Baleux, B. *Cr. Lebd. Séanc. Acad. Sci. Paris. C*, **274**, 1617 (1972).
77. Favretto, L. and Tunis, F. *Analyst (Lond.)*, **101**, 198 (1976).
78. Favretto, L., Stancher, B. and Tunis, R. *Analyst (Lond.)*, **103**, 955 (1978),
79. Favretto, L., Stancher, B. and Tunis, F. *Analyst (Lond.)*, **104**, 241 (1979).
80. Favretto, L., Stancher, B. and Tunis, F. *Analyst (Lond.)*, **105**, 833 (1980).
81. Stancher, B., Tunis, F. and Favretto, L. *J. Chromat.* **131**, 309 (1977).
82. Favretto, L. *Annali. Chim.*, **66**, 621 (1976).
83. Favretto, L. and Marletta, P. G. *Annali. Chim.*, **63**, 807 (1973),
84. Stancher, B., and Favretto, L. *J. Chromat.*, **150**, 447 (1978).
85. Stancher, B. and Tunis, F. *Analyst (Lond.)*, **104**, 241 (1979).
86. Crabb, N. T. and Persinger, H. F. *J. Am. Oil Chem. Soc.*, **41**, 752 (1964).
87. Patterson, S. J., Hunt, E. C. and Tucker, K. B. E. *J. Proc. Inst. Sew. Purif.*, 190 (1966).
88. Patterson, S. J., Scott, C. C. and Tucker, K. B. E. *J. Am. Oil. Chem. Soc.*, **44**, 407 (1967).
89. Favretto, L., Petroldi Marletta, G., and Fabrielli, L. *Annali. Chim.*, **62**, 478 (1972).
90. Dozanska, W. *Roczn.panst.Zakl. Hig.*, **20**, 137 (1969).
91. Hey, A. E. and Jenkins, S. H. *Wat. Res.*, **3**, 887 (1969).
92. Heatley, N. G. and Page, E. J. *Water Sonit.*, **3**, 46 (1952).
93. Rosen, M. J. and Goldsmith, H. A. *Systematic Analyses of Surface Active Agents*, 2nd edn, Wiley, New York (1972).
94. Huber, J. F., Kolder, F. F. M. and Miller, J. M. *Anal. Chem.*, **44**, 105 (1972).
95. Otsuki, A. and Shiraishi, H. *Anal. Chem.*, **51**, 2329 (1979).
96. Cassidy, R. M. and Niro, C. M. *J. Chromat.*, **126**, 787 (1976).
97. Nickless, G. and Jones, P. *J. Cheomat.*, **156**, 87 (1978).
98. Musty, P. R. and Nickless, G. *J. Chromat.*, **89**, 185 (1974).
99. Musty, P. R. and Nickless, G. *J. Chromat.*, **120**, 369 (1976).
100. Jones, P. and Nickless, G. *J. Chromat.*, **156**, 99 (1978).
101. Nadeau, H. G. and Waszeciak, P. H. *Non-ionic Surfactants*. Marcel Dekker, New York, p. 906 (1967).
102. Wickbold, R. *Tenside*, **9**, 173 (1972).
103. Tsuji, K., Tanaka, H. and Konishi, K. *J. Am. Oil Chem. Soc.*, **52**, 111 (1975).
104. McCracken, J. R. and Datyner, A. *Chem. Ind.* **6**, 266 (1974).
105. Sawyer, R., Stockwell, P. B. and Tucker, K. B. E. *Analyst (Lond.)*, **95**, 879 (1970).
106. Sawyer, R., Stockwell, P. B. and Tucker, K. B. E. *Analyst (Lond.)*, **95**, 284 (1970).
107. Linhart, K. *Tenside*, **9**, 241 (1972).
108. Wang, L. K., Panzardi, P. J., Pedro, J., Schuster, W. W. and Autenbach, D. B. *J. Environ. Hlth*, **38**, 159 (1957).

Chapter 3
Pesticides and PCBs

ORGANOCHLORINE INSECTICIDES

Work on the determination of chlorinated insecticides has been almost exclusively in the area of gas chromatography using different types of detection systems, although a limited amount of work has been carried out using liquid chromatography and thin-layer chromatography.

By its nature gas chromatography is able to handle the analysis of complex mixtures of chlorinated insecticides. It is not surprising therefore, that much of the published work, discussed in this section, is concerned with the analyses of mixtures of different types of chlorinated insecticides as found in environmental samples. Work on the determination of individual insecticides is reported at the end of this section. Work on the determination of polychlorinated biphenyls and mixtures of these with chlorinated insecticides is reported in the section on PCB and PCB–organochlorine pesticide mixtures.

Gas chromatography of mixtures of chlorinated insecticides

Wilson and Forester[1,2] have discussed the determination of aldrin, chlordane, dieldrin, endrin, lindane, o,p and p,p' isomers of DDT and its metabolites, mirex and toxaphene in sea water and molluscs. The US Environmental Protection Agency has also published methods for organochorine pesticides in water and waste water. This is the official US approved method for the determination of organochlorine pesticides in an aqueous medium. It includes advice on the collection of the sample, choice of correct glassware, preparation of stock standards, solvents, and reagents, concentration of samples, preparation of gas chromatographic columns, and injection devices, and gives details of the gas chromatographic analytical method. The Food and Drug Administration (USA)[3] has conducted a collaborative study of a method for multiple organochlorine insecticides in fish. Earlier work by Wilson et al.[2,4] in 1968 has indicated that organochlorine pesticides were not stable in sea water as indicated in Table 80. These conclusions were confirmed in the 1974 work.

Table 80 Stability of pesticides in natural sea water (salinity 29.8 ppt)

Pesticide	Days after start of experiment					
	0 ppb	6 ppb	17 ppb	24 ppb	31 ppb	38 ppb
p,p'-DDT	2.9	0.75	1.0	0.27	0.18	0.16
p,p'-DDE*		0.096	0.95	0.065	0.034	0.037
p,p'-DDD*			0.081	0.041	0.038	0.037
Aldrin†	2.6	0.58	0.096	<0.01	<0.01	<0.01
Dieldrin*		0.74	1.0	1.0	0.75	0.56
Malathion	3.0	<0.2	<0.2			
Parathion	2.9	1.9	1.25	1.0	0.71	0.37

*Metabolites of parent compound.
†From the seventeenth day onward, 2 unidentified peaks appeared on the chromatographic charts after aldrin had eluted.

Since petroleum ether was the solvent used in these earlier studies for extracting the DDT from sea water, Wilson and Forester[1] initiated further studies to evaluate the extraction efficiencies of other solvent systems. Duplicate one gallon bottles of clear glass, containing 3.5 litres of sea water, were fortified with 10.5 μg of p,p'-DDT in 350 ml of acetone to yield a concentration of 3.0 ppb. Duplicate 500 ml samples were taken from each bottle and extracted with one of the following solvents: three 50 ml portions of petroleum ether, two 50 ml portions of 15% ethyl ether in hexane followed by one 50 ml of hexane, or three 50 ml portions of methylene chloride. All solvents were dried with sodium sulphate, concentrated to an appropriate volume and analysed by electron capture gas chromatography using at least two 180 cm × 2 mm (i.d.) columns of different liquid phases. The following columns were used: DC-200, QF-1, EGS, OV-101, mixed DC-200/QF-1, and mixed OV-101/OV-17. Just prior to extraction, all samples were fortified with o,p'-DDE to evaluate the integrity of the analysis. The recovery rates of o,p'-DDE in all tests were greater than 89%, indicating no significant loss during analyses.

After initial sampling, the bottles were sealed and incubated at 20 °C under controlled light conditions (12 hours light, 12 hours dark). Duplicate samples of 500 ml were extracted at various time intervals. Tables 81–84 show the average percentage recovery of p,p'-DDT extracted from duplicate sea water (salinity 16–21 ppt) or distilled water samples up to 14 days after initiation of the experiment. p,p'-DDE was the only metabolite measured and since it never exceeded 2% of the parent compound it is not included in the percentage recoveries. Table 81 shows that immediately after the sea water was fortified with 3.0 ppb of DDT all solvent systems removed 93% of the DDT. After 6 days of incubation this level of recovery was not observed with any of the solvents tested. However, methylene chloride was more efficient than petroleum ether or 15% ether in hexane. Part of this experiment was repeated with sea water (salinity 16 ppt) and incubated for 4 days with similar results (Table 82).

Table 81 Percentage recovery of p,p'-DDT from sea water by different extraction solvents

	Extraction solvent		
Day	Petroleum ether	15% Ethyl ether in hexane	Methylene chloride
0	93	93	93
6	67	66	76

Table 82 Percentage recovery of p,p'-DDT from sea water by petroleum ether and methylene chloride

0	90	95
4	67	85

Table 83 Percentage recovery of p,p'-DDT from sea water and distilled water by petroleum ether and methylene chloride

	Sea water		Distilled water	
Day	Petroleum ether	Methylene chloride	Petroleum ether	Methylene chloride
0	90	94	90	91
7	58	78	90	91
14	46	68	94	92

Table 84 Percentage recovery of p,p'-DDT from sea water incubated under different light and temperature conditions

Day	12 hours light and 12 hours dark at 20 °C	Dark at 5 °C
0	87	88
7	69	81
14	68	86

An experiment was performed with sea water (salinity 21 ppt) and distilled water using petroleum ether and methylene chloride. Table 83 shows that immediately after fortification, recoveries were greater than 90% for water and sea water. After 14 days, similar recoveries were observed only in distilled water. In sea water however, there was 90% and 28% reduction in recovery with petroleum ether and methylene chloride respectively. Since distilled water is devoid of particulate matter, this study suggests that DDT may be absorbed or adsorbed to plankton or particulate matter in sea water and the sorbed material was not removed resulting in low recoveries of DDT. This would explain the initially high extraction efficiency of DDT followed by the decline in recovery as DDT was associated with the particulate phase. Since methylene chloride was the most polar solvent used, it would have a greater affinity for removing the sorbed DDT. In another test, duplicate bottled containing sea water (salinity 20 ppt) and DDT were incubated under controlled lighting conditions at 20 °C and another set incubated at 5 °C without light. Both were extracted with methylene chloride at various time intervals. Table 84 shows low recovery at 14 days under controlled lighting conditions. However, those samples incubated at 5 °C in total darkness did not show a significant decrease in recovery rate. Since the metabolic activity of plankton was probably inhibited under these temperature and lighting conditions, these results suggest that DDT may be absorbed rather than adsorbed by plankton.

These experiments support the work of other investigators in that DDT and other insecticides are extremely hydrophobic and can be easily absorbed or adsorbed by suspended matter from liquid solutions. Results obtained using petroleum ether as extraction solvent[2] support the concept that physical or chemical transformations of insecticides altered the extraction efficiencies of the solvent and prevented complete recovery of the compounds. It is difficult to relate laboratory findings directly to those of the estuary of open oceans. However, the laboratory data illustrate clearly some problems that could be encountered in monitoring sea water for insecticide pollution. The conventional analyses of water samples by liquid–liquid extraction techniques may provide invalid data if suspended matter is not considered. Standardized methods are needed to analyse the water column and suspended material separately.

Analysis of mollusc samples[1]

Mollusc samples were dehydrated by mixing them with a 9:1 mixture of anhydrous sodium sulphate and Quso (a micro fine silica). They could be held at room temperature for up to 15 days without loss or degradation of chlorinated hydrocarbon. The tissues of oysters were homogenized. Approximately 30 g of the homogenate was added to a second Mason jar and blended with a 9:1 mixture of sodium sulphate and Quso. By alternately chilling and blending a free-flowing powder was obtained. The blended sample was wrapped in aluminium foil and shipped to the laboratory. Upon receipt of the sample, it was weighed and extracted in a Soxhlet apparatus for 4 hours with petroleum ether. The extracts were then purified by concentrating and transferring the extract to 250 ml separatory funnels. The extracts were diluted to 25 ml with petroleum ether and partitioned with two 50 ml portions of acetonitrile previously saturated with petroleum ether. The acetonitrile was evaporated to dryness and the residue eluted from a Florisil column.[5] In this technique, increasing proportions of ethyl ether to petroleum ether were used to elute fractions containing increasingly polar insecticides. The extracts were analysed by gas chromatography as discussed above. Table 85 shows typical recovery rates of DDT and its

Table 85 Percentage recovery of insecticides from fortified oyster samples

Insecticide	Actual (ppm)	Found (ppm)	Per cent recovery
DDE	0.333	0.026	79
	0.033	0.026	79
	0.34	0.30	88
	0.34	0.29	85
DDD	0.067	0.061	91
	0.067	0.064	96
	0.70	0.67	96
	0.70	0.63	90
DDT	0.10	0.087	87
	0.10	0.094	94
	1.0	0.95	94
	1.0	0.92	92

metabolites from fortified oyster samples. The values were adjusted to account for naturally occurring DDT residues. The lower limit of detection for a 30 g mollusc sample was 0.010 parts per million (milligrams per kilograms). Residues were reported on a wet weight basis without adjustment for recovery rates.

Deubeurt[6] has discussed the sources of compounds which interfere in the analyses in water and soil extract for DDT and dieldrin by gas electron capture chromatography. Nitration of these insecticides eliminated their peaks so that background interference peaks could be studied.

Croll[7] has given details of the use of back-flushing with electron capture gas chromatography for the determination of organochlorine insecticides in water. Attention is paid to equalization of column resistances under operating and back-flushing conditions; baseline drift is thereby minimized. The system has been successfully used (with a variety of stationary phases, temperature ranging from 25 to 225° and nitrogen flow rates from 25 to 200 ml min^{-1}).

Kahanovitch and Lahav[8] used gas chromatography to study the occurrence of 12 organochlorine insecticides in water sources in Israel (αBHC, γBHC, o,p'-DDD, p,p'-DDT, dieldrin, endrin, heptachlor, heptachlor epoxide, Thiodan, Treflan, diazinon, and malathion).

Water samples of 10 litres were extracted at the sampling sites with about 100 ml of hexane. Special precautions were taken to prevent contamination during the extraction. After drying over anhydrous sodium sulphate, the extract was evaporated in a Buchi rotavapor concentration apparatus: 2–3 ml of hexane were then added to redissolve the insecticides. The clean-up procedure used was that described by Law and Goerlitz.[9] Effluent volumes of 0.5–2 ml were collected and their amounts determined gravimetrically. The total effluent volume was usually 10 ml. Pesticide mixtures of known concentrations were used to identify the eluent fraction in which each compound was eluted. The insecticide determination was made with a gas chromatograph equipped with an electron capture detector. The glass columns used were of 4 mm i.d. and 180 cm length and packed with (1) 5% OV-2% QF-1, and (2) 5% QF-1. Nitrogen was used as the carrier gas. The temperatures were: injection point, 210 °C; column oven, 190 °C; detector, 200 °C. Concentrations were calculated from the peak height, since a linear relationship was found between the two. Due to the large water volumes used in the method, insecticide concentrations lower than 1 ng l^{-1} could be detected without difficulty.

Solvent extraction in a continuose liquid–liquid mode[25,26] and batch sample extraction[27,28] have been used for the concentration of chlorinated insecticides from water samples.

Various workers[10-18] have investigated the use of activated carbon filters to concentrate chlorinated insecticides from water samples prior to gas chromatographic analysis. Quantitative values reported by these methods must be considered minimal because of lack of control on adsorption and desorption characteristics of the carbon filter bed and also because of bacterial breakdown of adsorbed organics during the sampling period. Booth et al.[19] using the carbon adsorption method determined that maximum organic adsorption efficiencies are achieved at a reduced flow rate (not greater than 120 ml min^{-1}) and total throughput volume of 1500 l or less. In a study designed to determine carbon adsorption–desorption efficiency and reproducibility, it was determined by Eichelberger and Lichtenberg[20] that the carbon is useful for some organochlorines, giving a 70–85% recovery rate, and less dependable for others, including DDT, which was determined to have an average 37% recovery rate.

Eichelberger and Lichtenberg[21] used carbon adsorption prior to gas chromatography for the determination of methoxychlor, lindane, endrin, dieldrin, chlordane, DDT, heptachlor epoxide and endosulfan in water samples. The eluate from the carbon were extracted with ethyl ether–hexane (3:17), and the carbon was dried and extracted with chloroform.

Aue et al.[22] used support bonded silicones for the extraction of 10 to 20 parts of lindane, heptachlor, aldrin, heptachlor epoxide and dieldrin from 10^{12} parts of water prior to gas chromatography. This absorbent was packed on to a glass tube (35 × 1 cm), and 10 litres of sample (or of pure water treated with known compounds) was passed through the column at 50–55 ml min^{-1}; the column was then dried by passage of nitrogen. The sorbed compounds were eluted with pentane (2 × 5 ml), and a portion of the elute was injected directly into a borosilicate-glass column (170 cm × 3.5 mm) packed with 1.5% of QF-1 plus 2% of OV-17 on Chromosorb W-HP (100–120 mesh). The column was operated at 185 °C for chlorinated insecticides with nitrogen (60 ml min^{-1}) as carrier gas and a ^{63}Ni electroncapture detector. Burnham[23] used ion-exchange resins (Amberlite XAD). However, initial results confirmed an earlier report[24] that, although those resins are very efficient for the absorption of organochlorine insecticides, the desorption procedure gave poor recoveries. Harvey[29] has also examined ion-exchange resins (Amberlite XAD-2) as an absorption medium for chlorinated insecticides in sea water.

Brodtmann[30] carried out a long-term study on the qualitative recovery efficiency of the carbon adsorption method vs. that of a continuous liquid–liquid extraction method for several chlorinated insecticides. Comparative results obtained by electron capture gas chromatography indicate that the latter method may be more efficient. Samples of river water for analysis by the carbon adsorption method were collected at a rate of approximately 120 ml min^{-1} for a 7 day period in each case. At the end of the weekly sampling period, the total throughput volume was recorded. The carbon chloroform extract was then obtained by chloroform extraction of the carbon in a modified Soxhlet apparatus[31] for 36 h using glass distilled, pesticide grade petroleum ether. The neutral fraction of the chloroform extract was then prepared for gas chromatography by the methods of Breidenbach.[31] Brodtmann[30] used a continuous liquid–liquid extraction apparatus as described by Kahn and Wayman[25] and Goldberg et al.[26] for the extraction of non-polar solutes from river water. Pesticide grade petroleum ether, used in all cases, was recycled internally (initial solvent charge of 350 ml), thereby continuously exposing essentially fresh solvent to the river water. The throughput rate of river water was set at 33 ml ml^{-1} for the sampling period of 7 days. To further enhance solute recoveries, three extractors were connected in series and solvent charges were pooled prior to concentration and clean-up. The pooled extracts were then reduced in volume over a steam bath in flasks equipped with three-ball Snyder columns. A Florisil clean-up step using sequential elutions with 6% and 15% ethyl ether–petroleum ether solutions was employed (the Florisil was activated at 130 °C). A 300 mm × 20 mm i.d. chromatographic tube was prepared by the addition of 20 mm × 120 mm plug of Florisil topped by a 15 mm × 20 mm plug of anhydrous sodium sulphate and a wad of extracted glass wool. The Florisil was settled by tapping the tube lightly. This packing was then prewetted with 35 ml of hexane which was then discarded. When the last of the hexane reached the top of the Florisil packing, the sample was quantitatively transferred to the column. The sample was then eluted with 200 ml of 6% v/v ethyl ether–petroleum ether solution. When the last of the 6% eluate had reached the top of the packing, 200 ml of 15% v/v ethyl ether–petroleum ether solution was added for the elution of dieldrin and endrin. Elution rates for both procedures were adjusted to approximately 5 ml min^{-1}. At this stage the 6% eluate was ready for gas chromatography, but the 15% eluate required a further clean-up step on a magnesium oxide–Celite 545 column. The magnesium oxide was prepared for use by forming a slurry of 200 g magnesium oxide in distilled water. The slurry was then heated on a steam bath for 30 min, vacuum filtered, and dried overnight at 130 °C. A blender was used to pulverize the dried filtrate which was then mixed in a 1:1 ratio (wt/wt) with Celite 545. A second 300 mm × 20 mm i.d. chromatographic tube was attached to a 250 ml vacuum flask. A plug of glass wool was placed in the bottom of this tube, then 10 g of 1:1 magnesium oxide–Celite 545 was added under full vacuum (approximately 640 mmHg) to pack it tightly. The vacuum line was then bled so that a 35 ml petroleum ether wash eluted at a rate of 15–20 ml min^{-1}. It was not necessary to discard this wash. After refluxed evaporation to 10 ml, the 15% eluate was quantitively transferred to the magnesium oxide–Celite 545 columns and eluted with 100 ml petroleum ether at the 15–20 ml min^{-1} rate with partial vacuum.

This eluate was again evaporated to 15 ml, by use of a Snyder column-equipped flask and was then ready for gas chromatographic analysis as described below.

A dual-column gas chromatograph equipped with two tritium electron capture detectors was employed by Brodtmann.[30] Both columns were acrylic glass, 1.83 m long by 0.32 cm i.d. Column A was packed with 5% DC-260 on 80/100 DCMS Chromosorb W. Flow rate of carrier gas (5% methane–argon) though this column was 80 ml min^{-1}. Column B was packed with 1.5% OV-17/1.95% QF-1 on 80/100 Chromosorb W DCMS support. Flow rate of carrier gas (5% methane–argon) though this column was 50 ml min^{-1}. Both pairs of injectors, columns, and detectors were maintained at, respectively, 212, 184, and 204 °C.

Quantitation of unknown samples was accomplished by the comparison of peak heights for peaks in the unknown samples, confirmed by retention times on two different columns, to those of calibration mixtures of reference standard pesticides on both columns. Pesticide recoveries obtained by Brodtmann[30] using the carbon adsorption method and the liquid–liquid extraction method are given in Table 86. These values were obtained by spiking carbon samples followed by extraction and clean-up for the carbon adsorption method and by spiking aliquots of petroleum ether after a 7 day sample period in the extractors.

Table 86 Overall recovery rates

	Percentage recovered	
Insecticide	Carbon adsorption method	Liquid–liquid method
α-BHC	75.8	82.2
γ-Chlordane	78.4	95.2
p,p'-DDD	84.8	91.5
p,p'-DDE	55.6	93.8
o,p-DDT	74.0	95.9
p,p'-DDT	83.9	85.8
Dieldrin	85.2	95.6
Endrin	55.0	97.5
Heptachlor	52.6	83.8
Heptachlor epoxide	73.8	96.8
Lindane	77.5	92.2

Reprinted from *Journal AWWA*, **67**, 10 (October 1975), by permission. Copyright © 1975, The American Water Works Association.

The overall conclusion of this work was that use of the liquid–liquid extractor method often finds low levels of chlorinated insecticides in water samples when the carbon adsorption method is insufficiently sensitive to do so.

Aspila *et al.*[32] reported the results of an interlaboratory quality control study involving five laboratories on the electron capture gas chromatographic determination of ten chlorinated insecticides in standards and spiked and unspiked sea water samples (lindane, heptachlor, aldrin, γ-chlordane,

Table 87 Insecticide concentration (pg μl^{-1}) provided by participants

Pesticide	Sample	Lab 1 A	Lab 1 B	Lab 2 A	Lab 2 B	Lab 3 A	Lab 3 B	Lab 4 A	Lab 4 B	Lab 5 A	Lab 5 B
Lindane	1	3	3	d	d	6	3	d	d	6	6
(8/12)	2	9	15	d	d	14	15	7.9	10	9	9
	3	7	10	4	11	9	13	5.4	12	3	8
	4	10	10	d	d	13	20	8.8	14	9	8
	5	10	11	17	20	14	19		13	11	14
	6	8.08	11.56	5.6	9.9	4	6	8.3	14	8.3	10.9
Heptachlor	2	4	5	d	d	4	4	d	d	d	d
(9.2/13.8)	3	3	4	d	d	4	4	d	d	d	d
	4	4	5	d	d	2	3	d	d	d	d
	5	6	7	d	d	2	4	d	d	d	d
	6	8.17	11.57	5.7	9.8	8.1	11.8	7.8	11	8.1	11.1
Aldrin	2	9	14	d	d	12	18	5.6	6.1	5	9
(16/24)	3	8	11	12	20	15	18	7.9	15	3	10
	4	15	17	d	d	21	31	12	14	10	11
	5	14	18	22	24	17	24		21	10	18
	6	16.6	22.98	14.6	23.3	16.4	25	15	25	12.7	15.7
α-Chlordane	2	17	25	d	d	20	27	11	15	17	16
(22/23)	3	13	16	14	21	17	22	16	20	5	20
	4	19	25	31	24	26	40	14	21	17	17
	5	21	29	22	31	24	33		26	22	28
	6	22.45	32.37	18.7	28.6	23.3	35.4	22	33	20.1	25.4
γ-Chlordane	2	19	24	d	d	20	29	11	15	16	17
(24/36)	3	13	18	16	23	16	23	14	19	6	20
	4	25	30	44	28	24	37	15	22	19	18
	5	22	30	26	36	22	32		28	23	28
	6	23.38	33.67	20.0	31.2	20	31.8	23	31	21	25.7
Dieldrin	2	24	37			32	48	19	33	25	25
(25/37.5)	3	23	32	24	34	28	38	23	32	10	28
	4	22	32	25	29	27	44	17	30	25	25
	5	24	35	27	40	26	39		33	32	38
	6	25	35	22.4	34.1	23.4	36.4	26	39	26.9	37.7
Endrin	1	d	d	78	92	d	d	d	d	d	d
(30/45)	2	29	52	89	102	31	52	30	41	24	25
	3	32	43	95	92	36	52	52	92	10	29
	4	28	42	88	66	35	56	27	44	26	28
	5	31	43	60	83	34	52		43	33	45
	6	32	46	25.1	37.1	29.6	46.3	32	53	26.2	34.4
p,p'-DDT	2	25	36	d	d	37	56	d	d	29	26
(41/61.5)	3	14	23	12	22	24	33	22	33	9	35
	4	38	52	31	34	50	84	21	35	41	43
	5	52	58	34	55	53	84		d	59	75
	6	42	61	34.7	58.3	39.9	65.5	40	64	42.7	59.3
Methoxychlor	2	111	165	21	46	93	143	40	70	78	69
(95/142.5)	3	99	160	86	120	112	157	93	150	39	117
	4	127	163	99	101	110	162	91	120	95	114
	5	108	165	88	151	113	169		120	113	138
	6	100	146	77.8	126	88.3	139.2	95	150	102	139

(continued)

Table 87 *(continued)*

Pesticide	Sample	Lab 1 A	Lab 1 B	Lab 2 A	Lab 2 B	Lab 3 A	Lab 3 B	Lab 4 A	Lab 4 B	Lab 5 A	Lab 5 B
Mirex	2	5	10	d	d	40	44	4.3	2.5	9	12
(23/34.5)	3	4	6	d	d	16	21	5.0	8	3	12
	4	4	5	d	d	27	40	6	8.1	17	15
	5	24	35	23	29	20	34		26	26	31
	6	23	34	21.5	30.2	22	32.9	24	32	24.9	35.7

a Concentrations (pg μl^{-1}) contained in 1 ml organic extract obtained by analysis of 1 l test sample.
b Values in parentheses are the designed concentrations for series A and B (i.e. the A/B ratio).
c Sample I (A and B) was a natural, unspiked water.
d Unreported value (either not present, not quantitated, or not recovered).
Reprinted with permission from Aspila et al.[32] Copyright (1977) Association of Official Analytical Chemists.

α-chlordane, dieldrin, endrin, p,p'-DDT, methoxychlor, and mirex). The methods of analyses used by these workers was not discussed, although it is mentioned that the methods were quite similar to those described in the water quality Branch Analytical Methods Manual.[33] Both hexane and benzene were used for the initial extraction of the water samples. In Table 87, results obtained by the five participating laboratories are presented for five paired water samples and the pair of standards (6A and 6B). Samples 1A and 1B were unspiked natural water. Given the design details, we can evaluate the raw data through various processes. The simpler approach taken was converting each value in Table 87 to a per cent recovery. A summary of per cent recovery is given in Table 88. The standard deviation values are large because they contain all the errors, bias, and deviations accumulated from the interlaboratory comparisons. Heptachlor revealed a poor recovery (Table 88), which confirmed its degradation.[34-38] The degradation product[34-36] is known to be 1-hydroxychlordene, but at the concentrations used in these samples, sensitivity was inadequate to recognize if it was indeed present. Mirex was also excluded from the mean recoveries because the methodology of its determination is questionable. The results for paired samples identify the capabilities of each laboratory in obtaining the designed concentration ratio between series A and B. The average values for each insecticide, if it was indeed present, are shown in Table 89, as well as the standard deviation for each series. It is significant to note that each sample (2, 3, 4, and 5) in each series, contained, by design, identical concentrations. Therefore, the variety of test samples could represent a typical average water sample analysed routinely by the participants. With these items noted, the relative within-laboratory precision of all five laboratories can easily be identified from their adherence to the design ratio (B/A) and from the standard deviations calculated for each insecticide in each series. The ratio in most cases tends to be biased low. A low bias could be anticipated because the lower concentration

Table 88 Average per cent recovery of insecticides

Insecticide	Sample: Statistic	1 (as is)	2 (stripped)	3 (filtered)	4	5 (distd water)	6 (std)
Lindane	x̄	113	83	117		130	89
	s.d.	31	25	34		30	25
Heptachlor	x̄	c	c	c		c	81
	s.d.						8
Aldrin	x̄	49	55	83		85	93
	s.d.	20	23	22		15	14
α-Chlordane	x̄	68	59	82		94	98
	s.d.	18	17	25		10	10
γ-Chlordane	x̄	64	53	80		87	88
	s.d.	17	13	21		8	8
Dieldrin	x̄	98	85	89		102	99
	s.d.	22	21	18		13	6
Endrin	x̄	95	114	96		105	100
	s.d.	20	55	19		8	13
p,p'-DDT	x̄	52	46	89		108	100
	s.d.	36	13	30		50	4
Methoxychlor	x̄	82	96	105		110	101
	s.d.	31	25	18		14	5
Mirex	x̄	56	31	54		95	100
	s.d.	60	23	43		13	5
Mean rec. (%)		78	74	93		103	96
Mean s.d.		24	24	23		19	11

a Mean recoveries exclude data for mirex and heptachlor. All data from laboratory 2 were excluded in calculating the average per cent recoveries.
b s.d. is the standard deviation for the average % recovery (X): x̄ is the mean value of 8 results (refer to Table 87).
c No results were used because heptachlor is too unstable.
Reprinted with permission from Aspila et al.[32] Copyright (1977) Association of Official Analytical Chemists.

in series A approach the detection limit, and could not be corrected for the blank (unspiked water). For laboratory 2 in Table 89, the magnitude of the standard deviation in relation to the mean values and the inability to obtain the required B/A ratio confirmed the operational difficulties that were evident from scrutiny of the gas chromatograms submitted by this laboratory. A low single laboratory standard deviation for the B/A ratio calculated for all insecticides is a strong indicator of internal laboratory precision and is consistent with good precision for each insecticide from a given small series of results.

The total error by each laboratory could be analysed by using the method outlined by McFarren,[39] since this study included sufficient values for estimating both the in-laboratory standard deviation and the accuracy. The equation applied was given above. Because two concentrations were used (series A and B), the individual total errors for each series were evaluated and the average was computed. A summary of these values is given in Table 90. Although the per cent errors appear somewhat high, such per cent errors calculated as

Table 89 Results for paired samples (2, 3, 4, and 5 in series A and B)

Insecticide	Statistic	Lab 1 A	Lab 1 B	Lab 2 A	Lab 2 B	Lab 3 A	Lab 3 B	Lab 4 A	Lab 4 B	Lab 5 A	Lab 5 B
Lindane	$\bar{x}$	9.0	11.5	5.25	7.75	12.5	16.8	7.37	12.3	8.0	9.95
	s.d.	1.4	2.4	8.06	9.67	2.3	3.3	1.76	1.7	3.46	2.87
B/A ratio		1.28		1.48		1.34		1.66		1.22	
Aldrin	$\bar{x}$	11.5	15.0	8.5	11.0	16.3	22.8	8.5	14.0	7.0	12.0
	s.d.	3.5	3.2	10.6	12.6	3.8	6.2	3.2	6.1	3.6	4.1
B/A ratio		1.30		1.29		1.4		1.65		1.71	
α-Chlordane	$\bar{x}$	17.5	23.8	16.8	19.0	21.8	30.5	13.7	20.5	15.3	20.3
	s.d.	3.4	5.5	13.2	13.2	13.3	4.0	7.8	25	45	5.0
B/A ratio		1.36		1.13		1.40		1.50		1.33	
γ-Chlordane	$\bar{x}$	19.8	25.5	21.5	21.8	20.5	30.3	13.3	21.0	16.0	20.0
	s.d.	5.1	5.7	18.4	15.5	3.4	5.9	2.1	5.5	7.3	5.0
B/A ratio		1.29		1.01		1.48		1.58		1.30	
Dieldrin	$\bar{x}$	23.3	34.0	19.0	25.8	28.3	42.3	19.7	32.0	23.0	29.0
	s.d.	1.0	2.5	12.7	17.8	2.6	4.6	3.1	1.4	9.3	6.2
B/A ratio		1.46		1.36		1.50		1.63		1.37	
Endrin	$\bar{x}$	30.0	45.0	83.0	85.6	34.0	53.0	36.3	55.0	23.3	31.8
	s.d.	1.8	4.7	15.6	15.8	2.2	2.0	13.7	24.7	9.2	9.0
B/A ratio		1.50		1.03		1.56		1.51		1.37	
p,p' DDT	$\bar{x}$	32.2	42.3	19.3	27.8	41.0	64.3	b		34.5	44.8
	s.d.	16.4	15.8	16.1	23.0	13.3	24.7			21.0	21.3
B/A ratio		1.31		1.44		1.57				1.30	
Methoxychlor	$\bar{x}$	111	160	73.5	104	107	158	74.7	115	86.8	110
	s.d.	11.7	8.0	35.4	44	9.4	11	30.0	33.2	36.3	29
B/A ratio		1.44		1.42		1.47		1.54		1.26	
Mean ratio (B/A)		1.368		1.270		1.465		1.581		1.344	
s.d. for mean ratio		0.086		0.089		0.081		0.067		0.155	

a $\bar{x}$ is the average value for series A and B and was obtained from Table 87; s.d. is the standard deviation for samples A and B.
b Insufficient data.
Reprinted with permission from Aspila et al.[32]

Table 90 Comparison of total errors (%)

Insecticide	Lab 1	Lab 2	Lab 3	Lab 4	Lab 5	Mean % total error	Mean concn of insecticides,[b] ($\mu g\,l^{-1}$)
Lindane	46	216	104	41	77	66	0.010
Aldrin	68	190	53	90	92	76	0.020
α-Chlordane	56	133	46	63	84	63	0.0275
γ-Chlordane	61	145	65	73	82	70	0.030
Dieldrin	15	147	34	29	66	36	0.031
Endrin	17	223	27	100	61	51	0.038
p,p'-DDT	93	130	58	—	107	86	0.051
Methoxychlor	29	93	29	75	80	53	0.119
Mean total for lab	48	160	52	67	81		

a Data from laboratory 2 were excluded in calculating the total error.
b Average concentration of insecticides in series A and B.
Reprinted with permission from Aspila et al.[32] Copyright (1977) Association of Official Analytical Chemists.

total error are considered acceptable for pesticide residue analysis at concentrations which approach the detection limit.

Thin polyethylene film has been employed for the adsorption of traces of chlorinated insecticides from water. Weil[40,41] used 20–25 μm thick film, and presents results for adsorption of γ-BHC, heptachlor expoxide, methoxychlor, dieldrin and DDT. Experiments[42] with '^{14}C"lindane and '^{14}C"DDT showed that adsorption of these substances on polyethylene film was not likely to be of practical use in river water analyses because of the variate effects of foreign matter. Mangani[43] used Carbopack B columns to recover chlorinated insecticides in water and soil samples. A continuous liquid–liquid extraction has been described for the monitoring of nanogram levels of eight chlorinated insecticides in River Missouri water.[44] Weil and Ernst[45] have compared three published methods, involving extraction with petroleum ether for obtaining concentrates of chlorinated insecticides from water. Compounds studied included: DDT, γ-BHC, heptachlor epoxide, dieldrin, and methoxychlor. The effect was examined of suspended matter on extraction efficiency. Direct cold extraction involving the use of a special glass microseparator was found to give the best results and was the only suitable method for samples containing suspended matter.

Engst and Knoll[46] used hexane extraction–gas chromatography to study the occurrence in rain water of down to 0.001 pp 10^9 of p,p'-DDT, p,p'-DDE, and p,p'-TDE. Samples (500 ml) were extracted by shaking with hexane (100, 50, and 50 ml) for 1 min each time. The combined hexane phases were dried with sodium sulphate and evaporated to between 1 and 5 ml in a rotary evaporator. The extract (1 ml) was injected, without further clean-up, on to gas chromatographic assembly consisting of one column (6 ft × 0.25 in.) packed with 5% of QF-1 on Varaport 30 (100–120 mesh) and a second column (3 ft × 0.25 in.) packed with 4% of OV-17 on AW-DMCS Chromosorb W (60–80 mesh), both operated at 180 °C with nitrogen as carrier gas (30 ml min^{-1}) and a tritium electron capture detector.

Thompson et al.[47] have described a gas chromatographic procedure for the multiclass, multiresidue analyses of organochlorine insecticides in water. It involves extraction with methylene chloride, separation into groups on a partially deactivated silica gel column, and sequential elution with different solvents. Final determinations of halogenated compounds and derivatized carbamates are made by gas chromatography with electron capture detection, and for organophosphorus compounds a flame photometric detector is used. This study included 42 organochlorine insecticides, 33 organophosphorus insecticides and 7 carbamate herbicides. These are discussed here as such mixtures are likely to be encountered in actual samples. The work on organophosphorous and carbamate herbicides types are referred to again in the appropriate chapters. Thompson et al.[47] point out that only few methods have been published which are ideally suited to the estimation of a wide variety of pesticidal compounds such as organochlorine and organophosphorus insecticides and herbicides. Table 91 is presented to illustrate the dearth of broadly applicable multiresidue

Table 91

Reference	Year	Author	Insecticides included	Extraction	Clean up	Recovery studies
49	1968	Pionke et al.	4 OGP and 8 OGC	Benzene	Act. silica gel, modified	Some data
50	1968	Kadoum	15 OGC compounds 4 OGP compounds		Aqueous acetonitrile partition	No data in abstract
51	1968	Kadoum	5 OGP compounds			
52	1969	Konrad et al.	9 OGC compounds 5 OGP compounds	Benzene	None used	Data reported on all compounds
53	1969	Askew et al.	40 OGP compounds	Used hexane, benzene, and chloroform. The latter preferred	Nuchar as needed but very rarely	Data reported on all compounds
54	1970	Johnson	8 OGC compounds	Pet. Ether	Silica gel column	None reported
55	1970	Law and Goerlitz	15 OGC compounds 3 OGP compounds	Hexane	Micro columns of alumina, Florisil and silica gel	Data reported for 6 OGC and 2 OGP compounds
56	1970	Herzel	Lindane, aldrin, DDT, and metabolites	Hexane	None	None reported in abstract
57	1970	Ahling and Jensen	BHS, lindane, DDE, DDD, DDT, PCB	Adsorption on undecane and Carbowax filter, then pet. ether	H_2SO_4	50 to 100%
58	1971	Ballinger	19 OGC compounds	15% ether in hexane	Florisil	Some data
59	1974	Ripley et al.	14 OGP compounds	Benzene	None	Data for all
60	1972	Thompson	15 OGC compounds	15% methylene dichloride in hexane	Florisil	Some data

Reprinted with permission from Thompson et al.[47] Copyright (1977) Springer-Verlag N.Y.

293

methods in the literature. Of these references, only five are intended as multiclass, multiresidue procedures outstanding amongst which is the work of Sherma and Shafik.[48] To correct this deficiency Thompson et al.[47] developed the multiclass multiresidue method described below which will provide the analyst with a means of simultaneously monitoring a water sample for a wide variety of pesticides. The silica gel columns are prepared as follows: the bottom of a 7 mm i.d. Chromaflex column is plugged lightly with a small wad of glass wool. To the column is added 1 g of deactivated silica gel, tapping firmly to settle, then 25 mm of anhydrous sodium sulphate, again tapping firmly. 10 ml of hexane are passed through columns and the eluate discarded. When the last prewash hexane just reaches the top surface of the sodium sulphate, a 15 ml empty conical centrifuge tube is placed quickly under the column. Using a disposable pipette, the 0.5 ml of concentrated extract are transferred to the column. When this has sunk into the bed, the walls of the tube are rinsed with 0.5 ml of hexane and, using the same disposable pipette, this wash increment is transferred to the column. The wash is repeated twice more and finally 8.0 ml of hexane are added to the column. This is fraction I. Another empty 15 ml centrifuge tube is immediately positioned under the column and 15 ml of the 60% benzane–hexane eluting solution are added. This is fraction II. A third elution is made with 15 ml of the 5% acetonitrile–benzene solution into one of the centrifuge tubes with 10 drops of keeper solution added. This elute is fraction III. A fourth fraction is necessary if there is reason to suspect the presence in the sample of crufomate, dimethoate, mevinphos, phosphamidon, or the oxygen analogues of diazinon or malathion. The elution solution is 25% acetone–methylene dichloride. This is fraction IV. Each tube of eluates is placed under a gentle nitrogen stream at ambient temperature and fractions I and II are concentrated to about 3.0 ml and fractions III and IV to 0.3 ml. The tube side walls are rinsed with hexane and each tube diluted to exactly 5.0 ml with hexane. If the presence of carbamates is suspected in the sample, the fraction II and III extracts, after completion of the electron capture and flame photometric detection gas chromatography, should be reconcentrated under a gentle nitrogen stream to 0.1 ml preparatory to derivatization of the carbamates.

Carbamate derivatization

0.5 ml of the fluorodinitrobenzene–acetone reagent and 5 ml of sodium borate buffer solution are added to the tubes containing the 0.1 ml concentrates of fractions II and III. The same reagents are added to an empty tube to serve as a reagent blank. The tubes are tightly capped and heated at 70 °C for one hour in a water bath. The tubes are removed from the heater, cooled to room temperature, 5 ml of hexane added to each tube and the tubes shaken vigorously for 3 min The layers are allowed to separate and 4 ml of the hexane (upper) layer are transferred carefully to a test tube and stoppered tightly.

Table 92 Recoveries of 42 organochlorine compounds

Compound	Concentration (ppb)	Extraction only	Recoveries (%) silica gel partitioning elution fraction				Total
			I	II	III	IV	
Aldrin	0.20	89	88				88
Atrazine	66.5	101			99		99
α-BHC	0.09	91	76	4			80
β-BHC	0.47	99		88			88
γ-BHC (lindane)	0.12	90	16	51			67
Captan[a]	6.50	100+			100+		100
CDEC	0.27	63		41			41
Chlorbenside	0.47	91	62				62
Chlordane	1.54	85	89	1			90
Chlordecone (Kepone)	3.64	72		18	8		26
2,4-D, butylester	4.08	93			90		90
2,4-D, butoxyethanol ether ester	8.65	109		9	90		99
2,4-D, iso-octyl ester	3.28	105		95			95
2,4-D, isopropyl ester	3.28	75		71			71
DCPA	0.50	98		84			84
p,p'-DDD	0.80	97	94				94
p,p'-DDE	0.45	96	101				101
o,p'-DDT	1.05	94	93				93
p,p'-DDT	1.58	104	98				98
Dichlone	13.4	85		79			79
Dieldrin	0.72	97		96			96
Dilan	2.42	97		94			94
Dyrene	8.70	95		77			77
Endrin	1.11	105		98			98
Endosulfan (Thiodan)	0.53	91	24	80			104
Folpet[a]	1.5	100		131			131
Heptachlor	0.18	90	79				79
Heptachlor epoxide	0.31	91		89			89
Hexachlorobenzene (HCB)	0.20	74	96				96
1-Hydroxchlordene	0.34	81			82		82
Methoxychlor	5.70	97		104			104
Mirex	2.35	83	83				83
PCNB	0.10	87	88				88
Perthane	66.5	89	80	15			95
Simazine	66.5	71			28		28
2,4,5-T, butyl ester	2.00	102		99			99
2,4,5-T, butoxyethanol ether ester	3.00	103		71	23		94
2,4,5-T, iso-octyl ester	6.05	109		97			97
Tetradifon (Tedion)	2.99	103		102			102
Toxaphene	22.3	103	93				93
Aroclor 1254	25.6	93	96				96
Aroclor 1260	25.6	93	92				92

[a]Non-linear response
Reprinted with permission from Thompson et al.[47] Copyright (1977) Springer-Verlag N.Y.

Table 93 Recoveries of 38 organophosphorus pesticidal compounds

Compound	Concentration (ppb)	Extraction only	I	II	III	IV	Total
Azinphos methyl (Guthion)	320	78			88		88
Carbophenothion (Trithion)	48	99	93				93
Carbophenoxon	80	94					0
Chlorpyrifos (Dursban)	4	99		87			87
Crufomate (Ruelene)	90	80				58	58
DEF	24	102			90		90
Diazinon	20	108			104	104	
Diazoxon	10	92				72	72
Dichlofenthion	1.6	102			102		102
Dicrotophos (Bidrin)	120	17				15	15
Dimethoate (Cygon)	24	40				60	60
Dioxathion (Delnav)	28	103		72	17		89
Disulfoton (Di-Syston)	2.6	92					0
EPN	60	99		96			96
Ethion	20	100		94			94
Ethoprop (Prophos)	2	97			96		96
Fenitrothion (Sumithion)	12	99		84			84
Fenthion (Baytex)	12	93		76			76
Fonofos (Dyfonate)	20	98		78			78
Leptophos (Phosvel)	200	107		91			91
Malaoxon	80	104				50	50
Malathion	4	100			78		78
Methamidophos (Monitor)	200	5					0
Mevinphos (Phosdrin)	6	69			32	33	65
Monocrotophos (Azodrin)	72	0					0
Naled (Dibrom)	56	92			45		45
Oxydemeton methyl (Metasystox R)	300	67					0
Paraoxon ethyl	40	99			90		90
Paraoxon methyl	36	98			93		93
Parathion ethyl	16	101	99				99
Parathion methyl	16	99		93			93
Phenkapton	60	99		98			98
Phorate (Thimet)	1.3	98		56			56
Phosalone (Zolone)	400	102		91			91
Phosmet (Imidan)	220	82			85		85
Phosphamidon	80	43				43	43
Ronnel (fenchlorphos)	4	100		96			96
Ronnoxon	120	94			92		92

Reprinted with permission from Thompson et al.[47] Copyright (1977) Springer-Verlag N.Y.

Gas chromatography

For multiresidue analysis of water of unknown pesticidal content, two gas chromatographic columns yielding quite divergent compound elution patterns are the 5% OV-210 and the 1.5% OV-17/1.95% OV-210. For electron capture detection, the column oven should be set at 200 °C for the mixture column and at 180 °C for the 5% OV-210. Carrier gas rates flow for electron capture detection should be adjusted to produce a retention time of 16–19 min for p,p'-DDT, generally from 50 to 70 ml min^{-1}. Response levels for both electron capture and flame photometric detectors should be carefully established before starting chromatography. By electron capture detection, the system should be capable of yielding a chromatographic peak on the recorder of at least 50% full scale deflection with an injection of 100 pg of aldrin. The flame photometric system should yield a minimum of 50% full scale deflection with the injection of 2.5 ng of parathion. The majority of the organochlorine insecticides will be detected in fractions I and II with a few of the more polar compounds in fraction III (Table 92). Most of the organophosphorus compounds will be found in fractions II and III, a very few in fraction IV, and none in fraction I (Table 93).

Thompson et al.[47] used a gas chromatograph with electron capture and flame photometric detectors, the latter operated in the phosphorus mode at 526 nm. Columns were 1.8 m × 4 mm i.d., borosilicate glass packed with 1.5% OV-17/1.95% OV-210 or 5% OV-210, both coated on Gas-Chrom Q, 80–100 mesh.

Reagents

Pesticide quality methylene chloride, hexane, benzene, acetone, acetonitrile, and methanol.

Silica gel, Woolm, activity grade I, activated for 48 h at 175 °C before use. Final deactivated material is prepared by adding 1.0 ml of water to 5.0 ml of silica gel in a tube with Teflon lined screw cap. Mix on rotating mixer for 2 h at about 50 rpm.

Sodium sulphate, granular, anhydrous is purified by Soxhlet extraction with methylene dichloride for 60 cycles.

1-Fluoro-2,4,-dinitrobenzene (FDNB): a 1% reagent solution in acetone is prepared.

Sodium borate buffer, $Na_2B_4O_7 \cdot 10H_2O$, 0.1 M solution, pH 9.4. Carborundum chips, approximately 12 mesh.

'Keeper' solution, 1% paraffin oil, USP, in hexane.

Eluting solutions

Fraction I: hexane.
Fraction II: 60% benzene–hexane, v/v.
Fraction III: 5% acetonitrile–benzene, v/v.
Fraction IV: 25% acetone–methylene dichloride, v/v.

Contaminant-free water: To 1500 ml of distilled water in a 2 l separatory funnel, is added 100 ml of methylene dichloride. The funnel is stoppered and shaken for 2 min After phase separation the methylene dichloride is discarded, another 100 ml portion is added, and the extraction repeated. The double-extracted water is transferred into a glass-stoppered bottle for storage.

Extraction and concentration

500 ml of water sample are transferred to a 1 l separatory funnel and 10 g of anhydrous sodium sulphate and 50 ml of methylene dichloride added. The funnel is shaken vigorously for 2 min and sufficient time allowed for complete phase separation. At this point, 500 ml of the pre-extracted water should be carried through all subsequent steps as a reagent blank. A small wad of glass wool is placed at the bottom of a 22 × 300 mm Chromaflex column and anhydrous sodium sulphate added to a 50 mm depth. The tip of the column is positioned over an assembly consisting of a 250 ml Kuderna Danish flask attached to a 10 ml evaporative concentrator tube containing 2 or 3 carborundum chips and 5–10 drops of 'Keeper' solution. The lower phase methylene dichloride is drained from the separatory funnel through the sodium sulphate column on to the Kuderna Danish flask. 50 ml more methylene dichloride is added to the aqueous phase, the flask stoppered, and the 2 min shaking repeated. This methylene dichloride portion is drained through the sodium sulphate column into the Kuderna Danish flask. It is not uncommon at this point to encounter emulsion problems at the methylene dichloride–water interface. With highly polluted water, it is often necessary to take some special procedural steps to break this emulsion sufficiently to effect complete phase separation. The Kuderna Danish flask is connected to an evaporator and the assembly inclined to an angle of about 20° from the vertical (Figure 139) with the concentrator tube about half immersed in a water bath at 35 °C. The rotator is turned on, adjusting the speed to a slow spin. The water bath heat is switched off and vacuum of about 125 mmHg applied. Evaporation is continued until the extract is condensed to about 4 ml when the assembly is removed from the bath and the walls of the flask rinsed down with 4 ml of hexane. The concentrator tube is disconnected from the Kuderna Danish flask, rinsing the joint with 2 ml of hexane. The tube is placed under a gentle stream of nitrogen at room temperature and the extract concentrated to 0.5 ml. Thompson *et al.*[47] discuss in detail some of the practical problems the analyst may encounter in carrying out these analyses such as the formation of persistent emulsions at the water–methylene dichloride interface and changes in activity of silica gel. They found that an initial water sample of 500 ml was suitable for the pesticide concentrations as shown in Tables 92–94. If lower concentrations are anticipated, the sample should be increased to 1000 or 1500 ml and the volume of extracting solvent increased to 75 ml. Air blowdown should be scrupulously avoided in this procedure for reducing the volume of extract or eluates. Oxidative effects would seriously hamper recoveries of a number of organophosphorus compounds and some of the carbamates,

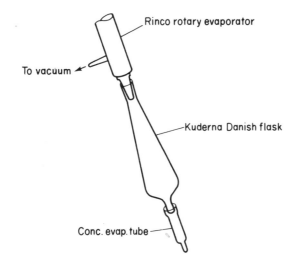

Figure 139 Evaporation assembly. Reprinted with permission from Thompson et al.[47]. Copyright (1977) Springer Verlag, N.Y.

and would even jeopardize recoveries of extremely low concentrations of the organohalogen compounds. Of the 42 halogenated compounds evaluated, reproducible recoveries of 80% or better were obtained for 31 (Table 92). Gas chromatographic linearity problems were encountered with captan and folpet, and surprisingly, a sizeable portion of lindane was lost during the silica gel fractionation. Recovery of lindane from the extraction step was 90%. In the fractionation, 51% eluted in fraction II and 16% in fraction I, totalling 67% recovery. Twenty-one of the 38 organophosphorus compounds were included in the 80% plus recovery range (Table 93). Six of these compounds fell in the 60–79% recovery range. The method proved wholly unsuitable for reproducible and satisfactory recoveries of carbophenoxon, disulfoton, methamidophos, monocrotophos, and oxydemeton methyl. Of these five compounds, excellent extraction efficiency was observed for carbophenoxon and disulfoton, but complete losses were experienced during silica gel fractionation. Six compounds gave partial recoveries in the range of 0–60%. Of the 17 organophosphorus compounds yielding total recoveries of less than 80%, six gave over 90% extraction recovery but losses occurred during silica gel column chromatography. Recoveries of the carbamates are shown in Table 94. No problems were encountered with metalkamate, carbofuran, methiocarb or propoxur. Final recoveries after fractionation fell in an acceptable range. Acceptably high recoveries were obtained for aminocarb and carbaryl by direct derivatization and gas chromatography of the concentrated extract resulting from the methylene dichloride extraction bypassing the silica gel fractionation step. Losses were noted during the fractionation for these two compounds. Recoveries of mexacarbate were highly inconsistent. Even by direct analysis of spiked methylene dichloride recovery variations were extreme. Expectedly, recovery data resulting

Table 94 Recoveries of 7 carbamate compounds

Compound	Concentration (ppb)	Recoveries (%) No partitioning		Silica gel partitioning elution fraction				Fraction total
		Extraction only	Fortified CH$_2$Cl$_2$	I	II	III	IV	
Aminocarb (Matacil)	10	89	90			59		59
Metalkamate (Bux)	10	102	101			93		93
Carbaryl (Sevin)	10	100	98			68		68
Carbofuran (Furadan)	10	96	95		4	94		98
Methiocarb (Mesurol)	10	99	94		55	57		112
Propoxur (Baygon)	10	96	100			99		99
Mexacarbate (Zectran)	10	72	71			58		58

Reprinted with permission from Thompson et al.[47] Copyright (1977) Springer-Verlag, N.Y.

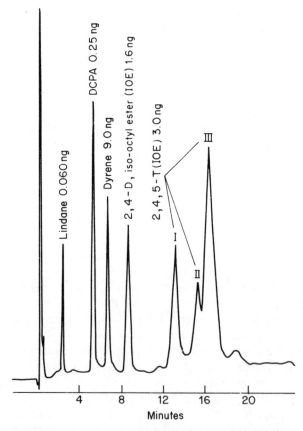

Figure 140 Five chlorinated compounds eluted in fraction II. G.c. column 1.5% OV-17/1.95% OV-210. Reprinted with permission from Thompson et al.[47] Copyright (1977) Springer Verlag, N.Y.

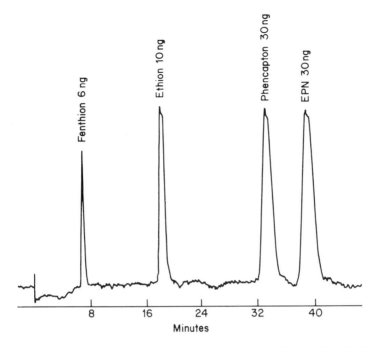

Figure 141 Four organophosphorous compounds eluted in fraction II. G.c. column 1.5% OV-17/1.95% OV-210. Reprinted with permission from Thompson et al.[47] Copyright (1977) Springer Verlag, N.Y.

from analysis of the direct water extraction were similarly inconsistent. Fractionation of this compound resulted in some further loss. Typical chromatograms of the electron capture gas chromatography of chlorinated compounds, and the FPD detection of the organophosphorus compounds are shown in Figures 140 and 141, respectively. Thompson et al.[47] emphasize that even the use of two dissimilar gas chromatographic columns does not ensure irrefutable compound identification. For example, if the retention characteristics of a given peak obtained from two dissimilar columns suggest the possibility of the presence of a compound which appears wholly out of place in a specific sample, further confirmation is clearly indicated by such techniques as specific detectors, coulometry, P values, or gas chromatography–mass spectrometry or thin-layer chromatography.

Sackmauerevá et al.[61] have described the method, given below, for the determination of chlorinated insecticides (BHC isomers, DDE, DDT, and hexachlorobenzene) in water, fish, sediments, and water weeds.

Isolation of chlorinated insecticides from waters

The water sample (1–3 l) is extracted with three portions of petroleum ether (boiling point 30–40 °C), using the glass apparatus shown in Figure 142.

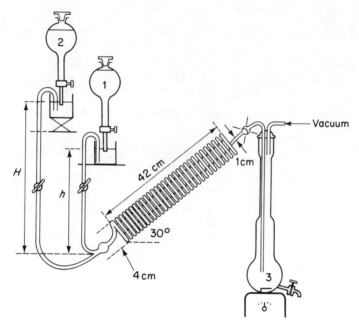

Figure 142 Apparatus for isolation of chlorinated insecticides from water samples. Reprinted with permission from Sackmauerevá et al.[61] Copyright (1977) Pergamon Press Ltd.

A vacuum is used to transfer the water sample (separation funnel no. 1) and petroleum ether (separation funnel no. 2) into a glass spiral filled with 8 mm glass balls. Phases are separated in the connected flask. The petroleum ether layer is then concentrated to a volume of about 0.5 ml using a vacuum, and purified on an alumina column (Woelm, neutral, activated by heating at 300 °C for 3 h and deactivated by adding 11% water). Thereafter, insecticides were eluted with 15% dichloromethane in petroleum ether. The eluate was concentrated in a vacuum rotary evaporator to a volume of 1 ml and then used for gas–liquid chromatography by previously described procedures.[62–65]

Isolation of chlorinated insecticides from fish

One hundred g of fish sample are weighed and homogenized. From the homogenate, 10 g are weighed and rubbed with cleaned sea sand. Water contained in the sample is bound by adding some anhydrous sodium sulphate to the sample in order to obtain a homogeneous powdery mixture. The mixture is shaken three times with portions of light petroleum (200, 100, and 100 ml) for one hour periods. The separate extracts are filtered into 500 ml flasks through a layer of anhydrous sodium sulphate and concentrated using a vacuum rotary evaporator. The concentrated extract is quantitatively transferred into a 250 ml separation funnel using 20 ml petroleum ether saturated with acetonitrile. The mixture is shaken three times with 40 ml acetonitrile saturated with petroleum ether. The acetonitrile extracts are combined in a one litre separating funnel, 500 ml of 5% sodium chloride in water solution is added, and the insecticides are extracted from the samples twice with 100 ml petroleum ether. The mixed ether extracts are concentrated using the vacuum rotary evaporator to a small volume and purified on a chromatographic column filled with 4 g Celite and a mixture of 8 g Celite with 6 ml oleum. The upper layer of the column consisted of a 15 mm layer of anhydrous sodium sulphate. The thickened extract is quantitatively transferred to the top of the washed column and the insecticides eluated with 250 ml petroleum ether. The elute is reduced to a volume of 1 ml and used for chromatography.[62]

Isolation of chlorinated insecticides from sediment

The sediment sample is allowed to dry in open air and then sieved. To 20 g of the sample 20% distilled water is added for deactivation purposes and the excess water is then bound to active silica (Siloxid), so that a powdery consistency is obtained. The insecticides studied are extracted with petroleum ether (b.p. 30–60 °C) in a Soxhlet apparatus. The extract is concentrated using the vacuum rotary evaporator and the coextractants are separated on a Celite oleum column (see isolation from fish). The petroleum ether eluate is then concentrated to a volume of 1 ml and used for chromatography.[63-66]

Isolation of chlorinated insecticides from water plants

From an average sample, dried at room temperature, 200 g are weighed and homogenized. From the pulverized sample 20 g are taken and extracted with petroleum ether in the Soxhlet apparatus (petroleum ether b.p. 30–60 °C) for 12 h. The concentrated extract is purified from coextracts on a Florisil filled column activated by heating at 120 °C for 48 h with addition of 5% water to the cooled column. Insecticides are eluted from the column with 15% dichloromethane in petroleum ether. The elute is concentrated to a volume of 1 ml and used for the chromatographic determination.[63-65,67]

Determination of chlorinated insecticides by gas chromatography

Working conditions: temperature of the column 180–200 °C, temperature of the injection port 210 °C, temperature of the electron capture detector (^{63}Ni) 200–225 °C, nitrogen flow rate 60–80 ml min^{-1}, EC detector voltage 20–70 V. The optimum operating voltage is to be found experimentally to obtain the highest response towards the components. One μl of the concentrated sample is injected into the gas chromatograph (Carlo Erba, type 452 GI) used. When necessary, the sample is diluted with hexane. Then, identical volumes of standard compound solution mixtures are placed to the apparatus under standard conditions. Under the above conditions, the insecticide concentration is in linear proportion to the peak height over the following range:

α-BHC	0.03–0.12 µg ml^{-1}
β-BHC	0.15–0.60 µg ml^{-1}
γ-BHC	0.04–0.18 µg ml^{-1}
δ-BHC	0.03–0.12 µg ml^{-1}
p,p'-DDE	0.15–0.60 µg ml^{-1}
o,p'-DDT	0.30–1.2 µg ml^{-1}
p,p'-DDD	0.30–1.2 µg ml^{-1}
p,p'-DDT	0.30–1.2 µg ml^{-1}

When the individual insecticides are present in the solution in such a concentration range, the electron capture responds nearly uniformly to all insecticides. A column filled with 1.5% silicone OV-17 plus silicone oil (fluoralchylsiloxane) on Chromosorb W (80–100 mesh) is used for separation of the BHC alpha, beta, gamma and delta isomers (hexachlorocyclohexane), o,p'-DDT, p,p'-DDE, p,p'-DDD, and p,p'-DDT. α-BHC and hexachlorobenzene (HCB) have a common peak. They can be separated on a column filled with 2.5% Silicone Oil XE-60 (β-cyanoethyl-methysilicone) on Chromosorb W (80–100 mesh).

Sackmauerevá et al.[61] used thin-layer chromatography on silica plates to confirm the identity of chlorinated insecticides previously identified by gas chromatography. The compounds can be separated by single or repeated one-dimensional development in n-heptane or in n-heptane containing 0.3% ethanol. The plate is dried at 65 °C for 10 min and detected by spraying with a solution of silver nitrate plus 2 phenoxyethanol.[68,69] Thereafter, the plate was dried at 65 °C for 10 min and illuminated with an ultraviolet light ($\lambda = 254$ nm) until spots representing the smallest amounts of standards were visible (10–15 min). The pesticide residues may be evaluated semiquantitively by simple visual evaluation of the size and of the intensity of spot coloration, and by comparing extracts with standard solutions.

Using the gas chromatography methods Sackmauerevá et al.[61] obtained from spiked samples the four BHC isomers at 93–103.5% recovery. Both DDT and DDE were yielded in 85.6–94%, 90–93.2%, 90–102.4%, and 92–105.8% from water, fish, plants, and sediment, respectively. Purification on a Florisil column was used in determining chlorinated insecticides unstable at low pH (aldrin, dieldrin). The type and activity of Florisil influence the yield and accuracy of the method. Therefore, the activity of this adsorbent had to be verified and adjusted.[70,71] From the results of the analyses of 92 drinking water samples the average content of γ-BHC was 0.069 µg l^{-1}, that of β-BHC 0.023 µg l^{-1}, and that of the other isomers ($\alpha + \delta$) of BHC + HCB (hexachlorobenzene) 0.018 µg l^{-1}. The average content of DDE in drinking water was 0.022 µg l^{-1} and that of DDT 0.042 µg l^{-1}. In the period 1971–4, 185 samples of surface water from the Danube were examined for the content of BHC, DDE, and DDT residues. The average content of γ-BHC was 0.117 µg l^{-1}, that of β-BHC

0.040 µg l^{-1}, and that of the other BHC isomers plus HCB was 0.049 µg l^{-1}. The average content of DDE was 0.050 µg l^{-1} and that of DDT 0.125 µg l^{-1}. The average content of γ-BHC in samples of herbivorous fishes (*Abramis ballerus* L., *Cyprinus carpio* L., *Chrondrosroma nastus* L.) was 0.054 mg kg^{-1}, that of β-BHC 0.009 mg kg^{-1}, and that of the remaining BHC isomers and HCB 0.049 mg kg^{-1}. The average DDE content was 0.133 mg kg^{-1} and that of DDT 0.094 mg kg^{-1}. From the analyses of 78 samples of carnivorous fishes (*Esox lucius* L., *Lepomis gibbosus* L., *Aspius fluviatilis* L.) these workers found the average content of γ-BHC 0.062 mg kg^{-1}, that of β-BHC 0.023 mg kg^{-1}, and that of the remaining BHC isomers plus HCB 0.060 mg kg^{-1}. The average DDE content was more than 10 times higher, 1.53 mg kg^{-1} and likewise that of DDT 1.175 mg kg^{-1} in comparison with the herbivorous fishes. Besides the BHC isomers, DDE, and DDT, the concentration of hexachlorobenzene was also studied in waters and fish since 1973. Its concentration in waters varied between 0.001 and 0.03 µg l^{-1}, while in fish it was from 0.001 to 0.26 mg kg^{-1}. Comparing the average content of the BHC isomers, of DDT and metabolites in fish it can be seen that DDT and DDE levels greatly exceed the levels of BHC, mainly as far as carnivorous fish is considered. Comparing the contents of the BHC isomers and of DDT and its metabolites in waters and fish, a 1000–10,000 fold higher concentration was detected in fish. The average concentration of the BHC γ and β isomer and $\alpha + \delta$ isomers, and of DDE and DDT in sediment was 0.010, 0.010, 0.016, 2.11, and 0.70 mg kg^{-1}, respectively. The average content of the γ isomer in water plants was 0.026 mg kg^{-1} while the β isomer was present at 0.00 mg kg^{-1}, $\alpha + \delta$ at 0.032 mg kg^{-1}, DDE at 0.003 mg kg^{-1} and DDT at 0.002 mg kg^{-1}. These results suggest that chlorinated insecticides, due to their physical and chemical properties, can accumulate and adsorb on to solid particles.

Suzuki *et al.*[72] studied the determination of chlorinated insecticides in river and surface waters and sediments and soils using high resolution electron capture gas chromatography with glass capillary columns. They compared resolution efficiencies of organochlorine insecticides and their related compounds with wall-coated open tubular (WCOT) and support-coated open tubular (SCOT) glass capillary columns with those of conventional packed glass columns. These columns were coated with silicone OV-101 as the liquid phase. Applicabilities of the glass capillary column to environmental samples were investigated. An all-glass system was used to prevent thermal decomposition. The 'resolution index', i.e. peak height/half-width of peak of standard injected, generally increased in the following order, conventional packed glass column $\ll$ WCOT glass capillary column $\leq$ SCOT glass capillary column. Excellent resolution of insecticides was obtained with SCOT glass capillary columns and WCOT glass capillary columns. Log–log plots of the resolution index *vs.* relative retention times compared to aldrin were linear. These workers used a Shimadzu GC-5AIEE glass chromatograph equipped with a dual electron capture detector (^{3}H, 300 mCi, foil type). Coiled support-coated open tubular glass capillary columns (23 m $\times$ 0·28 mm) were used. With SCOT glass capillary columns, the liquid phase was coated to the salt layer adhered to the inner wall of the glass capillary column. With WCOT glass capillary columns, the liquid phase was directly coated to the inner wall of the glass capillary column. Therefore, the surface area per unit of length was broader in SCOT glass capillary column than in the WCOT glass capillary columns. The glass capillary column was connected to a holder. An OV-101 PGC (3% on Gas Chrom Q, 80–100 mesh, U-shaped, 2 m × 3 mm) was used.[73] The gas chromatographic conditions used were as follows: temperatures of column, injector, and detector, for both conventional packed glass column chromatography and glass capillary column chromatography were 190, 210, and 200 °C, respectively. For the packed column a flow rate of carrier gas (highly purified nitrogen gas, 99.9999 + %) was 60 ml min^{-1}. The flow rates through the glass capillary column were both adjusted to 2 ml min^{-1}. The effluent from both glass capillary columns was scavenged at 60 ml min^{-1} and entered into the electron capture detector. The splitter ratio was 1:25 was SCOT glass capillary column and 1:40 for WCOT glass capillary column, respectively. All columns used were well conditioned before use. The chart drive was 2 cm min^{-1} for estimation of the 'resolution index', i.e., peak height/half-width of peak, which showed the sharpness of a peak and the degree of resolution efficiencies, and 1 cm min^{-1} for the analyses of environmental samples. Attenuations for the conventional packed glass columns, SCOT glass capillary column, and WCOT glass capillary columns were 8×10, 16×10^2, and 16×10^2, respectively. Under these conditions, duplicate injections of 1–5 µl of each standard showing 30% in full-scale deflection were made, and the resolution index was calculated. Also, 5 µl of extracts from samples were injected. Minimum detectable levels of α-BHC, β-BHC, γ-BHC, δ-BHC, heptachlor, heptachlor epoxide, aldrin, dieldrin, endrin, p,p'-DDE, p,p'-TDE, and p,p'DDT in 100 g samples of field soil and bottom sediment were 0.0005, 0.0032, 0.0014, 0.0040, 0.0012, 0.0020, 0,0014, 0.0020, 0.0056, 0.0032, 0.0080, and 0.0120 ppm, respectively, on SCOT glass capillary columns. Also, the levels of α-BHC, β-BHC, γ-BHC, and δ-BHC in a 1 l sample of river surface water were 0.002, 0.013, 0.006, and 0.016 ppb, respectively, on the glass capillary column.

Procedure

Materials

Reagent grade n-hexane and acetonitrile were used after distillation in an all-glass apparatus. Anhydrous sodium sulphate was heated at 625 °C for 2 h before use to eliminate the contaminants. Insecticide-free distilled water was prepared by distillation of tap water in an all-glass apparatus and washing of the distillate with n-hexane (1 vol. of n-hexane to 10 vol. of distillate). All glasswares were rinsed twice with n-hexane before use to remove any contaminants. Insecticides in river surface water were extracted successively with 100, 50, 50, 50, and 50 ml of n-hexane. The combined extracts and several rinsings were dried over anhydrous sodium sulphate. The dried extract was concentrated first to 2 ml by a Kuderna Danish evaporative concentrator with a three-ball Snyder column and finally to 0.2 ml with a gentle stream of nitrogen gas. An appropriate amount of heptachlor epoxide was added as an internal standard, and the extract was stored until gas chromatographic analysis. River bottom sediment and soil were dried at ambient temperature, pulverized with a mortar, and screened through a 20-mesh sieve to remove small gravel and other interfering materials. The screened sample was well mixed in a ball-mill jar rotating without balls for 5 h. Insecticides in these well mixed samples (100 g) were extracted for 5 min with 200 ml of acetonitrile in a high-speed mixer, 30 min after the addition of 70 ml of insecticide-free distilled water. The extract was passed through a glass filter (25G-4), and the filtrate and several rinsings were partitioned with 100 ml of n-hexane and 600 ml of insecticide-free distilled water. The resulting organic layer was dried over anhydrous sodium sulphate and concentrated to 5 ml using a Kuderna Danish evaporative concentrator, and stored in a refrigerator until gas chromatographic analysis. Heptachlor epoxide was added as an internal standard to the samples which did not contain its residue.

The relative retention times of the individual insecticides and their related compounds compared to aldrin, amounts injected, and 'resolution index' of conventional packed glass columns and, WCOT and SCOT glass capillary columns are shown in Tables 95–97, respectively. Among BHC isomers, greater enhancements in the 'resolution index' of γ-BHC were shown with WCOT glass capillary columns and/or SCOT glass capillary columns than with a conventional glass packed column. Moreover, the 'resolution index' with a SCOT glass capillary column was about twice as much as that with WCOT glass capillary columns. Similar results were obtained in the case of α-BHC and β-BHC. The 'resolution index' of δ-BHC with conventional packed glass columns, WCOT glass capillary columns, and SCOT glass capillary columns were 26.08, 36.29, and 44.86, respectively, supposing that the ratio of the 'resolution index' with SCOT glass capillary column/'resolution index' with SCOT glass capillary column was the least among those of BHC isomers. The 'resolution index' of heptachlor and heptachlor epoxide were 24.13 and 9.36 with a conventional packed glass column, 33.80 and 19.25 with WCOT glass capillary column, and 54.86 and 23.67 with SCOT glass capillary column, respectively. The 'resolution index' with WCOT glass capillary column and SCOT glass capillary columns were 140 and 227% of 'resolution index' with conventional packed glass columns in heptachlor and 206 and 252% in heptachlor epoxide, respectively. A marked increase in the 'resolution index' of heptachlor epoxide with SCOT glass capillary column was found. The 'resolution index' of aldrin, dieldrin, and endrin similarly increased in the following order: conventional packed glass column < WCOT glass capillary column < SCOT glass capillary column, and remarkable enhancement of the 'resolution index' with WCOT glass capillary column and SCOT glass capillary column, compared with that of a conventional packed glass column, was shown in the case of endrin.

Table 95 Relative retention times compared to aldrin (RRT),[a] ng injected, and 'resolution index' on OV-101 packed column

Insecticides	RRT	ng	RI
α-BHC	0.42	0.4	39.22
β-BHC	0.46	1.6	26.73
γ-BHC	0.50	0.8	25.58
δ-BHC	0.52	1.2	26.08
Heptachlor	0.81	0.8	24.13
Heptachlor epoxide	1.22	1.2	9.36
Aldrin	1.00	1.2	15.18
Dieldrin	1.79	1.6	7.20
Endrin	2.00	0.4	2.90
p,p'-DDE	1.80	0.4	10.77
p,p'-TDE	2.27	0.5	4.71
p,p'-DDT	2.49	1.2	5.27
o,p'-DDE	1.78	0.4	9.54
o,p'-TDE	1.84	1.0	7.61
o,p'-DDT	2.41	4.0	4.05
Mirex	5.53	2.2	2.03
Methoxychlor	4.54	4.0	2.35
Photodieldrin	4.38	2.0	2.82

[a]Retention time of aldrin was 4.19 min.
Reprinted with permission from Suzuki et al.[72] Copyright (1977) American Chemical Society.

Table 96 Relative retention times compared to aldrin (RRT),[a] and 'resolution index' on OV-101 WCOT glass capillary column

Insecticides	RRT	ng	RI
α-BHC	0.46	0.3	47.17
β-BHC	0.49	2.0	35.43
γ-BHC	0.53	0.8	38.86
δ-BHC	0.54	1.0	36.29
Heptachlor	0.82	1.0	33.80
Heptachlor epoxide	1.21	1.2	19.25
Aldrin	1.00	1.4	21.29
Dieldrin	1.73	2.0	11.16
Endrin	1.92	8.4	7.26
p,p'-DDE	1.73	3.2	12.70
p,p'-TDE	2.16	5.5	9.93
p,p'-DDT	2.81	8.4	7.72
o,p'-DDE	1.72	3.2	11.76
o,p'-TDE	1.77	7.5	13.79
o,p'-DDT	2.23	18.0	4.10
Mirex	5.04	8.0	2.53
Methoxychlor	4.04	20.0	4.02
Photodieldrin	3.90	6.0	1.95

[a]Retention time of aldrin was 6.25 min.
Reprinted with permission from Suzuki et al.[72] Copyright (1977) American Chemical Society.

Table 97 Relative retention times compared to aldrin (RRT),[a] ng injected, and 'resolution index' on OV-101 SCOT glass capillary column

Insecticides	RRT	ng	RI
α-BHC	0.46	0.26	85.20
β-BHC	0.49	1.60	65.67
γ-BHC	0.53	0.70	77.20
δ-BHC	0.56	2.00	44.86
Heptachlor	0.82	0.60	54.86
Heptachlor epoxide	1.21	1.00	23.67
Aldrin	1.00	0.70	34.56
Dieldrin	1.73	1.20	17.83
Endrin	1.92	2.80	15.74
p,p'-DDE	1.73	1.60	16.11
p,p'-TDE	2.24	4.00	13.67
p,p'-DDT	2.93	6.00	9.16
o,p'-DDE	1.79	1.60	15.16
o,p'-TDE	1.84	5.00	21.32
o,p'-DDT	2.37	15.00	8.36
Mirex	5.38	6.00	4.27
Methoxylene	4.33	20.00	7.57
Photodieldrin[b]			

[a]Retention time of aldrin was 4.99 min.
[b]No peak appeared.
Reprinted with permission from Suzuki et al.[72] Copyright (1977) American Chemical Society.

Suzuki et al.[72] found that endrin decomposes during high temperature gas chromatographic analysis forming aldehyde and/or ketone derivatives. However, it appears that the endrin does not decompose, or is not being adsorbed on the column wall of SCOT glass capillary columns. As for the results of DDT and its related compounds, no greater increases of the 'resolution index' in p,p'-DDE p,p'-DDT, o,p'-DDT, with a WCOT glass capillary column, in comparison with a conventional packed glass column, could be found. Significant enhancements of the 'resolution index' in p,p'-TDE and o,p'-TDE with WCOT glass capillary columns were shown. Also, similar results were obtained with SCOT glass capillary column. Considerable improvements in the 'resolution index' of mirex and methoxychlor with a SCOT glass capillary column were shown in comparison with those of conventional packed glass columns and WCOT glass capillary columns. However, the peak of photodieldrin, the photochemically and microbially formed isomer of dieldrin which was present in the field soil, could not be detected with a SCOT glass capillary column. This might be due to the thermal decomposition and/or adsorption of photodieldrin to the support of SCOT glass capillary column. The 'resolution index' with WCOT glass capillary columns was smaller than with conventional glass packed columns. The resolution of insecticides and their related compounds on conventional glass packed columns, WCOT glass capillary columns, and SCOT glass capillary columns is shown in Figures 143–145. With conventional glass packed columns, broad peaks with greater retention times

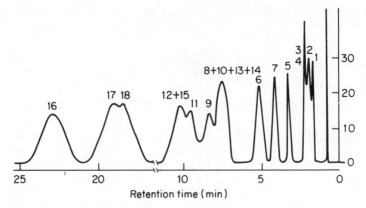

Figure 143 Resolution of standard insecticides and their related compounds on conventional packed columns. G.c. conditions shown in text. 1, α-BHC; 2, β-BHC; 3, γ-BHC; 4,6-BHC; 5, heptachlor; 6, heptachlor epoxide; 7, aldrin; 8, dieldrin; 9, endrin; 10, p,p'-DDE; 11, p,p'-TDE; 12, p,p'-DDT; 13, o,p'-DDE; 14, o,p'-TDE; 15, o,p'-DDT; 16, mirex; 17, methoxychlor; 18, photodieldrin. Reprinted with permission from Suzuki et al.[72] Copyright (1977) American Chemical Society

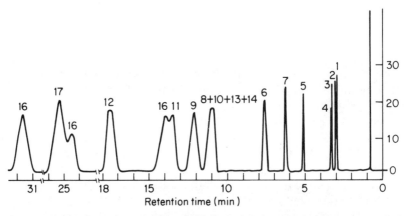

Figure 144 Resolution of standard insecticides and their related compounds on WCOT g.c.c. Peak numbers identical with those in Figure 143. Reprinted with permission from Suzuki et al.[72] Copyright (1977) American Chemical Society

were shown, and the BHC isomers could not be separated. However, with WCOT glass capillary columns and SCOT glass capillary columns, good resolution of the BHC isomers was obtained. Aldrin, heptachlor, heptachlor epoxide, and mirex could not be separated on all the columns used. On the contrary, resolution of dieldrin and p,p'-DDE could not be obtained. In general, better resolution was obtained with WCOT glass capillary columns. The relationships between relative retention time and 'resolution index' are shown in Figure 146. For the conventional packed glass column the equation was:

$$Y = 1.085 X^{-1.85} \tag{15}$$

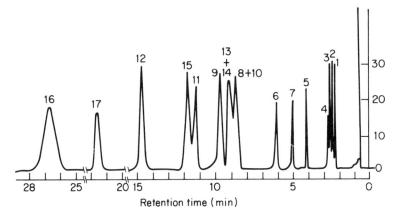

Figure 145 Resolution of standard insecticides and their related compounds on SCOT g.c.c. Peak numbers identical with those in Figure 143. Reprinted with permission from Suzuki *et al.*[72] Copyright (1977) American Chemical Society

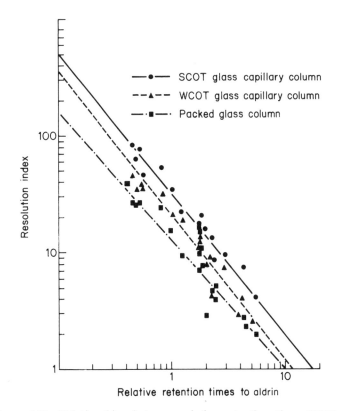

Figure 146 Relationships between relative retention time (RTT) and resolution index (RI). Reprinted with permission from Suzuki *et al.*[72] Copyright (1977) American Chemical Society

where X is the relative retention time to aldrin of insecticides and their related compounds employed, Y is 'resolution index' obtained. For WCOT glass capillary columns and SCOT glass capillary columns the equations were:

$$Y = 1.312 X^{-1.228} \qquad (16)$$
$$Y = 1.492 X^{-1.193} \qquad (17)$$

From these equations, the 'resolution index' can easily be calculated on the relative retention time to aldrin of all insecticides and their related compounds. Figures 147–149 show gas chromatograms of extracts from field soil, river surface water, and river bottom sediment samples, respectively, with SCOT glass capillary columns. SCOT glass capillary columns were chosen for analyses of environmental samples because of better resolution efficiency of insecticides. In field soil, 0.067, 1.172, 0.100, 0.194, 0.004, 0.506, 0.019, 0.246, 3.964, and 0.166 ppm of α-BHC, β-BHC, γ-BHC, δ-BHC, aldrin, dieldrin, heptachlor epoxide, p,p'-DDE, p,p'-DDT, and o,p'-DDT residue levels on dry matter bases, respectively, were determined. In river surface water, 0.039, 0.025, 0.010, and 0.012 ppb of α-BHC, β-BHC, γ-BHC, and δ-BHC, respectively, were estimated. However, insecticides other than BHC could not be detected. Also, 0.012, 0.008, 0.007, 0.004, 0.002, and 0.002 ppm of α-BHC, β-BHC, γ-BHC, δ-BHC, dieldrin, and endrin residues on dry matter bases, respectively, were found in river bottom sediment. From the sample with high residue levels such as in field soil (ppm) to that with significantly lower residue levels such as in river surface water (ppt to ppb), organochlorine insecticides and their related compounds in common environmental samples could be determined.

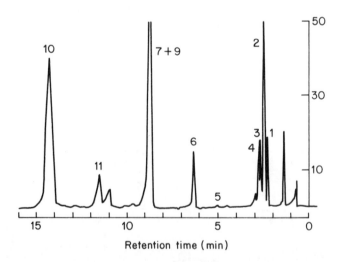

Figure 147 Gas chromatogram of residual organochlorine insecticides and their related compounds in spinach field soil on SCOT g.c.c. 1, α-BHC; 2, β-BHC; 3, γ-BHC; 4, δ-BHC; 5, aldrin; 6, heptachlor epoxide; 7, dieldrin; 8, endrin; 9, p,p'-DDE; 10, p,p'-DDT; 11, o,p'-DDT. Reprinted with permission from Suzuki et al.[72] Copyright (1977) American Chemical Society

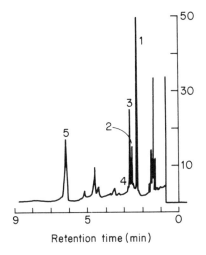

Figure 148 Gas chromatogram of residual organochlorine insecticides and their related compounds in river surface water from Onga River on SCOT g.c.c. Peak numbers identical with those in Figure 147. Reprinted with permission from Suzuki et al.[72] Copyright (1977) American Chemical Society

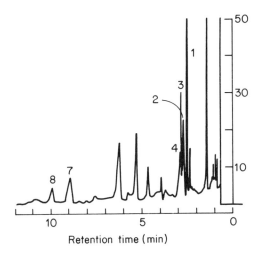

Figure 149 Gas chromatogram of residual organochlorine insecticides and their related compounds in river bottom sediment from Higashitani River on SCOT g.c.c. Peak numbers identical with those in Figure 147. Reprinted with permission from Suzuki et al.[72] Copyright (1977) American Chemical Society

Other workers[74] have used electron capture gas chromatography coupled with the use of glass capillary columns for the separation of chlorinated insecticides. Figure 150 shows separations achieved of 15 chlorinated insecticides on a WCOT glass capillary column containing polyphenyl methyl silicone gum (CPtm Sil7). Concentrations of one-tenth the amount depicted in Figure 150 can be detected. Specially purified hexane was used as the extractant. In Figure 151 is shown the linearity of response obtained for 1–5 μl injections of endrin. This curve shows the absence of adsorption even at the sub-picogram level. These workers emphasize that both the capillary column and the electron capture detector require extremely pure carrier gas. To avoid contaminations, especially oxygen, water, and organic volatiles from pressure regulators, Teflon couplings, etc., proper molecular sieve carrier gas filters and oxygen traps are absolutely necessary. To eliminate organic impurities a high capacity charcoal filter should be used. All filters are placed as near as possible to the injection port respectively the detector purge inlet, i.e. beyond the pressure/flow regulators. It is also advisable before installation to heat the carrier gas inlet tubing red hot, in order to thoroughly clean organic materials from the tubing. This can be done with a torch, whilst carrier gas is flowing through.

The American Public Health Association[75,76] published a gas chromatographic method for the solvent extraction and gas chromatographic determination of eleven chlorinated insecticides in water samples in amounts,

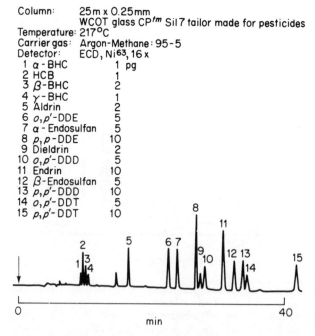

Figure 150 Chromatogram of 15 chlorinated insecticides (test mixture 50× diluted). Copyright (1979) Chromopak International B.V.

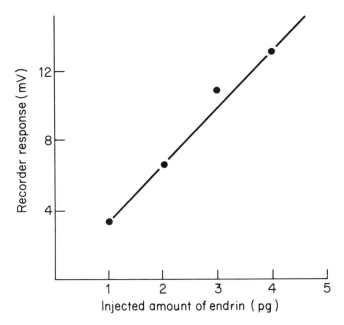

Figure 151 Calibration curve, endrin. Copyright (1979) Chromopak International B.C.

down to $0.005\,\mathrm{mg\,l^{-1}}$, p,p'-DDD, p,p'-DDE, p,p'-DDT, aldrin, dieldrin, endrin, heptachlor, heptachlor epoxide, lindane, isodrin, and methoxychlor. The insecticides carbophenothion, chlordane, dioxathion, diazinon, ethion, malathion, parathion methyl, methyl trithion, parathion, toxaphene, and VC-13 may be determined when present at higher levels. Also, the chemicals chlordene, hexachlorobicycloheptadiene, and hexachlorocyclopentadiene, which are pesticide manufacturing precursors, may be analysed by this method.

The insecticides are extracted directly from the water sample with n-hexane. After drying and removing the bulk of the solvent, the insecticides are isolated from extraneous material by microcolumn adsorption chromatography. The insecticides are then analysed by gas chromatography. This method is a modification and extension of the procedures developed by Lamar, Goerlitz, and Law.[77,78] For the analysis of insecticides in waters that are grossly polluted by organic compounds other than pesticides, a high-capacity clean-up procedure is used, as detailed in Federal Water Pollution Control Administration *Method for Chlorinated Hydrocarbon Pesticides in Water and Wastewater*.[79]

Recoveries obtained by this procedure for twelve insecticides from surface water samples are tabulated in Table 98.

A procedure involving ultraviolet irradiation followed by gas–liquid chromatographic detection of photodecomposition products has been used by Erney[80] to confirm the identity of selected organochlorine insecticides. Erney[80] studied the photodecomposition product patterns of 33 chlorinated insecticides.

Table 98 Insecticides in water: recovery of compounds added to surface-water samples. Insecticides and amount added ($\mu g\,l^{-1}$)

Sample no.	Aldrin 0.019	p,p'-DDD 0.080	p,p'-DDE 0.040	p,p'-DDT 0.081	Dieldrin 0.019	Endrin 0.040	Heptachlor 0.018	Heptachlor epoxide 0.021	Lindane 0.021	Malathion 0.181	Parathion methyl 0.082	Parathion 0.076
1	82.0	92.5	86.5	95.0	98.8	95.1	86.8	94.0	90.7	92.9	75.1	99.0
2	113	89.1	94.3	97.0	104	98.0	98.7	94.9	101	106	94.6	96.0
3	90.1	96.0	93.5	103	99.6	86.0	95.2	99.2	99.0	120	89.8	110
4	92.1	95.5	92.1	101	104	81.9	96.3	98.1	107	89.3	81.0	86.0
5	97.0	95.0	93.2	96.0	106	81.1	99.6	103	97.5	105	87.0	107
6	89.5	90.5	92.1	96.0	97.7	83.4	95.5	94.2	109	99.3	86.3	84.1
7	91.2	105	95.6	99.0	104	86.6	95.7	99.1	103	107	81.5	103
8	96.1	99.5	96.8	99.0	105	85.0	103	101	115	109	97.1	118
9	95.7	94.0	99.0	102	103	83.3	100	98.3	101	115	91.7	97.9
10	85.0	93.5	98.3	98.0	103	83.3	93.7	98.5	111	103	99.0	101
11	89.9	93.5	95.6	97.5	99.4	90.0	93.3	92.1	99.8	97.8	96.6	85.7
12	86.5	93.0	89.2	92.6	94.9	83.3	90.0	91.7	94.5	101	83.4	87.8
13	91.9	87.6	87.4	93.6	99.3	89.9	99.2	95.4	94.4	106	86.8	124
14	95.3	89.6	90.7	93.1	105	88.2	97.8	100	98.5	89.0	95.1	86.7
15	85.9	86.6	92.8	93.1	104	89.0	88.1	93.4	108	99.6	86.8	88.4
16	96.4	85.6	92.1	88.6	102	90.6	98.1	97.6	102	105	93.7	89.4
17	96.4	84.1	86.5	92.6	104	92.4	99.8	100	101	100	92.1	95.6
18	84.2	98.5	107	104	110	82.5	90.5	90.6	117	107	121	110
Mean	92.1	92.8	93.4	96.7	102	87.2	95.5	96.7	103	103	91.0	95.3
Variance	49.3	27.9	24.9	17.1	12.9	23.0	21.5	12.5	51.1	63.4	96.5	139.7
s.d.	7.02	5.28	4.99	4.14	3.59	4.80	4.64	3.54	7.15	7.96	9.82	11.8
Mean error	−7.9	−7.2	−6.6	−3.3	+2.0	−13.0	−4.5	−3.3	+3.0	+3.0	−9.0	−1.7
Total error	22	18	13	12	9.2	22	14	10	17	19	29	25

The attractiveness of this technique lies in the fact that it offers a means of independent identification of insecticides. The degree of photochemical reaction is dependent on wavelength, intensity, and time of irradiation as well as the physical state and chemical properties of the molecules involved. The conditions of irradiation and physical state must be controlled to give consistent reactions suitable for identification.

Erney[80] recommends iso-octane or hexane as solvent for electron capture gas chromatography. He used a Packard series 7800 equipped with electron capture detection and a 180 cm × 4 mm i.d. glass column packed with 10% OV-101 on Chromosorb WHP. Operation parameters: temperatures—column, 190 °C; inlet and outlet, 220 °C; detector 220 °C; sensitivity 1×10^{-9} amps mV^{-1}. Voltage was adjusted from 40 to 80 V to produce 50% full scale deflection for 1.0 ng heptachlor epoxide eluting in 5 minutes. For irradiation he used an ultraviolet light source at 254 nm. Samples were contained in a 2 cm pathway cylindrical quartz cells with tight fitting caps. The intensity of irradiation was standardized by determining the position relative to the ultraviolet source which resulted in equal gas chromatographic peak heights for the parent compound and major photodecomposition product after 8 minutes of exposure of a solution of heptachlor epoxide in iso-octane.

Chromatograms for 33 organochlorine insecticides are shown in Figures 152–184. Where possible, time of irradiation (I_t) to obtain approximately equal peaks heights for photodegradation product and parent compound is

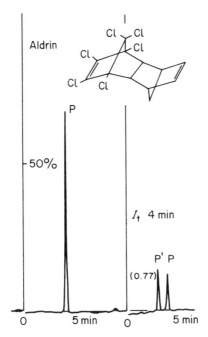

Figure 152 Gas chromatogram of aldrin. Reprinted from Erney[80] by courtesy of Marcel Dekker Inc.

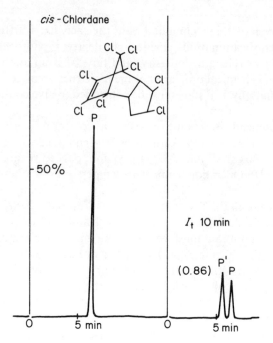

Figure 153 Gas chromatogram of *cis*-chlordane. Reprinted from Erney[80] by courtesy of Marcel Dekker Inc.

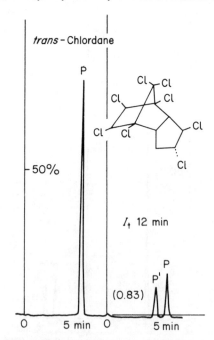

Figure 154 Gas chromatogram of *trans*-chlordane. Reprinted from Erney[80] by courtesy of Marcel Dekker Inc.

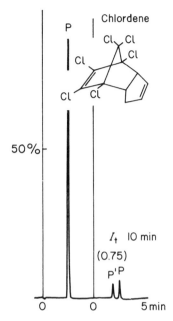

Figure 155 Gas chromatogram of chlordene. Reprinted from Erney[80] by courtesy of Marcel Dekker Inc.

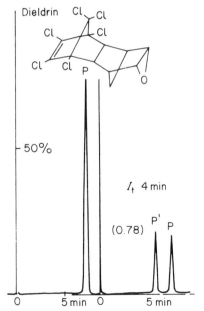

Figure 156 Gas chromatogram of dieldrin. Reprinted from Erney[80] by courtesy of Marcel Dekker Inc.

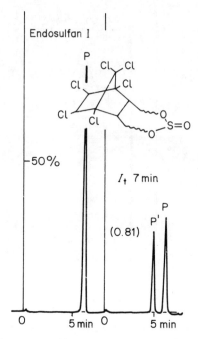

Figure 157 Gas chromatogram of endosulfan I. Reprinted from Erney[80] by courtesy of Marcel Dekker Inc.

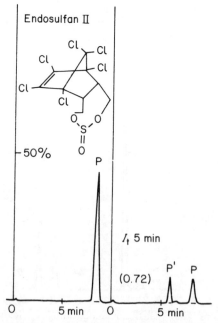

Figure 158 Gas chromatogram of endosulfan II. Reprinted from Erney[80] by courtesy of Marcel Dekker Inc.

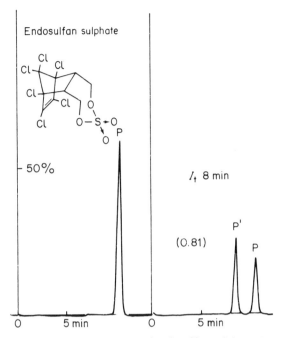

Figure 159 Gas chromatogram of endosulfan sulphate. Reprinted from Erney[80] by courtesy of Marcel Dekker Inc.

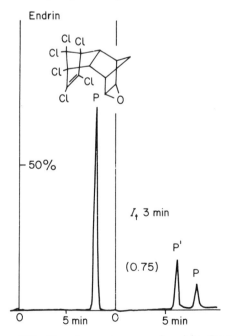

Figure 160 Gas chromatogram of endrin. Reprinted from Erney[80] by courtesy of Marcel Dekker Inc.

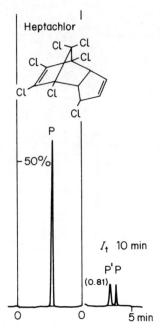

Figure 161 Gas chromatogram of heptachlor. Reprinted from Erney[80] by courtesy of Marcel Dekker Inc.

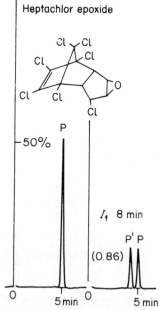

Figure 162 Gas chromatogram of heptachlor epoxide. Reprinted from Erney[80] by courtesy of Marcel Dekker Inc.

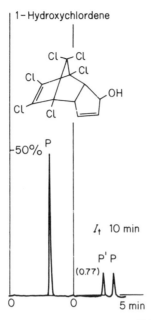

Figure 163 Gas chromatogram of 1-hydroxychlordene. Reprinted from Erney[80] by courtesy of Marcel Dekker Inc.

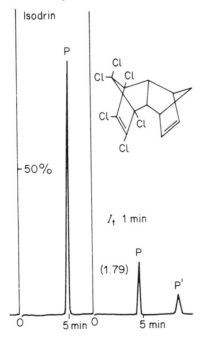

Figure 164 Gas chromatogram of isodrin. Reprinted from Erney[80] by courtesy of Marcel Dekker Inc.

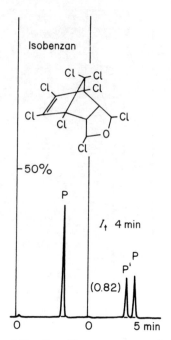

Figure 165 Gas chromatogram of isobenzan. Reprinted from Erney[80] by courtesy of Marcel Dekker Inc.

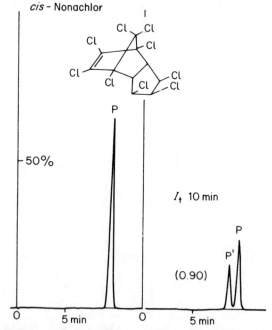

Figure 166 Gas chromatogram of *cis*-nonachlor. Reprinted from Erney[80] by courtesy of Marcel Dekker Inc.

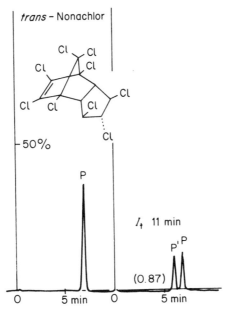

Figure 167 Gas chromatogram of *trans*-nonachlor. Reprinted from Erney[80] by courtesy of Marcel Dekker Inc.

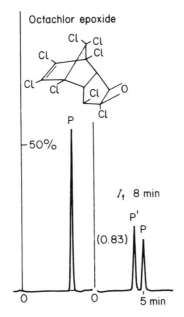

Figure 168 Gas chromatogram of octachlor epoxide. Reprinted from Erney[80] by courtesy of Marcel Dekker Inc.

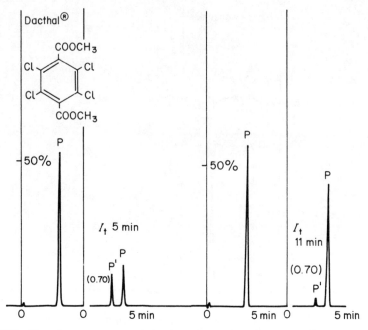

Figure 169 Gas chromatogram of dacthal. Reprinted from Erney[80] by courtesy of Marcel Dekker Inc.

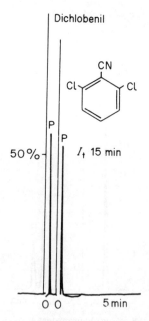

Figure 170 Gas chromatogram of dichlobenil. Reprinted from Erney[80] by courtesy of Marcel Dekker Inc.

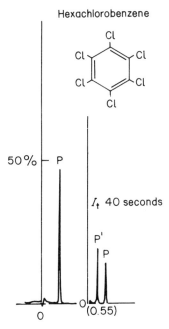

Figure 171 Gas chromatogram of hexachlorobenzene. Reprinted from Erney[80] by courtesy of Marcel Dekker Inc.

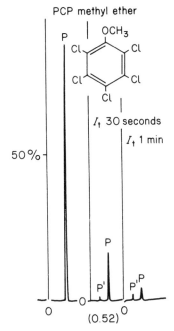

Figure 172 Gas chromatogram of PCP methyl ether. Reprinted from Erney[80] by courtesy of Marcel Dekker Inc.

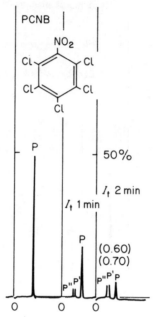

Figure 173 Gas chromatogram spectrum of PCNB. Reprinted from Erney[80] by courtesy of Marcel Dekker Inc.

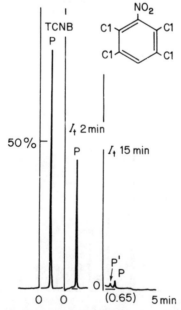

Figure 174 Gas chromatogram of TCNB. Reprinted from Erney[80] by courtesy of Marcel Dekker Inc.

given. Compounds in Figures 152–168 had a methanopthalene, methanoindene or similar type chemical structure. These compounds gave definite photodegradation patterns adequate for confirmation, and approximately a 1:1 peak height ratio for photodecomposition product and parent in the I_t reported. Conclusive identification generally is possible prior to the I_t reported; for example, 2 minutes is sufficient to confirm dieldrin. Chlorinated benzene derivatives depicted by Figures 169–174 also produced characteristic photodegradation patterns except for dichlobenil (Figure 170). Dichlobenil did not appear to degrade using the photochemical technique. Figure 169 shows confirmation of the presence of Dacthal is possible after 5 minutes of irradiation although 11 minutes is required to obtain an approximately equal gas chromatographic peak height for parent and product. Chromatographic evidence indicates photochemical reaction is affected by the composition and relative positions of the substituents on the ring. Results for DDT and somewhat similar compounds (Figures 175–182), show elimination of parent, but no significant observable photodegradation product. Similar observations are found for pesticides with chlorine adjacent to the sulphur–carbon bond outside the ring as shown in Figures 183 and 184. Thus, the photodegradation technique does not allow conclusive identification of these compounds.

Chau[81] has investigated photodegradation products of endrin including hexachloroketone and pentachloroketone. The residues obtained after clean-up on a Florisil column were heated with potassium t-butoxide in t-butyl alcohol,

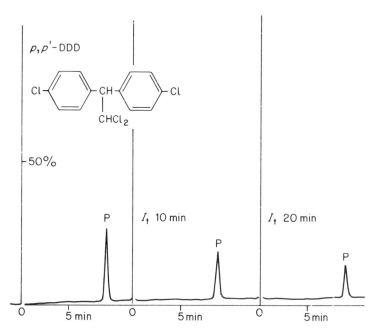

Figure 175 Gas chromatogram of p,p'-DDD. Reprinted from Erney[80] by courtesy of Marcel Dekker Inc.

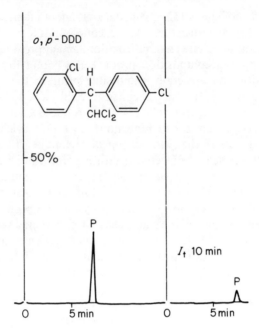

Figure 176 Gas chromatogram of o,p'-DDT. Reprinted from Erney[80] by courtesy of Marcel Dekker Inc.

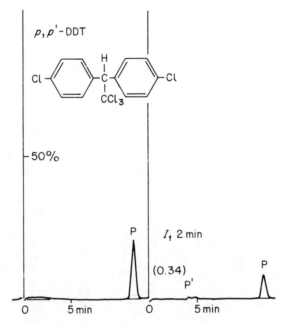

Figure 177 Gas chromatogram of p,p'-DDT. Reprinted from Erney[80] by courtesy of Marcel Dekker Inc.

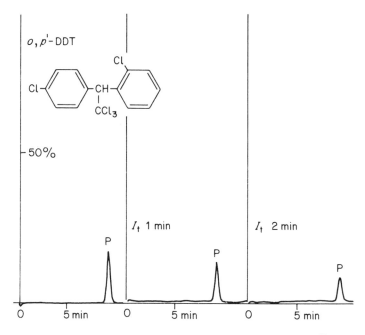

Figure 178 Gas chromatogram of *o,p'*-DDT. Reprinted from Erney[80] by courtesy of Marcel Dekker Inc.

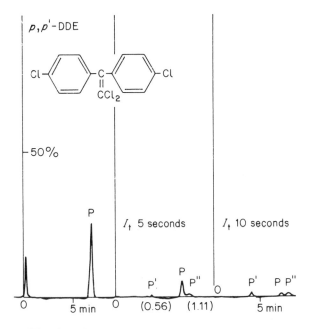

Figure 179 Gas chromatogram of *p,p'*-DDE. Reprinted from Erney[80] by courtesy of Marcel Dekker Inc.

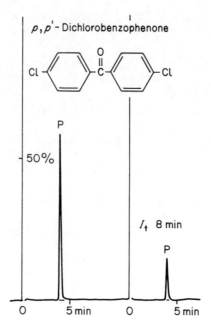

Figure 180 Gas chromatogram of *p,p'*-dichlorobenzophenone. Reprinted from Erney[80] by courtesy of Marcel Dekker Inc.

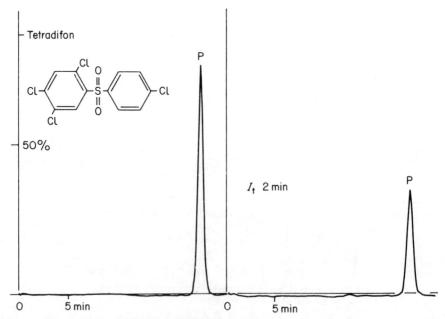

Figure 181 Gas chromatogram of tetradifon. Reprinted from Erney[80] by courtesy of Marcel Dekker Inc.

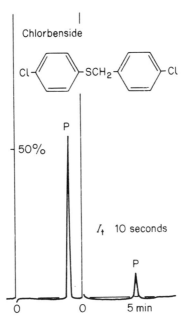

Figure 182 Gas chromatogram of chlorbenside. Reprinted from Erney[80] by courtesy of Marcel Dekker Inc.

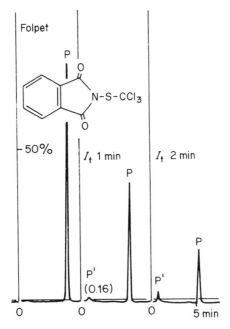

Figure 183 Gas chromatogram of folpet. Reprinted from Erney[80] by courtesy of Marcel Dekker Inc.

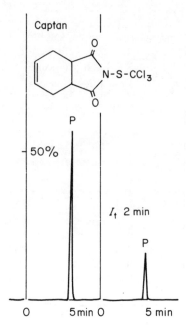

Figure 184 Gas chromatogram of captan. Reprinted from Erney[80] by courtesy of Marcel Dekker Inc.

and the products were silylated or acetylated. The respective volatile derivatives were analysed by conventional gas chromatography. Levels as low as 30 and 0.01 ng g^{-1} in fish and water, respectively, could be identified.

Hesselberg and Johnson[82] used Florisil column extraction followed by gas chromatography to determine DDT, dieldrin, endrin and, methoxychlor in fish and mud. Samples are prepared by blending with sodium sulphate (plus solid carbon dioxide for fish) until a free-flowing dry mixture is obtained. A glass column (400 mm × 20 mm) is packed with sodium sulphate (2 g), the sample mixture is added and tamped down, and the resulting column is washed with solvent (200 ml) at 3–6 ml min^{-1}. The solvents used are 1% methanolic phosphoric acid to elute 2,4-D, cyclohexane for DDT, dieldrin, endrin, and methoxychlor, ethyl ether for simazine and 10% of ether in light petroleum for parathion. The eluates are cleaned up by solvent partitioning and column chromatography on Florisil and the pesticides are then determined by gas chromatography on a packed OV-7 column. Recoveries of added pesticides (>5 ng g^{-1}) were 95–100%.

LeRoy and Goerlitz[83] have described the use of microcolumns (4 cm × 5 mm) with deactivated alumina, silica gel, and Florisil for the clean-up of chlorinated insecticides prior to gas chromatography. Weil and Quentin[84] and Lauren[85] have used silica gel for clean-up. Lauren[85] extracted the sample containing 10–12 g chlorinated insecticide with ethyl ether (50 ml). The extract is concentrated to 1 ml and placed on a column of deactivated silica gel (1 g, 60–200 mesh) moistened with hexane (pesticide grade), and elution is effected with benzene–hexane (1:9) (6.5 ml, followed by a further 6.5 ml), the two fractions being collected separately. Each fraction is concentrated in a stream of air and injected into a gas chromatography equipped with an electron capture detector and a glass column (4 ft × 3 mm) packed with 3% of QF-1 and 2% of OV-1. Organophosphorus insecticides, which are retained on the silica, can be eluted as a third fraction with ethyl acetate–benzene (1:9). Griffitt and Craun[86] have described a gel permeation chromatographic procedure using a column packed with Bio-Beads S-X2 to separate chlorinated insecticides from glycerides in a clean-up procedure preparatory to gas chromatography of the extract.

Södergren[87] used digestion with fuming sulphuric acid to clean up samples prior to the determination of chlorinated insecticides in algae by gas chromatography. A hexane extract of the sample is concentrated to 350 μl, and 50 μl is sealed in a glass tube with 50 μl of fuming sulphuric acid (10% SO_3). After mixing, the phases are then separated and the hexane layer is subjected to gas chromatography. A second portion of the original extract is mixed with an equal volume of 5% propanolic potassium hydroxide in a special pipette, which is sealed and heated in a water bath for 10 min. After heating, 5 ml of water is added and, after further mixing, the hexane fraction is allowed to separate for analysis. The remainder of the original extract is evaporated, and the residue of extractable lipids is weighed. Sample recoveries are 78–94%, losses occurring mainly in the extraction stage.

The removal of sulphur from water samples prior to the gas chromatographic determination of chlorinated insecticides has been discussed.[88,89] Woodham and Collier[88] used copper, whilst Baird et al.[89] used a column packed with a mixture of 4% of OV-17, 47% of QF-1, and 1% of DC-200 on Gas-Chrom G.

An all-glass apparatus for sampling water samples for chlorinated insecticide analyses has been described.[90]

Interference effects on the gas chromatogram have been reported when water samples are filtered through ordinary filter paper clamped in a PTFE gasket.[91] Levi and Nowicki[92] have overcome this interference by using a filtration apparatus constructed in stainless steel with a fritted borosilicate glass filtration disc.

Dolan and Hall[93] have described a Coulson electrolytic conductivity detector of enhanced sensitivity for the gas chromatographic determination of chlorinated insecticides in the presence of polychlorinated biphenyls. The detector was modified by the replacement of the silicone-rubber septum and stainless steel fitting at the exit of the pyrolysis furnace with a PTFE fitting, by the reduction in diameter of the PTFE transfer tube, and by the replacement of the 4 mm (i.d.) reaction tube with one of 0.5 mm. These modifications reduced hydrogen chloride adsorption and tailing, and improved sensitivity and reproducibility; sensitivity was also enhanced by increasing the cell voltage to 44 V.

Chlorinated insecticides in foods, fish etc.

Novikova[94] has reviewed the literature (209 references) covering the extraction, clean-up, and analyses of organochlorine and organophosphorus insecticides in food, soil, and water. Arias et al.[95] have described a method for the determination of organochlorine insecticide residues in molluscs, vegetables, rice, butter, and in river water. The method involves extraction, Florisil column clean-up, and analysis of the extract by thin-layer chromatography on silica gel G or alumina with hexane or hexane–acetone (49:1) as solvent, or gas chromatography on a polar column of 10% of DC-200 on Chromosorb W HMDS and on a semipolar column of 5% of DC-200 plus 7.5% of QF-1 on Chromosorb W, with electron capture detection.

Gas chromatography has been used for the determination of organochlorine insecticides residues in crabs, shrimp, and fish,[96] fresh water Triidad,[97] planarians, plankton,[98] and sediments and fish tissues.[99]

The methods for determining DDT, TDE, and dieldrin in sediments and fish tissues are reviewed below.

Air dried mineral sediments (25 g) were brought to within 50% of field storage capacity and left for 24 hours. Acetone and hexane in the ratio 1:1 (250 ml) were added to the sediment and the mixture was shaken for 2 hours. An aliquot (100 ml) was filtered off, water was added, and the organochlorine insecticides were partitioned into hexane. Air dried organic sediments (25 g) were blended with acetonitrile and water (2:1) for 5 minutes and an aliquot (10 g) was filtered off. The filtrate was partitioned into hexane. A sample (10 g) of the fish was mixed with granular anhydrous sodium sulphate (100 g) and standard Ottawa sand (25 g) by grinding in a mortar and pestle until all tissue was finely subdivided and homogenized. This mixture was transferred to a Soxhlet extraction apparatus and subjected to exhaustive extraction with hexane for 7 hours. Extracts from sediments and fish tissues were evaporated to dryness by rotary vacuum at 45 °C. In the case of fish the percentage of fat or oil was determined gravimetrically. A one-step column clean-up method[100] was used for the isolation of organochlorine insecticides. Florisil (60–100 mesh) activated at 650 °C was reheated at 135 °C for a minimum of 24 hours. After cooling, the adsorbent was partly deactivated by the addition of water at the rate of 5 ml 100 g^{-1} and allowed to equilibrate. Up to 1 g of fat from fish extracts and 10 g of sediment extract were mixed thoroughly with 25 g of conditioned Florisil. This was placed on top of a second 25 g portion of prewashed adsorbent in pyrex columns (25 mm × 300 mm) fitted with reservoirs (350 ml). The entire column was eluted with 300 ml of the 1:4 dichlormethane–hexane (v/v) solvent mixture at a percolation rate of approximately 5 ml min^{-1}. Eluates were concentrated just to dryness with rotary vacuum evaporation at 45 °C, the residue was redissolved in 5 ml hexane and used for subsequent chromatographic analysis. All solvents had been redistilled from glass. Varian Aerograph Models 204 and 1200 gas chromatographs, equipped with 250 millicurie tritium electron capture detectors, were used for qualitative and quantitative assays. Operating parameters were as follows:

Column: 5 ft × ⅛ in. pyrex packed with 4% SE-30 + 6% QF-1 on Chromosorb W preconditioned 72 h at 225 °C (13).
Temperatures: column, 175 °C; detector, 200 °C; injection block, 225 °C.
Carrier gas: nitrogen at 40 ml min^{-1}.

Injection volumes of 5 μl were used for both sample solutions and comparison standards. Qualitative residue confirmation was accomplished with thin-layer chromatography using silica gel. Plates were developed with 1% chloroform in n-heptane, and visualized with alkaline silver nitrate spray as the chromogenic agent. Alternatively, p,p'-DDT and p,p'-TDE were confirmed by treatment with 5% methanolic potassium hydroxide.[101] Partial confirmation of dieldrin was achieved by fractionating the analysis solution on a Mills column,[102] thus isolating dieldrin in the second eluate fraction. Per cent recoveries of pesticides from fish tissue were checked by fortification directly into the oil obtained as the result of the hexane extraction. Averaged recoveries were: p,p'-DDE, 97.8%; o,p'-DDT, 91.2%; p,p'-TDE, 94.6%; p,p'-DDT, 89.6%; dieldrin, 89.3%.

Other chromatographic methods

Liquid chromatography

Liquid chromatography coupled with an electron capture detector has been used[103] for the analyses of aldrin, lindane, dieldrin, also parathion, carbamate herbicides and 2,4-D in surface waters. These workers combined high performance liquid chromatography with electron capture detection by vaporizing the total eluant from the liquid chromatographic column and passing it directly into an electron capture detector. Contrary to expectations, the reduction in sensitivity (compared to standard gas chromatographic operation) is only modest so that the detector is still much more sensitive to organochlorine insecticides than currently available liquid chromatographic detectors. About 10^{-10} g aldrin can be detected by this method. A schematic diagram of the liquid chromatograph is given in Figure 185. The sampling valve was used to inject 3 μl of a sample directly onto the head of the column without interrupting flow of the mobile phase. The column consisted of 10 μm spherical particles of Spherisorb SLOW (UKAEA, Harwell) packed into a 100 mm × 2 mm glass-lined stainless steel tube (SGE Pty Ltd.). The column eluant passes into a stainless steel transfer tube (400 mm × 0.25 mm) enclosed in an oven, the temperature of which is such that the liquid is completely vaporized. The increase in volume involved in this transition forces the vapour into a ^{63}Ni electron capture detector which is contained in a separate oven. A purge of 30 ml min^{-1} nitrogen sweeps the vapour through the detector and prevents back diffusion of eluant into the purge conduit. The vapour was condensed in a coil of stainless steel tubing (2000 mm × 4.5 mm) and collected as a liquid. The signal from the

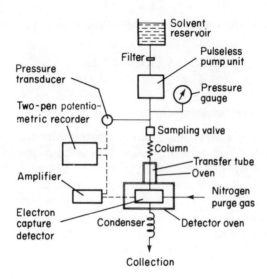

Figure 185 Schematic diagram of the liquid chromatograph. Reproduced from the *Journal of Chromatographic Science* by permission of Preston Publications, Inc.

electron capture detector was amplified using either a pulse mode amplifier or a constant current amplifier. The latter operates with 60 V pulses having a nominal width of 1 μs over a frequency range 0–130 kHz. The standing current can be varied between zero and 5×10^{-9} A.

Pulse sampling and constant current operator were used to establish the electron capture detection standing current. Table 99 compares detection sensitivities for aldrin obtained by various previously reported detector techniques and compares them with those achieved by Willmott and Dolphin[103] using an electron capture detector. It is seen that their detector is two orders of magnitude more sensitive than the only other reported combination of liquid chromatography and electron capture detector using the moving wire system.[104] The present system also shows a significant improvement in sensitivity compared with all available liquid chromatographic detectors, including the commonly used ultraviolet adsorption monitor. The magnitude of the expected reduction of sensitivity of the electron capture detector is evident when the minimum detectable concentration of 1.6×10^{-15} g ml^{-1} for gas chromatograph–electron capture detector[110] is compared with the Willmott and Dolphin[103] value of 3.2×10^{-13} g ml^{-1} (calculated with respect to flow of vapour phases through the electron capture detector.) Part of the overall reduction is due to a twofold increase in noise but the major contribution is due to the background of electron capturing species in the solvent vapour reducing the probability of electron capture by specific molecules of interest. Bleeding of gas chromatographic stationary phases has been shown to have a similar effect.[111] Figure 186 shows linearity plots for both the pulse sampling

Table 99 Comparison of detector sensitivities

System	Detector	Reference compound	Min. detectable concentration (g ml^{-1}) l.c.	g.c.	Min. det. quantity (g)	Reference
l.c.	Polargraphic	Parathion	4×10^{-9}			105
l.c.	Fluorimetric	Carbamates			$1-10 \times 10^{-9}$	106
l.c.	Refractometer	Aldrin			0.4×10^{-6}	107, 108
l.c.	u.v. (278 nm)	2,4-D			5×10^{-9}	174
g.c.	e.c.d. constant current	Dieldrin		4×10^{-14}		109
g.c.	e.c.d. pulse sampling	Aldrin		1.6×10^{-15}†		110
l.c.	e.c.d.	Aldrin	1.2×10^{-10}	3.2×10^{-13}*	4×10^{-11}	103
l.c.	u.v. (230 nm)	Aldrin	3.8×10		1.3×10^{-8}	103

*Noise—4×10^{-12} A.
†Noise—2×10^{-12} A.

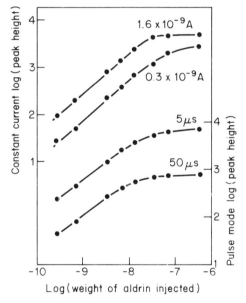

Figure 186 Comparison of linearity in pulse sampling and constant current modes. Flow rate, 1 ml min^{-1}; detector temperature, 300 °C; purge 30 ml min^{-1} N$_2$. Reproduced from the *Journal of Chromatographic Science* by permission of Preston Publications, Inc.

and constant current methods using aldrin. The upper region of the linear range extends further for constant current than for pulse sampling which is in agreement with gas chromatographic–electron capture detector work.[106] As the minimum detectable concentration values are comparable the constant current method gives the better linear range (approximately 500) and is the preferred technique. Although in the constant current mode the minimum detectable concentrations decreases monotonically with increasing standing current the upper limit of linearity also decreases (Figure 186). Both effects are of a similar order so that the net result on linear range is insignificant. In practice with such a sensitive detector it was found more useful to operate at a low standing current, hence extending the upper limit of linearity.

Willmott and Dolphin[103] also reported on other parameters which affect quality of results obtained by their technique, viz. solvent flow rate, purge flow and transfer tube temperature.

Willmott and Dolphin[103] applied their high performance liquid chromatography–electron capture detector approach to the determination of chlorinated insecticides in surface waters. Because the concentration of pesticides in most surface waters is less than 1 ppb some form of extraction and concentration from large volumes of water was necessary before analysis is possible. These workers applied a conventionally coated chromatographic support as a reverse liquid–liquid partitioning filter to the extraction of pesticides. They used uncoated polyether polyurethane foam; 0.5 g of the flexible foam were inserted into a 10 mm i.d. quartz tube and cleaned by washing with consecutive 100 ml aliquots of acetone, n-hexane, ethanol, and distilled water. Ten litres of water sample were passed through the column at a flow rate of about $2 l h^{-1}$ by applying slight suction. Excess water was removed and the adsorbed compounds were eluted with 20 ml of n-hexane which was subsequently dried with anhydrous sodium sulphate and reduced by slow evaporation in a stream of nitrogen to

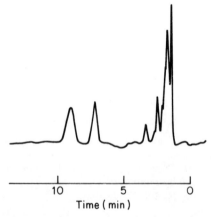

Figure 187 Chromatogram of sewage outfall extract. Flow rate, 0.2 ml min^{-1}; detector, ECD; attenuation, ×128; standing current 1.8×10^{-9} A; purge 30 ml min^{-1} N$_2$. Reproduced from the *Journal of Chromatographic Science* by permission of Preston Publications, Inc.

a volume of 1 ml. By applying the procedure to water spiked to 1 ppb with dieldrin the overall extraction efficiency for that pesticide was found to be about 80%. Figure 187 shows a chromatogram of an extract of the final outfall from a plant treating domestic sewage. The major components correspond to approximately 30 ng l^{-1} in the original sample based on an aldrin calibration. In the analysis of environmental samples by gas chromatography with an electron capture detector it is essential that polychlorinated biphenyls are previously separated from DDT and related compounds as both these groups have similar retention times. Separation is usually effected by column chromatography using alumina, Florisil, or silica gel columns. High performance liquid chromatography alleviates the requirement for such rigorous clean-up, as adequate separation of PCBs and most organochlorine pesticides can be obtained on the column. Figure 188 illustrates the separation of nanogram quantities of p,p'-DDT and Aroclor 1260.

Figure 188 Separation of p,p'-DDT from Aroclor 1260. Flow rate, 0.2 ml min^{-1}; detector, ECD; attenuation ×256; standing current 1.8×10^{-9} A; purge 30 ml min^{-1} N$_2$. Identification of peaks: 1,2,3, Aroclor 1260; 4, DDT impurity; 5, p,p'-DDT. Reproduced from the *Journal of Chromatographic Science* by permission of Preston Publications, Inc.

Thin-layer chromatography

Thin-layer chromatography has been used by several workers to confirm the identity of chlorinated pesticides.[114-117] Armstrong and Terrill[118] used aqueous solutions of sodium sulphate with polyamide to separate p,p'-DDT, p,p'-DDD and p,p'-DDE and decachlorobiphenyl on alumina coated thin-layer sheets.

Suzuki et al.[112,113] separated chlorinated insecticides into two groups by column chromatography and then into three further groups by thin-layer chromatography. Individual insecticides were then isolated and determined by gas chromatography. The members of the two groups separated by column chromatography were: first division insecticides—aldrin, DMC ethylene

[1,1-*bis*-(4-chlorophenyl)ethylene], chlorfensulphide, DDS [*bis*-(4-chlorphenyl)-disulphide], *o,p'* and *p,p'*-DDT, heptachlor, quintozene, isobenzan and tetrasul; second division insecticides—alpha-, beta-, and gamma-BHC, trifluralin, nitrofen, endrin, dieldrin, and dicofel.

Achari et al.[119] used thin-layer chromatography to confirm the identity of organochlorine pesticides in ground water. Bevenue et al.[120] point out that when a comparatively small sample of water (one gallon or less) is extracted for detection of pesticides and submitted to thin-layer chromatography, the use of spray reagents may not be practical. The spots are therefore extracted, and the concentrated extracts are submitted to gas chromatography. Extraneous interferences are magnified on the recorder chart unless special precautions are taken to exclude impurities from the organic solvents or from the apparatus used. The necessary pretreatment of glassware and silica gel is discussed and described. Plastic ware and filter paper should not be used.

N.m.r. spectroscopy

Keith and Alford[121] have reviewed the applications of n.m.r. spectroscopy to the determination of organochlorine pesticides (57 references).

Yeast hexokinase assay

Sadar and Guilbauelt[124] have described an assay method in which down to one micromole of chlorinated insecticides, organophosphorus insecticides, and herbicides and fungicides can be determined with an error of about 5%.

Determination of individual organochlorine insecticides
DDT, DDE, and DDD

Ernst et al.[122,123] have determined by gas chromatography–mass spectrometry residues of DDT, DDE and DDD and polychlorinated biphenyls in birds and scallops from the English Channel. The procedure used for working up the tissue samples has been described by Ernst et al.[123] and is summarized in Figure 189. Two gas chromatographs were used under the conditions stated in Table 100. Calibration was done daily using five different concentrations of either PCB (Clopen A60) + DDE or DDT + DDD. PCB values were calculated from the three major peaks and then averaged.

Method

Reagents Alumina, neutral (Woelm), heated 3 h at 850 °C, deactivated with 5% water, Florisil 60–100 mesh heated 2 h at 650 °C, addition of 0.4–0.6%

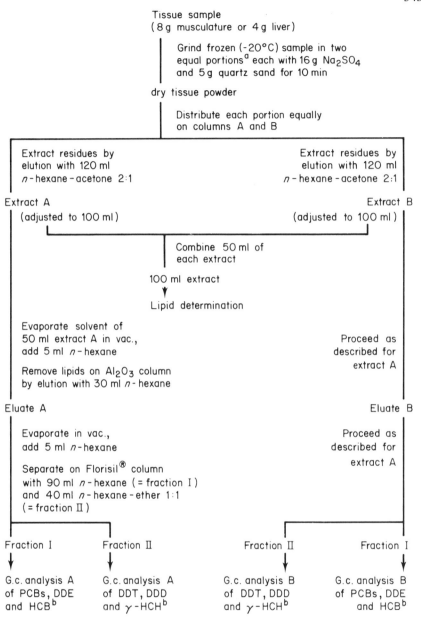

Figure 189 Procedure for two parallel residue analyses of a tissue sample. Reprinted with permission from Ernst et al.[123] Copyright (1974) Springer Verlag

Table 100 Conditions of gas chromatographic analysis

	Beckman GC 5	Varian 2740
Column	1.8 m, all glass, 2 mm i.d., 5% SE 30 on Chromosorb W 80–100 mesh	1.8 m, all glass, i.d., 1.5% SP-2250 (OV-17) + 1.95% SP-2401 (QF-1) on Supelcopot 100–120 mesh
Column Temperature	200 °C (DDD + DDT) 220 °C (DDE + PCB)	190 °C
Inlet Temperature	10 °C above column temperature	225 °C
Detector	Beckman ECD	^{3}H-ECD
Detector Temperature	300 °C	225 °C
Carrier gas	Helium, 40 ml min^{-1}	Nitrogen, 30 ml min^{-1}

Reprinted with permission from Ernst et al.[123] Copyright (1974) Springer Verlag.

water after 24 h, ready for use after additional 24 h, storage in well closed glass-stoppered flasks; sodium sulphate anhydrous GR, heated 2 h at 650 °C; quartz sand; acetone GR, distilled over 65 cm Widmer column; n-hexane for synthesis, distilled over 50 cm packed column; ether GR, threefold distilled.

Apparatus Mortar mill, type Pulverisette 2, with agate equipment. Glass columns for extraction 13 × 280 mm, solvent reservoir on top 50 × 100 mm. Alumina and Florisil columns 5 × 480 mm and 10 × 480 mm, solvent reservoirs on top 35 × 120 mm. Column fillings 8 g Al_2O_3 and 13 g Florisil respectively. Gas chromatograph conditions: see Table 100. Mass Spectrometer: Varian MATCH7 coupled to Varian-Aerograph, packed column.

Taylor et al.[125,126] isolated DDT from water and determined it by gas chromatography. This method is based on the extraction of pesticides from water with light petroleum, purification of the concentrated extract on a Florisil column followed by gas chromatography at 195 °C on a column (160 × 3 mm) packed with 5% of DC-11 on Chromosorb W AW DMCS, with argon (45 ml min^{-1}) as the carrier gas and a tritium electron capture detector. For water containing appreciable amounts of suspended solids, the sample was filtered, and the solids were collected on a filter, dried, and extracted in a Soxhlet apparatus for 5 h with acetone. For samples of water containing appreciable amounts of municipal sewage, a preliminary separation of the insecticides by thin-layer chromatography on silica gel G (0.25 mm thick), with carbon tetrachloride as solvent, is necessary. The mean recoveries of DDT and its metabolites were 76%.

Balayanis[127] used surface modified adsorbents in a clean-up technique for analysis of DDT and γ-BHC in water prior to gas chromatography. The adsorbents comprised silica gel modified by treatment with trichlorooctadecylsilane or with dichlorodimethylsilane. These two products effected substantial clean-up, but also displayed different selectivities towards the pesticides. p,p'-DDE was strongly retained by the first product, requiring a second elution with 10 ml of hexane to recover it all, whereas with the second product, almost the whole of the added p,p'-DDE was recovered in the first 10 ml of hexane·p,p'-DDT was strongly retained by both products, as was γ-BHC. With 10 ml of hexane–acetone (19:1), γ-BHC could be eluted readily, without interfering peaks, from the first product.

Gooding et al.[128] used DDT-dehydrochlorinase for the identification of DDT in soils. The enzyme converts DDT to DDE which is then determined gas chromatographically on Chromosorb WHMDS at 190 °C using an electron capture detector.

Gillespie et al.[129] have pointed out that static bioassays for DDT and other toxins give relatively imprecise data and suggest improvements in methodology to overcome this.

Wilson[130] showed that liquid–liquid extraction of estuarine water immediately after addition of DDT gave acceptable recovery levels with all solvent systems tested, but analyses carried out several days later gave only partial recovery owing to adsorption of DDT on suspended matter.

Liquid scintillation counting of [^{14}C]DDT has been used to study the pick-up and metabolism of DDT by fresh water algae and to determine DDT in sea water.[132] Neudorf and Khan[131] investigated the uptake of ^{14}C-labelled DDT, dieldrin, and photodieldrin by *Ankistrodesmus amalloides*. The results of liquid scintillation spectrometric analyses show that the total pick-up of DDT during a 1–3 hour period was 2–5 times higher than that of dieldrin, and 10 times higher than that of photodieldrin. The algae metabolized 3–5% of DDT to DDE and 0.8% to DDD. The metabolism of DDT by *Daphnia pulex* was also monitored by exposing 100 organisms to 0.31 ppm of the labelled pesticide for 24 hours without feeding. The metabolites were then extracted and separated by thin-layer chromatography and the R_f values of radioactive spots were compared to R_f values for non-radioactive DDD and radioactive DDE. The results show a conversion of DDT to DDE of about 13.6%.

Picer et al.[132] described a method for measuring the radioactivity of labelled DDT contaminated sediments in relatively large volumes of water using a liquid scintillation spectrometer. Various marine sediments, limestone, and quartz in sea water were investigated. External standard ratios and counting efficiencies of the systems investigated were obtained as was the relation of efficiency factor to external standard ratio for each system studied.

Aldrin, dieldrin, endrin

Woodham et al.[133] converted dieldrin and endrin to chemical derivatives prior to gas chromatographic determination of these substances in crops, soil sediments, and water. An aliquot of extract after appropriate clean-up and evaporation to dryness, is treated with 1 ml of conversion reagent (10% soln of boron trichloride in 2-chloroethanol) in a centrifuge tube, which is then placed, unstoppered, in a water bath at 90 °C and left for 2 h or for 10 min for a sample that may contain dieldrin, or endrin, respectively, or for 2 h for a mixture. The tube is cooled, 5 ml of hexane and 10 ml of 7% aqueous sodium sulphate are added, and the contents are mixed and left for the phases to separate. The hexane phase is analysed on a column of OV-17-QF-1 on Gas-Chrom Q with electron capture detection (tritium source). Down to 0.01 ppm of either pesticides (0.01 part per 10^9 in water) could be detected.

Lester and Smiley[134] developed a rapid method for identifying aldrin in the presence of sulphur by electron capture gas chromatography. The presence of elementary sulphur in hexane extracts of water or sediment produces a response during gas chromatography with electron detection that is similar to the response produced by aldrin. It has been established that sulphur and aldrin can be separated on a glass column (6 ft × 0.25 in. o.d.) packed with 3% of OV-17 on Chromosorb W-HP. The column is operated at 180 °C and the carrier gas flow is 90 ml min^{-1}.

Dieldrin can be adsorbed from river water with XAD-2 ion-exchange resin.[135] A 4 litre sample is passed at 50 ml min^{-1} through a column (20 cm × 2 cm) containing XAD-2 resin (Rohm and Haas), and the pesticides are eluted with 200 ml of hot acetonitrile. The eluate is then diluted with 700 ml of water and the solution is extracted with pesticide grade light petroleum (boiling range 30–60 °C) (3 × 60 ml). The combined extracts are washed with water (2 × 15 ml) and dried on a column of sodium sulphate then concentrated to 1 ml for gas chromatographic analysis. The method is more convenient than conventional solvent extraction procedures and gives comparable results for dieldrin.

Simal et al.[136] have described solvent extraction and gas chromatographic methods for estimating dieldrin in fish, mollusc, and river water samples originating from contamination of the environment resulting from the wreck of a ship carrying dieldrin. The insecticide was extracted from samples of water with light petroleum, and from fatty materials by partition between light petroleum and acetonitrile or by direct homogenization with acetonitrile. The extracts were dried over sodium sulphate, evaporated, purified on columns of Florisil, and analysed on a glass column (2 m × 3.5 mm) with a non-polar packing of 5% DC-200 on Chromosorb W HMDS (80–100 mesh) or with a semipolar packing of 5% DC-200 plus 7.5% QF-1 on the same support, operated at 200 °C with nitrogen as carrier gas (180 ml min^{-1}) and with flame ionization and electron capture detectors.

Johnson and Starr[137] and Chiba and Morley[138] have studied factors affecting the extraction of dieldrin and aldrin from different soil types; ultrasonic extraction was recommended by these workers.

BHC

Konrad et al.[139] and Pionke[140] have described a method for determining γ-BHC, methoxychlor, and organophosphorus insecticides involving the extraction of 500 ml of water sample with 25 ml benzene, and concentration to 1 ml by a stream of air followed by gas chromatography using an electron capture or a KCl thermionic detector. Down to $0.06\,\mu\text{g}\,\text{l}^{-1}$ of γ-BHC was detected.

Mahel'ova et al.[141] determined BHC isomers in soil by gas–liquid and thin-layer chromatography after extraction with light petroleum. An air-dried, ground sample (18 mesh) (20 g) was deactivated by addition of 25% of water and set aside for 24 h. Siloxid (active silica) was added to form a powdery mixture, which was extracted with 250 ml of light petroleum (boiling range 35–50 °C) in a Soxhlet apparatus for 12 h. The extract was evaporated to 5 ml, and purified on a column of Celite 545 mixed with fuming sulphuric acid. The insecticides were eluted with light petroleum (250 ml), which was evaporated to dryness; the residue was dissolved in hexane and an aliquot of the solution was subjected to gas chromatography on a column packed with 1.5% OV-17 and 2% QF-1 on Chromosorb (80–100 mesh). The column was operated at 190 °C, nitrogen (60–80 ml min^{-1}) was used as carrier gas with an electron capture detector. Recoveries of the isomers, of p,p'-DDT and of 1,1-dichloro-2,2-bis-(4-chlorophenyl)ethylene at the 0.1 ppm level ranged from 82 to 106% with a coefficient of variation between 4.3 and 11.9%. The purified extracts were also examined by thin-layer chromatography.

Yamato[142] studied the persistence of BHC in river water using a solvent extraction gas chromatographic technique.

Strachel et al.[143] have reported on methods for on-site continuous liquid–liquid extraction from water samples of chlorinated compounds. The limits of detection were: γ-BHC (lindane) $0.5\,\text{ng}\,\text{l}^{-1}$, α-BHC $0.7\,\text{ng}\,\text{l}^{-1}$ and hexachlorobenzene $0.2\,\text{ng}\,\text{l}^{-1}$.

Trichlorphon

Mosinaka[144] has described a semiquantitative thin-layer chromatographic method for the determination of trichlorphon in drinking water. The sample (100–500 ml) is extracted with redistilled chloroform (3 × 100 ml). The extract is dried with sodium sulphate, reduced in volume to 4 ml *in vacuo* and then evaporated to dryness in a stream of air. The residue is dissolved in 1 ml of acetone, and 0.01–0.03 ml portions are used for the thin-layer separation on chloride-free silica gel G plates (activated at 100 °C for 1 h) with benzene–methanol (17:3) as solvent. The spots, revealed with ammoniacal silver nitrate in acetone, are compared with those of standards for semiquantitative determination. The detection limit is $0.2\,\mu\text{g}$ ($= 0.02$ ppm in the sample) and the efficiency of extraction is 70%.

Trichlorphon has been determined[145] in water, acid soil, leaves etc., by solvent extraction followed by gas chromatography on a glass column (6 ft × 0.375 in.) packed with 16% of XF-1150 on Chromosorb W-AW operated at 125 °C, with a carrier gas flow of 60 ml min^{-1} and a flame photometric detector operated in the phosphorus mode. Average recoveries were 96%, and down to 0.002 ppm of trichlorphon could be determined in water and 0.05 ppm in all other samples.

α-, β-Endosulfan, and heptachlor

Gorbach et al.[146] and Chau and Terry[147-149] have described a procedure involving hexane extraction followed by gas chromatography on a steel column (1.5 metres) packed with 3% of SE-30 on Chromosorb W, operated at 190 °C, with nitrogen as carrier gas and electron capture detection. The technique was applied to water from rivers, ponds, and canals and sea water.

Chau and Terry[147] in a series of papers have discussed gas chromatographic methods for the determination of α- and β-endosulfan. The procedure involved heating the sample on a 2 cm column of alumina impregnated with concentrated sulphuric acid and heated 1.5 hours at 100 °C to convert endosulfan to an ether which was examined by gas chromatography. Down to 0.003 ng of the insecticide could be detected per ml of water extract. The method was subsequently applied to the determination of 10 pg of heptachlor in one litre of water.[150]

An automatic sampler has been used[151] to observe an endosulfan wave in the river Rhine. The automatic sampler at the gauging station was programmed to provide 2 hour mixed samples composed of 125 ml portions taken every 10 min. The resulting samples, together with the 24 hour mixed samples for the preceding and succeeding periods were analysed for endosulfan by gas–liquid chromatography. The results show a distinct peak in the endosulfan concentration with a maximum recorded value of 276 ng l^{-1} or about 100 times the normal background level. Estimation of the total mass flow with the aid of flow measurements indicated that 40–50 kg endosulfan passed Koblenz during August, compared with a baseline value of only 5 kg.

Hexachlorophane

In a method[152] for determining this insecticide, the hexachlorophane is converted into its di-*p*-anisate ester, which is subjected to chromatography on a column (2 ft × 2.3 mm) packed with Sil-X silica (36–40 μm) with 1-chlorobutane as solvent and detection by u.v. adsorption. Calibrations are made over the range 40–280 ng of the derivative. The coefficient of variation at the 20 ng level is 1.0%.

Chlorobromuron, (3,4-(bromo-3-chlorophenyl)1-1 methoxy 1-methyl urea)

Katz and Strusz[153] have described gas and thin-layer chromatographic methods for the determination of traces of chlorobromuron and its metabolites in soils which with minor modifications could be applied to water samples. The samples were extracted in a Soxhlet apparatus with 250 ml of ethyl acetate for 6 h. Each extract was then acidified with 2 ml of anhydrous acetic acid and evaporated to 5 ml on a steam bath; 15 µl portions were applied to the 0.1 mm layer of silica gel on an Eastman Chromagram sheet, and the chromatogram was developed twice with hexane–ethyl acetate (15:2). After drying, the chromatogram was developed with chloroform–pyridine (10:1), dried again, and sprayed with Mitchell reagent (20 ml of 2-phenoxyethanol added to a solution of 1.7 g silver nitrate in 5 ml of water, the mixture is diluted with acetone to 200 ml, then 1 drop of 30% aqueous hydrogen peroxide is added), and the spots were located under u.v. radiation. For gas chromatographic confirmation, standards were treated as above and corresponding unsprayed sample areas were removed, extracted with 5 ml of ethyl acetate and evaporated to 1 ml; 5 µl aliquots were injected on to a glass column (16 in. × 0.25 in.) packed with 1.5% of XE-60 on Gas-Chrom Q (80–100 mesh) (previously aged at 240 °C for 48 h). The column was temperature programmed from 75 to 230 °C at 50° per minute. Nitrogen was the carrier gas and detection was by flame ionization.

Kepone (chlordecone)

Gas chromatography–mass spectrometry involving chemical ionization–mass spectrometry has been used to detect and confirm the presence of Kepone (chlordecone), Kepone photoproducts, and a conversion product of Kepone in environmental samples.[155]

Mirex (dechlorane, $C_{10}Cl_{12}$)

Andrade and Wheeler[154] have studied the biodegradation of this substance by sewage sludge organisms utilizing ^{14}C-labelled mirex. They did not succeed in identifying the metabolites.

Laseter et al.[156] and Kaiser[157] have both utilized gas chromatography and mass spectrometry for the determination of mirex in fish. Kaiser[157] has pointed out that under standard gas chromatographic conditions, the mirex peak is superimposed on that of the polychlorinated biphenyls and, as a result, the presence of mirex may have been interpreted by several workers as a PCB isomer. He used a computer controlled gas chromatographic–mass spectrometric system to positively identify mirex and distinguish it from other highly chlorinated insecticides that could have been present in the samples including aldrin, chlordane, dieldrin, endrin, endosulfan, heptachlor, Kepone (chlordecone), pentac, and toxaphene.

Determination of chlorine

It is sometimes useful to be able to determine very low amounts of chlorine in insecticide-containing extracts of water and other environmental samples. Various insecticides and industrial chemicals such as polychlorinated biphenyls (PCB) may occur at concentration levels of about $1\ ng\ l^{-1}$ in water and about $1\ \mu g\ g^{-1}$ in marine biological material. Organically bound bromine and iodine are known to occur in water and in marine organisms, but very little is known about the composition of such substances. Compared to chlorine, much smaller amounts of organic bromine and iodine are introduced to the environment by man. Various techniques have been used for these measurements and some of these which have been applied to the aqueous environment are discussed below.

Neutron activation analysis

Gether *et al.*[158] have reported on the determination of chlorine, bromine, and iodine present as non-polar, hydrophobic, hydrocarbons in environmental samples. The organohalogen compounds are separated from water into an organic phase by on-site liquid–liquid extraction, of the type described by Ahnoff and Josefsson.[159] This method is described in detail as it emphasizes the precautions necessary to avoid contamination by apparatus and chemicals.

Method

Purification of solvents The solvents used, i.e. cyclohexane, *n*-hexane, isopropanol, and acetone (for cleaning apparatus) are purified by distillation in the apparatus shown in Figure 190. Column (1) is a rectification column (1500 mm length, 25 mm diameter) filled with stainless steel turnings, and insulated with a 30 mm thick layer of expanding polystyrene foam. The solvent to be purified is introduced through the inlet of a 1 l flask (2) equipped with a Teflon stopcock in the bottom, which allows the apparatus to be drained and cleaned without dismantling. A condenser (3) is fitted with an adapter (4) to the 1 l funnel (5) used as receiver. This funnel may be detached from the apparatus, in which case a 'swan-neck' adapter (6) prevents dust from entering. An activated carbon trap (7) may be inserted behind the condenser. Heating is done with a mantle (8). When 80–90% of the solvent has been distilled off, the residue is drained through the bottom stopcock of the flask (1). The apparatus is completely closed to the atmosphere, and may be operated with an inert gas such as highly purified nitrogen. If heat-labile solvents or benzene are to be purified, the heating mantle is replaced by an aluminium block (see below), and the apparatus is purged with nitrogen. When *n*-hexane or cyclohexane is to be distilled, 10 ml of a 10% w/v solution of sodium metal in anhydrous ethanol per litre of solvent are added prior to distillation. In the case of isopropanol, 0.1 g of sodium metal is added directly. Acetone is purified without any additions. Technical grade solvents are used as starting materials. Blanks are run on all solvents before use. Although the apparatus may be rinsed from the top of the rectification column, the lowest and most reproducible blank values are obtained when rinsing and dismantling of the apparatus are reduced to a maximum. Distilled water may be prepared in an all-glass apparatus over alkaline potassium permanganate. Small sliced pieces of Teflon sheet efficiently prevent bumping.

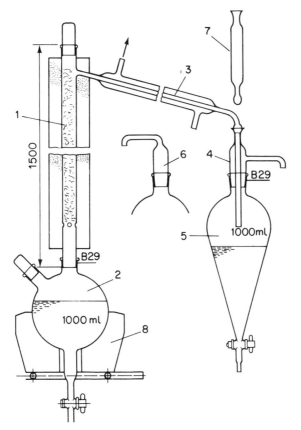

Figure 190 Distillation apparatus. See text for key to numbering. Reprinted with permission from Gether et al.[158] Copyright (1979) Elsevier Scientific Publishing Co.

Cleaning of apparatus

Two suitable procedures are as follows. Centrifuge tubes and flasks of a volume less than about 1 l are kept overnight at 550 °C. The glassware is placed in large beakers with raised lids, where it is exposed to the atmosphere but protected from dust. It is subsequently stored in the same beakers with a large petri dish as cover, and treated as if sterile. All flasks are stoppered as soon as they have cooled sufficiently. For apparatus that cannot easily be placed in the oven, thorough rinsing with purified acetone is employed.

Sample preparation

On-site liquid–liquid extraction The apparatus used consists of three main parts: an extraction unit, a magnetic stirring device, and a pump. The extraction unit has been described by Ahnoff and Josefsson.[159] The water is continuously pumped into an 800 ml extraction chamber and mixed vigorously with 200 ml of an organic solvent less dense than water. The resulting emulsion is separated in the lower part of the chamber and the water extracted is drained off at constant flow. The flow rate is about $4 l h^{-1}$ and the total volume of water extracted is typically about 200 l. The organic extract, usually in n-hexane, is transported to the laboratory in glass flasks with glass stoppers.

Lipid phase extraction About 50 g of tissue are homogenized in a Waring-type blender, transferred to a glass-stoppered flask to which twice the weight of a 1:1 isopropanol–hexane mixture is added, and placed in an ultrasonic bath for 15 min. The flask is subsequently placed in a shaking apparatus or on a ball mill (100–200 rpm) for 24 h at 20 °C. The mixture is allowed to settle and the hexane phase is separated.

Codistillation About 500 g of tissue are homogenized in a Waring-type blender with 1 l of distilled water and transferred to a distillation apparatus equipped with a condenser and a receiving flask. Then 10 ml of cyclohexane are added, the sample is carefully brought to boiling, and about 25 ml of water are distilled over along with the cyclohexane. The apparatus is cooled to 20 °C, another 3 ml of cyclohexane are added, and distillation is continued for 4–5 min to flush out traces of volatile compounds remaining in the condenser. To obtain a better idea of the volatility of the halogenated hydrocarbons, these distillation sequences may be repeated, and the second distillate analysed separately.[160]

Concentration of extracts

In most cases, the extracts are evaporated to a volume of a few ml in a modified Kuderna Danish apparatus (Figure 191). The modifications provide better retention of volatile material through an efficient column with provision for regulating the reflux rate, an aluminium heating body of known and controllable temperature to prevent overheating of the glass walls and possible destruction of sample material, and facilities for operation under vacuum or in inert gas. The distillation column (1,2) is constructed from two pieces of concentric glass tubing with a glass rod wound as a coil in the space between them; the diameter of the inner tube is about 20 mm, and the space between the inner and the outer tube is about 2 mm. The column is insulated with expanded polystyrene foam. The inserted tube (3) is an inverted condenser, the length of which may be varied by lifting or lowering the central tube and the bottom closure attached to it. The sample is distilled from a 500 ml round bottom flask (4) to which a 10 ml centrifuge tube may be attached by means of a B14 ground glass joint. Heating is effected by an aluminium block (6) machined to fit the flask, and fitted with three 150 W heating elements connected to an adjustable mains supply. The block also has a spring under the centrifuge tube to keep it in position (neither is shown in Figure 191). A second ground glass joint allows introduction of nitrogen gas into the flask, or application of vacuum for cleaning. An ordinary condenser

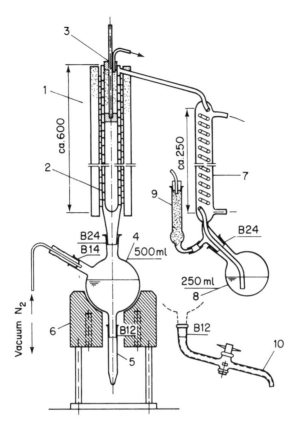

Figure 191 Modified Kuderna Danish apparatus. See text for key to numbering. Reprinted with permission from Gether et al.[158] Copyright (1979) Elsevier Publishing Co.

(7) is fitted to a 250 ml receiving flask (8), and a tightly packed glass wool filter (9) prevents access of particulate material. This filter may be connected to a vacuum line for operation at reduced pressure. In operation, the sample to be concentrated is introduced into the evaporation flask through a funnel attached to the B14 ground glass joint (for complete closure to the atmosphere a separating funnel equipped with a Teflon stopcock and a male glass joint may be used). The aluminium block is set at a suitable temperature (60–70 °C for hexane), and the reflux condenser is adjusted to give a suitable reflux ratio. A flow of nitrogen (30–50 ml min^{-1}) is started simultaneously. With most of the sample removed, the round bottom flask is rinsed by applying vacuum momentarily to the nitrogen inlet, so that a portion of the distillate is sucked back into the distillation column. This is repeated twice as the surface of the liquid becomes visible beneath the aluminium block. Evaporation of the sample will stop at this stage, and the sample is left in an unheated state until removed from the apparatus. This is done simply by removal of the centrifuge tube. The

concentration of one 200 ml sample to 4 ml requires about 90 min. Smaller final volumes can be achieved by using centrifuge tubes with greater taper at the bottom. Accurate determination of the concentrated sample is best achieved by weighing. The apparatus is rinsed by substituting the draining attachment (10) for the centrifuge tube and a flask containing purified acetone for the receiver, and applying vacuum from a water aspirator to the nitrogen inlet. Two such rinses are sufficient. The acetone is drained and the draining attachment replaced by a clean centrifuge tube to prepare the apparatus for a new sample.

Removal of inorganic halides

Traces of inorganic halides are estimated by washing the extracted phase with distilled water. This is always done before, and in some cases also after, the concentration step. Washing of sample volumes larger than a few millilitres with water is done in glass-stoppered separatory funnels equipped with Teflon taps and with tapered outlets. For smaller volumes glass-stoppered (B12) centrifuge tubes are used. About equal volumes of water and organic phase are employed. Separatory funnels are placed for 10–20 min in a reciprocating shaker, the contents of centrifuge tubes are intimately mixed by means of a 'Whirlmixer'. Extractions are repeated three times. Emulsions should be avoided. They normally cause no problems with hexane extracts, but may be encountered with oils that contain much phospholipid; in such cases, the addition of some isopropanol will help.

Neutron activation analysis

After the final work-up of samples, the part of the sample intended for the determination of total organically bound halogens is sealed in a glass ampoule and stored in a refrigerator until neutron activation analysis can be done. For irradiation, 1–2 ml of the sample is sealed in an ampoule made from polyethylene tubing. These ampoules are cleaned batchwise: washing with distilled water, washing with cyclohexane, storage under cyclohexane overnight, and drying in an oven at 60 °C. The cleaned ampoules are stored in airtight containers until used. Testing for blanks with pure cyclohexane is carried out separately for four ampoules from each cleaned batch. Immediately after sealing, the samples are packed in pairs in polythene 'rabbits' and then stored at -20 °C until the start of the irradiation. This procedure is to prevent appreciable diffusion of solvent or organohalogen compounds through the walls of the vial during the pre-irradiation and irradiation period. The irradiations were performed in the pneumatic-tube facility of a JEEP-II reactor (Norway) at a thermal neutron flux of about 1.5×10^{13} n cm^{-2} s^{-1}. The analyses are based on γ-peaks from ^{38}Cl ($t_{1/2} = 37.2$ min, $E_\gamma = 1643$ keV), ^{80}Br ($t_{1/2} = 17.6$ min, $E_\gamma = 617$ keV), and ^{128}I ($t_{1/2} = 25.0$ min, $E_\gamma = 443$ keV). Calibration is done with an aqueous solution (1 M NH$_4$OH) of the halides in question, irradiated and counted under the same conditions as the samples. Simultaneous detection of ^{24}Na ($t_{1/2} = 15.0$ h,

$E_\gamma = 2754$ keV) induced in sodium may serve as a check for contamination by inorganic halides.

Results from radiotracer experiments with ^{38}Cl as chloride and ^{82}Br as bromide to check the efficiency of the washing procedure for the removal of inorganic halides from hexane extracts are summarized in Table 101. In some of the experiments the hexane had been treated with concentrated sulphuric acid, which is a procedure used in the analyses of chemically persistent chlorinated hydrocarbons. In all cases a single washing was found to reduce the level of chloride and bromide by a factor of more than four decades, even in the presence of fairly high concentrations of oil-soluble amines. These results are in accordance with those reported by Schmitt and Zweig,[161] and with all previous experience from work wtih marine oils.

Table 101 Removal of inorganic chloride by washing of hexane extracts with water, indicated by the ratio of the activity in hexane extract to the activity in the aqueous phase (H:aq)

Organic phase	Aqueous phase	H:aq($\times 10^{-3}$)
n-Hexane	0.05 M Cl$^-$	18
	0.05×10^{-3} M Cl$^-$	8
	0.05 M Br$^-$	<10
n-Hexane + H$_2$SO$_4$	0.05 M Cl$^-$	8
n-Hexane + TMA[a,b]	0.05 M Cl$^-$	10
	0.05 M Br$^-$	<10
n-Hexane + TMA + conc. H$_2$SO$_4$	0.05 M Cl$^-$	10
n-Hexane + DPA[a,b]	0.05 M Cl$^-$	<5
	0.05 M Br$^-$	<10
n-Hexane + DPA + H$_2$SO$_4$	0.05×10^{-3} M Cl$^-$	8
	0.05 M Cl$^-$	7
	0.05 M Br$^-$	<10
Water extracted from polluted river	0.05 M Cl$^-$	11
	0.05×10^{-3} M Cl^{-1}	14
n-Hexane + herring oil[b]	0.05 M Cl$^-$	75
	0.05 M Br$^-$	<10
n-Hexane + herring oil + H$_2$SO$_4$	0.05 M Cl$^-$	11

[a]TMA, trimethylamine; DPA, diphenylamine. [b]100 μg ml^{-1} added in each case.
Reprinted with permission from Gether et al.[158] Copyright (1979) Elsevier Science Publishers B.V.

Gether et al.[158] found that the polyethylene ampoules used for irradiation of samples were a serious source of contamination with chlorine. When samples of pure hexane were stored in polyethylene ampoules not subjected to special pretreatment and subsequently analysed, values up to several μg Cl ml^{-1} of hexane were observed, evidently because of extraction of chlorinated compounds from the polyethylene. After introduction of the washing procedure described, the blank from this part of the analysis became a minor problem. For chlorine, values obtained by testing the ampoules with pure cyclohexane are typically 50 ng ml^{-1} and negligible for bromine and iodine. The role of treatment of

Table 102 Effects of purification procedures on organically bound chlorine in hexane and cyclohexane

Solvent[a]	Cl in final extract (μg ml^{-1}) A	B
Merck analytical grade cyclohexane (glass bottle)	2.1	1.4
Merck puriss grade cyclohexane (can)	0.94	
Merck puriss grade n-hexane	0.66	
Merck puriss grade n-hexane (can) + Na-ethylate	0.18	0.14
Merck puriss grade n-hexane (can) + Na-ethylate, equilibrated with puriss isopropanol (glass bottle)	0.31	0.26
As above, isopropanol purified over sodium	0.16	0.13
Merck puriss grade n-hexane (can) + Na-ethylate treated also with concentrated sulphuric acid	0.11	0.13

[a]Samples were prepared in a room where chlorinated solvents were not used, and were concentrated from 200 to about 4 ml. For further details of analysis, see method.
Reprinted with permission from Gether et al.[158] Copyright (1979) Elsevier Science Publishers B.V.

solvents is indicated from the results given in Table 102. The use of sodium ethylate in distillation of the solvents seems to be crucial.

It was found that the recovery of the on-site liquid–liquid extraction is normally quite satisfactory in clean waters, but may be poor in waters with a high content of particulate material. Losses may also occur in the concentration step, particularly with volatile material. Some results to indicate the degree of such losses are given in Table 103; losses are quite low for compounds with boiling points above 200 °C.

Table 104 lists typical blank levels obtained in between actual samples, relating to both water and biological material. The values are given in terms of the contribution they would make to an actual sample. Values for PCB, determined by gas chromatography on aliquots of the same extracts, are included. The

Table 103 Recovery of various compounds during the concentration step (experimental conditions: 200 ml of hexane concentrated to 4 ml; distillate flushed back 3 times for washing; reflux ratio ca. 50%, final volume obtained in ca. 90 min; 25 ml N$_2$ min^{-1} passed through the apparatus)

	B.p. (°C)	Amount added (μg)	Amount found (μg)	Recovery (%)
Dichloromethane	40	138	14	11
Chlorobenzene	132	110	83	75
1,2-Dichlorobenzene	180.5	127	103	81
Hexachlorobenzene	322	160	147	92
DDT	260	118	107	91
n-Alkanes C$_{10}$	174	75	53	71
C$_{16}$	287	75	69	92
C$_{23}$		75	72	96

Reprinted with permission from Gether et al.[158] Copyright (1979) Elsevier Science Publishers B.V.

Table 104 Blank levels obtained for different types of separation procedures

Type of separation	Element/ compound	concentration in final extract (ng ml^{-1})	concentration, calculated to normal sample (ng kg^{-1})[a]
Water/extraction	Total Cl	100-150	4-6
	Total Br	2-5	0.08-0.2
	Total I	0.5-1	0.2-0.04
	PCB	1-2	0.05-0.1
Herring/extraction	Total Cl	100-200	1000-2000
	Total Br	1-5	
	Total I	0.5-3	
	PCB	1-2	
Herring/codistillation	Total Cl	200-300	1000-2000

[a] 200 l for water; 100 g wet weight for herring tissue.
Reprinted with permission from Gether et al.[158] Copyright (1979) Elsevier Science Publishers B.V.

relative standard deviation of the chlorine blank values is generally about 20% for both water and biological material. For bromine and iodine, the blank levels are insignificant. Table 105 shows results from activation analyses of solutions with known concentrations of chlorinated hydrocarbons. The precision attained at the 1-5 μg ml^{-1} level of chlorine is of the order of 1.5-3%. Experiments with 1,2-dichlorobenzene show that there is a loss of about 20%, referred to an inorganic chloride standard solution. Similar results were obtained in other experiments, where solutions of carbon tetrachloride in hexane gave results in the range 76-84% of the nominal values when analysed at different times. Corresponding analyses of hexane solutions of bromoform analysed simultaneously showed no similar loss. The detection limits of the activation analysis were about 0.02 μg Cl ml^{-1}, 0.005 μg Br ml^{-1}. At high levels of chlorine, the

Table 105 Recovery and reproducibility of the neutron activation analysis method for organically bound chlorine

	No. of samples	Mean (μg ml^{-1})	Relative standard deviation (%)	Recovery (%)
1,2-Dichlorobenzene in hexane[a]	10			
with ammonium chlorine as reference		3.73	1.7	78
with 1,2-dichlorobenzene as reference		4.70	1.17	97
PCB in hexane[b]	2	0.95		103

[a] Corresponding to 4.83 μg Cl ml^{-1}. [b] corresponding to 0.92 μg Cl ml^{-1}.
Reprinted with permission from Gether et al.[158] Copyright (1979) Elsevier Science Publishers B.V.

detection limits for bromine and iodine are somewhat higher because of the Compton continuum from ^{38}Cl in the γ-spectra.

Jager and Hagenmaier[162] have described a simple method for the determination of the total extractable organochlorine content of waste water samples. Chlorine is determined by extraction with petroleum ether, followed by reductive hydrolysis using elemental sodium and isopropanol, and determination of the liberated chloride ions either by titration (Mohr's method) or by amperometric procedures. Volatile and non-volatile organochlorine compounds can be determined, the detection limit being 0.1 mg chlorine per litre by titration against silver nitrate, or 0.01 mg chlorine per litre using a coulometer. Recoveries are between 95% and nil, depending on the type of organochlorine compound being investigated. Interference due to sulphide ions can be eliminated by addition of hydrogen peroxide.

Van Steenderen[163] has described the construction of a total organohalogen analyses system, which is applicable to potable water and other samples. It is based on a coulometric titration system. In this method an aliquot of sample extract is injected into a pyrolysis tube and mixed with reagent gas, from which the gas is eluted through a heated capillary inlet tube into a titration cell which is specific for the measurement of chloride, bromide, and/or iodine derived from the combusted halogenated organic compounds present in the sample. Halides entering the titration cell form insoluble precipitates with silver ions in a 13.2 mol dm^{-3} acetic acid solution. The most sensitive cell conditions occur when the silver ion concentration is equal to the chloride ion concentration. For water the chloride ion concentration is 10^{-5} mol dm^{-3} while for acetic acid this is at less than 10^{-7} mol dm^{-3}. The titration cell is designed to operate at a constant titrant ion concentration which is determined by the bias set of a microcoulometer. The cell contains four electrodes immersed in acetic acid solution agitated by means of a magnetic stirrer rotating at *ca.* 17 rps. A change in titrant ion concentration within the titration cell due to halides entering the cell from the combusted sample, is potentiometrically indicated by a reference and sensor electrode combination. The e.m.f. of each of these half cells is a direct function of the ion concentration immediately upon the surface of the metal portion of the electrodes. The reference electrode produces a constant e.m.f. at constant temperature and pressure while the sensor electrode in the stirred electrolyte is at a lower titration ion concentration. The half cell potentials of these two electrodes are summed and compared to an opposing e.m.f. (bias setting) obtained from the coulometer. Any change in titration ion concentration due to halides entering the cell or by shifting the bias setting will result in a voltage imbalance. This imbalanced voltage is the input to an amplifier supplying current to a pair of generator electrodes. This current generates titrant ions until the electrolyte has been restored to its original concentration. A recorder monitoring the voltage drop appearing across a precise resistor in series with the generator electrode circuit records the current needed to restore the electrolyte to its original equilibrium. The number of coulombs required to complete a titration are calculated by measuring the area under a current–time peak

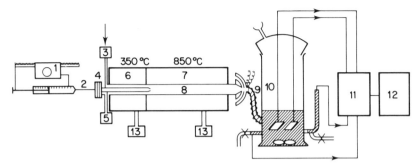

Figure 192 Microcoulometric total organic halogen analysing system. 1 A fractionating push button dispenser utilizing a 10 mm³ syringe. 2 The syringe needle must be at least 65 mm long to deposit the sample into the 350 °C heating zone. 3 Regulating oxygen supply via a 5 mm pore size molecular sieve. 4 Silicon septum. 5 Regulated argon carrier gas supply via 5 nm pore size molecular sieve. 6 Low temperature (350 °C) combustion furnace for sample volatilization. 7 High pyrolysis combustion furnace (850 °C). 8 Quartz combustion tube. 9 Electric heating tape (35 V) wrapped round the titration cell inlet capillary. 10 Dohrmann T-300-S titration cell. 11 Coulometer, incorporating the instrument control panels. 12 Recorder and/or integrator. 13 Furnace temperature controls. Reprinted with permission from Van Steenderen.[163] Copyright (1980) *Laboratory Practice*, United Trade Press

registered on a recorder. The amount of halogen is equated to the number of coulombs obtained by applying Faraday's Law. A schematic configuration of the microcoulometric analyser is given in Figure 192. Argon was used as sample carrier gas (40 cm³ min⁻¹) and oxygen as the oxidant gas (120 cm³ min⁻¹). Petroleum ether (40–60 °C) was used as extraction solvent. *n*-Hexane (spectroscopic, 69 °C) was found to produce three times the background noise to that of petroleum ether (40–60 °C). *n*-Hexane could be purified to the same level as petroleum ether by passing it through a glass column packed with aluminium oxide (chromatographic type). Samples were extracted in 125 cm³ Wheaton serum bottles previously baked at 500 °C to remove any organic contamination. The bottles were completely filled (no head-space) and capped with a Teflon coated rubber septum. Two ml solvent were introduced to the bottle while simultaneously removing an equivalent amount of water. Samples were extracted for 30 minutes by clamping the bottles on to a tumbling machine rotating at 0.6 rps. Water has an adverse effect on the equilibrium of the silver potential in the titration cell due to a diluting effect resulting in the generation of silver ions to restore the original equilibrium. Under oxidative conditions a standard solution containing 100 ng mm⁻³ Cl⁻ as chlorobenzene was spiked with 100 ng sulphur as butylsulphide and 100 ng nitrogen as pyridine. At these concentrations nitrogen and sulphur did not interfere with the halogen determination. It is suggested that 3 mg of ammonium sulphamate per 20 ml cell electrolyte could be used to combat interference from excess amounts of nitrous oxides derived from the combustion of nitrogen-containing compounds causing negative errors in measurement due to reduction effects at the indicator electrode. Phosphorus, arsenic, and fluoride compounds did not interfere with measurement. The minimum practical limit of detection in water (after extraction) was approximately 10 μg dm^{-3} Cl$^-$. An imperative requirement of the analyser is that sample injection is carried out stepwise, i.e. 0.25 mm^3 s^{-1} in order to ensure complete combustion of the solvent and maximum recovery of the liberated chloride atoms in the titration cell liquid.

In Table 106 are some recoveries obtained in this procedure when applied to known solutions of various organohalogen compounds in water. Samples were extracted with head-space (extraction containers not filled completely) and without head-space (extraction containers filled completely). The recovery of the organohalogen compounds from water was most efficient using the technique of extraction without head-space. This is of special significance where the recovery of tri- and tetrachloromethane was concerned since recoveries for both these compounds was in the order of 20% or more above those obtained by extraction with head-space.

Table 106 The recovery of known amounts of organohalogen compounds added to water ($n = 3$)

Compound	ng Cl$^-$ and/or Br$^-$ calculated	weight recovered (ng)						Boiling point (°C)	Vapour pressure in mmHg at 20 °C
		with head-space	%	CV	Without head-space	%	CV		
Trichloromethane	89	58	65	1.6	74	83	5.4	61.2	10×1.89
Tetrachloromethane	92	62	67	2.3	92	100	0.4	76.8	10×9.1
Bromodichloromethane	92	76	82	0	76	83	6.0	90.0	
Chlorodibromomethane	90	79	88	2.4	77	86	0.1	119.0	
Tribromomethane	95	93	97	1.8	85	89	3.8	149.5	4.5
Hexachlorobenzene	100	80	80	2.3	72	72	5.3		10×10^{-3}
Chlorobenzene	250	61	24	1.0	152	61	2.0	132.0	9

CV, coefficient of variation; n, number of determinations.
Reprinted with permission from Van Steenderen.[163] Copyright (1980) Laboratory Practice, United Trades Press.

A method has been described[164] for the determination of organically combined chlorine (total organochlorine, TOCl) in high molecular weight aquatic organics. This technique involves photochemical oxidation. TOCl is measured after preconcentration of the sample, photolysing the organics to mineralize the chlorine, and measuring inorganic chloride. In chlorinated sewage effluent TOCl was found to increase after chlorination, so that from 0.002 to 0.009 mol of TOCl per mol of total organic carbon (TOC) was present. Chlorination was shown to produce small amounts of organochlorine compounds of molecular weight greater than 1000 as defined by ultrafiltration of the sample.

Kühn et al.[165] have conducted studies of the determination of organically bound chlorine using a novel concentration process. The procedure involves flocculation and sedimentation of the solution with powdered activated carbon under controlled conditions. By a two-stage application of this technique about 0.97% of the total chlorine is adsorbed and subsequently detected by analysis. The detection limit of the process is about 20 µg organic chlorine per litre. Separate determinations of chloride ion by means of nitrate ion exchange is possible.

CARBAMATE INSECTICIDES

Carbaryl (Sevin) (1-naphthyl-N-methylcarbamate)

Probably the most common of insecticides is carbaryl. This substance has obtained wide acceptance due to its effectiveness and low mammalian toxicity.

Crosby and Bowers[166] have described a method for determination of carbaryl in which the sample (0.5 g) is heated under reflux for 1 h with 2-chloro-$\alpha\alpha\alpha$-5-nitrotoluene (1 millimole) or 4-chloro-$\alpha\alpha\alpha$-trifluoro-3,5-dinitrotoluene (0.7 millimole), acetone (20 ml) and 0.1 M $Na_2B_4O_7$ (20 ml) to convert its amine moiety into an N-substituted nitro-(trifluoromethyl) aniline derivative. The derivative is subjected to gas chromatography on a stainless steel column (10 ft × 0.125 in. o.d.), packed with 3% SE-30 or 3% FFAP on HMDS treated Chromosorb G at a temperature from 150–205 °C, with nitrogen (30 ml min^{-1}) as carrier gas and electron capture or flame ionization detection. Down to 50 pg of the more volatile or 200 pg of the less volatile derivatives can be determined by electron capture detection. Methods also described are the determination of molinate (S-ethyl hexamethylenecarbamate) in water.

Frei et al.[167] have described an in situ fluorimetric method for the determination of carbaryl and 1-naphthol by thin-layer chromatography. The compounds are extracted from the water sample with dichloromethane, and are separated by thin-layer chromatography on activated silica gel (without binder) with chloroform as solvent (ascending); the R_f values are 0.43 and 0.60 for carbaryl and its hydrolysis product (1-naphthol), respectively. The chromatogram is sprayed with sodium hydroxide to convert carbaryl into 1-naphthol and to make the spots fluorescent; they are located in u.v. radiation, marked, and measured by fluorimetry on the plate at 488 nm within 1.5 h after spraying. Visual and instrumental detection limits are about 6 and 1 ng per spot, respectively. Calibration graphs are rectilinear from 0.01 to at least 0.3 µg per spot. Standard addition techniques can be used for small amounts (1–5 ng per spot). Recovery of carbaryl added to tap water was 90–94%; that of 1-naphthol was only 50%. Other carbamates do not interfere with the determination.

Handa and Dikshit[168] have described a spectrophotometric method, described below, for the determination of carbaryl in lake, stream, and pond waters.

Method

Apparatus

Spectrophotometer: Perkin-Elmer, Model 402, with 1 cm silica cells or equivalent Colorimeter with 12×100 mm tubes. Test-tubes: 30 ml capacity, with B19 sockets and stoppers. Reagents: potassium hydroxide solution, 0.5 M solution in methanol; vanillin solution, 0.6 g of vanillin dissolved in 100 ml of analytical reagent grade glacial acetic acid; orthophosphoric acid; potassium carbonate solution, 0.1 M.

Preparation of standard solution

100 g of pure carbaryl are dissolved in acetone and the volume made up to 100 ml with acetone in a calibrated flask. 2 ml of the solution are transferred into a 100 ml calibrated flask and the volume made up to the mark with acetone, thus obtaining a solution containing $20 \,\mu g \, ml^{-1}$ of carbaryl.

Preparation of standard graph

Into clean, dry test-tubes is transferred 0, 0.5, 1.0, 1.5, 2.0, 2.5, and 3.0 ml portions of standard solution. The volume is diluted to 3 ml with acetone in all of the test-tubes, then the solvent is evaporated under a current of dry air. Into each of the tubes is pipetted 0.2 ml of 0.5 M methanolic potassium hydroxide solution. The test-tubes are rotated in order to wet the sides with solution for 5 min and to evaporate the methanol from each test-tube. Next 1 ml of vanillin solution and 4 ml of orthophosphoric acid are pipetted into each tube. The test-tubes are placed in a water bath at 60 °C for 20 min, then removed from the bath and cooled in a beaker of cold water. The absorbance of each solution in the spectrophotometer at 575 nm is then read.

Extraction of water samples

After collection of the water samples (minimum volume l) pH values are adjusted to below 5 with 20% sulphuric acid, and 10 g of sodium sulphate are dissolved in each 1 l sample. Next each sample is extracted in a 2 l separating funnel with 150 ml of methylene chloride, shaking the funnel for 2–3 min. The methylene chloride extract is transferred into a 1 l funnel and the aqueous phase re-extracted (2–3 min) with a further 100 ml methylene chloride. The second methylene chloride extract is added to the first and the combined extract washed with 0.1 M potassium carbonate solution. The methylene chloride is then dried by passing it through 15–20 g of sodium sulphate in a filter funnel and the extract collected in a 500 ml flask fitted with a ground glass stopper. The methylene chloride extract is concentrated to 100 ml.

Determination

A suitable portion of the methylene chloride extract (expected to contain 10–60 μg of carbaryl) is placed in a test-tube. The solution in the test-tube is evaporated and the procedure as described under 'Preparation of standard graph' is repeated exactly.

Recoveries of carbaryl of 92–96% were obtained from water samples by the above procedure.

Various workers[169-174] have discussed gas chromatographic methods for the determination of carbaryl in water samples. Lewis and Paris[174] determined carbaryl and its hydrolysis product 1-naphthol by direct injection of a benzene or iso-octane extract into a glass column (0.3 m × 4 mm) packed with 3% SE-30 on Gas-Chrom Q (80–100 mesh) and operated at 145 °C with nitrogen (120 ml min^{-1}) as carrier gas. Considerable preconditioning of the column was necessary to obtain optimum sensitivity.

Ueji[175] determined carbaryl and propoxur (*o*-isopropylphenyl *N*-methylcarbamate) in crops in amounts down to 0.005 ppm. The carbamates were reacted with trifluoroacetic anhydride solution in ethyl acetate by heating in the dark at 50 °C. This reaction was quantitative and reproducible and the stability of the *N*-trifluoroacetyl derivatives was high. The derivatives of these insecticides were subjects to gas chromatography with electron capture detection on 5 ft columns packed with Chromosorb W coated with 5% OV-17 or OV-25, 2% poly(ethanediol adipate). The most efficient stationary phases were OV-17 and OV-25, 5% OV-17 being particularly good. To determine *m*-tolyl methylcarbamate residues in unpolished rice grain or rice straw, the sample of powdered rice (10 g) or chopped straw (5 g) was extracted with dichloromethane in a Soxhlet apparatus, the extract was partitioned between acetonitrile and hexane, and the acetonitrile phase was diluted with 4% sodium chloride solution and extracted with dichloromethane. This extract was passed through a column of Florisil (5 g), water-saturated dichloromethane being used as eluant, and the eluate was cleaned up on a column of activated alumina (10 g), with acetone–hexane (1:9) as eluant. The residue from evaporation of the eluate was dissolved in ethyl acetate, and, after the trifluoroacetylation reaction, the solution was gas chromatographed. Recoveries ranged from 91.2 to 98.8% for samples fortified with *m*-tolyl methylcarbamate at the 0.1–0.4 ppm level. Other spectrophotometric methods for determining carbaryl have been described by various workers.[176-178]

Methomyl (*S*-methyl-*N*-[(methyl carbamoyl)oxy] thioacetimidate)

Reeves and Woodham[179] have described a gas chromatographic method for the determination of this insecticide in water, soil sediments, and crops. The residues were extracted from soil, sediment, and water with dichloromethane, and the extracts were purified on a column of Florisil. The residues were extracted from tobacco with dichloromethane–benzene (39:1), and the extracts were purified by a coagulation procedure with ammonium chloride–phosphoric acid. The purified and concentrated extracts were then analysed by gas chromatography on a glass column (6 ft × 3/16 in.) packed with 10% DC-200 on Chromosorb W HP (80–100 mesh) and operated at 140 °C, with nitrogen as carrier gas (80 ml min^{-1}) and a 394 nm S-interference filter. The limits of detection were 0.05 ppm for soil, sediment, and tobacco and 0.01 ppm for water; the recoveries were 90.8, 80.1, 78.0, and 75.1% respectively.

Zectran ((4-dimethylamine-3,5-xylyl)methylcarbamate)

Hasler[180] has studied the rate of degradation of this substance in alkaline water. High performance liquid chromatography was used to identify xylenol as a degradation product of this herbicide.

N-methylcarbamate and N,N'-dimethylcarbamates

These substances were determined in water and soil samples by hydrolyses with sodium bicarbonate and the resulting amines reacted with 4-chloro-7-nitrobenzo-2,1,3-oxadiazole in isobutyl methyl ketone solution to produce fluorescent derivatives.[181] These derivatives were separated by thin-layer chromatography on silica gel G or alumina with tetrahydrofuran–chloroform (1:49) as solvent. The fluorescence is then measured *in situ* (excitation at 436 nm, emission at 528 and 537 nm for the derivatives of methylamine and dimethylamine, respectively). The method was applied to natural water and to soil samples containing parts-per-10^9 levels of carbamate. The disadvantage of the method is its inability to differentiate between carbamates of any one class.

m-*S*-butylphenyl methyl-(phenylthio)carbamate (RE 11775)

Westlake *et al.*[182] determined this substance in water, soil, and vegetation by a gas chromatographic procedure. The sample is extracted with dichloromethane, chloroform, or acetonitrile, followed by clean-up, if necessary, on a column of Florisil, silica gel, or alumina. The purified residue is submitted to gas chromatography on either a stainless steel column (3 ft × 0.25 in.) packed with 5% OV-225 on Gas-Chrom Q (60–80 mesh) and operated at 242 °C, with nitrogen as carrier and a flame photometric detector operated in the S mode, or on a glass column (3 ft × 6 mm o.d.) with identical packing and operated at 195 °C, with hydrogen as carrier gas (100 ml min^{-1}) and an electrolytic conductivity nitrogen detector. Recoveries of added RE 11775 from water, soil, and mud samples were about 100% and from grass and lucerne about 80%. Down to 0.01 and 0.1 ppm could be determined in water, soil, and grass and in lucerne, respectively.

Mixtures of carbamates

Methods for analysing mixtures of carbamate insecticides have been described by various workers.[183-185] Cohen *et al.*[183] used an electron capture gas chromatograph to determine carbamate insecticides as their 2,4- dinitrophenyl derivatives in water and plant material. After isolation from the sample material by the appropriate extraction method, the carbamates were reacted with 1-fluoro-2,4-dinitrobenzene. The resulting 2,4-dinitrophenyl ethers were determined by subjecting their solution in hexane to gas chromatography on a glass column (140 cm × 15 mm) packed with 1% GE-XE 60 and 0.1% Epikote 1001 on AW-DMCS Chromosorb G (60–80 mesh) operated at 211 °C; nitrogen was used as carrier gas. The recovery of carbamates added to water and to peas, lettuce, and apples was in the range 82–100%, except for carbaryl (63%) and butacarb (40–65%) added to water. The limits of determination of carbamates in water and plant materials were 0.005 ppm and 0.1 ppm respectively. The method is inapplicable to soil.

Sundaram et al.[185] have described a rapid and sensitive analytical technique to quantify carbamate insecticides at nanogram levels using sorption on an Amberlite XAD-2 resin column and desorption followed by N–P gas–liquid chromatographic analysis. The carbamates were extracted from natural water by percolation through a column of Amberlite XAD-2, followed by elution with ethyl acetate. The carbamate residues were then analysed directly. Recoveries between 86 and 108% were obtained for the following insecticides; aminocarb (4-dimethylamino)*M*-tolyl-*N*-methylcarbamate), carbaryl (1-naphthyl-*N*-methylcarbamate), carbofuran (2,3-dihydro-2,2-dimethylbenzofuran-7-yl-methylcarbamate), mexacarbate (4-dimethylcarbamate), and propoxur (*o*-isopropoxyphenyl-*N*-methylcarbamate). Only 41–58% recovery was obtained for methomyl (*S*-methyl-*N*-[(methyl carbamoyl)oxy] thioacetimidate).

Extraction procedure

A Tracor model 550 gas chromatograph, equipped with a Hall 310 electrolytic conductivity detector and a model 702 N–P detector was used for the analysis of carbamate residues present in the acetone or ethyl acetate extracts of water samples. Three Pyrex glass columns (75 cm × 4.0 mm i.d.) were used: (1) 1.95% QF-1 plus 1.5% OV-17; (2) 3% OV-25; and (3) 3% Carbowax 20M TPA; all on Chromosorb W HP, 80–100 mesh. The operating conditions are given in Table 107.

Table 108 shows the results obtained in a study of the effect of the pH of the water sample on the Amberlite XAD-2 column on the recovery of one particular carbamate insecticide. The recoveries of aminocarb at 0.50 and 5.00 ppm levels above pH 4.0 were 91–104%. However, at acidic pH, namely 3.0 and 4.0, the Amberlite XAD-2 did not absorb all aminocarb residues present in the water. At concentration levels of 0.5 and 5.0 ppm, only 62 and 52% of fortified aminocarb were extracted by the resin columns. The unabsorbed aminocarb residues present in the water samples were quantitatively recovered by neutralizing the sample with saturated aqueous sodium carbamate solution, repercolating the water through the same resin columns and eluting with ethyl acetate.

Table 107 Operating conditions for g.l.c.

Parameter	Hall 310 electrolytic conductivity detector	Tracor model 702 N-P detector
Detector temperature	850 °C	240 °C
Inlet temperature	210 °C	210 °C
Outlet temperature	300 °C	210 °C
Column temperature	160 °C	160 °C
Carrier gas (helium) flow rate	80 ml min^{-1}	80 ml min^{-1}
Reaction gas (hydrogen) flow rate	20 ml min^{-1}	
Solvent flow rate	50% isopropanol in distilled deionized water at 1 ml min^{-1}	
Plasma gas flow rate		Hydrogen 2.5 ml min^{-1} Air 120 ml min^{-1}

Reprinted with permission from Sundaram et al.[185] Copyright (1979) Elsevier Science Publishers B.V.

Table 108 Recovery of aminocarb from Amberlite XAD-2 at various pH levels

	Aminocarb fortification (ppm)	
	0.50	5.00
pH	$\bar{X} \pm$ s.d. (%)	$\bar{X} \pm$ s.d. (%)
3.0	61.8 ± 5.99	51.8 ± 5.12
4.0	89.1 ± 1.84	83.8 ± 6.03
5.0	91.5 ± 4.82	98.0 ± 5.41
6.0	101 ± 1.15	103 ± 2.45
7.5	104 ± 3.23	97.2 ± 4.00
9.0	98.5 ± 1.91	104 ± 4.03

$\bar{X}$, mean recovery; s.d., standard deviation ($n = 4$)
Reprinted with permission from Sundaram et al.[185] Copyright (1979) Elsevier Science Publishers B.V.

Table 109 shows the recoveries obtained for eight carbamate insecticides when put through the whole analytical procedure. With the exception of methomyl, recoveries are reasonably good.

Coburn et al.[186] reported a procedure for the extraction and analysis of N-methylcarbamates in natural water. Their procedure involved the extraction at pH 3–4 by solvent partitioning with methylene chloride, salting out with sodium sulphate, hydrolysis of the extracts by methanolic potassium hydroxide to form the corresponding phenols, re-extraction of the phenols at a pH of 2 or lower with methylene chloride, chemical derivatization with pentafluorobenzyl bromide to form the ether derivatives, clean-up on a silica gel microcolumn, and analysis by gas chromatography with electron capture detection. Their recoveries were from 87 to 98% for several carbamates, namely propoxur, carbofuran, 3-ketocarbofuran, metmercapturon, carbaryl, and mobam. The

Table 109 Recovery of carbamate pesticides from natural water

	Fortification (ppm)	
	1.0	0.01
Compound	$\bar{X} \pm$ s.d. (%)	$\bar{X} \pm$ s.d. (%)
Aminocarb	103 ± 2.00	98.0 ± 3.00
Mexacarbate	91.4 ± 1.69	101 ± 2.73
Carbaryl	91.5 ± 1.32	108 ± 2.86
Propoxur	102 ± 2.22	94.5 ± 11.0
Carbofuran	95.3 ± 0.95	97.8 ± 9.74
Pirimicarb	106 ± 7.76	88.3 ± 6.24
Methiocarb	105 ± 1.63	86.4 ± 3.25
Methomyl	58.4 ± 5.91	41.7 ± 3.05

$\bar{X}$, mean recovery; s.d., standard deviation ($n = 4$)
Reprinted with permission from Sundaram et al.[185] Copyright (1979) Elsevier Science Publishers B.V.

determination of mexacarbate and aminocarb was not possible with this procedure as the phenolic products obtained on hydrolysis were not extracted from the acidified media.

Thompson et al.[187] have devised a multiresidue scheme of analyses based on silica gel column chromatography followed by gas chromatography for the analyses of mixtures of organochlorine, organophosphorus and carbamate types of insecticides. Details of this procedure are given in the section on organochlorine insecticides. The relevant work on seven carbamate insecticides is discussed below. The mixtures of the three types of compounds is fractionated into groups on a partially deactivated silica gel column with three sequential elutions. Final determinations were made by gas chromatography using the electron capture detector for the halogenated compounds and derivatized carbamates, and the flame photometric detector for organophosphorus compounds. Derivatization of the carbamate fraction is carried out with 1-fluoro-2,4-dinitrobenzene as follows. To the tubes containing the 0.1 ml concentrates of fractions 11 and 111 is added 0.5 ml of 1-fluoro-2,4-dinitrobenzene (1% in acetone) and 5 ml of sodium borate buffer solution ($Na_2 B_4O_7 \cdot 10H_2O$ 1 M solution at pH 0.4). The same reagents are added to an empty tube to serve as a reagent blank. The tubes are tightly capped and heated at 70 °C for one hour in a water bath. The tubes are cooled to room temperature, and 5 ml of hexane added to each tube. The tubes are shaken vigorously for 3 min, either manually or on a wrist action shaker. The layers are allowed to separate and 4 ml of the hexane (upper) layer are transferred carefully to a tube and stoppered tightly. The results obtained in applying electron capture gas chromatography to a mixture of seven carbamate insecticides is shown in Figure 193. The peaks eluted significantly later than those resulting from reagent impurities or other contaminants. Recoveries of the carbamates are shown in Table 110. No problems were encountered with

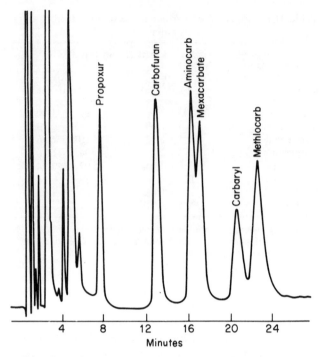

Figure 193 Six carbamates eluted in fraction III (portion of carbofuran and methiocarb in fraction II); 5.0 ng of each compound; g.c. column 5% OV-210. Reprinted with permission from Thompson et al.[187] Copyright (1977) Springer Verlag, N.Y.

Table 110 Recoveries of seven carbamate compounds

Compound	Concentration (ppb)	No partitioning		Recoveries (%) Silica gel partitioning Elution fraction				
		Extraction only	Fortified CH_2Cl_2	I	II	III	IV	Fraction total
Aminocarb (Matacil)	10	89	90			59		59
Metalkamate (Bux)	10	102	101			93	68	
Carbaryl (Sevin)	10	100	98			68		68
Carbofuran (Furadan)	10	96	95		4	94		98
Methiocarb (Mesurol)	10	99	94		55	57		112
Propoxur (Baygon)	10	96	100		99			99
Mexacarbate (Zetran)	10	72	71		58			58

Reprinted with permission from Thompson et al.[187] Copyright (1977) Springer Verlag, N.Y.

metalkamate, carbofuran, methiocarb or propoxur. Final recoveries after fractionation fell in an acceptable range. Acceptably high recoveries were obtained for aminocarb and carbaryl by direct derivatization and gas chromatography of the concentrated extract resulting from the methylene dichloride extraction, bypassing the silica gel fractionation step. Losses were noted during the fractionation for these two compounds. Recoveries of mexacarbate were highly inconsistent. Even by direct analysis of spiked methylene dichloride solution recovery variations were extreme. Expectedly, recovery data resulting from analysis of the direct water extraction were similarly inconsistent. Fractionation of this compound resulted in some further loss.

Liquid chromatography with an electrochemical detector has been used[188] to estimate 2–7 ppm of various carbamate insecticides in water.

Thin-layer chromatographic methods have been described[189] for determining N-arylcarbamate and urea pesticides in water and waste water (barban, chloropropham, diuron, fenuron, fenuron TCA, linuron, monouron, monoruron TCA, neburan, propham, siduron, and swep) and for determining O-arylcarbamate pesticides in water and waste water (aminocarb, carbaryl, methiocarb, mexacarbate, and propoxur). To determine N-arylcarbamates a measured volume of water sample is extracted with methylene chloride and the concentrated extract is cleaned up with a Florisil column. Appropriate fractions from the column are concentrated and portions are separated by thin-layer chromatography. The insecticides are hydrolysed to primary amines, which in turn are chemically converted to diazonium salts. The layer is sprayed with 1-naphthol and the products appear as coloured spots. Quantitative measurement is achieved by visually comparing the response of sample extracts to the responses of standards on the same thin-layer plate. Direct interferences may be encountered from aromatic amines that may be present in the sample. These materials react with the chromogenic reagent and yield reaction products similar to those of the insecticides. In cases where amines are suspected of interfering with a determination, a different solvent system should be used to attempt to isolate the insecticides on the layer. Indirect interferences may be encountered from naturally coloured materials whose presence masks the chromogenic reaction.

Table 111 lists R_f values obtained for N-arylcarbamate and urea insecticides for several migration solvent systems. Table 112 lists column responses and detection limits for the quoted N-arylcarbamates and ureas.

To determine O-arylcarbamate insecticides a measured volume of water is extracted with methylene chloride. The concentrated extract is cleaned up with a Florisil column. Appropriate fractions from the column are concentrated and portions are separated by thin-layer chromatography. The carbamates are hydrolysed on the layer and the hydrolysis products are reacted with 2,6-dibromoquinone chlorimide to yield specific coloured products. Quantitative measurement is achieved by visually comparing the responses of sample extracts with the responses of standards on the same thin layer. Identification is confirmed by changing the pH of the layer and observing colour changes of the reaction products. Direct interferences may be encountered from phenols that may be present in the sample. These materials react with the chromogenic reagent and yield reaction products similar to those of the carbamates. In cases where phenols are suspected of interfering with a determination, a different

Table 111 R_f values of N-arylcarbamate and urea insecticides in several solvent systems

	A	B	C	D	E	F	G
Carbamates							
Propham	0.49	0.54	0.73	0.48	0.36	0.68	0.69
Chloropropham	0.57	0.60	0.73	0.49	0.37	0.70	0.73
Barban	0.61	0.59	0.72	0.41	0.28	0.70	0.74
Swep	0.48	0.44	0.70	0.41	0.28	0.67	0.66
Ureas							
Fenuron	0.03	0.04	0.38	0.22	0.10	0.41	0.30
Fenuron TCA	0.03	0.04	0.36	0.22	0.10	0.41	0.30
Monuron	0.04	0.05	0.37	0.24	0.10	0.47	0.34
Monuron TCA	0.04	0.06	0.34	0.24	0.10	0.46	0.34
Diuron	0.05	0.09	0.38	0.28	0.13	0.54	0.44
Linuron	0.40	0.43	0.62	0.39	0.24	0.66	0.64
Neburon	0.21	0.28	0.64	0.41	0.26	0.68	0.65
Siduron	0.02	0.7	0.68	0.39	0.25	0.62	0.55

Solvent systems: A, methylene chloride; B, chloroform; C, ethyl acetate; D, hexane–acetate (2:1); E, hexane–acetone (4:1); F, chloroform–acetonitrile (2:1); G, chloroform–acetonitrile (5:1).

Table 112 Colour responses and detection limit for the N-arylcarbamates and ureas

	Colour	Detection limit (μg)
Carbamates		
Propham	Red-purple	0.2
Chloropropham	Purple	0.1
Barban	Purple	0.05
Swep	Blue-purple	0.2
Ureas		
Fenuron	Red-purple	0.05
Fenuron TCA	Red-purple	0.1
Monuron	Pink-orange	0.05
Monuron TCA	Pink-purple	0.1
Diuron	Blue-purple	0.1
Linuron	Blue-purple	0.1
Neburon	Blue-purple	0.1
Siduron	Red-purple	0.05

solvent system should be used to attempt to isolate the carbamates. Indirect interferences may be encountered from naturally coloured materials whose presence masks the chromogenic reaction.

Table 113 quotes R_f values of O-arylcarbamates obtained with several migration solvent systems. Table 114 lists column responses and detection limits.

Table 113 R_f values of O-arylcarbamate insecticides in several solvent systems

	A	B	C	D	E	F
Carbaryl	0.26	0.22	0.48	0.41	0.58	0.24
Aminocarb	0.26	0.02	0.46	0.52	0.54	0.04
Mexacarbate	0.34	0.22	0.54	0.53	0.60	0.24
Methiocarb	0.31	0.31	0.55	0.55	0.59	0.28
Propoxur	0.27	0.10	0.53	0.59	0.60	0.13

Solvent systems: A, hexane–acetone (3:1); B, methylene chloride; C, benzene–acetone (4:1); D, benzene–cyclohexane–diethylamine (5:2:2); E, ethyl acetate; F, chloroform.

Table 114 Colour responses and detection limit for O-arylcarbamates

| | Colours | | Detection limit |
	Before buffer	After buffer	(μg)
Carbaryl	Brown	Red-purple	0.1
Aminocarb	Grey	Green	0.1
Mexacarbate	Grey	Green	0.1
Methiocarb	Brown	Tan	0.2
Propoxur	Blue	Blue	0.1

ORGANOPHOSPHORUS INSECTICIDES

Introduction

The most commonly used insecticides of this type are listed below in alphabetical order. From this list can be seen the wide variety of these phosphorus based compounds now in use. Gas chromatography has played a major part in the development of suitable methods for the determination of submicro amounts of such substances in environmental samples such as crops, animal tissues and water.

Abate	O,O,O',O'-Tetramethyl-O,O'-(thiodi-p-phenylene diphosphorthioate)
Accothion (sumithion, fenitrothion, folithion)	O,O-Dimethyl-O-4-nitro-m-ethyl phosphorothioate
Amidithion	S-(N-2-Methoxyethylcarbamoylmethyl)dimethyl phosphorothiolothionate.
	S-(N-2-methoxyethylcarbamoylmethyl)O,O-dimethyl phosphorodithioate
Amiton	S-(2-(Diethylamino)ethyl) diethyl phosphorothiolate
	S-(2-(Diethylamino)ethyl) O,O-diethyl phosphorothionate

Azinphos-ethyl	S-(3,4-Dihydro-4-oxobenzo(d)-[1,2,3]triazin-3-ylmethyl) diethyl phosphorothiolothionate
	S-(3,4-Dihydro-4-oxobenzo(d)-[1,2,3]triazin-3-ylmethyl) O,O-Diethyl phosphorodithioate
Azinphos-methyl	S-(3,3-Dihydro-4-oxobenzo(d)-[1,2,3] triazin-3-ylmethyl) dimethyl phosphorothiolothionate
	S-(3,4-Dihydro-4-oxobenzo(d)-[1,2,3] triazin-3-ylmethyl) O,O-dimethyl phosphorodithioate
Azothoate	4-(4-Chlorophenylazo)phenyldimethyl phosphorothionate
	O-4-(4-Chlorophenylazo)phenyl-O,O-dimethyl phosphorothioate
Bensulide	N-2-(O,O-Di-ispropylphosphorothiolothioyl)-ethyl benzenesulphonamide
	di-Isopropyl S-(2-phenylsulphonylaminoethyl) phosphorothiothionate
	O,O-Di-isopropyl S-(2-phenylsulphonylamino-ethyl)phosphorodithioate
Bromophos	4-Bromo-2,5-dichlorophenyldimethyl phosphorothionate
	O-(4-Bromo-2,5-dichlorophenyl)-O,O-dimethyl phosphorothioate
Bromophos-ethyl	4-Bromo-2,5-dichlorophenyldiethyl) phosphorothionate
	O-(4-Bromo-2,5-dichlorophenyl)-O,O-diethyl phosphorothioate
Butonate	Dimethyl-1-butyryloxy-2,2,2-dichloroethylphosphonate
Carbophenothion	S-(4-Chlorophenylthiomethyl)diethyl phosphorothiolothionate
	S-(4-Chlorophenylthiomethyl)-O,O-diemethyl phosphorodithioate
Chlorfenvinphos	2-Chloro-1-(2,4-dichlorophenyl)vinyldiethyl phosphorodithioate
Chlorphonium	Tributyl-2,4-dichlorobenzylphosphonium
Chlorphoxim	(2-Chloro-α-cyanobenylideneamion) diethyl phosphorothionate
	O-2-Chlorocyanobenzylideneamino-O,O-diethyl phosphorothioate
	2-Chloro-α-(diethoxyphosphinothioyloxyimino) phenylacetonitrile
Coumaphos	3-Chloro-4-methyl-7-coumarinyldiethyl phosphorothionate
	O-(3-Chloro-4-methyl-7-coumarinyl)-O,O-diethyl phosphorothionate

Coumithoate	Diethyl-7,8,9,10-tetrahydro-6-oxobenzo(c) chroman-3-yl phosphorothionate
	O,O-Diethyl-O-(7,8,9,10-tetrahydro-6-oxobenzo(c) chroman-3-yl) phosphorothioate
	O,O-Diethyl-O-(7,8,9,10-tetrahydro-6-oxo-6-didibenzo(bd)pyran-3-yl) phosphorothionate)
Crotoxyphos	Dimethyl *cis*-1-methyl-2-(1-phenylethoxycarbonyl) vinyl phosphate
	1-Methylbenzyl-3-(dimethoxyphosphinyloxy) isocrotonate
Cyanophos	4-Cyanophenyldimethyl phosphorothionate
	O-4-Cyanophenyl-O,O-dimethyl phosphorothionate
Cyclophosphamide	
Dasanite	O,O-Diethyl-O-(4 methylsulphinyl)phenyl) phosphorothioate
Demephion	A mixture of demephion-O and demephion-S (see below)
Demephion-O	Dimethyl-2-(methylthio)ethyl phosphorothionate
	O,O-Dimethyl-O-(2-(methylthio)ethyl phosphorothioate
Demephion-S	Dimethyl-S-(2-(methylthio)ethyl) phosphorothiolate
	O,O-Dimethyl-S-(2(methylthio)ethyl) phosphorothioate
Demeton	A mixture of demeton-O and demeton-S (see below)
Demeton-methyl	A mixture of demeton-O-methyl and demeton-S-methyl (see below)
Demeton-O-methyl	2-(Ethylthio)ethyldimethyl phosphorothionate
	O-(2-(ethylthio)ethyl)-O,O-dimethyl phosphorothioate
Demeton-O	Diethyl-2-(ethylthio)ethyl phosphorothionate
	O,O-Diethyl-O-(2-(ethylthio)ethyl) phosphorothioate
Demeton-S	Diethyl-S-(2-(ethylthio)ethyl) phosphorothiolate
	O,O-Diethyl-S-(2-(ethylthio)ethyl) phosphorothioate
Demeton-S-methyl	S-(2-Ethylthio)ethyl)dimethyl phosphorothiolate
	S-(2-Ethylthio)ethyl)-O,O-dimethyl phosphorothioate
Diazinon	Diethyl-2-isopropyl-6-methyl-4-pyrimidinyl phosphorothionate
	O,O-Diethyl-O(2-isopropyl-6-methyl-4-pyrimidyl) phosphorothioate

Dichlorvos	2,2-Dichlorvinyldimethyl phosphate
Dimethoate (rogor)	
Disulfoton	*O,O*-Dimethyl-*S*-[2(ethylthio)ethyl] phosphorodithioate
Dursban (chlorpyrifos)	*O,O*-dimethyl-*O*-(3,5,6-(triethylchloro-2-pyridyl) phosphorothioate
Edifenphos	Ethyl-*S,S*-diphenyl phosphorodithiolate thionate *O*-Ethyl-*S,S*-diphenyl phosphorodithioate
Ethion	*S,S*-Methylene-*O,O,O',O'*-tetraethyl phosphorodithioate
Fenchlorphos (ronnel)	*O,O*-Dimethyl-*O*-2,4,5-trichlorophenyl phosphorothioate
Fenthion	*O,O*-Dimethyl-*O*-[4(methylthio)-*m*-tolyl] phosphorothioate
Fonofos	Ethyl-*S*-phenylethyl phosphonthiolothionate Ethyl-*S*-phenylethyl phosphonodithioate
Iodophos	
Malathion	
Maretin (napthalphos)	
Methidathion (supracide)	
Methocrotophos	Dimethyl-*cis*-2(*N*-methoxy-*N*-methyl-carbamoyl) 1-methylvinyl phosphate 3-(Dimethoxyphosphinyloxy)-*N*-methoxy-*N*-methylisocrotonamide
GS-13005	*S*-(5-Methoxy-2-oxo-1,3,4-thiadazolin-3-yl)-*O,O*-dimethyl phosphorodithioate
Monitor	*O,S*-Dimethyl phosphoramidothioate
Nemacur	Ethyl-4-(methylthio)-*m*-tolyl isopropyl phosphoramidate
Paraoxon	The oxygen analogue of parathion
Paraoxon methyl	
Parathion	*O*-(4-Nitrophenyl phosphorothioate)
Parathion-ethyl	
Phenthioate	
Phorate	*O,O*-Dimethyl-*S*-(ethylthio)methyl phosphorodithioate
Phosalone (zolone)	
Phosdrin (mevinphos)	
Phosphamidon	
Pirimiphos-ethyl	2-Diethylamino-6-methypyrimidin-4-yl-diethyl phosphorothionate *O*-(2-diethylamino-6-methylpyrimidin-4-yl)-*O,O*-diethyl phosphorothioate
Pirimiphos-methyl	2-Diethylamino-6-methylpyrimidin-4-yl-dimethyl phosphorothionate

	O-(2-Diethylamino-6-methylpyrimidin-1-yl)-O,O-dimethyl phosphorothioate
Prothoate	O,O-Dimethyl-S-N-isopropyl carbamoyl methyl phosphorothiolothionate
Salithion	2 Sulphide of 2-methoxy-4-H-benzo-1,2,3,-dioxaphosphrin
Tetrachlorvinphos	trans-2-Chloro-1-(2,4,5-trichlorophenyl)vinyldimethyl phosphate
Velsicol VCS-506	O-(4-Bromo-2,5-diethylchlorophenyl)-O-methylphenyl phosphorothioate
Zyron	O-2,4-Dichlorophenyl-O-methylisopropyl phosphoramidothioate

Gas chromatography

Detector systems

Much of the development work on the determination of organophosphorus insecticides which has been carried out in recent years has hinged on the development of suitable detectors which are ultrasensitive and which are specific for phosphorus in the present of other elements such as carbon, hydrogen, oxygen, halogens, nitrogen, and sulphur and, indeed in some cases can be used to determine compounds containing these other elements. Several types of organophosphorus insecticides also contain halogens, nitrogen, or sulphur. Electron capture, flame ionization, flame photometric, microcoulometric, thermionic, and electrolytic conductivity detectors have all been studied in this application. The advantages and disadvantages of these types of detectors are discussed in general proceeding to a discussion of the application of gas chromatography to the determination of particular organophosphorus insecticides or mixtures, thereof, in water. The quantitative analysis of organophosphorus pesticide residues has become much easier since the development of the alkali flame and flame photometric detectors, both of which are now commercially available.

The Karmen-Guiffrida detector (the so-called thermionic detector) has proved extremely useful. It is basically a flame ionization detector with an alkali salt ring placed on the flame tip. Not only does this enhance the sensitivity but also the selectivity. This may be of the order of 10^4 to 10^5 for phosphorus-containing compounds as compared with the equivalent carbon compounds. The selectivity for halides and nitrogen is between 10^2 and 10^3 and 10^2 for sulphur and about 10 for arsenic. The responses for phosphorus increases with increasing hydrogen flow, but some increase in the background current is also observed. A variety of alkali salts are used including potassium chloride, caesium bromide, and rubidium sulphate. The type and shape of the tip and anode, together with the flow rates, are very much a matter of personal choice. De Loach and Hemphill[191,192] have discussed the design of a rubidium sulphate detector and with optimum conditions claim a sensitivity of 1 pg. Often a charcoal column clean-up is used before injection to prevent the column becoming contaminated.

The flame photometric detector was patented as early as 1962. It consists basically of a flame ionization detector fitted with optical filters and photomultiplier tubes. When organophosphorus compounds burn in the flame two optical bands of primary interest can be produced, at 526 nm due to HPO species and, when sulphur is present, at 394 nm due to SS species. Brody and Chaney[193] developed the method of filtering and measuring the emission at these wavelengths. A highly selective detector, which was reported as early as 1965 is the helium plasma microwave emission spectrometric detector. Bache and Lisk[194] pioneered this type of detector and have successfully used it at the residue level for both organophosphorus insecticides and carbamates. The detector is based on the emission resulting from the atoms of a specific element in a helium or argon stream eluting from a gas chromatograph when passed into a microwave sustained plasma discharge at either atmospheric or reduced pressure. Reduced pressure permits the use of helium as the carrier gas with the advantages of less background radiation and higher excitation energy than argon plasmas. All elements or organic compounds can be detected, so it is possible to examine for phosphorus and sulphur simultaneously. For most elements, including phosphorus, the detection limits are of the order 0.1 to 1.0 ng s^{-1}, but for oxygen and nitrogen it is about 3 ng s^{-1}. Spiked samples have been satisfactorily recovered at the 0.05–0.5 mg kg^{-1} level. Another promising detector makes use of the piezoelectric effect. A quartz piezoelectric crystal coated with an inorganic salt can selectively adsorb organophosphorus compounds. A detector based on this principle has been developed; the crystal is incorporated in a variable oscillator circuit and by measurement of frequency changes caused by the increase in mass on adsorption, organophosphorus compounds can be determined at the mg kg^{-1} level.

Hartmann[195] described a specific, highly sensitive detector for phosphorus-containing compounds for use with single and dual channel gas chromatographic columns. It is a modified electron capture detector in which the flame burns above a compressed caesium bromide tip. The detector is simple in construction, rapidly interchangeable and easy to operate, has high reliability and shows good rectilinearity of response. Hartmann gives examples of the use of this method in the identification of phosphorothioate pesticides in mixtures with chlorinated hydrocarbon pesticides. The limit of detection is 3×10^{-12} for phorate and 12×10^{-12} for malathion. Kanazawa and Kawahara[196] applied electron capture gas chromatography to the determination of pesticides. They investigated retention time, peak area sensitivity, and rectilinearity of the calibration graph for various pesticides under two sets of operating conditions. The columns used were 5% of Dow silicone II at 170 °C and 2% of polyoxyethylene glycol adipate at 180 °C. For organophosphorus compounds the polyoxyethylene glycol adipate is the more satisfactory stationary phase, peak area sensitivities on this column being higher than those on the silicone column.

Guiffrida et al.[197] studied the effects of varying the operating conditions for electron capture and flame ionization detectors. Use of a hydrogen flame burning in an atmosphere of alkali-metal salt (e.g. potassium chloride) results in increased sensitivity and complete specificity in the detection of organophosphorus insecticides. Typical conditions used by these workers include a column (6 ft × 4 mm) of 10% DC-200 silicone fluid on 80–100 mesh Gas Chrom Q, operated at 190–230 °C, with nitrogen as carrier gas. A technique involving the parallel use of an electron capture detector and a modified flame ionization detector is described. A method for specific detection of subnanogram quantities of organophosphorus compounds by flame ionization has been described by Kamen.[198] This detector is specific for phosphorus and halogens.

The flame photometric detector, which is specific for phosphorus and sulphur has been used[193] for the detection of subnanogram quantities of pesticides containing these elements. The detector can be constructed by modifying existing flame ionization equipment. The detector is sensitive to parts per 10^9 of phosphorus and to less than 1 ppm sulphur but is insensitive to hydrocarbons and to compounds containing chlorine, nitrogen, and oxygen which are burned inside the burner tip, which is shielded from the photomultiplier tube. Response to phosphorus is rectilinear for up to 63 ppm; that to sulphur varies exponentially with up to 100 ppm. Nitrogen is used as carrier gas, the flow of hydrogen plus air is kept at 200 ml min^{-1}, and the ratio of nitrogen to oxygen is maintained at 4:1. The application of the microcoulometric gas chromatograpic detector[199] to the selective detection of phosphorus, sulphur, and halogen in pesticides has been studied by Burchfield et al.[200] after the separation of phosphorus-, sulphur-, or chlorine-containing pesticides on a column (1 m × 3 mm) of Anakrom ABS supporting 10% DC-200 silicone oil, with hydrogen (15 ml min^{-1}) as carrier gas and temperature programming from 175 to 230 °C in 10 min. The column effluent is passed through a small furnace at 950 °C in which the compounds are converted into phosphine, hydrogen sulphide, or hydrogen chloride. These gases are measured with a microcoulometric titration cell equipped with silver electrodes. The insertion of a short tube containing alumina between the furnace and the titration cell removes hydrogen sulphide and hydrogen chloride whereas a short tube containing silica gel similarly placed retains hydrogen chloride and separates phosphine and hydrogen sulphide. In a further paper Burchfields et al.[201] reported work on the selective detection of phosphorus-, sulphur-, and halogen-containing pesticides. They discussed ionization detectors and the possibility of increasing their specificity. They also describe microcoulometric methods for the analysis of pesticides containing the above three elements. These methods are briefly reviewed below.

Halogens and sulphur

The sample solutions in an organic solvent are carried through the chromatograph column by means of nitrogen and oxygen added to the effluent before combustion at 800 °C. To determine halogens, the combustion products are passed through a titration cell in which halides (except fluoride) cause precipitation of silver halide and the current required to generate sufficient silver ions to replace those consumed is recorded. Down to 0.001 g of chloride ion can be detected. To determine sulphur the combusted products are passed through a cell in which the sulphur dioxide is automatically titrated with iodine, the current required to generate sufficient iodine being recorded also.

Phosphorus

The sample of organophosphorus compounds are first separated on a chromatographic column and are then reduced at 950 °C to phosphine by means of hydrogen. The reduction products pass through a titration cell in which silver phosphides are precipitated and the current necessary to regenerate silver ions equivalent to those precipitated is recovered. To prevent interference by hydrogen chloride and hydrogen sulphide a 'subtraction' column of alumina or silica gel is inserted between the reduction tube and the titration cell. The determination of chloride plus sulphur plus phosphorus in a sample mixture is possible by omitting the subtraction column. The method has been applied to the determination of mevinphos, diazinon, and carbophenothion. In 1967, Ives and Guiffrida[202] described their investigations into the development of a thermionic detector for compounds containing phosphorus, nitrogen, arsenic, and chlorine. They found that phosphorus could be distinguished from nitrogen or arsenic by comparing the responses of thermionic and flame photometric detectors. They found that potassium chloride was the preferred source for organo-phosphorus compounds and rubidium chloride for nitrogen compounds. They recommended thermionic detection for compounds containing several nitrogen atoms and examined the applicability of the technique to various types of pesticides. Dressler and Janak[203,204] studied the use of a thermionic detector to discriminate between volatile phosphorus-, chlorine-, nitrogen-, sulphur-, and carbon-containing compounds. These compounds were subjected to chromatography on a 68 cm column packed with 15% of polyoxyethylene glycol 400 on Strechamol and operated at 120 °C. The salt tip was compressed from a 1:1 mixture of alkali-metal salt and molecular sieve 5A. Variables were studied as a function of the background current. The effect of changing the cation of the salt (each as the chloride) was to increase the response in the order sodium, potassium, rubidium, and caesium for the nitrogen and phosphorus compounds and to decrease it for the others. Thus, the response to the phosphorus compound was 4 to 5 times as high with a caesium chloride tip as with a sodium chloride tip. By using sodium salts, the response to the phosphorus and chlorine compounds also varied with the anion used, increasing the order chloride, bromide, sulphate, carbonate, and nitrate. In both these series the flow rate of hydrogen necessary to obtain a given background current varied in the same way. These results can be related to the vapour pressures of the salts and the ionization potentials of the cations. The noise level also increased with increasing atomic number of the cation, but, up to a certain background current level, the increase was less than the increase in response. Dressler and Janak[203,204] found that the difference effects of background current and salt-tip material changes can be used for qualitative differentiation between types of compounds.

Wooley[205] described the mechanism of the determination of organophosphorus compounds using a flame ionization thermionic detector. He investigated the changes in concentration of free electrons in a hydrogen flame into which caesium nitrate is aspirated and organophosphorus compound is introduced. This work showed that it is not essential for the alkali-metal salt to be in the solid state to achieve a sensitive detection of phosphorus. Wooley[205] proposed a mechanism for the release of electrons under these conditions.

Novak and Malmstadt[206] used an alkali-metal salt flame thermionic detector coupled with a photomultiplier and interference filter to monitor the emission from organic compounds containing phosphorus and chlorine. They found that under some operating conditions the visible radiation caused by the alkali metal is also enhanced, in addition to that due to phosphorus and chlorine. Subnanogram quantities of halogen compounds were detected but the sensitivity to phosphorus was two orders of magnitude lower.

Bowman and Beroza[207] incorporated two photomultipliers into the central burner assembly of a flame thermionic detector for phosphorus and sulphur. One photomultiplier had a 526 nm filter to detect phosphorus and the other a 394 nm filter to detect sulphur. The phosphorus response was linear but proportional to the square of the quantity of phosphorus present. Bowman and Beroza[207] applied this detector to the determination and identification of organophosphorus–sulphur pesticides.

Bowman et al.[208] studied the applicability of the Melpar flame photometric detector in the phosphorus (526 nm filter) and sulphur (394 nm filter) modes to the multicomponent determination of a wide range of pesticides in foods. They used OV-101 or OV-210 columns and concluded that operation of the detector in the phosphorus mode was preferable as most of the samples examined showed less interference using this detector.

Kamen[209] has discussed the question of differential specificity in the detection by alkali flames of phosphorus, nitrogen, and halogens.

De Loach and Hemphill[192] have described modifications to the Barber-Coleman and Varian Aerography rubidium sulphate flame detectors to effect improvement in response to organophosphorus insecticides. Optimization of anode design and position were important factors in achieving an improvement in response. Scolnick[210] detected organophosphorus insecticides by their gas-phase ionization reaction with caesium bromide vapour in an electrically heated inert gas atmosphere. Response factors measured in coulombs per mole for a number of pesticides are lower by factors of 20 to 60 than those obtained with an alkali flame detector. Although this chemi-ionization detector is insensitive to hydrocarbons, their presence reduces its response to phosphorus compounds.

Grice et al.[211] studied the response characteristics of the commercially available Melpar (caesium/rubidium) flame photometric detector for organophosphorus and sulphur compounds and showed that the minimum detectable quantities with this detector are 200 pg (sulphur) and 40 pg (phosphorus).

These workers also carried out a comparison between the flame ionization and flame photometric types of detectors. Craven[212] has described a simplified version of alkali flame detector for phosphorus and nitrogen. This detector utilizes a bead of fused rubidium sulphate and potassium bromide placed on the tip of the jet of a flame ionization detector. Svojanovsky and Nebola[213] have described a combined flame ionization and photometric detector for the specific determination of organophosphorus compounds. This detector system is provided with a silica lens that magnifies the image of the flame 10-fold and directs it on to the photomultiplier. An interference filter in front of the photomultiplier ensures the selective properties of the photometric system. The signal of this system is 200 times greater than that of the normal flame ionization detector.

Greenhalgh and Cochrane[214] compared the responses of alkali flame (rubidium chloride) and electroltic conductivity detectors under these gas chromatographic conditions to phosphates, phosphorothioates containing nitrogen, phosphoramidates, triazines, amides, and a carbamate. The columns were operated with nitrogen (alkali flame) or helium (electrolytic) as carrier gas at 40 ml min^{-1}. The sample size was limited to 1 ng for the organophosphorus compounds and to 50 ng for the others. The coefficient of variation for the alkali flame detector was 3.1% and that for the conductivity detector was 4.0%. Structural factors affect the response of the rubidium chloride detector, which gives a much greater response for phosphorus than for nitrogen. When the molecule contained both phosphorus and nitrogen not bonded together, the response was greater than for a similar compound containing nitrogen, but when phosphorus–nitrogen bonds were present the response was decreased. In the absence of phosphorus the response of the two detectors to nitrogen was similar. If the compound contains nitrogen and no phosphorus the electrolytic conductivity detector is preferred as it is simpler in operation and its response to nitrogen is specific. If the compound contains phosphorus also, the alkali flame detector is preferred because of its greater sensitivity (by a factor of 100).

Mixtures of organophosphorus compounds

Gas chromatography

Askew et al.[215] have developed an early general method for the determination of organophosphorus insecticide residues and their metabolites in river waters and sewage effluents utilizing gas chromatography. The organophosphorus pesticides vary greatly in their polarity, and the extent of their extraction from aqueous samples is markedly dependent on the nature of the solvent used. Table 115 gives a comparison of the efficacies of three solvents in removing these insecticides from water, and it is apparent that a polar solvent such a chloroform is the most generally useful.

Table 115 Extraction of insecticides from aqueous solution by organic solvents

Insecticides	Insecticide extracted (%)		
	Hexane	Benzene	Chloroform
Chlorfenvinphos	92	95	87
Demeton-S-methyl	0	76	98
Dimethoate	0	0	41
Pyrimithate	92	93	90

Results obtained by extracting 1 litre of 0–0.01 ppm insecticide solution with 20 ml of solvent. Determination made by gas chromatography. Reprinted with permission from Askew et al.[215] Copyright (1959) Royal Society of Chemistry.

Askew et al.[215] found that most water extracts are sufficiently low in coextractive to require no clean-up for gas chromatographic purposes. The nature of the coextractive will undoubtedly vary with the location from which the samples are taken, and no complete comprehensive clean-up procedure can be recommended. A column containing 1 g of Nuchar carbon eluted with chloroform gave the most useful clean-up where this was found to be necessary. The following insecticides are exceptional in being retained by the column described: azinophos-ethyl, azinphos-methyl, coumaphos, dichlorvos, maloxon, menazon, phosalone, and vamidothion. When these insecticides are encountered, a clean-up on alumina[216] or magnesium oxide[217] would be preferable. In Figure 194 are shown gas chromatographs obtained by Askew et al.[215] on solvent extracts of river Thames water unspiked and spiked with various organophosphorus insecticides. Figure 194 gives a comparison of the responses of an electron capture detector and a phosphorus detector (flame thermionic), an Apiezon L column was used for the separation. Note the enhanced response for the chlorine-containing insecticide, chlorfenvinphos, on electroncapture and the poor response for demeton-S-methyl. The variation in the relative response of the electron capture detector for organophosphorus compounds is a disadvantage for general screening.[218,219] The interpretation of gas chromatograms obtained when using a phosphorus-specific detector of the Hartmann type[220] must be made with caution. A chromatographic peak may often be caused by the presence of relatively large amounts of compounds that do not contain phosophorus; volatile nitrogen-containing compounds, in particular, can be detected at the microgram level. Additional confirmation of a characterization needs to be made with another technique, such as thin-layer chromatography. The pesticides amenable to gas chromatography can be separated on a variety of stationary phases; the separating times shown in Table 116 are based on columns that are commonly used for separating organochlorine insecticides.[221] The retention times may vary greatly, and it is necessary to use either temperature programming or isothermal runs at more than one temperature for successful screening. The technique can be rendered quantitative down to subnanogram levels, and the comparison of relative times on two or more columns enables a tentative pesticide characterization to be made.

Table 116 Usage and chromatographic results of organophosphorus insecticides

Insecticide	Usage	Gas chromatographic results			Thin-layer chromatographic results			Gel chromatographic results
		Apiezon L	SE30 × E60		solvent (i)	solvent (ii)	solvent (iii)	
Azinphos-ethyl	AH	995*	970*	870*	0.33	0.90		113
Azinphos-methyl	AH	840*	870*	870*	0.19	0.88		113
Bromophos	V	135	102	63	0.85	0.93		100
Carbophenothion	V	385*	270*	187*	0.83	0.96		98
Clorfenvinphos	AHV	129	140	102	0.24	0.79		77
Coumaphos	V	ND	ND	ND	0.33	0.90		99
Crufomate	V	ND	164	141	0.06	0.43	0.86	73
Demeton-S	AH	19	21	13	0.33	0.93		79
Demeton-S-ethyl	AH	22	33	27	0.17	0.73		86
Diazinon	AHFV	38	41	18	0.61	0.95		77
Dibrom	AH	26	6	25	0–0.22[a]	0 to 0.89[a]		86
Dichlofenthion	V	66	58	29	0.77	0.96		89
Dichlorvos	AHFV	3*	4*	3.5*	0.22	0.27	0.73	83
Dimefox	AH	1*	2*	2*	0.08	0.44	0.66	71
Dimethoate	AH	43	65	95	0.05	0.37	0.59	95
Disulfoton	AH	47	45	26	0.82	0.97		87
Ethion	AH	224*	221*	156*	0.77	0.97	0.75	87
Fenchlorphos	AHFV	86	70	40	0.84	0.93		101
Fenitrothion	F	81	88	95	0.49	0.91		114
Haloxon	V	ND	ND	ND	0.04	0.71	0.86	112
Malathion	AHFV	66	85	75	0.37	0.95		86

Mecarbam	AH	117	127	226	0.42	0.95		85
Menazon	AH	ND	ND	ND	0	0.02	0.38	79
Mevinphos	AH	8.10≠	13.16≠	12.15≠	0.10	0.64	0.69, 0.82	106
Morphothion	AH	212*	285*	356*	0.06	0.49	0.77	79
Oxydemeton-methyl	AH	ND	ND	ND	0	0.05	0.20	100
Parathion	AH	100	100	100	0.57	0.91		104
Phenkapton	AH	640*	420*	290*	0.74	0.97		89
Phorate	AH	31	29	18	0.80	0.97		104
Phosalone	AH	730*	600*	685*	0.39	0.97		104
Phosphamidon	AH	41, 55	67, 91	73, 110	0.04	0.34	0.60	72
Pyrimithate	AH	78	79	42	0.62	0.96		81
Schradan	AH	78	130	74	0	0.02	0.16	59
Sulfotep	AH	19	26	18	0.75	0.92		75
TEPP	AH	ND	ND	26	0	0.50^a	0.03, 0.62	73
Thionazin	AH	21	25	16	0.45	0.92		89
Trichlorphon	AHV	2≠	4≠	ND≠	0.03	0.18	0.61	80
Vamidothion	AH	ND	ND	ND	0.01	0.16	0.30	73

Usage: A denotes agricultural, V veterinary, F food storage, and H horticultural.

Reprinted with permission from Askew et al.[215] Copyright (1959) Royal Society of Chemistry.

Gas chromatography results: the values shown are retention times relative to that of parathion = 100. The columns used contained the following stationary phases (i) Apiezon L, 2% and Epikote 1001, 0.2%; (ii) SE30, 4% and Epikote 1001, 0.4%; and (iii) XE 60, 2% and Epikote 1001, 0.2%, coated on acid-washed, dimethyl dichlorosilane-treated 80–100 mesh chromosorb G. All columns were 150 cm in length with 0.3 dm o.d. Retention times were determined at 195 °C except where marked * = 220 °C and ≠ = 150 °C. The retention times of parathion on the three columns were (i) 220 °C, 1.70 min; 195 °C, 4.00 min; and 150 °C, 19 min; (ii) 220 °C, 1.80 min; 195 °C, 4.50 min; and 150 °C 22 min; and (iii) 220 °C, 1.60 min; 195 °C, 4.25 min; and 150 °C, 21 min. ND denotes not detected. Thin-layer chromatographic results: the values shown are the R_f values in the solvent (i) hexane–acetone (5+1), (ii) chloroform–acetone (9+1) and (iii) chloroform–acetic acid (9+1)^a denotes streaking. Gel chromatographic results: the values shown are the elution volumes relative to that of parathion = 100 when eluting the ethanol from a Sephadex LH 20 column.

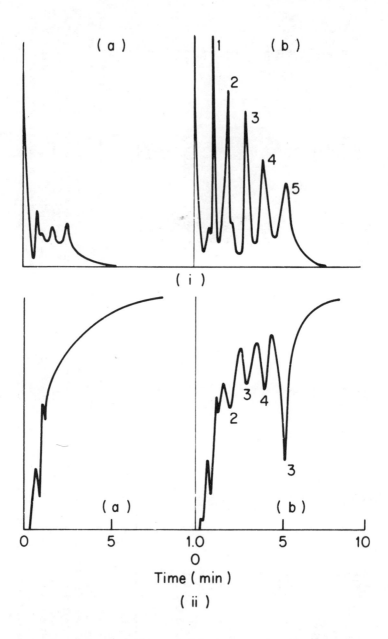

Figure 194 Comparison of gas chromatogram obtained with I a phosphorus detector and II an electron capture detector: (a) an extract equivalent to 1 l of Thames water and (b) the same extract fortified with pesticides. Peaks correspond to 5 ng of each compound: 1, demeton-S-methyl; 2, dimethoate; 3, pyrimithate; 4, parathion; 5, chlorfenvinphos. Reprinted with permission from Askew et al.[215] Copyright (1959) Royal Society of Chemistry

The general screening technique developed by Askew et al.[215] is described below in full. Stage (a) is applied as a general screening technique, and stage (b) is incorporated only when suspected pesticides are encountered.

Stage (a)

Sample
|
Chloroform extraction*
|
Clean-up
(only if coextractives required removal)
|
Chloroform removed and replaced by ethanol or acetone
|
Gas chromatographic examination
(see Table 116)
|
Thin-layer chromatography
(gel plates and solvent (i))

*If menazon is suspected remove by acidic extraction at this stage.

Stage (b)

Separate the extract before gas or thin-layer chromatography as follows:

Ethanol solution
Separation on Sephadex LH 20
(column of 75 ml bed volume)

Fraction 1	Fraction 2	Fraction 3
Elution volume 40 to 60 ml, RV 50–80)	(Elution volume 60 to 70 ml, RV 81–93)	(Elution volume 71 to 110 ml RV 94–145)
Chlorfenvinphos	Demeton-O-methyl	Azinphos-ethyl
Crufomate	Demeton-S-methyl	Azinphos-methyl
Demeton-S	Dibrom	Bromophos
Diazinon	Dichlofenthion	Carbophenthion

Dimefox	Dichlorvos	Coumaphos
Mevinphos	Disulfoton	Dimethoate
Oxydemeton-methyl	Ethion	Fenchlorphos
Phosphamidon	Ethoate-methyl	Fenitrothion
Schradan	Malathion	Haloxon
Sulfotep	Macarbam	Morphothion
TEPP	Phorate	Parathion
Vamidothion	Pyrimithate	Phenkapton
also	Thionazin	Phosalone
Trimethyl phosphate	Trichlophon	
Triethyl phosphate and		
Tributyl phosphate		

Examined by gas liquid and thin-layer chromatography.

Apparatus

Gas chromatography equipment: a single or dual column instrument, fitted with a phosphorus-specific detector (Varian Aerograph, model 205–8 was used).

Thin-layer chromatographic equipment: suitable for the preparation of 250 μm thick, silica gel chromatoplates 20×20 cm.

Ovens: 120 °C and 180 °C.

Development chamber: a glass tank, $22 \times 21 \times 9$ cm with well fitting lid.

Kuderna-Danish evaporator: this is fitted with a 10 ml pear-shaped flask, glass chromatographic column 2.5×30 cm.

Reagents

All reagents should be of recognized analytical grade whenever possible.
Chloroform
Ethanol
Ammonia solution sp. gr. 0.88
Hydrochloric acid, M
Sodium hydroxide solution, 2 M
Silica gel — for thin-layer chromatography
Nuchar carbon — a grade with particles of about 1 mm diameter was used (obtainable from Eastman Kodak Ltd)
Sodium sulphate, granular, anhydrous
Hydriodic acid spray — 25 ml of hydriodic acid (sp. gr. 1.7), 25 ml of glacial acetic acid and 50 ml of water are mixed together. The solution is stable for several weeks.

Ammonium molybdate spray: 2 g of ammonium molybdate, $NH_4MO_7O_{24} \cdot 4H_2O$, are dissolved in 20 ml of water. Concentrated hydrochloric acid $(1+1)$ is added with gentle heating, and the volume adjusted to 100 ml with water. The solution is stable for several weeks.

Tin (II) chloride spray: 1 g of tin(II) chloride, $SnCl_2 \cdot 2H_2O$, is dissolved by heating with 10 ml of concentrated hydrochloric acid, and 40 ml of water and 50 ml of acetone are added. This should be prepared freshly each day.

Extraction

One litre of water or sewage effluent is extracted with three portions each of 50 ml of chloroform (the emulsified portion of the sample is centrifuged if stable emulsions form; the extent of emulsion formation can often be reduced by dissolving about 20 g of sodium sulphate in the sample solution before extraction). The combined chloroform extracts are dried by passing through a 10×1 cm column containing granular anhydrous sodium sulphate, and collecting the eluate in a Kuderna-Danish evaporator fitted with a 10 ml pear shaped flask. The column is washed with a further 25 ml of chloroform and the solution evaporated to a small volume on a steam bath.

When menazon is suspected to be present in a sample the procedure is as follows. The combined chloroform extracts are shaken with two portions of 25 ml of M hydrochloric acid; both phases are retained. The acidic extract is washed with 25 ml of chloroform adding this to the previous extracted chloroform. To the acidic solution is added 30 ml of the 2 M sodium hydroxide and the menazon is back extracted by shaking with the two portions of 25 ml of chloroform. The chloroform is dried using the procedure described above. The extract containing menazon should be examined by thin-layer chromatography. The procedure for the other solution is as follows.

Clean-up

This is necessary only when coextractives interfere in the subsequent chromatographic determinations.

One gram of Nuchar carbon is degassed by evacuating in the presence of about 25 ml of chloroform, and the slurry is added to a 1 cm diameter $\times 30$ cm long column containing about 0.5 g of Celite 545 (the latter prevents charcoal fines from eluting from the column). The chloroform sample solution (1 ml) is placed on the column and eluted with 100 ml of chloroform, collecting the eluate in a Kuderna-Danish evaporator and evaporating to a small volume on a steam bath.

Gas chromatography

All traces of chloroform are removed from the sample solution by subjecting it to a gentle stream of air or nitrogen until dry. To prevent losses by evaporation of volatile insecticides, such as dichlorvos, dimefox, and mevinphos, the sample is removed from the air stream immediately after the last traces of solvent have evaporated; alternatively, a micro-snyder column is used, evaporating successively with acetone or ethanol to remove the final traces of chloroform.

To the residue is added 1 ml of ethanol or acetone, swirling it to effect dissolution, and 5 μl aliquots are injected on to suitable gas chromatographic columns. An injection liner (e.g. Pyrex glass) is used if uncleaned extracts are injected, to minimize column contamination. The elution time of any chromatographic peaks is determined relative to a suitable standard insecticide, such as parathion, and the results are compared with those previously obtained with insecticide standards (see Table 116).

Thin-layer chromatography

250 μg thick layers of silica gel G on 20 × 20 cm glass carrier plates are prepared and activated by heating at 120 °C for at least 2 hours. The sample solution is spotted on and the plate developed in a suitable solvent system by ascending chromatography. The plates are removed from the tank when the solvent has travelled about 10 cm from the origin.

When the plate has dried, it is sprayed uniformly with the hydriodic acid spray. A similar glass carrier plate is clipped over the sprayed surface and heated in an oven at 180 °C for 30 minutes preferably with the plates standing in a vertical plane to aid heat flow over the surface. The plates are removed from the oven and the cover-plate unclipped (this is carried out under a flame hood, as iodine fumes are evolved). When the plate is cool, it is sprayed with ammonium molybdate solution and replaced in the oven for 5 minutes. It is then removed from the oven and, when cool, sprayed with tin(II) chloride solution. The background is bleached by placing the plate in a tank containing an atmosphere of ammonia vapour. Organophosphorus pesticides appear as blue spots on a buff background. Compounds containing no phosphorus do not give this reaction, but large amounts of some coextractives can appear as light brown, charred areas, which may sometimes mask the blue coloration. The R_f values of any blue spots are determined and a characterization attempted by reference to standards (see Table 116).

Gel chromatography

This technique only need be applied as an adjunct to identification when the presence of pesticides is indicated by one or both of the previously mentioned chromatographic technique.

Beads of Sephadex LH 20 are soaked in absolute ethanol for 24 hours and the swollen gel slurry poured into a column 2.5 cm long, until a packed bed volume of 75 ml is obtained. The ethanol is run from the column until the meniscus just touches the top of the gel bed and the 1 ml of ethanol sample solution is added without disturbing the column surface. Ethanol is again run from the column until the meniscus again touches the surface, a further 1 ml of ethanol is added and the process is repeated. 5 ml of ethanol are added to the column, a reservoir containing about 150 ml of ethanol is attached and allowed to elute at a rate of about 2 ml min^{-1}, using pressure if necessary. The fractions eluting between 40–60, 61–70 and 71–110 ml are collected. The column should not be allowed to run dry, as it can be used repeatedly. Note that the elution fractions are based on an elution volume for parathion of 75.5 ml, which has been found to be constant on several different columns; should parathion show a different elution volume, the new fractions to be collected can readily be evaluated from the relative elution volumes shown in Table 116 and stage (b) of the analytical scheme. Each fraction is evaporated to a small volume in a Kuderna Danish evaporator and examined by gas and thin-layer chromatography as previously described.

The analytical scheme described by Askew et al.[215] is essentially a qualitative procedure but can readily be rendered quantitative. Those pesticides giving gas chromatographic peaks can be quantitatively determined by reference to a standard peak height — concentration curve. A standard curve should be prepared the same day as the analysis is made; dilute substandards, which decompose fairly readily at room temperature and on exposure to light, are prepared from a refrigerated stock solution (about $1000\,\mu g\,ml^{-1}$).

Suffet and Faust[222] applied the p-value approach to the liquid-liquid extraction of diazinon, parathion, malathion, and fenthion and their oxygen anologues and hydrolysis products from water samples prior to their analyses by gas chromatography on Reoplex-400 with electron capture and flame ionization detection.

Thompson et al.[223, 230] have developed a multiclass, multiresidue gas chromatographic method for the determination of insecticides (organophosphorus, organochlorine, carbamate types) and herbicides in water samples. The compounds are extracted from water with methylene chloride, and the extract is concentrated by an evaporative technique utilizing reduced pressure and low temperature. Compounds are segregated into groups using a column of partially deactivated silica gel and sequential elution with four different solvent systems. Carbamate residues, converted to their 2,4-dinitrophenyl ether derivatives, are gas chromatographed via electron capture detection as are the parent compounds of the organohalogen compounds. Organophosphorus compounds are determined by gas chromatography using a flame photometric detector. Recovery studies were conducted on 42 halogenated compounds, 38 organophosphorus compounds, and 7 carbamates. Fifty-eight of 87 compounds tested produced recoveries in excess of 80%; another 13 compounds yielded recoveries exceeding 60%, while the recoveries on the remaining 16 compounds feel below 60%. The methods for determining organochlorine insecticides are discussed in more detail in the section on Organochlorine insecticides.

Daughton et al.[224] have described a procedure for isolating and determining, in large volumes of aqueous media, ionic diethyl phosphate, diethyl thiophosphate, dimethyl phosphate, and dimethyl thiophosphate suitable for application in environmental monitoring. Procedures for eliminating interference due to inorganic phosphate are also discussed. In this approach the aqueous sample containing ionic dialkyl phosphates and thiophosphates are passed down a column of Amberlite XAD-4 resin. Recoveries for diethylphosphoric acid and diethylthiophosphoric acid at 0.01–0.1 ppm in 500–400 ml of aqueous media were 100 and 85%, respectively; recoveries for dimethylphosphoric acid and dimethylthiophosphoric acid at 0.1 ppm in 500 ml of aqueous media were 50 and 97%, respectively. Following a clean-up procedure the effluents are gas chromatographed using a gas chromatograph equipped with a phosphorus thermionic detector and a glass column, 1.8 m × 2 mm i.d., packed with equal parts of 15% QF-1 and 10% DC-200 on Gas-Chrom Q (80–100 mesh) at 140 °C; injector, 200 °C detector, 250 °C nitrogen, 22 ml min^{-1}; air, 23 ml min^{-1}; hydrogen, 55 ml min^{-1}. The effectiveness of the clean-up procedure is shown

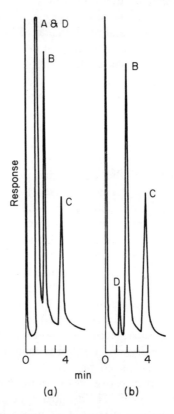

Figure 195 Gas chromatogram (500 ml sample, 0.1 ppm each of DEP and DETP, and 20 ppm of K$_2$HPO$_4$): (a) prior to clean-up with molybdic acid and safranine; (b) after clean-up (A), trimethyl ester of inorganic phosphate; (B), O-methyl ester of DEP; (C), S-methyl ester of DETP; (D), O-methyl ester of DETP. Reprinted with permission from Daughton et al.[224] Copyright (1976) American Chemical Society

in Figure 195 which shows gas chromatographs of a sample containing 0.1 ppm diethylphosphoric acid and potassium diethylthiophosphate before and after clean-up to remove inorganic phosphate.

Blanchet[225] used hexane to extract organophosphorus insecticides from pond and marsh water prior to flame photometric gas chromatographic analysis. He also studied sample preservation techniques. Sixteen different organophosphorus insecticides were included in this study. To extract the water[226] 10 ml of hexane (pesticide grade) was added to a glass-stoppered volumetric flask containing 300 ml or 1000 ml of solution, to which 1 ml of concentrated hydrochloric acid had been added. The mixture was stirred rapidly for at least 15 min. After separation of the two phases, 5 ml of the hexane layer was recovered for analysis. The samples were dried with sodium sulphate and analysed by the gas chromatographic procedure given below. A Varian Model 3700 gas chromatograph equipped with a flame photometric detector was operated with an interference filter for spectral isolation of phosphorus emission at 526 nm. The 60 cm glass column (2 mm i.d.) contained 4% SE 40/6% OV-210 on 60–80 mesh Gas-Chrom Q and was operated isothermally at various temperatures as shown in Table 117, after being conditional for 24 h at 275 °C. Except for the nitrogen (carrier gas) as shown in Table 117, the flow rates of the gases were

Table 117 Experimental conditions for g.c. analysis

Insecticide	Temperature (°C)			Flow rate carrier gas (ml min^{-1} ±1)	Retention time* (min)
	Injection	Detection	Column		
Abate (temephos)	240	240	280	80	3.90
Amidithion	230	230	200	30	1.98
Diazinon	220	220	180	30	1.31
Dursban (chlorpyrifos)	220	220	180	30	2.95
Dowco 214 (chlorpyrifos-methyl)	220	220	180	30	1.95
Ethion	220	220	210	30	2.95
Fenitrothion (folithion)	220	220	190	30	2.95
Guthion (azinphosmethyl)	230	230	220	30	1.96
Iodofenphos	230	230	200	30	2.25
Malathion	220	220	190	30	2.50
Methidathion	230	230	200	30	2.57
Ronnel (fenchlorphos)	220	220	190	30	2.55
Trithion	230	230	200	30	1.58
Vapona (dichlorvos)	160	160	100	30	2.09
Phenkapton	240	240	220	30	2.45
Phosphamidon	240	240	220	30	2.75

Reprinted with permission from Blanchet.[225] Copyright (1979) Elsevier Science Publishers B.V.

constant throughout. The conditions for the analysis of specific insecticides are listed in Table 117.

Recoveries of organophosphorus insecticides from spiked pond and marsh water samples are shown in Table 118. For all the solutions, the recovery ranged from 89 to 102% for the majority of the insecticides used.

Table 118 Recovery of insecticides from aqueous solution with hexane

Insecticide	Volume extracted (ml)*	ng μl^{-1} added	Recovery (%)[†]
Abate (temephos)	300 or 1000	1.0, 0.1, 0.01 (0.001)	91–98 (45)
Amidithion	1000	1.0	3.3–4
Diazinon	300 or 1000	1.0, 0.1, 0.01, 0.001	93–97.5
Dursban (chlorpyrifos)	300 or 1000	1.0, 0.1, 0.01, 0.001	97–101
Chlorpyrifos-methyl	300 or 1000	1.0, 0.1, 0.01, 0.001	96–101
Ethion	300 or 1000	1.0, 0.1, 0.01, 0.001	98.5–102
Fenitrothion	300 or 1000	1.0, 0.1, 0.01, 0.001	91–99
Guthion (azinphosmethyl)	300 or 1000	1.0, 0.1, 0.01, 0.001	88–95.5
Iodofenphos	300 or 1000	1.0, 0.1, 0.01, 0.001	96–99.5
Malathion	300 or 1000	1.0, 0.1, 0.01, 0.001	91–97.5
Methidathion	300 or 1000	1.0, 0.1, 0.01, 0.001	93.5–98
Ronnel (fenchlorphos)	300 or 1000	1.0, 0.1, 0.01, 0.001	96.5–100
Trithion	300 or 1000	1.0, 0.1, 0.01, 0.001	97.5–101.5
Vapona (dichlorvos)	1000	1.0	11–16
Phenkapton	300 or 1000	1.0, 0.1, 0.01, 0.001	87–103
Phosphamidon	1000	1.0	<5

*The results were the same using natural or distilled water.
[†]The recovery percentages reported here are the lowest and highest values of five extractions at each concentration.
Reprinted with permission from Blanchet.[225] Copyright (1979) Elsevier Science Publishers B.V.

Table 119 Stability of the extracts

Insecticide	Recovery (%)*				
	0 hours	24 hours	1 week	4 weeks	6 weeks
Abate (temephos)	96	96.5	95	94	92.5
Amidithion					
Diazinon	93	93	93.5	93.5	90
Dursban (chlorpyrifos)	99	97	99.5	98	91
Chlorpyrifos-methyl	98.5	98	97	97.5	90
Ethion	100	100.5	99	99.5	98
Fenitrothion	92	95	93.5	94	92
Guthion (azinphosmethyl)	98	97	96	96	89
Iodofenphos	96	98.5	97	98.5	88
Malathion	98	97	96.5	94	91
Methidathion	95	94.5	96.5	95	91.5
Ronnel (fenchlorphos)	100	102	99	98	99.5
Trithion	98	97.5	97.5	97.5	89
Vapona (dichlorvos)					
Phenkapton	89	93	93.5	92	89
Phosphamidon					

*The results shown are an average of five extracts.
Reprinted with permission from Blanchet.[225] Copyright (1979) Elsevier Science Publisher B.V.

Table 119 shows the excellent stability of hexane extracts containing various organophosphorus compounds kept in an airtight container for up to six weeks.

Although Blanchet[225] found that hexane was not a good extractant for amidithion, vapona, and phosphamidon (Table 118), he found that these three compounds extracted with near 100% recovery into chloroform. Chloroform gave good results as an extractant for most of the other insecticides mentioned above.

Verweij et al.[227] have described a procedure for the determination of PH_3-containing insecticides in surface water. In this procedure the insecticide is hydrolysed to methylphosphonic acid, and the acid is concentrated by anion exchange and converted to the dimethyl ester. After clean-up on a microsilica gel column the ester is analysed by gas chromatography using a thermionic phosphorus-specific detector. Detection limit is 1 nmol per litre.

Method

The hydrolysis was carried out in sealed 750 ml Carius tubes containing at most 500 ml water samples adjusted to pH 3 using 0.5 M hydrochloric acid. The tubes are heated in an oil bath at 160 °C for 24 h. After filtration through glass fibre paper (Whatman, GF/A) the hydrolysed sample was passed through an anion-exchange column (length 185 mm, i.d. 15 mm) packed with 15 ml of AG 1-X8 (formate form, BIO-RAD) at a flow rate of 1–2 ml min^{-1}. After the passage of the sample the exchange column was washed with 30 ml of methanol in order to remove the water. Methylphosphonic acid and other acids adsorbed on the resin were eluted at a flow rate of 0.5–1 ml min^{-1} with 20 ml of acidified (with dry gaseous hydrochloric acid up to 3 M) methanol. Interstitial air bubbles were removed by carefully stirring the exchanger in the column. The eluate, collected in a pear-shaped flask, was concentrated to a volume of 1 ml by evaporation in a water bath at 50 °C, using a gentle stream of air. To methylate the products a solution of diazomethane in ether was added to the residue of the eluate until a yellow colour persists. The mixture was allowed to stand for 15–20 minutes. Excess diazomethane was removed by means of a few drops of acetic acid. After the addition of 10 ml of benzene the methylated solution was concentrated by boiling water under reflux using a Vigreux column (length 19 cm, i.d. 11 mm) to a residue volume of 3–4 ml. During boiling the pear-shaped part of the flask was immersed in an oil bath, which was gently heated from room temperature up to 160 °C over 45 minutes with an air bleed to prevent bumping. Silica gel (70–230 mesh), after pretreatment by heating for 48 h at 135 °C, was partially deactivated by shaking with 3% (w/w) distilled water. After 4 hours the gel was ready for use. To a column (length 19 cm, i.d. 8 mm) plugged with glass wool, 1 g of the silica was added, followed by 2 g of anhydrous sodium sulphate. The column was prewashed with 10 ml of hexane. The sample solution is transferred to the silica gel column which was successively rinsed with 16 ml of benzene, 24 ml of ethyl acetate and 1.5 ml methanol at a flow rate of 0.2–0.4 ml min^{-1}. The gas chromatographic analysis was carried out on a

Becker gas chromatograph, type 409, equipped with a thermionic detector, type 712. The coiled glass column (length 2 m, i.d. 1.5 mm) was packed with Chromosorb W AW/DMCS 80–100 mesh coated with Triton X-305 (25% w/w after sieving in the particle range from 149 to 177 μm. The column, injector and detector were maintained at 150, 200, and 200 °C respectively. The carrier gas flow rate is 40 ml min^{-1} of nitrogen. Because of the use of a splitter at the end of the column (ratio 3:1) only 30 ml of nitrogen per minute reaches the thermionic detector. The remaining part is led to a flame ionization detector. In Figure 196 a representative gas chromatogram is shown as obtained on

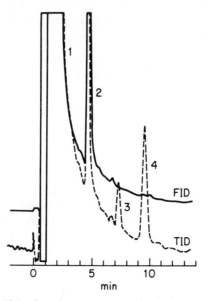

Figure 196 Gas chromatogram of a Rhine river water sample taken 20-6-77. 1, methanol; 2, acetic acid; 3, dimethyl methylphosphonate; 4, trimethyl phosphate; FID, flame ionization detector (att. 2×10^{-12} A); TID, thermionic detector (att. 2×10^{-11} A); injected volume: 2.5 μl

application of the analytical procedure to a Rhine river water sample. Peaks 3 and 4 represent nanogram amounts of dimethyl methylphosphonate and trimethyl phosphate respectively. The latter is due to a small part of the thousand-fold excess of trimethyl phosphate which is collected in the methanol fraction. The bulk of trimethyl phosphate proved to be present in the ethyl acetate fraction. The analytical procedure was checked on Rhine river water samples doped with microgram quantities of methylphosphonic acid, dimethyl methylphosphonate or O-ethyl-S-2-di-isopropylaminoethyl methylphosphonothioate. Based on methylphosphonic acid a mean recovery of 75 ± 12% was obtained. The mean lowest amount of dimethyl methylphosphonate detectable based on four times the noise level) was 0.23 ng (range 0.25–0.30 ng) of dimethyl methylphosphonate introduced into the column in 5 µl of solvent. This detection limit corresponds with 95 ng of methylphosphonic acid or more generally with about 1 nmol of PCH_3 per litre of water, being corrected for a mean recovery of 75% and an original water sample of 0.5 litre, which is concentrated during the complete analytical procedure to a volume of 1 ml.

Between 0.2 and 1.2 µg^{-1} PH_3 was found by this method in rivers flowing through Holland. The source of this contamination has not been identified. Mecarphon, the only PH_3-containing insecticide is supposedly no longer produced. Detergents and flame retardants are other possibilities. To prove that peak 3 ascribed to dimethyl methylphosphonate is not due to the presence of a non-phosphorus compound in relatively high concentrations, the thermionic detector was used in combination with a flame ionization detector. In case of a non-phosphorus compound the last-mentioned detector will give a relatively high response. The identity of the compound giving rise to peak 3 was confirmed by simultaneous injection of a reference sample of dimethyl methylphosphonate as well as by mass fragmentography.

Bargnoux et al.[229] used preconcentration at low temperatures as a means of extracting organophosphorus insecticides from water, prior to gas chromatography.

Addition of chloroform to water samples containing traces of pesticides, as reported by Bourne,[231] is not only a technique with numerous and tedious steps in the extraction procedure, but recovery percentages are lower for natural water samples. Moreover, traces of moisture in the extracts increase the degradation of the organophosphorus insecticides markedly and the recovery percentage decreases rapidly. The use of Amberlite XAD-2 resins as reported by Mallet and co-workers[232] is a breakthrough in the search for an adequate technique for the preservation of aqueous samples containing traces of organophosphorus insecticides.

Gas chromatography has been used[190] to determine the following organophosphorus insecticides at the microgram per litre level in water and waste water samples: azinphos-methyl, demeton-O, demeton-S, diazinon, disulfoton, malathion, parathion-methyl, and parathion-ethyl. This method is claimed to offer several analytical alternatives, dependent on the analyst's assessment of the nature and extent of interferences and the complexity of the pesticide mixtures found. Specifically, the procedure uses a mixture of 15% v/v methylene chloride in hexane to extract organophosphorus insecticides from the aqueous sample. The method provides, through use of column chromatography and liquid-liquid partition, methods for the elimination of non-pesticide interference and the pre-separation of pesticide mixtures. Identification is made by selective gas chromatographic separation and may be corroborated through the use of two or more unlike columns. Detection and measurement are best accomplished by flame photometric gas chromatography using a phosphorus-specific filter. The electron capture detector, though non-specific, may also be used for those compounds to which it responds. Confirmation of the identity of the compounds should be made by gas chromatography–mass spectrometry when a new or undefined sample type is being analysed and the concentration is adequate for such determination. Detailed instructions are given for clean-up of reagents, solvents, and glassware to avoid the occurrence of discrete artefacts and/or elevated baselines. The interferences in industrial effluents are high and varied and often pose great difficulty in obtaining accurate and precise measurement of organophosphorus insecticides. Sample clean-up procedures are generally required and may result in the loss of certain organophosphorus insecticides. Therefore, great care should be exercised in the selection and use of methods for eliminating or minimizing interferences. Compounds such as organochlorine insecticides, polychlorinated biphenyls, and phthalate esters interfere with the analysis of organophosphorus insecticides by electron capture gas chromatography. When encountered, these interferences are overcome by the use of the phosphorus-specific flame photometric detector. Elemental sulphur will interfere with the determination of organophosphorus insecticides by flame photometric and electron capture gas chromatography. The elimination of elemental sulphur as an interference is discussed in detail.

Typical gas chromatograms obtained with various column packings are shown in Figures 197–200. Retention data is reproduced in Table 120.

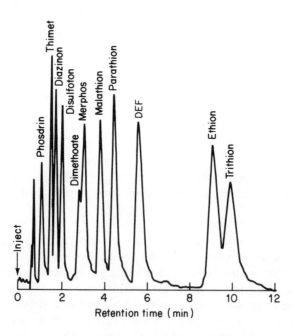

Figure 197 Column packing: 1.5% OV-17 + 1.95% QF-1. Carrier gas: nitrogen at 70 ml min^{-1}. Column temperature, 215 °C. Detector, flame photometric (phosphorus)

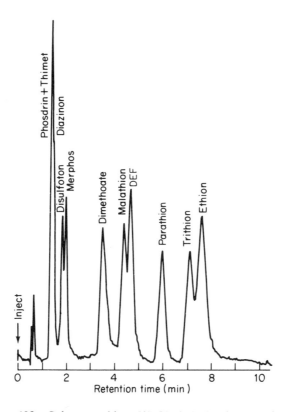

Figure 198 Column packing: 5% OV-210. Carrier gas, nitrogen at 60 ml min^{-1}. Column temperature, 200 °C. Detector, flame photometric (phosphorus)

Table 120 Retention times of some organophosphorus insecticides relative to parathion

Liquid phase[1]	1.5% OV-17 + 1.95% QF-1[2]	6% QF-12 + 4% SE-30	5% OV-210	7% OV-1
Column temp. (°C)	215	215	200	200
Nitrogen carrier flow	70 ml min^{-1}	70 ml min^{-1}	60 ml min^{-1}	60 ml min^{-1}
Pesticide	RR	RR	RR	RR
Demeton[3]	0.46	0.26	0.20	0.74
		0.43	0.38	
Diazinon	0.40	0.38	0.25	0.59
Disulfoton	0.46	0.45	0.31	0.62
Malathion	0.86	0.78	0.73	0.92
Parathion-methyl	0.82	0.80	0.81	0.79
Parathion-ethyl	1.00	1.00	1.00	1.00
Azinphos-methyl	6.65	4.15	4.44	4.68
Parathion (min absolute)	4.5	6.6	5.7	3.1

[1] All columns glass, 180 cm × 4 mm i.d., solid support Gas-Chrom Q, 100–120 mesh.
[2] May substitute OV-210 for QF-1.
[3] Anomalous, multipeak response often encountered.

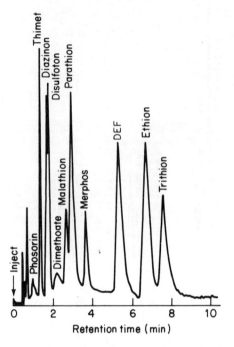

Figure 199 Column packing: 3% OV-1. Carrier gas, nitrogen at 60 µl min^{-1}. Column temperature, 200 °C. Detector, flame photometric (phosphorus)

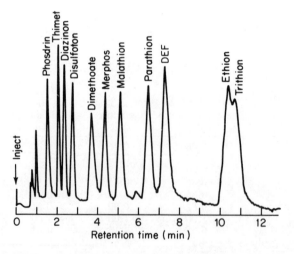

Figure 200 Column packing: 6% QF-1 + 4% SE-30. Carrier gas, nitrogen at 70 ml min^{-1}. Column temperature, 215 °C. Detector, flame photometric (phosphorus)

Thin-layer chromatography

This technique has been found to be useful by several workers for examining mixtures of organophosphorus insecticides. Rodica[233] applied the technique to traces of malathion, dimethoate, and ethion insecticides in chloroform or dichloroethane extracts of water samples. The insecticides were subjected to thin-layer chromatography on 0.25 mm layers of silica gel G activated at 130 °C: the solvents are: for malathion, hexane–ethyl ether, benzene–ether, or benzene–ethyl acetate (each 9:2); for dimethoate, hexane–ether or benzene–ether (each 5:1); and for ethion, benzene or hexane. The R_f values are respectively 0.4–0.45; 0.3–0.35; 0.8–0.85. The spots were located by exposure of the chromatograms to bromine vapour followed by spraying with (1) 0.5% 3,5-dibromo-p-benzoquinonechlorimine solution in dimethylformamide; (2) 0.1% bromophenol blue solution in acetone followed by a 3:100 ethanolic dilution of 0.01% fluorescein solution in M sodium hydroxide. Limits of detection were 0.2–1 μg. There was no interference from dimethyl phosphorodithioate or from malaoxon.

Malathion, foschlor (trichlorphon), and dichlorvos[234] have been determined by extracting a 1 l sample with chloroform (200 ml) drying the extract with sodium sulphate and concentrating to 1 ml. Trichlorphon and dichlorvos were separated on silica gel G containing silver nitrate and activated for 30 min at 110 °C with chloroform–acetone (9:1) as solvent; the spots are located in u.v. radiation after spraying with a methanolic solution of butter yellow. Malathion, which is separated on plates containing flourescein instead of silver nitrate, is revealed by exposure to bromine vapour. The R_f values were trichlorphon (0.1–0.12), dichlorvos (0.55–0.6), and malathion (0.6–0.5); 1–2 μg of each insecticide could be determined.

Leoni and Puccetti[235] have reviewed the thin-layer chromatography of organophosphorus insecticides in their study of the environmental pollution of Italian surface waters. The water samples were extracted and the extracts concentrated; 40 μl portions (equivalent to 500 ml of water) were applied to 0.25 mm layers of Alumina G (Merck type E) and a chromatogram was developed with acetone–hexane (7:93) for 10 cm. Pesticides were located by an enzymic cholineesterase inhibition method[236] and by a chemical method[237] based on spraying successively with tetrabromophenolphthalein solution in acetone, silver nitrate solution in aqueous acetone, and citric acid solution in aqueous acetone. The migration values relative to parathion-methyl and sensitivities (mainly in the submicrogram range) by both methods are tabulated by these workers for 23 organophosphorus compounds.

Bidleman *et al.*[238] used fluorescent indicators as spray reagents for the *in situ* determination of 16 phosphorothioate insecticides on thin-layer chromatograms. After separation on layers of silica gel with hexane–acetone or hexane–ethyl ether as solvent, the insecticides are located as fluorescent spots by spraying with an aqueous solution of the Pd(II)–calcein or Pd(II)–calcein blue complex. Limits of detection (in the 10–50 ng range) and R_f values are reported for 16 phosphorothioate insecticides. Quantitative methods (based on peak area measurements) are described for azinphos-methyl (10–400 ng) and dimethoate (5–150 ng). Recoveries of dimethoate from spiked lake water samples were 87–113%.

Jovanovic and Prosic[239] also used palladium chloride as a detection reagent for determining down to 0.1 ppm of malathion, parathion, parathion-methyl, bromophos, diazinon, fenthion, and phenkapton in water samples. After separation from water samples by extraction with dichloromethane, pesticides are subjected to thin-layer chromatography on silica gel F previously activated at 130 °C. The plate is developed with hexane–dichloromethane (1:1), and sprayed with 0.5% palladium chloride solution; the pesticides appear as yellow spots.

Guven and Aktugla[240] identified parathion, parathion-methyl, phosphamidon, malathion, azinphos-methyl, dichlorvos, carbaryl, diazinon, methidathion, carbophenothion, fenitrothion and endosulphan by thin-layer chromatography on silica gel G containing 10% aqueous copper sulphate and 2% aqueous ammonia (5:1). After application of the test solution and development of the chromatogram with hexane–ethyl ether (7:3) the pesticide spots were made visible by heating for 10 minutes at 100 °C. Detection limits for some of the compounds were between 4 and 25 μg.

Detection of organophosphorus compounds using cholinesterase enzyme

Gauson et al.[241] have described a method using a source of cholinesterase enzyme impregnated on paper for the rapid and simple detection of organophosphorus inhibitors in water in concentrations at or about the level of acceptability for drinking contaminated water. The papers are immersed in a suspect water and after a suitable incubation time, a chromogenic substrate is added to the paper. The appearance of a blue colour, different from that of the substrate is evidence that the enzyme is active and that the water does not contain sufficient anticholinesteratic material to inhibit the enzyme. Conversely, no change in the colour is evidence that an anticholinesteratic material is present in a concentration equal or greater than that acceptable for ingestion. These workers give experimental data on studies with the organophosphorus esters isopropyl methylphosphonofluoridate (Sarin), O-ethyl-S(2-di-isopropyamino)ethyl methylphosphonothioate, and O-(4-nitrophenyl phosphorothioate) (parathion). Comparison of inhibition data with rate constants indicates that the sensitivity of the method to any given inhibitor can be estimated if the rate constant value is known for that inhibitor with horse serum cholinesterase.

Cranmer and Peoples[242, 243] have also described a method for determining anticholinesterase pesticides. Compounds that are direct inhibitors can be differentiated from those that must be activated with bromine before they possess anticholinesterase activity. The method involves incubation of various amounts of non-treated and bromine treated sample with acetycholinesterase for 1 h at 37 °C, addition of substrate (3,3-dimethylbutyl acetate) and, after a further 30 min at 37 °C, stopping the reaction by adding formic acid and then extracting with carbon disulphide. The reaction product (3,3-dimethylbutanol) in an aliquot of the extract is determined by gas chromatography and the extent of inhibition is calculated from the results for the uninhibited and the inhibited samples. The method was applied to potable water.

Karlhuber and Eberle[244] have discussed the determination of organophosphorus and carbamate insecticides by cholinesterase inhibition.

Goodson et al.[245] used an immobilized cholinesterase product for the detection of organophosphorus and carbamate insecticides in water. In this method pads in which horse-serum cholinesterase is entrapped in starch gel on the surface of open-pore polyurethane foam[246] were used to detect enzyme inhibitors in air or water. Pads with improved enzyme activity and better resistance to loss of such activity are prepared by adsorption of cholinesterase on aluminium hydroxide gel during the precipitation of aluminium hydroxide from a solution of aluminium chloride followed by suspension of the gel enzyme mixture in a slurry of starch for application to the urethane foam. Pads so prepared retain their activity during the monitoring of water for certain organophosphorus and carbamate insecticides.

The US Environmental Protection Agency[247] has also reported on a rapid detection system based on the use of immobilized cholinesterase for the detection of organophosphorus and carbamate insecticides in water.

Determination of individual organophosphorus insecticides

Much work has been published which discusses the determination of individual organophosphorus insecticides. Whilst, as has already been discussed, in general it is advisable to adopt multicompound approaches when analysing for this class of compounds or mixtures of these with organochlorine and carbamate insecticides and herbicides and PCBs, there are circumstances when a simple direct method for a particular organophosphorus insecticide might be useful. Such methods are discussed in this section.

Parathion, O-(4 nitrophenyl phosphorothioate)·Parathion methyl (dimethyl-p-nitrophenyl phosphorothioate)·Parathion ethyl (diethyl-p-nitrophenyl phosphorothioate)

Kawahara et al.[248] described a procedure for the determination of phosphorothioate insecticides including parathion and parathion-methyl. It consists of solvent extraction, clean-up by thin-layer chromatography on silica gel G (0.25 mm layer), and identification by gas chromatography on an aluminium column (1.2 m × 0.6 cm o.d.) packed with equal portions of acid-washed Chromosorb P supporting 5% of DC 200 silicone oil, and unwashed Chromosorb W supporting 5% of Dow-11 silicone; the column is operated at 180 °C, with argon–methane (9:1) as carrier gas (120 ml min^{-1}) and electron capture detection. If the sample volume is sufficient, identification can be confirmed by infrared spectrophotometry. This method was used to follow accidental contamination of river water by pesticides. Other workers[249-251] have used gas chromatography to determine parathion in water samples.

Moye[252] has discussed the application of liquid chromatography to the determination of parathion-ethyl.

Paschal et al.[253] have discussed the determination of parathion-ethyl and parathion-methyl in run-off water using high performance liquid chromatography. The organic compounds are concentrated on an XAD-2 resin before analysis by reverse phase, high performance liquid chromatography. Detection limits were found to be approximately $2-3$ mg^{-1}. These workers examined the possible interferences to the method from other agricultural chemicals and organic compounds commonly occurring in water. This method is based on the use of Rohm and Haas XAD macroreticular resins.[254] Organics in water can be sorbed on a small column of resin, and the sorbed organics then eluted by diethyl ether. After evaporation of the eluate, the concentrated organics can be determined by chromatography. In addition to the obvious benefit of 100- to 1000-fold concentration, this method offers the possibility for on-site sampling.

Method

Apparatus A modular chromatographic system was used consisting of a six-port valve injector, a Whatman prepacked microparticle reverse phase column (Partisil ODS), and a variable wavelength detector.

Reagents and materials

The macroreticular resin XAD-2 was purified by Soxhlet extraction as described by Junk,[254] and stored under pure methanol. Pesticide grade acetonitrile was used as received; diethyl ether was glass distilled before use. Glass distilled water was used throughout.

Procedure

Preparation of standards: Microlitre amounts of 100 ppm stock solutions of organophosphorus insecticides made up in methanol were diluted volumetrically with distilled water to 100 ml. The diluted standards were then passed through a 10 cm column of purified XAD-2 resin, at a rate of $4-6$ ml min^{-1}. After the last of the dilute aqueous standards were passed through the column, most of the water clinging to the resin was removed by gentle vacuum aspiration. Thirty ml of glass distilled diethyl ether were passed through the column at $2-3$ ml min^{-1}, after which the last of the ether was removed by passing dried purified nitrogen through the column. The ether was dried by shaking with 2 g of anhydrous sodium sulphate, and evaporated to dryness using a rotary evaporator at temperatures not exceeding 35 °C. The residue was then dissolved in 100 ml of pesticide grade acetonitrile, and the resulting solution chromatographed on a Partisil-ODS reverse phase column at 2.40 ml min^{-1} with 50% acetonitrile water mobile phase. An average recovery of 99% was obtained through the procedure.

Extraction of run-off water

Extracts of 2 litre samples were prepared as in the above procedure, and chromatograms were evaluated to establish calibration curves for the parathions.

To evaluate the efficiency of extraction of XAD-2 resin for trace organics in run-off water, Paschal *et al.*[253] ran a sample of water through the procedure and obtained a large number of peaks in the 2-3 minute region. The chromatograph of run-off water is shown in Figure 201. A number of relatively polar compounds elute early in the chromatograph, with relatively few peaks in the 3-10 min region of the chromatogram. On changing from 50% acetonitrile to 100% acetonitrile to regenerate the column, several more peaks were eluted, apparently consisting of less polar materials strongly adsorbed under the conditions of the procedure. No interference was obtained from these strongly adsorbed compounds, although it was found to be useful to regenerate the column with 100% acetonitrile after every five to six runs to ensure reproducibility. The retention times for methyl and ethyl parathion, obtained from volumetric dilutions of methanolic standards with acetonitrile, were 3.45 and 4.65 min, indicating no interference from naturally occurring organics in the run-off water. Spiked samples of run-off water were prepared containing parathion-ethyl and parathion-methyl. A typical chromatogram for such a spiked sample is shown in Figure 202. The parathions are well separated, with no observed interference from organics already present in the water. Calibration

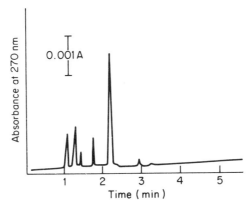

Figure 201 Chromatogram of run-off water extract. Separation of organics in run-off water. Eluant, 50:50 (v/v) acetonitrile water, Partisil ODS 4.6 mm × 25 cm. Detector at 270 nm, 0.02 AUFS. Reprinted with permission from Paschal *et al.*[253] Copyright (1977) American Chemical Society

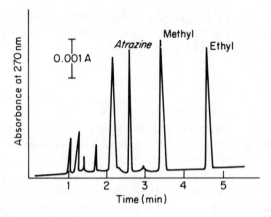

Figure 202 Chromatogram of spiked run-off water extract. Conditions as in Figure 201. Reprinted with permission from Paschal et al.[253] Copyright (1977) American Chemical Society

curves were prepared from a set of standards containing 10–120 ppb parathions in methanol. Atrazine was added as an internal standard to the concentrated extract. Ratio of peak heights or areas of parathions to those of atrazine were plotted vs. concentration. Good linearity was obtained over the range of concentration examined for both parathions. In order to evaluate the accuracy and reproducibility of the method, a series of solutions was prepared in run-off water with concentrations of parathions in the range of the calibration curve. The results of this study are given in Table 121. The lower limit of detection was calculated from those data to be 3.1 and 2.9 ng for parathion-methyl and parathion-ethyl, respectively.

Table 121 Reproducibility of method

Taken $\mu g\, l^{-1}$	Found $\mu g\, l^{-1}$	s.d.	Relative s.d. (%)
Parathion-methyl			
15.0	14.8	0.45	3.0
37.5	37.1	1.07	2.8
75.0	75.9	0.73	1.0
112.5	112.7	2.56	2.3
Parathion-ethyl			
10.0	9.9	0.37	3.7
25.0	24.6	1.40	5.6
50.0	49.3	0.97	1.9
75.0	75.0	2.40	3.2

[a]Average of six determinations.
Reprinted with permission from Paschal et al.[253] Copyright (1977) American Chemical Society.

Table 122 shows the effects of potential interference by other agricultural chemicals and organics commonly occurring in natural water. Wavelengths chosen for measurement were at or near the absorbence maxima for the compounds as determined by u.v. scans from 350 to 200 nm. If a potentially interfering compound showed a retention time near one of the parathions, then chromatography was performed with detection at 270 nm. Of the compounds investigated, only fonofos (ethyl S-phenylethyl phosphonothiolothionate) interferes at 270 nm. However, if the wavelength of detection is changed to 280 nm, the interference is overcome.

Table 122 Interference study

Compound	Relative retention (parathion-methyl = 1.00)	Wavelength measured (nm)
Aroclor 1260	3.94–5.88 multiple peaks	225
Atrazine	0.75	265
Azinphos-ethyl	1.14	285
Alachlor	0.89	235
Carbaryl (Sevin)	0.69	280
Carbofuran	0.61	270
Chloramben	0.26	240
Chlorpyrifos	2.01	290
p,p'-DDT	2.78	235
DEHP	1.59	235
Dialifor	1.61	290
Diazinon	1.30	245
Dyfonate (fonofos)	1.36	240
Fenitrothion	1.18	265
Methoxychlor	1.72	225
p-Nitrophenol	0.72	310
Phosmet	0.93	230
Phorate	1.30	220
Propachlor	0.67	260
2,3,5-T	0.28	250
Trifluralin	0.58	270

Reprinted with permission from Paschal et al.[253] Copyright (1977) American Chemical Society.

Venkataraman and Sathyamurthy[255] developed a simple and direct spectrophotometric method for the determination of parathion in water in which a benzene extract of parathion is hydrolysed and is reduced with zinc dust in acid solution. The resulting amino derivative is diazotized with sodium nitrite and hydrochloric acid. The diazo compound is coupled with naphthylethylenediamine hydrochloride and a magenta dye is produced, which is evaluated spectrophotometrically. Whilst it is unlikely that this method is specific for parathion it is reproduced here.

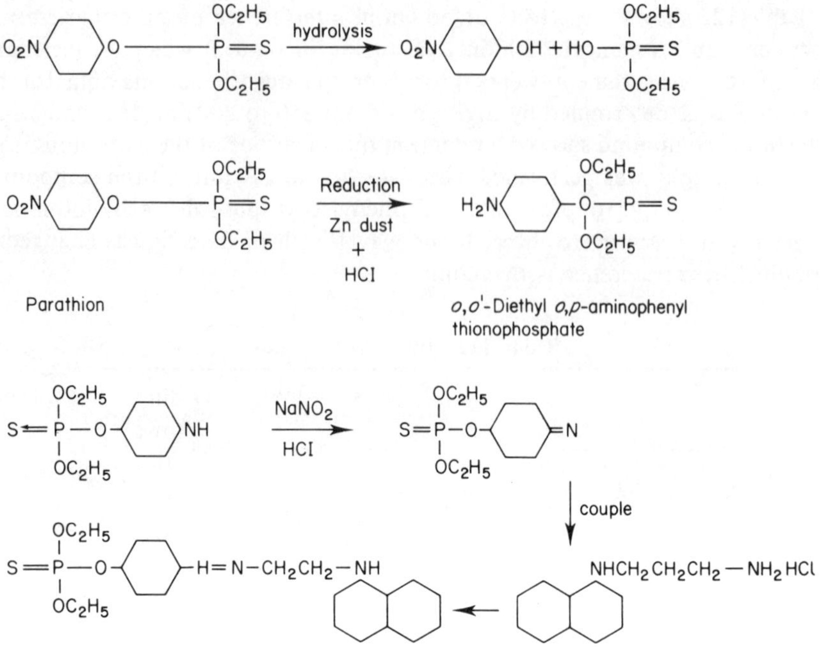

$O_2N-\langle\rangle-O-P(=S)(OC_2H_5)(OC_2H_5) \xrightarrow{\text{hydrolysis}} O_2N-\langle\rangle-OH + HO-P(=S)(OC_2H_5)(OC_2H_5)$

$O_2N-\langle\rangle-O-P(=S)(OC_2H_5)(OC_2H_5) \xrightarrow[\text{HCl}]{\text{Reduction Zn dust}} H_2N-\langle\rangle-O-P(=S)(OC_2H_5)(OC_2H_5)$

Parathion

o,o'-Diethyl o,p-aminophenyl thionophosphate

$S=P(OC_2H_5)(OC_2H_5)-O-\langle\rangle-NH \xrightarrow[\text{HCl}]{\text{NaNO}_2} S=P(OC_2H_5)(OC_2H_5)-O-\langle\rangle=N$

↓ couple

$S=P(OC_2H_5)(OC_2H_5)-O-\langle\rangle-H=N-CH_2CH_2-NH-\text{[naphthyl]}$ ← $\text{[naphthyl]}-NHCH_2CH_2CH_2-NH_2 \cdot HCl$

(Magenta red)
N(4-o,p-phenylazo-O-O-diethyl dithiophosphate)
-1-naphthylethylenediamine

Method

Reagents Benzene, redistilled and thiophene free. Standard parathion solutions: 1 ml = 1 mg and 1 ml = 0.02 mg parathion. Hydrochloric acid, approximately 0.5 M and 3 M. Sodium nitrite solution, 0.25%, prepared weekly. Ammonium sulphamate solution, 2.5%, prepared weekly. *N*-(1-Naphthyl)ethylenediamine dihydrochloride solution, 1%, filtered through Whatman no. 41 filter paper or equivalent into an amber coloured bottle, prepared daily. Zinc dust, pure.

Procedure

A 250 ml quantity of the shaken sample is placed in a separating funnel and extracted successively four times with 120 ml benzene, drawing the benzene extract in another separating funnel. The benzene extract is treated with 20 ml of 3 M hydrochloric acid twice to eliminate any interfering amines and the lower acid layers are discarded. The benzene extract is collected in an Erlenmeyer flask, 200 mg zinc dust are added to the treated benzene extract and then 20 ml of 0.5 M hydrochloric acid. The solution is reduced for 10 minutes over a steam bath, the contents transferred to the distillation flask, and the benzene rapidly distilled off. Boiling stops temporarily after all the benzene is eliminated and the vapour temperature falls. The aqueous solution must not boil as this contains the reduced parathion. As soon as all the benzene has distilled off, the flask is disconnected and 10 ml absolute alcohol are added to the reduced parathion in the distillation flask. A funnel is introduced such that the stream is inside the neck of the flask. The flask is placed over the steam bath and refluxed for 5 minutes. The flask is removed and cooled. The solution is then filtered through Whatman no. 42 filter paper and the filtrate collected in a 50 ml Nessler tube, which is cooled to 10 °C and kept as such until the additions of the following are completed. To the filtrate in the Nessler tube is added 1 ml of sodium nitrite solution. This is mixed well and allowed to standard for 10 minutes. Then 1.5 ml of ammonium sulphamate are added, stirred briskly and allowed to remain for 10 minutes. Next 2 ml of N-(1-natphthyl)ethylenediamine dihydrochloride are added, the solution is diluted to volume, mixed and allowed to stand for 5 minutes. The magenta colour produced is compared with the appropriate standards visually or in a photoelectric colorimeter.

Report as mg l^{-1} parathion Calculation:

$$\text{mg l}^{-1} \text{ parathion} = \frac{\text{mg parathion} \times 1000}{250}$$

Weber[256] has described a kinetic method for studying the degradation of parathion in sea water. Weber observed two pathways whereby parathion is hydrolysed. The first reaction proceeds via dearylation with loss of *p*-nitrophenol. This reaction is well known:

$$(C_2H_5O)_2 - \overset{\overset{S}{\|}}{P} - O - \underset{}{\bigcirc} - NO_2 \xrightarrow{+H_2O}$$

$$(C_2H_5O)_2 - \overset{\overset{S}{\|}}{P} - OH + HO - \underset{}{\bigcirc} - NO_2$$

Additionally they observed a second main pathway, hydrolysis through dealkylation leading to a secondary ester of phosphoric acid which still contains the *p*-nitrophenyl moiety, i.e. de-ethylparathion (*O*-ethyl-*O*-*p*-nitrophenyl-monothiophosphoric acid):

$$(C_2H_5O)_2 - \overset{\overset{S}{\|}}{P} - O - \underset{}{\bigcirc} - NO_2 \xrightarrow{+H_2O}$$

$$O_2N - \underset{}{\bigcirc} - O - \underset{C_2H_5O}{\overset{}{\underset{}{P}}} \overset{S}{\underset{}{\|}} - OH + C_2H_5OH$$

Biochemical degradation of parathion was also observed although the degradation products were not identified.

By the process shown in Table 123 aliquots of an ethanolic solution of parathion were separated into undecomposed insecticides and decomposed insecticide products. Among the products free p-nitrophenol (B) was detected photometrically at 410 nm in alkaline solution. p-Nitrophenol, chemically bound in acidic phosphorus compounds (C) and in non-hydrolysed neutral phosphorus compounds (A) were detected in the same way after saponification. As a control an aliquot of 10 ml was taken at the same time as the 30 ml aliquots and extracted with ether in the way described in Table 123. Saponification after removal of ether without separation of neutral and acidic compounds yielded total p-nitrophenol equivalents (D). Saponifications were conducted in closed and calibrated reaction tubes; a 2 h reaction time at 100 °C in 2 ml 3.5 M potassium hydroxide was used. Thereafter the solutions were diluted, if necessary, to ensure accurate measurement at 410 nm.

Table 123 30 ml aliquot

acidified with 2 M HCl to pH 1

extracted with Et$_2$O
(3 times with 20 ml) (consecutively)

aqueous phase
(basic compound)
either discarded or
checked for amino
compounds

organic phase
(neutral and acid compounds)
extracted with 0.1 M NaOH
at 5 °C (once with 10 ml,
twice with 5 ml consecutively)

aqueous phase
(acidic compounds)

photometric detection of
p-nitrophenolate:
FREE p-NITROPHENOL (B)

reacidified with 2 M HCl to pH 1

extracted with Et$_2$O (as above)

organic phase
(acidic compounds)

ether evaporated

saponification of the residue

photometric detection of
p-nitrophenolate:
p-NITROPHENOL
EQUIVALENTS
IN ACIDIC COMPOUNDS
(C + B)

aqueous phase
(discard)

organic phase
(neutral compounds)
checked for organophosphates
other than parathion

ether evaporated

saponification of the residue

photometric detection of
p-nitrophenolate:
p-NITROPHENOL EQUIV-
ALENTS IN NEUTRAL
COMPOUNDS (A)

Reprinted with permission from Weber.[256] Copyright (1976) Pergamon Press Ltd.

Malathion

Bargnoux et al.[257] carried out a comparative chromatographic study, utilizing thin-layer and gas chromatography with phosphorus and sulphur dual detectors for the detection of parathion and malathion residues in water. They discuss the application of two low temperature methods, lyophilization and cryoconcentration for the recovery of these insecticides from water. The use of lyophilization and cryoconcentration represents an improvement in the analysis of some particularly labile phosphorus and sulphur pesticides, over previous methods by significantly reducing the risks of degradation in aqueous solution. Moreover lyophilization offers two other advantages, by allowing on the one hand the simultaneous analysis of many water samples irrespective of their mineralization, and on the other, optimum conservation either in the freezing state before their treatment or after the lyophilization step.

Wolfe et al.[258] studied the kinetics of the chemical degradation of malathion in water. They investigated potential chemical and photochemical pathways for the degradation of malathion. The results of acid-catalysed degradation and of oxidation indicate that these pathways would be too slow to be of significance in the aquatic environment, but photolysis could be important under acid conditions. Malathion was much more susceptible to alkaline degradation, and underwent two competing reactions at 27 °C. The persistence of the malathion monoacids formed was also studied, and they proved to be 18 times more stable than malathion under the same conditions.

Dursban (O,O-diethyl-O-(3,5,6-trichloro-2-pyridyl phosphorothioate)

A gas chromatographic procedure using electron capture detection has been described for the determination of dursban in water and silt (Rice and Dishberger[259]). In this method, water samples are extracted with dichloromethane, the extract is evaporated, and a solution of the residue is cleaned up on a column of silicic acid, dursban being eluted with hexane (37:3). The eluate is evaporated to dryness under reduced pressure, and a solution of the residue in hexane is subjected to gas chromatography on a glass column (4 ft × 0.25 in. o.d.), packed with 5% of SE-30 on Anakrom ABS (80–90 mesh) operated at 200 °C (or 215 °C with methane–argon (1:19) (60 ml min^{-1}) as carrier gas, or, 5% SF-96 on Chromosorb W (80–100 mesh), operated at 215 °C with nitrogen (90 ml min^{-1}) as carrier gas. Dried silt samples, finely powdered, are blended with dichloromethane and Celite, the filtered extract is evaporated to dryness and a solution of the residue in hexane is extracted with acetonitrile. The concentrated solution plus added hexane is evaporated to dryness under reduced pressure, and dursban in a solution of the residue in hexane is subjected to clean-up and gas chromatographed as for water samples. Down to 10^{-4} ppm of dursban in water and down to 5×10^{-3} ppm in silt could be determined; average recoveries from water and silt were 92% and 83% respectively.

Deutsch et al.[260] determined dursban in water, mud, vegetation, fish, ducks, insects, and crustacea. After a preliminary clean-up the extract is chromatographed on a column packed with 3% Carbowax 20 m on Gas-Chrom (60–80 mesh), which gives excellent separation of dursban from other organophosphorus insecticides. Both thermionic and flame photometric detectors are satisfactory. Recoveries range from 75 to 105% depending on the nature of the sample. This procedure will detect as little as 0.5 ng of dursban corresponding to a level of 0.01 ppm in a 10 g sample.

Abate OOO'O'-tetramethyl-OO'-(thiodi-p-phenylene)diphosphorothioate)

Dale and Miles[261] showed that Abate and its sulphoxide were well separated from various other organophosphorus insecticides on a column of XE-60 on silanized Chromosorb W 240 °C, using nitrogen as carrier gas and a flame photometric and electron capture detector. Shafik[263] has described a method for the determination of this insecticide in water. Abate is first converted to its hydrolysis product, 4,4'-thiodiphenol, which is then silated by reaction with chlorotrimethylsilane and hexamethyldisilazane. Separation was achieved on an aluminium column (4 ft × 0.25 in. o.d.) containing 2.5% E-301 or 0.25% Epon 1001 on Chromosorb W (80–100 mesh) operated at 190 °C with nitrogen as carrier gas (100 ml min^{-1}) and a flame photometric detector equipped with a sulphur filter. Miller and Funes[262] used alkali flame gas chromatography to determine Abate. Separation was achieved on a column packed with 2.5% of E-301 plus 0.25% of Epon 1001 on Gas-Chrom W (AW-DMCS)HP, the column and detector being operated at 235 °C. The column was conditioned by injecting 20 ng standards of Abate solution between samples. The detector response was rectilinear for up to 200 ng of Abate. The recover of 0.01 to 1 ppm of added Abate to water was 97%, after extraction by the method of Dale and Miles.[261]

Thin-layer chromatography has also been applied to the determination of Abate residues in water.[264] The sample of surface water or sewage, acidified with sulphuric acid, was extracted with chloroform. The extract was evaporated under nitrogen at 60–70 °C and the residue was dissolved in acetone; 15 μl of this solution was applied to a layer (0.25 mm) of silica gel G, previously activated for 1 h at 90 °C, together with Abate standards. The plate was developed for 1 h in hexane–acetone (10:1), air dried for 1 min, exposed to bromine vapour for 1 min, exposed to air for 3 min, and sprayed with 1% N,N-dimethyl-p-(phenylazo)aniline (CI solvent yellow 2) solution in 95% isopropyl alcohol. Abate gave distinct red spots (R_f 0.11 ± 0.01) on a bright yellow background. Spot areas were related to concentration by a calibration graph. Down to 9 μg of Abate could readily be determined.

Otsuki and Takaku[265] have described a method for the determination of Abate down to 5–150 μg using reversed phase adsorption liquid chromatography. Only one ml of sample is required for this determination.

S-methyl fenitrothion
(O,S-*dimethyl*-O-*(3-methyl-4-nitrophenyl)phosphorothiolate*)

Coburn et al.[266] and others[267,268] have reported the use of XAD-2 for the extraction of fenitrothion from water.

Zitko and Cunningham[269] investigated the hydrolysis rate of this compound and prepared ultraviolet, infrared and mass spectra.

Fenitrothion

Various workers[270-272] have used macroreticular resin XAD-2 to recover Fenitrothion from river water samples and compared this method with solvent extraction techniques.

Method

Mallet et al.[270] used an automated gas chromatographic system which consisted of a gas chromatograph mounted with an automatic sampler interfaced to an integrator. A Melpar flame photometric detector (phosphorous mode) was connected with the flame gas inlets in the reverse configuration to prevent solvent flame-out. The detector was maintained at 185 °C and flame gases were optimized with flow rates (ml min^{-1}) as follows: hydrogen, 80; oxygen, 10; air, 20. A 1.8 m × 4.0 mm i.d. U-shaped glass column packed with 4% (w/w) OV-101 and 6% (w/w) OV-210 on Chromosorb W AW DMCS, 80–100 mesh, was used. Nitrogen was used as carrier gas at a flow rate of 70 ml min^{-1}. A column temperature of 195 °C sufficiently resolved fenitrooxon from its parent compound. The injection port temperature was set at 225 °C.

The water sample was passed through an XAD-2 column, which was subsequently eluted with ethyl acetate. The ethyl acetate extract was examined by gas chromatography.

In an alternative method involving t.l.c. and fluorimetric analysis a 1 l water sample was extracted with two 50 ml portions of chloroform, which were collectively dried through an anhydrous sodium sulphate (50 g) column. The chloroform was replaced with ethyl acetate on a flash evaporator, carefully reduced to 4–5 ml and made up to the mark with ethyl acetate in a 10 ml volumetric flask. Aminofenitrothion was recovered by this method from environmental water using an XAD-2 column. The results, given in Table 124, indicate good recoveries at an average flow rate of 153 ml min^{-1}. The relative error, *ca.* 10%, is normal at a concentration of 50 ppb when using *in situ* fluorimetry. Under similar conditions, fenitrooxon can also be recovered with good yields. The procedure was adapted to the simultaneous analysis of the parent compound and its two derivatives by gas–liquid chromatography. The data in Table 125 indicate that with XAD-2 resin, conditions such as flow rate and column length are crucial to obtain good recoveries. If a 10 × 1.9 cm i.d. column is used the maximum flow rate is limited to *ca.* 50 ml min^{-1}, which is easily sustained by gravity flow.

Table 124 Recovery of aminofenitrothion and fenitrothion from natural water at a concentration of 50 ppb

Aminofenitrothion		Fenitrooxon	
Experiment no.	Recovery (%)	Experiment no.	Recovery (%)
1	89	1	89
2	87	2	90
3	103	3	83
4	103		
5	103		
6	118		
7	98		
8	95		
Average	100	Average	87
Relative standard deviation	9.8%		

Method: t.l.c. and *in situ* fluorimetry; eluting solvent: hexane–acetone (6:1)
Reprinted with permission from Mallet *et al.*[270] Copyright (1978) Elsevier Science Publishers B.V.

Table 125 Calibration of XAD-2 columns for the analysis of fenitrothion (F), fenitrooxon (FO) and aminofenitrothion (AF)

Sample flow rate (ml min^{-1})	Column length (cm)	Recovery (%)		
		F	AF	FO
238	5	0	0	68
218	5	0	0	56
135	5	53	17	72
129	5	50	10	70
71	5	73	43	85
63	10	81	83	89
60	10	91	95	84
50	10	102	117	114

Reprinted with permission from Mallet *et al.*[270] Copyright (1978) Elsevier Science Publishers B.V.

Various good recoveries are illustrated in Table 126 when using XAD-2 for extracting spiked lake water with three compounds. Relative standard deviations of 5.1–6.4% are very good. The overall average recovery (99.7%) of the three compounds by the conventional serial solvent extraction procedure is somewhat better (Table 127) than by the XAD method (95.3%). Reproducibilities are all better for the exception of fenitrooxon.

Methods[273,274] to determine fenitrothion in water using gas–liquid chromatography with a flame photometric detector have been reported. An *in situ* fluorimetric method[275] to detect simultaneously fenitrothion, fenitrooxon, aminofenitrothion and nitrocresol on a thin-layer chromatogram has been developed. Fenitrothion, fenitrooxon and aminofenitrothion have been analysed

Table 126 Recovery of a mixture of fenitrothion (F), fenitrooxon (FO) and aminofenitrothion (AF) at various concentrations from natural water with XAD-2

Experiment no.	Recovery (%)		
	F	FO	AF
1	96	98	98
2	95	95	94
3	98	102	97
4	91	109	99
5	85	95	98
6	84	95	86
Average	91.5	99	95.3
Relative standard deviation (%)	6.4	5.6	5.1

Reprinted with permission from Mallet et al.[270] Copyright (1978) Elsevier Science Publishers B.V.

Table 127 Recovery of a mixture of fenitrothion (F), fenitrooxon (FO), and aminofenitrothion (AF) from natural water by the conventional serial solvent extraction procedure

Experiment no.	Recovery (%)		
	F	FO	AF
1	90	96	106
2	97	95	101
3	95	92	110
4	96	95	105
5	99	103	110
6	96	99	99
7	97	106	108
8	90	82	112
9	96	89	112
10	99	97	115
11	93	98	114
12	101	90	110
Average	95.7	95.1	108.5
Relative standard deviation (%)	3.5	6.7	4.5

Method: g.l.c. with FPD; concentrations: F, 10 ppb; FO, 100 ppb; AF, 30 ppb. Reprinted with permission from Mallet et al.[270] Copyright (1978) Elsevier Science Publishers B.V.

simultaneously by gas chromatography using an SE-30 plus QF-1 column and a flame photometric detector in the phosphorus mode.[276]

Coumaphos (Coral) and Bayrusil

Mallet and Brun[277] have described an *in situ* spectrofluorimetric method for the analysis of these substances in lake and sewage water. This procedure does not necessitate a preliminary cleaning up of samples, and interfering coextractives which fluoresce only in solution are also avoided. An advantage over other *in situ* fluorimetric techniques is that fluorogenic spray reagents are not used and the fluorescence produced is selective. In earlier papers Brun and Mallet[278] had shown that Coral (I) and Bayrusil (II) could be made to fluoresce on silica gel layers simply by heating the chromatogram at a specific temperature for a definite period of time and this is the basis of the method described below.

(I)	(II)
Excitation: 344 nm	356 nm
Emission: 440 nm	440 nm

Method

The visible ultraviolet chromatogram analyser was equipped with a photomultipler detector tube along with 230–420 nm and 405–800 nm interference filters in the excitation and analyser legs, respectively. Reducing apertures of 0.005 in. were used with both filters. A Turner Fluormeter Model 111 with t.l.c. attachment was used for quantitative measurements. Filters (300–400 nm) and >415 nm) were placed in the entrance and exit slits, respectively. The pesticide was extracted from 1000 ml of water by shaking with three portions of 50 ml of *n*-hexane. The combined organic phases were dried over 20 g of anhydrous sodium sulphate and the solvent evaporated with a flash evaporator to approximately 10 ml at 25–35 °C. The remaining solvent was transferred to a tube with two portions (5 ml) of *n*-hexane and further evaporated to around 10 μl. The concentrate is spotted 2 cm from the bottom of a thin-layer chromatogram (250 μm) with a 10 μl microsyringe. The chromatogram is also spotted with the appropriate standards. The plate is then eluted at a distance of 10 cm using a 7:2 (v/v) solution of *n*-hexane–acetone. For Coral the fluorescence is produced by heating the chromatogram at 200 °C for 20 minutes. In the case of Bayrusil, the plate is first sprayed with a solution of aqueous potassium hydroxide (0.1 M) and then heated at 100 °C for 30 minutes before measurement. Figure 203 illustrates a linear relationship between concentration and relative intensity with Coral. Sensitivity is not very good in the region between 0.002 and 0.01 μg. A similar behaviour is observed with Bayrusil. Recovery data for Coral and Bayrusil in water samples are given in Table 128. Coral can be detected quite readily at the 0.1 ppb level in ordinary lake water. In sewage effluents, problems are encountered at concentrations lower than 0.1 ppb. The problem is even worse with Bayrusil in sewage and it is thought that the loss is due in part to the formation of emulsions during the solvent

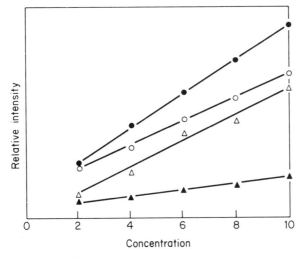

Figure 203 Calibration curves with Coral ( ): 2–10 μg; (●): 0.2–1 μg; (Δ): 0.02–0.1 μg; (▲): 0.002–0.01 μg. Reprinted with permission from Mallet and Brun.[277] Copyright (1974) Springer Verlag, N.Y.

Table 128 Recovery of Coral and Bayrusil from lake and sewage water

concentration (ppb)	% Recovery (A) Coral		Bayrusil	
	L.W.	S.W.	L.W.	S.W.
10	86	80	94	55
1.0	100	90	86	54
0.5	104	92	D	D
0.1	108	72	D	
0.01	86	D		

A, Average of three extractions; D, Detectable visually at that concentration; L.W., lake water; S.W., sewage water.
Reprinted with permission from Mallet and Brun.[277] Copyright (1974) Springer Verlag, N.Y.

extraction. In the case of Coral, recoveries are less at 10 ppb than they are at 1 ppb because of solvent saturation at the greater concentration.

Determination of phosphate equivalent of organophosphorous insecticides

Vogler[279] has described a method for the separate determination of orthophosphoric acid esters ('COP-phosphate') and condensed phosphates ('POP-phosphate') in natural water samples. The sample (100 ml) of filtered natural water is treated with 30% aqueous hydrogen peroxide (0.2 ml) and exposed to u.v. radiation at below 23 °C under conditions of maximum irradiation intensity and minimum time. Excess of hydrogen peroxide and orthophosphates are determined by standard procedures. The result obtained is a measure of COP-phosphate. The other half of the sample is treated in a polyethylene flask with 10 N sulphuric acid (0.4 ml) on a steam bath for 4 h. Excess of hydrogen peroxide is then removed with potassium permanganate solution and the orthophosphate is again determined, with a waiting time of 20 min after the addition of reagent. The value obtained is for total dissolved phosphate; POP-phosphate is obtained by subtraction.

Goossen and Kloosterboer[280] have described a method for the determination of organophosphorous derived orthophosphates in natural and waste water after photochemical ultraviolet decomposition and acid hydrolysis.

PCB AND PCB–ORGANOCHLORINE INSECTICIDE MIXTURES

Again, as with the chlorinated insecticides, gas chromatography possibly coupled with mass spectrometry predominates as a method of choice for PCBs. First, the work which discusses PCBs only is discussed, moving on to work covering the coanalysis of PCBs with chlorinated insecticides.

Polychlorinated biphenyls (PCBs) have been prepared industrially since 1929 by chlorination of biphenyl with anhydrous chlorine using either iron filings or iron(III) chloride as a catalyst. The product obtained is a complicated mixture of several PCBs. In the UK, PCBs are marketed as Arochlors by Monsanto. All Arochlors are characterized by a four digit number; the first two digits represent the type of molecule (e.g. 12 represents biphenyl, 54 terphenyl and 25 and 44 are mixtures of biphenyl and terphenyl); the last two digits give the percentage by mass of chlorine, e.g. Aroclor 1260 is a 12-carbon system with 60% m/m of chlorine. Polychlorinated biphenyls are sold under a variety of trade names, of which Arochlor is one. The following is a list of the principal trade names used for PCB-based dielectric fluids which are usually classified as Askarels: Aroclor (UK, USA), Pyoclor (UK); Inerteen (USA); Pyanol (France); Clophen (Germany), Apirolio (Italy); Kaneclor (Japan); Solvol (USSR). Other names were used for PCB products intended for different applications no longer in current use; these include: Sanothern FR (UK, prior to 1972 for heat transfer); Therminol FR (USA, prior to 1972 for heat transfer); Pydraul (USA, prior to 1972 for hydraulic applications); Phenoclor (France) and Fenclor (Italy). The trade names Santothern, Therminol, and Pydraul are still in use but they are now used to refer to non-chlorinated products.

PCBs

Gas chromatography

Albro and Fishbein[281] in 1972 carried out early work on the quantitative and qualitative analysis of PCBs using gas chromatography with flame ionization detection. Webb and McCall[282] studied the question of quantitative PCB standards for electron capture gas chromatography.

Polychlorinated biphenyls are often extracted from water by solvent extraction. However, the relatively small sample size, *ca.* 1–2 l, that can be handled restricts the quantitation limits of PCB analyses. Recently charcoal (Chriswell *et al.*[283]) polyurethane foams (Musty and Nickless[284]), Carbowax–undecane on Chromosorb W (Musty and Nickless[285]), as well as macroreticular resins (Chriswell *et al.*,[283] Musty and Nickless,[285] Lawrence and Tosine,[286] Coburn *et al.*[287]) have been used to isolate PCBs from water. Amberlite XAD-2 and XAD-4 macroreticular resins have been used by several workers (Chriswell *et al.*,[283] Musty and Nickless,[285] Coburn *et al.*[287]) to analyse PCBs from a variety of water sources but on relatively small sample size (1–2 l) and recovery studies were carried out at relatively high fortification levels (*ca.* 250 ng l^{-1}). However, Coburn *et al.*[287] suggested that the XAD-2 method could be used to sample large volumes of water to decrease the quantitation limits for PCB analysis. Following from this suggestion Le'Bel and Williams[288] described a method for determining PCBs in drinking water in which they obtained a concentrate by passing the water sample through Amberlite XAD-2 macroreticular resin cartridges. Following clean-up on a Florisil column the extract was gas chromatographed using an electron capture detector. In potable well water samples interference was minimal and the detection limit was about 0.04 ng l^{-1} for Aroclors 1016, 1232, 1242, and 1254. In potable water from a river source, interference in the chromatogram from other organic compounds made quantitative detection difficult at the 1 ng l^{-1} level. Aroclor 1254 was detectable at this level but 1232, 1016, and 1242 were only detectable at 10 ng l^{-1}.

Method

Reagents Solvents: distilled in glass, redistilled in an all-glass system.

PCB Standards: analytical standards were obtained from Monsanto Co. (Aroclor 1016, 1242, and 1254) or from Supelco (Supelco Inc., Bellefonte Pa, Aroclor 1232). Stock of each Aroclor is prepared at 1000 μg ml^{-1} in hexane. Spiking solutions are prepared at 10 and 1 ng μl^{-1} in acetone and standard solutions for gas chromatographic analysis at 0.4 and 0.1 ng μl^{-1} in 2,2,4-trimethylpentane.

Macroreticular resin: Amberlite XAD-2 (Rohm and Haas) is purified as described by McNeil *et al.*[289]

Anhydrous sodium sulphate: reagent grade, granular, is heated at 400 °C overnight, cooled and successively washed with methylene chloride, acetone, and hexane. It is stored in a glass bottle with a Teflon-lined cap.

Purified water: Millipore super-Q system water is passed through a XAD-2 cartridge at *ca.* 150 ml min^{-1}, collected, and stored in clean glass bottles.

Glass wool is washed with methylene chloride, acetone, and hexane and stored in a clean glass jar with a Teflon-lined cap.

Florisil is heated at 275 °C overnight, cooled, and deactivated with purified water (2% w/w).

Apparatus for PCB analysis

Mini-Florisil column: custom made 10 cm × 10 mm i.d. column is connected to a Teflon stopcock and a 100 ml round bottom flask as reservoir.

Gas chromatograph: Perkin-Elmer Model 910 equipped with ^{63}Ni electron capture detector is used; standing current setting 2.0; electrometer setting × 64– × 128. Operating conditions: temperatures (°C)—injector 250, interface 270, detector 300, oven programmed from 160 (hold 2 min) at 4° per min to 200 (hold 5 min), post program 225 (hold 5 min); carrier gas nitrogen at pressure of 15 psig; make-up gas—nitrogen at 45 ml min^{-1}. The glass capillary column was a 11 m × 0.25 mm i.d. WCOT, OV-17 contained in a cage connected to oven injector and detector ports with 0.3 mm i.d. glass lined ¹/₁₆ in. o.d. stainless steel tubing. The transfer lines in the g.c. interface were also 0.3 mm i.d. glass lined tubing. The injection port was modified by a splitless injection system with a removable 0.75 mm i.d. glass lined injector tube and a septum purge valve. The sample was injected using a Hamilton syringe with a 3 in. needle.

Sampling with macroreticular cartridges

The cartridge and sample tap water are prepared as described by Lebel *et al.*[290] Elution is carried out as previously described, except the sodium sulphate drying columns are rinsed with two volumes of 10 ml each of methylene chloride, acetone, and hexane respectively prior to percolating the XAD-2 organic layer eluate. The filtrate and rinsings are concentrated to near dryness on a rotary flask evaporator (water bath 37 °C). The solution is transferred with several 1 ml hexane rinsings to a small graduated centrifuge tube and concentrated to 1 ml with a gentle stream of dry nitrogen.

Florisil column chromatography

A plug of clean glass wool is inserted into the mini-column and pack with 3 g of deactivated Florisil (*ca.* 8 cm deep), topped by *ca.* 0.5 cm anhydrous sodium sulphate. The Florisil column is washed with *ca.* 25 ml hexane. The flow is stopped when hexane just reaches the top of the sodium sulphate. A clean 15 ml graduated centrifuge tube is placed under the column as receiver and the concentrated extract is transferred with a clean disposable pipette to the head of the column. The flow is resumed. The tube that contained extract is rinsed with two ml hexane washes, successively adding rinsings to column when the previous rinse just reaches the sodium sulphate layer. Elution is performed with hexane, collecting *ca.* 13 ml eluate as determined by calibration run to elute PCBs. To the eluate is added 1 ml of 2,2,4-trimethylpentane. The solution is concentrate to *ca.* 1 ml under a gentle stream of nitrogen and made up to required volume (2–5 ml) with 2,2,4-trimethylpentane. The extract is analysed by capillary column gas chromatography using an electron capture detector.

Field samples

Tap water is passed through the cartridge at a steady flow of 140 ml min^{-1} (*ca.* 3 bed volume per min) until 200 l have been sampled as determined by calculation or by passing effluent through a volumetric measuring counter (Lebel *et al.*[290]).

Le'Bel and Williams[288] found that low procedural blank values, equivalent to 0.04 ng l^{-1} for a 200 l potable water sample were attainable only by using doubly distilled solvents and by exhaustive washing of all reagents and glassware with these solvents. A gas chromatogram of a procedural blank is shown in Figure 204. Gas chromatographic analysis of concentrated extracts of potable water samples without Florisil column clean-up gave off off-scale peaks at instrument settings suitable for low ng l^{-1} PCB analysis. However, fractionation of the extract by Florisil column chromatography gave a PCB fraction sufficiently clear of interfering organics to permit PCB analysis at 1–10 ng l^{-1}. The gas chromatographic analyses were done on a relatively short (11 m)

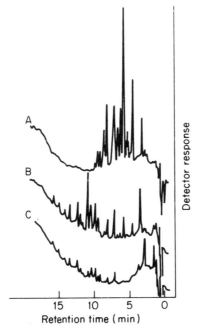

Figure 204 G.c.–e.c. chromatograms. A, 1 ng l^{-1} equivalent conc. of Aroclor 1016; B, suspected Aroclor 1016 contaminated well water extract; C, 2× conc. of blank. Reprinted with permission from Le'Bel and Williams.[288] Copyright (1980) Springer Verlag, N.Y.

capillary column, thus providing a compromise of short analysis time and good resolution with minimum interferences from other substances present in the water extract. These workers quantitated recovery runs, by comparing peak heights of selected peaks from the sample extracts with the corresponding peaks from standard PCB solutions. Low organic content Millipore Super Q water or laboratory tap water, after passage through XAD-2 resin, was used for fortification studies. Recoveries of Aroclors 1232, 1016, 1242, and 1254 from water samples fortified at the equivalent of 1 and 10 ng l^{-1} for a 200 l sample are reported in Table 129. Recoveries at the 10 ng l^{-1} fortification level ranged from 86 to 99% and at the 1 ng l^{-1} fortification level ranged from 91 to 110%.

The analysis of 200 l potable well water samples, suspected of contamination of Aroclor 1016, are reported in Table 130 and illustrated in Figure 204B. The interfering compounds were minimal except for an amplified series of peaks

Table 129 Recoveries of PCBs from treated XAD-2 resin extract[a]

PCBs	Fortification level (ng l^{-1})	Recoveries (%)
Aroclor 1254	10	89,99
	1	100,110
	1[b]	91
Aroclor 1242	10	86,98
	1	94,96
	1[b]	92
Aroclor 1016	10	88
	2	91
Aroclor 1232	10	97,96
	1	
	1[b]	109

[a]XAD-2 filtered tap water source.
[b]XAD-2 filtered Millipore Super Q system water as water source.
Reprinted with permission from Le'Bel and Williams.[288] Copyright (1980) Springer-Verlag N.Y.

Table 130 PCB levels in drinking water samples

Source	PCBs (ng l^{-1}) (calculated as Aroclor 1016)
Ottawa tap water	ND
ground water no. 1	0.2
no. 2	0.1
no. 3	0.1

Reprinted with permission from Le'Bel and Williams.[288] Copyright (1980) Springer-Verlag N.Y.

similar to the 2-fold concentration procedural blank shown in Figure 204C. The higher background was due to analysis before the method was fully optimized for minimum interferences. Quantitation of the Aroclor 1016 was done by comparing peaks from the interference-free region between retention times 3–7 min, where several of the Aroclor 1016 components (Figure 204A) are eluting. The detection limit was estimated to be *ca.* 0.04 ng l^{-1} of Aroclor 1016 from this source of water with Aroclors 1232, 1242, and 1254 having similar levels of detection. When the method was applied to potable water form a river source, the interference in the gas chromatogram (Figures 205C and 206C) from other organic compounds present in the sample made quantitation difficult at the 1 ng l^{-1} level. Aroclor 1254 would still be detectable at the 1 ng l^{-1} level (Figures 205A and 205C) but 1 ng l^{-1} levels of Aroclors 1232, 1016, and 1242 would not have been distinguishable from the background peaks. However, at the 10 ng l^{-1} level, it was possible to distinguish peaks due to Aroclor 1232 (Figure 206B), Aroclor 1242 (Figure 206A), and Aroclor 1016 from the background chromatogram due to other organics (Figure 206C). The analysis of a 200 l sample of potable water from a river source showed no detectable levels of Aroclor 1254 (<1 ng l^{-1}) or of Aroclor 1232, 1242, or 1016 (<10 ng l^{-1}). It is concluded that the XAD-2 macroreticular resin method can be used to analyse large volumes of potable water for low ng l^{-1} levels of PCB's. The detection

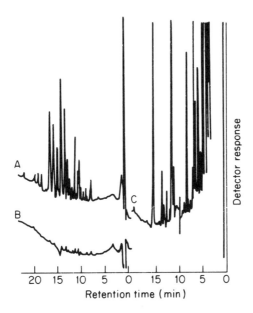

Figure 205 A, 1 ng l^{-1} equiv. conc. of Aroclor 1254; B, blank on regenerated XAD-2 resin cartridge; C, Ottawa tap water at 1 ng l^{-1} conc. level. Reprinted with permission from Le'Bel and Williams.[288] Copyright (1980) Springer Verlag, N.Y.

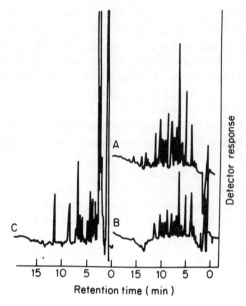

Figure 206 A, $10\,\text{ng}\,l^{-1}$ equiv. conc. of Aroclor 1242; B, $10\,\text{ng}\,l^{-1}$ equiv. conc. of Aroclor 1232; C, Ottawa tap water extract at the $10\,\text{ng}\,l^{-1}$ conc. level. Reprinted with permission from Le'Bel and Williams.[288] Copyright (1980) Springer Verlag, N.Y.

limits, however, are limited by the interference from other organics in the water sample but are typically in the 1–$10\,\text{ng}\,l^{-1}$ range from potable water from a river source and *ca.* $0.04\,\text{ng}\,l^{-1}$ from underground water sources low in organic content.

Bauer[291] has described a gas chromatographic procedure for the determination of PCBs in water. The PCBs were extracted from water with hexane, and the extract was dried and concentrated before gas chromatographic analysis. Extraction and concentration of PCBs were also effected by the use of slow filtration through sand or algae columns. The sand was extracted with 60% acetone–hexane, acetone was removed by washing the extract with water, and the washed hexane was passed through a column filled with layers of sodium sulphate, alumina, and Florisil. The eluate was analysed by gas chromatography. Algal material was dried, ground with sand, and treated in a manner similar to that used for sand alone. The gas chromatography was carried out on glass columns ($3\,\text{m} \times 2.8\,\text{mm}$) packed with 2.5% QF-1, 2.5% silicone rubber and 0.5% Epikote 1001 on Chromosorb W AW-DCMS (80–100 mesh) and operated at $200\,°\text{C}$ with nitrogen ($90\,\text{ml}\,\text{min}^{-1}$) as carrier gas and an electron capture detector.

The US Environmental Protection Agency[292] and the Inland Waters Directorate, Canada[293] have described a method for the determinations of PCBs in water.

Various methods have been described for the extraction and recovery of PCBs prior to gas chromatographic analysis. Coburn et al.[294] have shown that XAD-2 macroreticular resin can be successfully used to analyse PCBs in 2 l natural water samples fortified at the 250 ng l^{-1} level.

Porous polyurethane foam has also been used to preconcentrate water samples. Gesser et al.[295] found that the compounds could be adsorbed on a column composed of two polyurethane plugs (each 38 mm × 22 mm) inside a glass tube. The sample water was poured through at 250 ml min^{-1}, then the plugs were removed and squeezed free from water; the PCBs were then extracted by treating the plugs with acetone and hexane. The concentrated extract was analysed by gas chromatography on a glass column (6 ft × 0.25 in.) packed with Chromosorb W HMDS supporting 2% SE-30 and 3% QF-1 and operated at 200 °C; the carrier gas helium and a ^{63}Ni electron capture detector was used. In tests on 20 μg of polychlorinated biphenyls added to 1 litre of water, the recovery (based on measurements of 13 peaks on the chromatogram) ranged from 18.2 to 19.6 μg. Bedford[296] also used polyurethane foam plugs to extract PCBs from natural waters; these results indicated that Aroclor is probably adsorbed on to small particles in unfiltered lake water which can pass through the foam. Although this makes the foam method somewhat impractical for enriched and turbid waters, it is still a valuable technique for clear water containing low quantities of PCB.

Ahnoff and Josefsson[297] tested various clean-up procedures for PCB analysis on river water extracts. Extracted water samples were divided into a number of portions and used in testing the following clean-up treatments: with sulphuric acid, with sulphuric acid and activated Raney nickel, with a Florisil column, with a Florisil column and activated Raney nickel, and with a Florisil column and potassium hydroxide. The results of subsequent gas chromatographic analyses indicate the Florisil column procedure is more effective in removing contaminants than the sulphuric acid procedure. However, neither procedure could remove sulphur, which could be removed by the activated Raney nickel or potassium hydroxide procedures.

Several methods have been used to quantitate PCB levels in environmental samples such as perchlorination (Berg et al.[298]) and electron capture response factor to various isomers (Webb and McCall,[300] Sawyer[299]). Various other workers have discussed the gas chromatography of PCBs and Aroclor in various matrices. Whilst these methods do not specifically discuss the analysis of water samples the work may, nevertheless, be of interest to the water chemist. Stalling and Huckins[301] discussed gas chromatography–mass spectrometry characterization of PCBs Aroclors and ^{36}Cl labelling of Aroclors 1248 and 1254. These workers mention the separation of PCBs from polyterphenyls. Webb and McCall[302] isolated 27 isomers found in Aroclors 1221, 1242, and 1254 by preparative gas chromatography and identified them by comparison of their retention data and infrared spectra with those of known synthetic compounds. Analytical gas chromatography was carried out on a support coated open-tubular column (100 ft × 0.2 in.) of SE-30 at 190 °C with helium as carrier gas (6.5 ml min^{-1}) and flame ionization detection.

Tas and Kleipool[303] discuss the gas chromatographic identification of 2,2',3,3',6,6'-hexachlorobiphenyl, 2,2',3,4',5',6-hexachlorobiphenyl, and 2,2'3,3',4,5,6'-heptachlorobiphenyl. Albro and Fishbein[304] determined the retention indices on six stationary phases for some mono-, di-, tri-, tetra-, and hexachlorobiphenyls. They confirmed the additivity of the half-retention values in predicting retention indices. Webb and McCall[305] determined the weight of PCB producing each peak in chromatograms of several Aroclors by gas chromatography–mass spectrometry and the use of an electrolytic conductivity detector. Electron capture detector response factors were derived, and these were applied in association with rules for the division of chromatograms to the quantitative analyses of environmental samples containing one or more Aroclors.

Beezhold and Stout[306] studied the effect of using mixed standards on the determination of PCBs. Mixtures of Aroclors 1254 and 1260 were used as comparison standards and gas chromatograms of these mixtures were compared with those obtained from a hexane extract of the sample after clean-up on a Florisil column. Polychlorinated biphenyls were separated from DDT and its analogues on a silica gel column activated for 17 h and with 2% (w/w) of water added. The extracts were analysed on a silanized glass column packed with 5% DC-200 and 7.5% QF-1 on Gas Chrom Q (80–100 mesh) operated at 195 °C with nitrogen as carrier gas (50–60 ml min^{-1}) and a tritium detector.

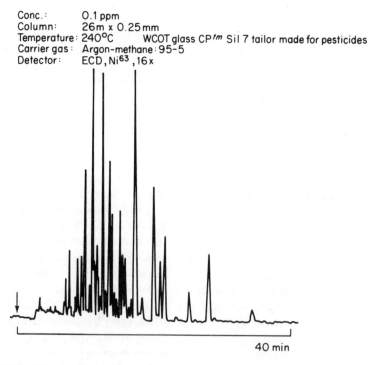

Figure 207 Chromatogram of Aroclor 1260. Copyright (1979) Chromapak International B.V.

Sissons and Welti[307] and Albro et al.[308] have studied the gas chromatography of PCBs using an electron capture detector.

Glass wall coated open tubular capillary columns have been used for the gas chromatography of PCBs.[309,310] Figure 207 shows a chromatogram obtained by this technique[309] for a mixture of Aroclors in hexane. The column comprised 25 m × 0.25 mm of WCOT glass GPtm Sil 7. It can be seen that 0.1 ppm of these substances is easily detected. Schulte and Acker[310] gas chromatographed PCBs with a glass capillary column at temperatures up to 320 °C. They used a 60 m column impregnated with SE 30-SC and helium as carrier gas with a flame ionization detector. The column was mantained briefly at 80 °C after injection of the sample, heated to 180 °C and then temperature programmed to 260 °C at 20 per minute.

Sullivan et al.[311] studied the loss of PCBs from sea water samples during storage.

Gas chromatography–mass spectrometry

Skinner[314] showed for the first time the application of mass fragmentography to residue analysis, namely the determination of PCBs along with DDE in extracts of a sewage effluent. Ahnoff and Josefsson[312] carried out confirmation studies on PCBs from river waters using mass fragmentography, a form of gas chromatography–mass spectrometry in which the mass spectrometer is focused at one or more individual mass numbers while the components emerge from the column. This technique is a more specific and sensitive technique than electron capture gas chromatography for detecting and identifying traces of PCB in river waters. The specificity is further increased by using the known intensity ratios of the isotopic peaks of chlorine-containing species. Webb and McCall[313] determined the weight of PCB producing each peak in chromatograms of several Aroclors by gas chromatography–mass spectrometry and the use of an electrolytic conductivity detector. Electron capture detector response factors were derived, and these were applied in association with rules for the division of chromatograms to the quantitative analyses of environmental samples containing one or more Aroclors. Eichelberger et al.[315] applied gas chromatography–mass spectrometry with computer controlled repetitive data acquisition from selected specific ions to the analyses of PCBs in lake sediments. The polychlorinated biphenyl mixtures were separated by gas chromatography at 180 °C in a coiled glass column (6 ft × 0.78 in.) packed with 1.5% OV-17 plus 1.95% QF-1 on Gas-Chrom Q (100–120 mesh), with helium (30 ml min^{-1}) as carrier gas. Effluent is passed via a glass jet enrichment device into a quadrupole mass spectrometer, controlled by a mini-computer in such a way that only selected ions of specific m/e pass through the quadruple field. There is a substantial gain in sensitivity, without loss of qualitative information contained in the complete mass system. This technique provides a basis for a sensitive qualitative and quantitative (from ion-abundance chromatograms obtained from subset scanning) analysis for polychlorinated biphenyls. Karlruber et al.[316] and

Ahnoff and Josefsson[317] have discussed the application of mass fragmentography to the detection of traces of PCBs and herbicides in water.

PCBs in sea water

Elder[318] determined PCBs in Mediterranean coastal waters by adsorption on to XAD-2 resin followed by electron capture gas chromatography. The overall average PCB concentration was 13 ng l^{-1}.

PCBs in sewage

Gaffney[319] showed that PCBs were present in certain municipal sewages. He showed that various chlorobiphenyl isomers are produced during chlorination of sewage and sewage effluents.

PCBs in paper mill effluents

Various workers have described methods for determining PCBs in paper mill effluents. The method described by Delfino and Easty[320] is capable of detecting down to 2 μg l^{-1} PCB. Easty and Wabers[321,322] have studied the effect of suspended solids on the determination of Aroclor 1242 in paper mill effluents.

Column chromatography

Berg *et al.*[323] separated PCBs from chlorinated insecticides on an activated carbon column prior to derivatization and gas chromatographic separation on the column. Separation is based on the observation that PCBs adsorbed on activated charcoal cannot be removed quantitatively with hot chloroform but can be with cold benzene. Insecticides of the DDT group and a variety of others (e.g. γ-BHC, aldrin, dieldrin, endrin, and heptachlor and its epoxide) can be eluted from the charcoal with acetone–ethyl ether (1:3). Typical recoveries from a mixture of *p,p'*-DDE (1,1-dichloro-2,2-*bis*-(4-chlorophenyl)ethylene), p,p'-TDE, and *o,p'*-DDT and a PCB (Aroclor 1254) by successive elution with 90 ml of 1:3 acetone–ether and 60 ml of benzene were 91, 92, 92, 94, and 90%, respectively. Identification and determination of the PCBs was effected by catalytic dechlorination to bicyclohexyl or perchlorination to decachlorobiphenyl, followed by gas chromatography. For bicyclohexyl, gas chromatography was carried out on a column (8 ft × 0.25 in.) of 10% DC-710 on Chromosorb W, operated at 90 °C for 2.5 min then temperature programmed at 10° per min, with flame ionization detection. For decachlorobiphenyl, the column (2 ft × 0.125 in.) is 5% SE-30 on Chromosorb W, operated at 215 °C, with nitrogen as carrier gas and a tritium detector. Hanai and Walton[324] have also studied the separation of PCBs on a column of pyrolytically deposited carbon.

Earlier work on the application of high performance liquid chromatography

Table 131 H.p.l.c. retention time and u.v. spectral responses for individual PCBs

PCB	Retention time (min)	Spectral response (relative to biphenyl = 1.000)	Analysis conditions
Biphenyl	7.62	1.000	Aroclor 1221 assay
2-	9.39	0.255	
3-	9.91	0.725	
4-	11.12	0.917	
2,2'-	10.24	0.047	
2,6-	11.09	0.075	
2,3-	12.71	0.278	
2,4'	13.57	0.465	
2,5-	13.81	0.295	
2,4-	14.31	0.508	
3,3'-	14.80	0.562	
3,4-	15.11	0.917	
4,4'-	15.43	1.087	
3,5-	16.63	0.685	
2,5,2'-	23.59	0.046	Aroclor 1016 assay
2,3,6-	23.82	0.042	
2,4,6-	28.67	0.093	
2,3,4-	30.05	0.394	
3,4,2'-	31.08	0.327	
2,5,3'-	32.12	0.190	
2,5,4'-	33.61	0.506	
2,4,5-	34.02	0.309	
2,4,4'-	36.31	0.674	
2,6,2',6'-	14.61	0.007	Aroclor 1254 assay
2,3,2',3'-	18.41	0.173	
2,3,2',5'-	19.80	0.034	
2,5,2',5'-	21.10	0.037	
3,4,3',4'-	21.15	0.086	
2,3,5,6-	21.64	0.092	
2,4,2',5'-	22.31	0.093	
2,4,2',4'-	22.80	0.156	
2,3,4,5-	24.95	0.352	
2,5,3',4'-	24.96	0.364	
2,4,3',4'-	25.80	0.564	
2,4,5,2',3'-	24.91	0.088	
2,3,6,2',5'-	25.48	0.063	
2,3,4,2',5'-	26.09	0.088	
2,3,4,5,6-	27.21	0.176	
2,4,5,2',5'-	27.42	0.089	
2,3,6,2',3',6'	25.81	0.007	
2,3,5,6,2',5'-	28.64	0.019	
2,3,4,2',3',4'-	30.83	0.163	
2,4,6,2',4',6'-	31.44	0.047	
2,3,4,2',4',5'-	31.85	0.254	
2,4,5,2',4',5'-	32.88	0.146	
2,3,4,5,6,2',5'-	34.10	0.096	
2,3,5,6,2',3',5',6'-	37.36	0.024	
2,3,4,5,6,2',3',4',5'-	41.96	0.204	
2,3,4,5,6,2',3',4',5',6'-	45.12	0.140	

U.v. detector at 254 nm wavelength (injected as 5 mg l^{-1} solutions in acetonitrile).
Reprinted with permission from Kaminsky and Fasco.[327] Copyright (1978) Elsevier Science Publishers B.V.

to the analysis of PCBs was carried out by Brinckmann et al.[325,326] These workers used a silica gel column which elutes the higher chlorinated PCBs in the normal phase. This system produced a reasonable separation of the lower chlorinated PCBs present predominantly in the commercial mixture Aroclor 1221, but was less efficient in separating the more highly chlorinated PCBs present in Aroclors 1254 and 1260. Kaminsky and Fasco[327] investigated the potential of reversed phase liquid chromatography to the analysis of PCB mixtures in environmental samples. They used mixtures of water and acetonitrile as the mobile phase to achieve analysis of 49 different PCBs and of samples of Aroclor 1221, 1016, 1254, and 1260.

These workers used a Waters Associates Model 244 liquid chromatograph equipped with a recording integrator. All solutions were monitored at 254 nm. The column (30 cm × 4 mm i.d.) was packed with reversed phase, microparticle silica (μ Bondapak C_{18}; Waters Assoc). The solvents used were water–acetonitrile (9:1) and acetonitrile–water (9:1). Elution conditions were varied to obtain best resolution. For example, for Aroclor 1221 elution was initiated with a solvent mixture of 60% of 9:1 acetonitrile–water and 40% of 9:1 water–acetonitrile and the solvent composition was altered in a non-linear gradient over a period of 30 min to achieve 100% of 9:1 acetonitrile–water. Elution was continued for a further 6 min Solvent flow rates were 2 ml min^{-1} throughout. The absolute retention times and u.v. detector adsorbance responses at 254 nm of 49 individual PCBs are presented in Table 131. In general, increasing the chlorine content of PCBs is a major factor in increasing the retention time. Consequently, isomeric groups tend to be eluted at similar times. The chromatograms of Aroclor 1221, 1016, and 1254 and a 1:1:1 (w/w/w) mixtures of the three Aroclors are depicted in Figures 208–211. Some of the major constituents of the Aroclor are tentatively assigned. For Aroclor 1221, (21% by weight of chlorine), 31 components were distinguishable chromatographically (Figure 208). The seven major components were tentatively identified and quantitated (Table 132). 3-Chlorobiphenyl is not completely resolved from an excess of 2-chlorobiphenyl and a trace amount of the former may be present and thus increase the reported value for the latter. It is apparent that the patterns of minor components of Aroclor 1221 closely resemble those

Table 132 Composition of Aroclor 1221

Peak no.	PCB	Content (wt. %)
1	Biphenyl	12.6
2	2-	38.2
3	2,2'-	5.8
5	4-	21.7
7	2,4'-	13.4
8	2,4-	1.6
10	4,4'-	4.7

Reprinted with permission from Kaminsky and Fasco.[327]
Copyright (1978) Elsevier Science Publishers B.V.

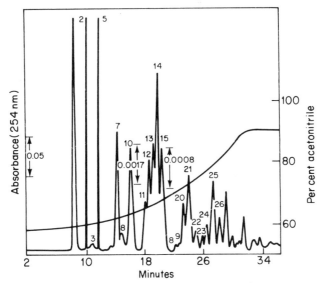

Figure 208 H.p.l.c. separation of Aroclor 1221 (5 mg ml^{-1} in tetrahydrofuran) on a μ Bondapak C$_{18}$ column monitored at 254 nm. Initial conditions, 40% water–acetonitrile (9:1) and 60% water–acetonitrile (1:9); final conditions, 100% water–acetonitrile (1:9); gradient period 30 min; flow rate, 2 ml min^{-1}; injection volume, 5 μl; amount injected 25 μg. Reprinted with permission from Kaminsky and Fasco.[327] Copyright (1978) Elsevier Science Publishers B.V.

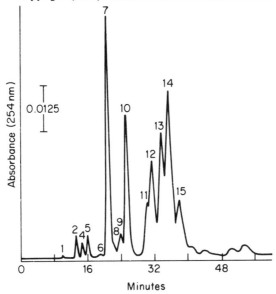

Figure 209 H.p.l.c. separation of Aroclor 1016 (5 mg ml^{-1} in tetrahydrofuran) on a μ Bondapak C$_{18}$ column monitored at 254 nm. Conditions, 45% water–acetonitrile (9:1) and 55% water–acetonitrile (1:9); flow rate 2 ml min^{-1}; injection volume 5 μl; amount injected, 25 μg. Reprinted with permission from Kaminsky and Fasco.[327] Copyright (1978) Elsevier Science Publishers B.V.

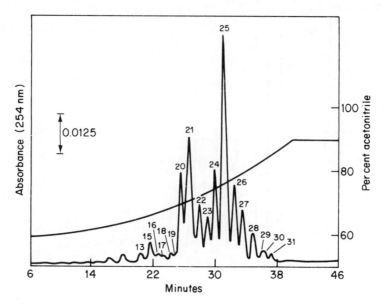

Figure 210 H.p.l.c. separation of Aroclor 1254 (5 mg ml^{-1} in tetrahydrofuran) on a μ Bondapak C$_{18}$ column monitored at 254 nm. Initial conditions, 40% water–acetonitrile (1:9) and 60% water–acetonitrile (1:9); final conditions, 100% water–acetonitrile (1:9); gradient time, 40 min; flow rate, 2 ml min^{-1}; injection volume, 5 μl; amount injected, 25 μg. Reprinted with permission from Kaminsky and Fasco.[327] Copyright (1978) Elsevier Science Publishers B.V.

of the major components of Aroclor 1016 and 1254 (compare Figures 208, 209 and 210).

For Aroclor 1016 (41% by weight of chlorine), 18 components were distinguishable chromatographically (Figure 209). The resolution of the components was inferior to that obtained with Aroclor 1221 and 1254. The peaks up to and including no. 10 were tentatively identified and together make up approximately 30% by weight of the sample (Table 133).

For Aroclor 1254 (54% by weight of chlorine), 22 components were distinguishable chromatographically (Figure 210). Those components which were

Table 133 Composition of Aroclor 1016

Peak no.	PCB	Content (wt. %)
1	Biphenyl	0.03
2	2-	1.1
5	4-	0.4
7	2,4'-	12.7
9	2,5,2'-	12.7
10	4,4'-	3.4

Reprinted with permission from Kaminsky and Fasco.[327] Copyright (1978) Elsevier Science Publishers B.V.

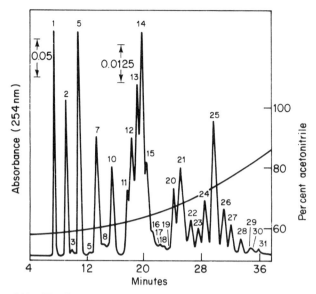

Figure 211 H.p.l.c. separation of Aroclor 1221–Aroclor 1016–Aroclor 1254 (1:1:1, w/w/w) (1.67 mg ml^{-1} of each in tetrahydrofuran) on a μ Bondapak C$_{18}$ column monitored at 254 nm. Initial conditions, 40% water–acetonitrile (9:1) and 60% water–acetonitrile (1:9); final conditions, 100% water–acetonitrile (1:9); gradient time, 40 min; flow rate, 2 ml min^{-1}; injection volume, 10 μl; amount injected 50 μg total. Reprinted with permission from Kaminsky and Fasco.[327] Copyright (1978) Elsevier Science Publishers B.V.

Table 134 Composition of Aroclor 1254

Peak no.	PCB	Content (wt. %)
13	2,3,2',5'-	5.3
15	2,5,2'5'-	10.3
20	2,5,3',4'-	3.3
22	2,4,5,2',5'-	11.7
26	2,3,4,2',4',5'-	4.9
27	2,4,5,2',4',5'-	5.3

Reprinted with permission from Kaminsky and Fasco.[327] Copyright (1978) Elsevier Science Publishers B.V.

tentatively identified and quantitated are shown in Table 134. The major peak (based on integrated areas) is no. 25 and probably represents a pentachloro- or hexachlorobiphenyl which has less than two ortho-substituents. Higher numbers of ortho-substituents would have responses that would preclude the possibility of a peak of such magnitude.

Polyurethane foam has been used as a liquid–liquid partitioning filter for the concentration of PCBs from water samples.[328,329]

Thin-layer chromatography

de Vos and Peet[330] applied mixtures of PCBs in hexane solution (20 µl) to a 0.25 mm layer of Kieselguhr G previously saturated with light petroleum containing 8% of liquid paraffin and dried. The chromatograms were developed (×3) with acetonitrile–acetone–methanol–water (20:9:20:1) previously saturated with liquid paraffin. After drying in air, the plates are sprayed with 0.85% ethanolic silver nitrate and exposed to u.v. radiation for detecting the spots (black on a white background).

Sackmauer et al.[331] determined PCBs in water and fish by thin-layer chromatography on silica gel plates impregnated with 8% paraffin oil. As a mobile phase a mixture of acetonitrile, acetone, methanol and water (20:9:20:1) was used. For detection a solution of silver nitrate and 2-phenoxyethanol was used, followed by irradiation with u.v. light. The detection sensitivity for Aroclor 1242 is 0.5–1.0 µg.

Polarography

PCBs, chlorinated insecticides, and polychlorinated naphthalenes and benzenes have been identified, volumetrically[332] using three electrode potentiostatic control circuitry with interruptable linear voltage sweep control and a normal voltage scan rate of $142\,\text{mV}\,\text{s}^{-1}$. The reference electrode was a standard calomel electrode, the auxiliary electrode was a platinum wire and the stationary electrode was a mercury coated platinum electrode. Dimethyl sulphoxide is used as solvent with $0.1\,\text{M}$ tetraethylammonium bromide as the supporting electrolyte, and nitrogen is used for deaeration.

Neely[333] determined PCBs in fish and water in lake Michigan whilst Frederick[334] measured by gas chromatography the comparative uptake of PCBs and dieldrin by the white sucker. Jan and Malversic[335] determined PCBs and polychlorinated terphenyls in fish using acid hydrolysis of the fish tissue and destructive clean-up of the extract. Olsson et al.[336] studied the seasonal variation of PCB levels in roach. Szelewski et al.[337] have also determined PCBs in fish samples.

Biodegradability of PCBs

Kaiser and Wong[338] studied the bacterial degradation of Aroclor1242 in water and identified aliphatic and aromatic hydrocarbon metabolites using gas chromatography and mass spectrometry. No evidence was found for chlorine containing metabolites. ^{14}C-labelled PCBs have been used,[339] to ascertain the fate of PCBs in activated sludge in a sewage works. It was found that tri- and pentachlorobiphenyl were practically unaffected by bacterial attack, and were partitioned, in unchanged form, between the water and the sludge, and major portion being retained by the sludge. Furukawa et al.[340] studied the effect of chlorine substitution on the biodegradability of 31 different PCBs. Their biodegradability using *Alcaligenes* and *Acinetobacter*, showed that degradation varied inversely with chlorine substitution. PCB with chlorine in the ortho position showed poor degradability; those with all chlorine atoms on a single ring degraded faster than those with the same number on both rings; preferential ring fission occurred with non-chlorinated or less chlorinated rings; and only in the degradation of 2,4,6-trichlorobiphenyl were significant differences observed between the two organisms.

Mixtures of chlorinated PCBs and chlorinated insecticides

The previous pages dealt with methods for PCBs alone. In actual practice, environmental samples which are contaminated with PCBs are also highly likely to be contaminated with chlorinated insecticides. Many reports have appeared discussing cointerference effects of chlorinated insecticides in the determination of PCBs and vice versa and much of the more recent published work takes account of this fact by dealing with the analysis of both types of compounds. This work is discussed below.

Gas chromatography

PCBs have gas chromatographic retention times similar to the organochlorine insecticides and therefore complicate the analysis when both are present in a sample. Several techniques have been described for the separation of PCB from organochlorine insecticides. A review of these methods has been presented by Zitko and Choi.[341] These techniques are time-consuming and, in general, semiquantitative. In addition, differential adsorption or metabolism of the Aroclor isomers in marine biota prevent accurate analysis of the PCBs. The gas chromatographic determination of chlorinated insecticides together with PCBs is difficult. Chlorinated insecticides and PCBs are extracted together in routine residue analysis, and the gas chromatographic retention times of several PCB peaks are almost identical with those of a number of peaks of chlorinated insecticides, notably of the DDT group. The PCB interference may vary, because the PCB mixtures used have different chlorine contents, but it is common for PCBs to be very similar to many chlorinated insecticides and the complete separation of chlorinated insecticides from PCBs is not possibly by gas chromatography alone.[342-346] Figure 212 illustrates the possibility of the interference of DDT-type compounds in the presence of PCBs. In an early paper on the determination of PCBs in water samples which also contain chlorinated insecticides, Ashling and Jensen[347] pass the sample through a filter containing a mixture of Carbowax 4000 monostearate on Chromosorb W. The adsorbed insecticides are eluted with light petroleum and then determined by gas chromatography on a glass column (160 cm × 0.2 cm) containing either 4% SF-96 or 8% QF-1 on Chromosorb W pretreated with hexamethyldisilazane, with nitrogen as carrier gas (30 ml min^{-1}) and a column temperature of 190 °C. When an electron capture detector is used the sensitivity was 10 ng of lindane

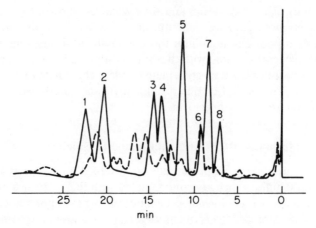

Figure 212 Interference of DDT-type compounds in the presence of PCBs. G.c. column: 5% QF-1 on Gas Chrom Q, 100–120 mesh. Solid line: 1 = p,p'-DDT; 2 = p,p'-DDD; 3 = o,p'-DDT; 4 = o,p-DDD; 5 = p,p'-DDE; 6 = p,p'-DDMU; 7 = o,p'-DDE; 8 = o,p-DDMU. Broken line: PCB Chlophen A 50. Reprinted with permission from Zitko and Choi.[341] Copyright Fisheries Research Board, Canada

per cubic metre, with a sample size of 200 litres. The recoveries of added insecticides range from 50 to 100%; for DDT the recovery is 80% and for PCB, 93–100%.

Elder[348] evaluated mixtures of PCBs and DDE in terms of mixtures of commercial preparations from peak heights of packed column gas chromatograms using a programmable calculator. They propose a method for evaluating gas chromatograms of multicomponent PCB mixtures and superimposed single components simultaneously. Apparent concentrations relative to calibration mixtures are assigned to a number of suitable peaks, and the apparent concentrations are related to the true concentrations by a set of linear equations, which are solved by least squares approximation.

Musty and Nickless[349] used Amberlite XAD-4 for the extraction and recovery of chlorinated insecticides and PCBs from water. In this method a glass column (20 cm × 1 cm) was packed with 2 g of XAD-4 (60–85 mesh) and 1 litre of tap water (containing 1 part per 10^9 of insecticides) was passed through the column at 8 ml min^{-1}. The column was dried by drawing a stream of air through, then the insecticides were eluted with 100 ml of ethyl ether–hexane (1:9). The eluate was concentrated to 5 ml and was subjected to gas chromatography on a glass column (5.5 ft × 4 mm) packed with 1.5% OV-17 and 1.95% QF-1 on Gas-Chrom Q (100–120 mesh). The column was operated at 200 °C, with argon (10 ml min^{-1}) as carrier gas and a ^{63}Ni electron capture detector (pulse mode). Recoveries of BHC isomers were 106–114%; of aldrin, 61%; of DDT isomers, 102–144%; and of polychlorinated biphenyls 76%. Girenko et al.[350] noted that PCBs interfered in the electron capture gas chromatographic identification and determination of chlorinated insecticides in water and fish. Prior to gas chromatography, water samples and biological material were extracted with *n*-hexane. A mixture of organochlorine insecticides was completely separated (Figure 213a). Girenko *et al.*[350] noted that it was difficult to analyse samples of sea water because they are severely polluted by various coextractive substances, chiefly chlorinated biphenyls. To determine organochlorine insecticide residues by gas chromatography with an electron capture detector, the chlorinated biphenyls were eluted from the column together with the insecticides. They produce unseparable peaks with equal retention time, thus interfering with the identification and quantitative determination of the organochlorine insecticides. The presence of chlorinated biphenyls is indicated by additional peaks on the chromatographs of the water samples and aquatic organic organisms (Figure 213b). Some of the peaks coincide with the peaks of the *o,p'* and *p,p'* isomers DDE, DDD, and DDT and some of the constituents are eluted after *p,p'*-DDT.

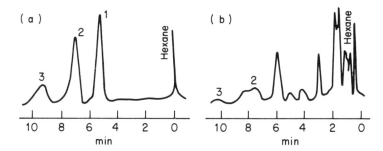

Figure 213 Chromatograms of a synthetic mixture of organochlorine insecticides (a) and of a water extract containing chlorinated biphenyls (b): 1 = *p,p'*-DDE, 2 = *p,p'*-DDD; 3 = *p,p'*-DDT. Reprinted with permission from Musty and Nickless.[349] Copyright (1974) Elsevier Science Publishers B.V.

The Department of the Environment (UK)[351] has used a tentative method for the determination of organochlorine insecticides and PCBs in natural and drinking waters and sewage effluents. The first part of this method is concerned with the extraction and determination of the amounts of individual substances present; the second part is concerned with methods for verifying the identity of the various substances quantified in the first part. Representative data are presented for a number of typical insecticides using different chromatographic stationary phases (Table 135).

Information on blanks and limits of detection are given in Table 136.

Szelewski *et al.*[352] claim that some loss of PCB homologues occurs during the chromium trioxide extraction of fish tissue. The biphenyl-free PCB extract is then perchlorinated using antimony pentachloride in a special apparatus at 200 °C. Following acidification and toluene extraction, the aqueous phase remaining is extracted with hexane and this extract passed down an anhydrous sodium sulphate microcolumn and concentrated in a Kuderna Danish evaporator prior to gas chromatography. Comparison of the gas chromatograms of the

Table 135 Retention lines of organochlorine insecticides relative to dieldrin, on some g.l.c. columns

Insecticide	1% Apiezon Mor L	2.5% Methyl silicone (e.g. OV-1)	2.5% Phenyl methyl silicone (e.g. OV-17)	2.5% Cyano silicone gum rubber XE-60	5% Trifluoropropyl silicone oil QF-1 (FS-1265)	1% Neopentylglycol succinate (NPGS)	1% FFAP	1.5% QF-1 + 1% OV-1 (2)	2.0% OV-1 + 3.0% QF-1 (1)
α-HCH	0.20	0.19	0.17	0.23	0.17	0.23	0.20	0.19	0.22
γ-HCH (lindane)	0.26	0.23	0.22	0.35	0.22	0.37	0.32	0.24	0.26
β-HCH	0.31	0.21	0.22	1.11	0.28	1.63	0.79	0.25	0.28
δ-HCH	0.35	0.26	0.43	1.09	0.31	1.34	0.27	0.29	0.31
Chlordane									0.34*
Heptachlor	0.36	0.41	0.36	0.23	0.23	0.22	0.22	0.35	0.36
Aldrin	0.49	0.53	0.46	0.26	0.28	0.22	0.25	0.43	0.44
Heptachlor epoxide	0.60	0.66	0.63	0.64	0.60	0.64	0.60	0.64	0.65
Endosulfan A	0.88	0.85	0.80	0.69	0.79	0.73	0.67	0.83	
Endosulfan B	1.45	1.16	1.35	2.33	1.66	2.69	2.39	1.32	
Dieldrin	1.00	1.00	1.00	1.00	1.00	1.00	1.00	1.00	1.00
p,p'-DDE	1.25	1.06	1.01	0.83	0.65	1.01	1.09	0.88	0.85
Endrin	1.49	1.28	1.55	0.71	1.22	1.12	1.07	1.26	1.15
o,p'-TDE	1.23	1.04	1.07	1.37	0.87	1.81	1.72	0.96	0.95
p,p'-TDE	1.83	1.30	1.40	2.44	1.21	3.39	3.03	1.30	1.24
o,p'-DDT	1.62	1.38	1.42	1.12	0.94	1.40	1.69	1.17	1.12
p,p'-DDT	2.40	1.72	1.85	2.07	1.32	2.84	2.99	1.60	1.51

*Chlordane is a multipeak compound with other smaller peaks at RRT 0.37, 0.69, 0.76, and 1.23. Glass columns 1 m × 3 mm internal diameter 60–80 mesh acid washed DMCS treated Chromosorb W support at 180 °C and 20–40 ml nitrogen min^{-1}.

Table 136 Blanks and limits of detection

Insecticides	mean (ng l^{-1})	Sw	Sb	St	Limit of detection (ng l^{-1})	Criterion of detection (ng l^{-1})
γ-HCH	0.21	0.320	NS	0.320	12	0.9
Aldrin	0.04	0.032	0.600	0.63	3	0.1
Dieldrin	0.28	0.205	0.430	0.480	14	0.6
p,p'-TDE	0.42	0.570	NS	0.70	8	1.5
p,p'-DDT	0.03	0.125	NS	0.125	15	0.3
PCB	2.8	1.8	3.2	3.7	106	4.8

Sw, within batch standard deviation.
Sb, between batch standard deviation.
St, total standard deviation.

polychlorobiphenyls thus obtained with those obtained for the perchlorination product of an authentic sample of PCB (e.g. Aroclor 1260) enables identifications to be made of the types of PCBs in the sample extract.

Workers at the US Environmental Protection Agency[353] have evaluated protocols for chlorinated insecticides and PCBs in raw waste water and sewage effluents. They concluded that the gas chromatographic method performed satisfactorily at the parts per billion level.

Södergren[354] investigated the simultaneous detection of PCBs, chlorinated insecticides, and other compounds by electron capture and flame ionization detectors combined in series using an open tube capillary column. Södergren[354] points out that much of the earlier work on combining detectors have utilized splitting the sample before a parallel detection system. As a result the sensitivity of the system is decreased, since only part of the injected sample is fed to each detector. Furthermore, it may be difficult to quantify the detected compounds since the splitting ratio, usually estimated by measurement of the carrier gas stream, may be affected by the size of the molecules eluted from the column. Södergren[354] combined the electron capture detector and the flame ionization detection in series to obtain a dual detection system capable of simultaneous detection of environmental pollutants of different character, e.g. organochlorine residues and oil and lipid constituents in samples from aquatic environments. In order that the limit of detection should not be adversely affected when using capillary columns, a splitless system without a scavenging gas was used.

Figure 214 Serial ECD–FID dual detection system. Reprinted with permission from Södergren.[354] Copyright (1979) Elsevier Science Publishers B.V.

Södergren[354,355] connected a modified all-glass electron capture detector to the flame ionization detector of a gas chromatograph (Figure 214). Since electron capture is a non-destructive process, the effluent from the column passes undisturbed through the electron capture detector. The effluent was then directed to the jet-tip of the flame ionization detector by means of a glass capillary tube. The detector system was connected to the column by means of a PTFE tube. Glass columns (190 cm × 1.5 mm i.d.) were used, and the stationary liquid phases (SF-96–QF-1, 3:1) were supported on Gas-Chrom G HP AW DMCS (100–120 mesh). The capillary column (30 m × 0.4 mm) i.d.) was coated with OV-101 by a dynamic procedure.[356] Unless the temperature of the column oven was programmed, the temperatures of the injector, column, and detector were 225, 185, and 220 °C, respectively. With packed columns the flow rate of the carrier gas (nitrogen) was adjusted to *ca.* 20 ml min^{-1}. With capillary columns a minimum flow rate of 1.9 ml min^{-1} was required to operate the electron capture detector. The flows of hydrogen and air to the flame ionization detector were *ca.* 25 and 250 ml min^{-1}, respectively. To prepare the detection system for open tube capillary columns, a low volume detector was developed. The electron capture detector is sensitive to solute concentration, and the use of a scavenging gas dilutes the sample and lowers the sensitivity. By modification of the arrangement of the tritium foil in the detector cell (Figure 215), a cell volume of 0.15 ml was obtained. The detector was operated with comparable sensitivity to that of a conventional cell at a carrier gas flow rate of 1.9 ml min^{-1}.

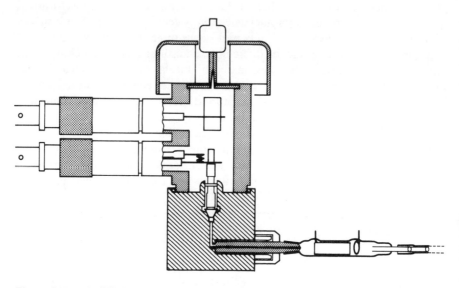

Figure 215 Modified arrangement of the tritium foil in the ECD cell. Reprinted with permission from Södergren.[354] Copyright (1979) Elsevier Science Publishers B.V.

The gas chromatograms in Figure 216 show that organochlorine insecticides and methyl esters of fatty acids were detected simultaneously using this system. The electron capture detector was apparently not affected by the fatty acids, and the flame ionization detector showed no response for the organochlorine compounds at the concentrations used. No interferences or abnormal behaviours were observed in the detection system when the two classes of compounds were injected separately. Obviously, the responses of the detectors are equal to those obtained when the detectors are operated singly. Södergren[354] used his detection system to study the degradation and fate of persistent pollutants in aquatic model ecosystems. Usually these pollutants are closely associated with lipids. Therefore, it is an advantage to be able to study the occurrence and amount of both lipids and, for example, organochlorine residues. A cell extract from a continuous flow culture of the green alga *Chlorella pyrenoidosa* to which polychlorinated biphenyls had been added was hydrolysed by treatment with

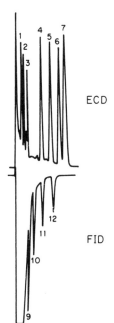

Figure 216 Simultaneous response of ECD–FID to a mixture of organochlorine insecticides and methyl esters and fatty acids. Peaks: 1 = lindane; 2 = BHC; 3 = aldrin; 4 = p,p'-DDE; 5 = dieldrin; 6 = p,p'-DDD; 7 = p,p'-DDT; 8 = lauric acid; 9 = myristic acid; 10 = palmitic acid; 11 = stearic acid; 12 = oleic acid. Reprinted with permission from Södergren.[354] Copyright (1979) Elsevier Science Publishers B.V.

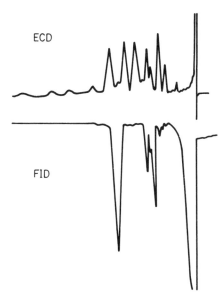

Figure 217 Fatty acids and PCBs in cells of the green alga *Chlorella pyrenoidosa*. For peaks see Figure 216. Reprinted with permission from Södergren.[354] Copyright (1979) Elsevier Science Publishers B.V.

a solution of acetyl chloride in methanol and then presented to the detection system (Figure 217). The PCBs added to the culture were efficiently taken up by the algal cells: the lipids detected in the extract were palmitic acid and stearic acid. Lipids and substances of lipophilic character tend to accumulate in aquatic environments at the interface between water and air. To assess the ability of the electron capture flame ionization detector system to simultaneously detect mineral oil and PCBs, a mixture of these substances was injected onto the water of an aquarium below the surface. The surface film thus created was sampled,[357] extracted, and an aliquot of the extract injected into the gas chromatograph equipped with a capillary column and a low volume electron capture detector and a flame ionization detector (Figure 218). The mineral oil was eluted before the main PCB components appeared. Due to the high sensitivity of the electron capture detector to changes in temperature, programming the column resulted in severe baseline drift at the beginning of the run. However, neither class of components affected the detection of the other. Thus, both the mineral oil and halogenated compounds can be conveniently analysed and quantified simultaneously after a single injection.

Bacaloni *et al.*[358] also used capillary column gas chromatography for the analysis of mixtures of chlorinated insecticides, PCBs, and other pollutants. Graphitized capillary columns were used in this work. These columns, introduced by Goretti and co-workers,[359-363] have the interesting feature that according to the amount of stationary phase coated on the walls, they operate in gas–solid, gas–liquid and gas–liquid–solid chromatographic modes. Bacaloni[358] used columns coated with PEG 20M and it is shown that selective columns

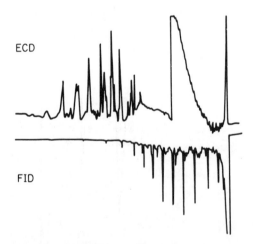

Figure 218 Simultaneous detection of PCBs and mineral oil in a surface film. Glass capillary column (30 m); stationary phase, OV-101. Column temperature programmed from 80 to 180 °C at 4 °C min^{-1}. Reprinted with permission from Södergren.[354] Copyright (1979) Elsevier Science Publishers B.V.

can be obtained that give different performances suitable for particular applications.

Bacaloni et al.[358] gas chromatographed a mixture of 15 chlorinated pesticides on columns of different lengths and loadings of stationary phase under the same operating conditions at 160 °C. The values of the capacity ratio (K') and the retention volumes relative to dieldrin are given in Table 137. The columns give the same elution sequence for most insecticides, with the exception of α-endosulfan, and γ-chlordane, and endrin and p,p'-DDE. These compounds are eluted in the order given on columns 1 and 2, whereas the order is reversed on other columns with higher loadings of stationary phase.

Table 137 Capacity ratios (K') and retention volumes (r) relative to dieldrin at 160 °C

No.	Insecticide	column 1		column 2		column 3		column 4	
		K'	r	K'	r	K'	r	K'	r
1	α-BHC	1.49	0.18	2.50	0.20	8.81	0.19	36.4	0.18
2	Heptachlor	1.70	0.21	2.63	0.21	9.02	0.20	40.0	0.20
3	Aldrin	1.81	0.22	2.85	0.23	9.80	0.21	42.1	0.21
4	γ-BHC	2.51	0.31	3.80	0.31	16.8	0.36	62.9	0.31
5	α-Endosulfan	5.41	0.66	7.92	0.65	31.2	0.66	140	0.69
6	γ-Chlordane	5.65	0.69	8.21	0.67	30.0	0.64	130	0.64
7	α-Chlordane	5.82	0.71	8.67	0.70	31.8	0.67	142	0.70
8	o,p'-DDE	6.84	0.84	10.3	0.84	40.5	0.86	172	0.84
9	Dieldrin	8.18	1.00	12.3	1.00	47.2	1.00	204	1.00
10	Endrin	8.94	1.09	13.6	1.09	51.6	1.09	236	1.16
11	p,p'-DDE	9.35	1.14	14.1	1.14	50.0	1.06	219	1.07
12	Perthane	11.8	1.44	17.3	1.41	62.0	1.31	278	1.36
13	o,p'-DDD	17.4	1.90	23.5	1.91	93.4	1.98	393	1.93
14	β-Endosulfan	20.0	2.70	33.2	2.69	125	2.65		
15	p,p'-DDT	26.1	3.51	43.8	3.59	138	2.94		

Reprinted with permission from Bacaloni et al.[358] Copyright (1979) Elsevier Science Publishers B.V.

In order to evaluate the feasibility of these columns for the separation of basic and acidic compounds that might occur in the environment, a mixture of 2,4-dichloroaniline, 2,4-dichlorophenyl, and an insecticide (γ-BHC) was analysed on column 1. The insecticide and the amine gave symmetrical peaks whereas tailed peaks were obtained with the phenol. By analysing a similar mixture on column 4 at 160 °C, it was found that all of the compounds tested gave symmetrical peaks. On both columns plots of HETP against linear velocity were obtained by using nitrogen and hydrogen as the carrier gas; the plot for column 4 is shown in Figure 219. By comparing the HETP values for the same compounds, it can be concluded that the use of hydrogen gives only a slight decrease in efficiency compared with the use of nitrogen but the analysis time is drastically reduced. Bacaloni et al.[358] concluded that a glass capillary column loaded with a large amount of PEG 20M is suitable for the analysis of volatile chlorinated compounds, phenols and amines (Figure 220), whereas for the

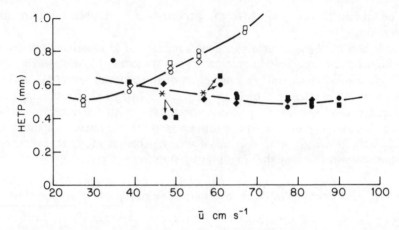

Figure 219 HETP versus linear velocity of carrier gas at 160 °C in column 4. Carrier gas: open symbols, N_2; closed symbols, H_2. ○,● = α-BHC, $k' = 38.0$; □,■ = 2,4-dichloroaniline, $k' = 16.6$; ◇,◆ = 2,4-dichlorophenol, $k' = 11.2$. * represents ■ + ●. Reprinted with permission from Bacaloni et al.[358] Copyright (1979) Elsevier Science Publishers B.V.

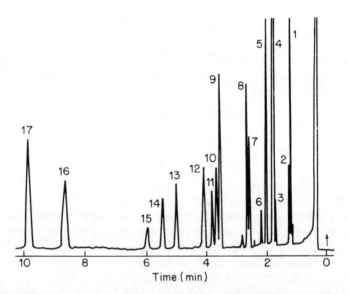

Figure 220 Gas chromatogram of a mixture of phenols and aniline on column 4 at 160 °C. Carrier gas H_2; flame ionization detector. Peaks: 1 = o-nitrophenol; 2 = o-chlorophenol; 3 = o-chloroaniline; 4 = 5-methyl-2-chloroaniline; 5 = o-cresol; 6 = 4-chloro-2-nitrophenol; 7 = p-cresol; 8 = m-cresol; 9 = p-chloroaniline; 10 = m-chloroaniline; 11 = 2-methyl-3-chloroaniline; 12 = 2-methyl-4-chloroaniline; 13 = 2,4-dichloroaniline; 14 = 2,5-dichloroaniline; 15 = 2,4,6-trichlorophenol; 16 = p-chlorophenol; 17 = p-chloro-m-cresol. Reprinted with permission from Bacaloni et al.[358] Copyright (1979) Elsevier Science Publishers B.V.

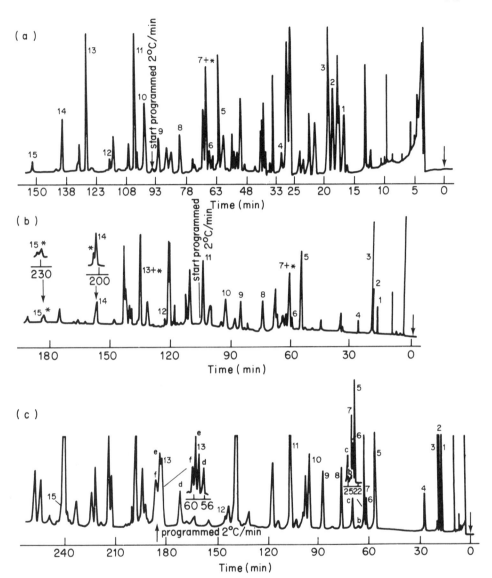

Figure 221 Gas chromatograms of mixtures of pesticides and Aroclors on column 5. Carrier gas, H$_2$; electron capture detector. Numbers of peaks correspond to those in Table 137. (a) Aroclor 1242; temperature programme, 140 °C for 95 min, then increased from 140 to 160 °C at 2 °C min^{-1}; *interfering peaks. (b) Aroclor 1254; temperature programme 140 °C for 105 min, then increased from 140 to 160 °C at 2 °C min^{-1}. Alternative programme: 140 °C for 170 min then increased from 140 to 160 °C at 2 °C min^{-1} (see inset peaks); *interfering peaks. (c) Aroclor 1260: temperature programme, 140 °C for 185 min, then increased from 140 to 160 °C at 2 °C min^{-1}. Alternative programme: isothermal at 160 °C (see inset peaks): a, b, c, d, e, and f are peaks of Aroclor. Reprinted with permission from Bacaloni et al.[358] Copyright (1979) Elsevier Science Publishers B.V.

analysis of chlorinated compounds including the PCBs a column with a low loading of stationary phase is desirable. On the basis of these results, the separation of mixtures of chlorinated insecticides and PCBs (Aroclor 1242, 1256, and 1260) was tried on column 5. This 70 m column had 210,000 theoretical plates for hexadecane at 100 °C (K' 2.5) and 160,000 for γ-BHC at 160 °C (K' 3.5). Figure 221 shows chromatograms of complex mixtures of chlorinated organic compounds. Although the complexity of the mixture prevents the separation of all of the components, by changing the operating conditions separations of specific compounds might be achieved. As an example, the interference of some PCBs with β-endosulfan and p,p'-DDT is eliminated by varying the temperature programme (Figure 221(b)). A similar situation is observed in Figure 221(c) for chlordane and o,p'-DDD and is resolved by operating isothermally at 160 °C. The results reported indicate that the thickness of the coating in graphitized capillary columns plays a predominating role. Probably columns with a low loading of stationary phase operate as gas–liquid–solid chromatographic columns and, because of the specific characteristics of graphitized carbon black, separations of chlorinated compounds of similar structure are obtained. Apparently the surface should be basic so that amines yield symmetrical peaks (Figure 222). The behaviour of capillary columns with a high loading of stationary phase should be considered typical of gas–liquid chromatography.

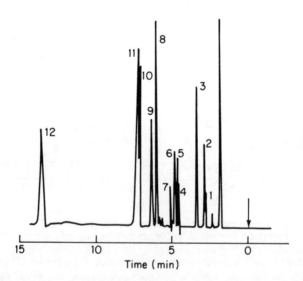

Figure 222 Separation of halogenated anilines on column 1 at 130 °C. Carrier gas, N_2; electron capture detector. Peaks: 1 = 2-chloroaniline; 2 = 2-chloro-5-methylaniline; 3 = 2-bromoaniline; 4 = 3-chloroaniline; 5 = 4-chloroaniline; 6 = 3-chloro-2-methylaniline; 7 = 4-chloro-2-methylaniline; 8 = 2,4-dichloroaniline; 9 = 2,5-dichloroaniline; 10 = 4-bromoaniline; 11 = 3-bromoaniline; 12 = 4-iodoaniline. Reprinted with permission from Bacaloni et al.[358] Copyright (1979) Elsevier Science Publishers B.V.

A method has been described[365] for the determination of the following PCBs (Aroclors) at the nanogram level in water and waste water:

PCB-1016
PCB-1221
PCB-1232
PCB-1242
PCB-1248
PCB-1254
PCB-1260

This method is an extension of the method for chlorinated hydrocarbons in water and waste water (described by Georlitz and Law.[364] It is designed so that determination of both the PCBs and the following organochlorine insecticides may be made on the same sample.

Aldrin	DDT	Mirex
BHC	Heptachlor	Pentachloronitrobenzene
Chlordane	Heptachlor epoxide	Strobane
DDD	Lindane	Toxaphene
DDE	Methoxychlor	Trifluralin

The PCBs and the organochlorine insecticides are coextracted by liquid–liquid extraction and, insofar as possible, the two classes of compounds separated from one another prior to gas chromatographic determination. A combination of the standard Florisil column clean-up procedure and a silica gel microcolumn separation procedure are employed. Identification is made from gas chromatographic patterns obtained through the use of two or more unlike columns. Detection and measurement is accomplished using an electron capture, microcoulometric, or electrolytic conductivity detector. Techniques for confirming qualitative identification are suggested by these workers.

The interference in industrial effluents are high and varied and pose great difficulty in obtaining accurate and precise measurement of PCBs and organochlorine insecticides. Separation and clean-up procedures are generally required and may result in the loss of certain organochlorine compounds. Therefore, great care should be exercised in the selection and use of methods for eliminating or minimizing interferences. Phthalate esters, certain organophosphorus insecticides, and elemental sulphur will interfere when using electron capture for detection. These materials do not interfere when the microcoulometric or electrolytic conductivity detectors are used in the halogen mode. Organochlorine insecticides and other halogenated compounds constitute interferences in the determination of PCBs. Most of these are separated by the method described. However, certain compounds, if present in the sample, will occur with the PCBs. Included are: sulphur, heptachlor, aldrin, DDE, chlordane, mirex, and to some extent, o,p'-DDT and p,p'-DDT.

Various gas chromatographic column packings are used including SE-30 or OV-1 (3%), OV-17 (1.5%), or OV-210 (1.95%) on Gas Chrom Q (100–120 mesh).

In the method 60 ml of 15% v/v_u methylene chlorine in hexane is added to up to one litre of water sample in a separatory funnel and the mixture shaken vigorously for two minutes. The mixed solvent is allowed to separate from the sample, then the water is drawn into a one litre Erlenmeyer flask. The organic layer is poured into a 100 ml beaker and then passed through a column containing 3–4 inches of anhydrous sodium sulphate, and collected in a 500 ml Kuderna Danish flask equipped with a 10 ml ampoule. The water phase is returned to the separatory funnel. The Erlenmeyer flask is rinsed with a second 60 ml volume of solvent, the solvent is added to the separatory funnel and the extraction procedure completed a second time. A third extraction is performed in the same manner. The extract is concentrated in the Kuderna Danish evaporator on a hot water bath. The sample is quantitatively analysed by gas chromatography with an electron capture detector. From the response obtained the following are decided:

(1) if there are any organochlorine pesticides present.
(2) if there are any PCBs present.
(3) if there is a combination of 1 and 2.
(4) if elemental sulphur is present.
(5) if the response is too complex to determine 1, 2, or 3.
(6) if no response, the sample is concentrated to 1.0 ml or less, as required, and the analysis repeated, looking for 1, 2, 3, 4, and 5. Samples containing Aroclors with a low percentage of chlorine, e.g. 1221 and 1232, may require this concentration in order to achieve the detection limit of $\mu g \, l^{-1}$. Trace quantities of PCBs are often masked by background which usually occurs in samples. Various clean-up steps are incorporated into the method depending on which of conditions (1) to (6) apply. Figures 223–228 show gas chromatograms obtained by this procedure for various PCB's.

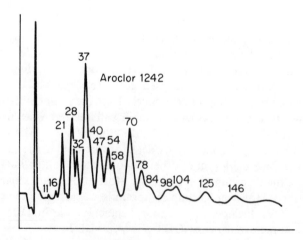

Figure 223 Column: 3% OV-1, carrier gas, nitrogen at 60 ml min^{-1}; column temperature, 170 °C; detector, electron capture

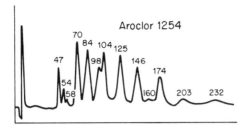

Figure 224 Column, 3% OV-1; carrier gas, nitrogen at 60 ml min^{-1}; column temperature, 170 °C; detector, electron capture

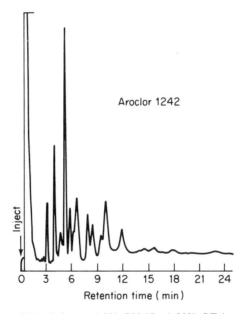

Figure 225 Column, 1.5% OV-17 + 1.95% QF-1; carrier gas, nitrogen at 60 ml min^{-1}; column temperature, 200 °C detector, electron capture

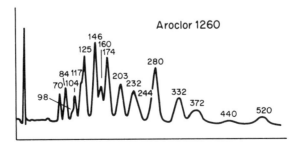

Figure 226 Column, 3% OV-1, carrier gas, nitrogen at 60 ml min^{-1}; column temperature, 170 °C; detector, electron capture

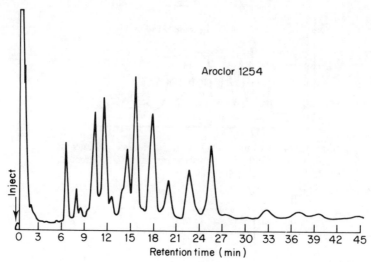

Figure 227 Column, 1.5% OV-17 + 1.95% QF-1; carrier gas, nitrogen at 60 ml min^{-1}; column temperature, 200 °C; detector, electron capture

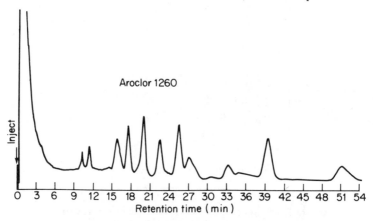

Figure 228 Column, 1.5% OV-17 + 1.95% QF-1; carrier gas, nitrogen at 60 ml min^{-1}; column temperature, 200 °C, detector, electron capture

Using the electron capture detector

Miller *et al.*[336] have described a gas chromatographic method for the determination of 25 organochlorine insecticides and PCBs in water. Data is presented on extractability, recovery from clean-up columns, recovery from spiked waste waters, and preservation of samples for analysis.

Chlorinated insecticides/PCBs in bottom deposits

Goerlitz and Law[367] determined chlorinated insecticides in sediment and bottom material samples which also contained PCBs by extracting the sample with acetone and hexane. The combined extracts were passed down an alumina column. The first fraction (containing most of the insecticides and some polychlorinated biphenyls and polychlorinated naphthalenes) is eluted with hexane and treated with mercury to precipitate sulphur. If the polychlorinated hydrocarbons interfere with the subsequent gas chromatographic analysis, further purification on a silica gel column is necessary.

Chlorinated insecticides/PCBs in sewage sludge

Jensen *et al.*[368] have described a procedure for the determination of organochlorine compounds including PCBs and DDT in sediments and sewage sludge in the presence of elemental sulphur. The method can also be used for a search for both volatile and/or polar pollutants. The sulphur interfering in the gas chromatographic determination is removed in a non-destructive treatment of the extract with tetrabutylammonium sulphite. This lipophilic ion pair rapidly converts the sulphur to thiosulphate in an organic phase. The recovery of added organochlorines was above 80% and the detection limit in the range of 1–10 ppb from a 10 g sample. Elemental sulphur present in most sediment and digested sludge has caused significant problems in residue analysis.[369,370] If the sulphur level is high, the electron capture detector will be saturated for a considerable period of time, and if the level of sulphur is low, it gives three or more distinct peaks on the chromatogram which can interfere with BHC isomers and aldrin. Treatment of the crude extract with potassium hydroxide in ethanol[371] or Raney nickel[372] will quantitatively destroy all sulphur, but will at the same time convert DDT and DDD to DDE and DDMU (1-chloro-2,2-*bis*(4-chlorophenyl) ethane), respectively, and most BHC isomers are lost. Metallic mercury has also been used for removal of sulphur.[373] Jensen *et al.*[368] described an efficient, rapid, non-destructive method to remove the sulphur according to the reaction:

$$(TBA^+)_2SO_3^{2-} + S(s) \longrightarrow 2TBA^+ + S_2O_3^{2-}$$

where TBA^+ = tetrabutylammonium ion.

Method

Reagents

Acetone, hexane, undecane (analytical grade), diethylether, 2-propanol, and 2,2,4-rimethylpentane (analytical grade) were used without further purification. Sulphuric acid monohydrate (concentrated sulphuric acid–water 98:18 w/w), 0.2 M sodium chloride in 10 mM phosphoric acid, phosphoric acid–water (1:1), 0.1 M sodium hydroxide, and potassium dihydrogen phosphate (0.5 M) (analytical grade).

TBA–sulphite reagent

A solution of 3.39 g (0.1 mol) tetrabutylammonium hydrogen sulphate in 100 ml of water is extracted with three 20 ml portions of hexane (to remove impurities) and then saturated with sodium sulphite 25 g (0.2 mol). The purities of all reagents were tested in blank procedures.

Extraction

An aliquot of the wet sediment sample (2–3 g) is dried at 110 °C for sediment dry weight estimation. Wet sediment (15–20 ml) is transferred into a weighed centrifuge tube (60 ml) with a Teflon-sealed screw cap and centrifuged for 10 min at 2000 rpm (1200 × g). The water is discarded, the tube weighed again, and the sediment wet weight calculated. Acetone (40 ml) is added to the tube, and the tube and the sample are stirred. The tube is then rotated for 60 min. After centrifugation, the liquid phase is transferred into a separatory funnel, containing the sodium chloride solution in phosphoric acid (50 ml). The extraction procedure is repeated with acetone–hexane (10 + 30 ml) for 30 min. After centrifugation, the liquid phase is transferred into the separatory funnel and then shaken. The aqueous phase is re-extracted with a 90:10:2 mixture of hexane–diethyl ether–undecane (10 ml). If the aim is to include more polar substances such as the chlorophenols, the acetone–water phase is re-extracted twice with diethyl ether (2 × 20 ml). The combined organic phases are shaken with 0.1 M sodium hydroxide (10 ml) for isolation of chlorophenols and the aqueous phase is treated as described below for isolation of chlorophenols. The solvents are finally removed in a rotary evaporator (vacuum) and the residue dissolved in trimethylpentane (2–5 ml). All partition steps in the following procedures were carried out in small test tubes with Teflon-sealed caps. The separation of phases can conveniently be carried out by freezing the lower phase in a dry ice–acetone bath followed by decanting the upper phase to another test tube. To remove sulphur the trimethylpentane extract (2 ml) is shaken with 2-propanol (1 ml) and the TBA-sulphite reagent (1 ml) for at least 1 min. If the precipitated sodium sulphite disappears, more is added in 100 mg portions until a solid residue remains after repeated shaking. Water (5 ml) is added and the test tube is shaken for another minute, followed by centrifugation and the trimethylpentane phase is transferred to a test tube. For further clean-up the trimethylpentane phase extract is shaken with an equal volume of sulphuric acid monohydrate. After centrifugation, the trimethylpentane phase is injected into the gas chromatograph. For samples also containing substances not stable against sulphuric acid monohydrate, other fat removing treatments should be carried out using aliquots such as potassium hydroxide treatment[371] or gel permeation chromatography.[374] To isolate chlorophenols the sodium hydroxide phase above is acidified with the phosphoric acid–water mixture (1:1) to pH 1–2 and shaken with trimethylpentane (3 ml) containing a suitable internal standard. The trimethylpentane phase is transferred to another test tube and suitable derivatives of the chlorophenols are then made before the gas chromatographic determination.[375,376]

The samples were run with an electron capture detector (^{63}Ni). The 240 × 0.18 (i.d.) cm glass column was filled with a mechanical mixture of 2 parts of 8% QF-1 and 1 part of 4% SF-96 on acid washed silanized Chromosorb W (100–120 mesh).[369] For all substances including the methylated and acetylated chlorophenols used in the recovery experiments, the column was normally programmed from 150 to 210 °C at a rate of 1 °C min^{-1}. Injector and detector were held at 250 °C. The extraction procedure described above will also extract elemental sulphur. Since the sulphur is soluble in the final trimethylpentane extract and sodium sulphite in water, the well known reaction between them to give thiosulphate proceeds very slowly. If the tetrabutylammonium ion is present, however, a lipophilic ion pair is formed with the sulphite. As higher alcohols are good solvents for such ion pairs, 2-propanol is added to the trimethylpentane extract. This will favour the solution of the pair in the non-polar phase and cause a momentary reaction with the sulphur. When 2-propanol, the TBA reagent, and the extract are mixed in the sulphur-removing step, some sodium sulphite precipitates. In an excess of sulphur, all sulphite is converted to the more water-soluble sodium thiosulphate. More sodium sulphate is therefore added until the precipitate remains, which implies that all sulphur has reacted. The sulphite present in the TBA reagent is, however, normally sufficient for sediment, but not for digested sewage sludge (see Figures 229 and 230).

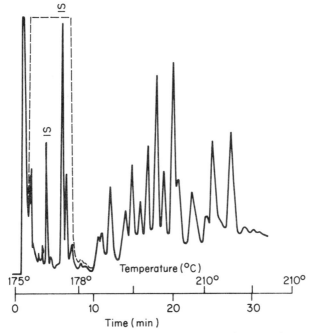

Figure 229 Gas chromatogram of extract from sediment before (– – –) and after (———) TBA–sulphite treatment (the PCB level is 240 ppb on a wet weight basis). · = PCB components. IS = internal standard. Reprinted with permission from Jensen et al.[368] Copyright (1977 American Chemical Society

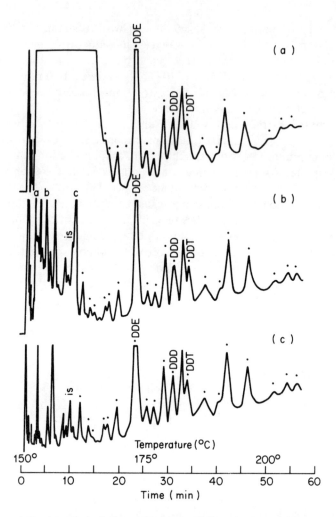

Figure 230 (a) Typical digested sewage sludge chromatogram, severely contaminated with sulphur. (b) The same sample after a normal TBA–sulphite treatment, showing that most of the sulphur has disappeared. A number of peaks (a,b,c) originating from traces of sulphur appear in the BHC–aldrin region. (c) The final chromatogram after additional treatment with sodium sulphite. · = PCB components. IS = internal standard. Reprinted with permission from Jensen et al.[368] Copyright (1977) American Chemical Society

Mattsson and Nygren[377] have described a glass capillary column gas chromatographic procedure for the determination of PCBs and some chlorinated insecticides in sewage sludge. The capillary column is coated with silicone oil SF 96.

The sample is extracted with a mixture of hexane, acetone, and water. After separation, the hexane phase is reduced in volume and divided into two aliquots,

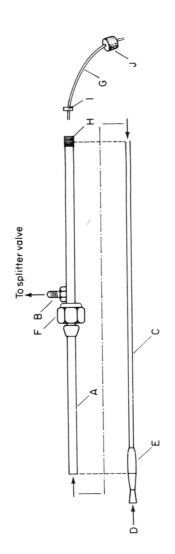

Figure 231 Equipment for injection of samples into the capillary column. The splitter consists of a stainless tube (A) with a Swagelok reducer SS-100-R-2 (B) soldered to it as outlet to the splitter valve. Inside the steel tube there is a glass tube (C), which is funnel-shaped at the injection end (D) and tightened to the steel tube by PTFE tape (E). The splitter is connected to the original injection port with a ¼ in. Swagelok nut (F). The capillary column (G) is connected to the splitter at H with a silicone rubber septum (I) and a screw cap (J). Reprinted with permission from Mattson and Nygren.[377] Copyright (1976) Elsevier Science Publishers B.V.

one of which is first shaken with 7% fuming sulphuric acid to remove lipids, and then with cyanide to eliminate interference by elemental sulphur. The other aliquot is evaporated to dryness and heated with ethanolic potassium hydroxide. The two aliquots are injected into a gas chromatograph fitted with a glass capillary column and an electron capture detector. Hexabromobenzene is used as an internal standard. Polychlorinated biphenyls are determined quantitatively by comparing the peaks of the sample with those of Clophen A 50 or A 60. The individual percentage composition of the chlorobiphenyls in the polychlorinated biphenyl oils is used. Figure 231 shows the details of the tritium source electron capture gas chromatograph with capillary columns used by Mattson and Nygren.[377]

The concentration levels of chlorinated hydrocarbons in sewage sludge allowed the use of a splitter injection technique.[378] These impurities, together in the first part of the column, cause poor separations and lower sensitivity. The effect of changing the glass tubes of the splitter is shown in Figure 232. Nearly 70 peaks were detectable when a PCB oil (Clophen A 50) was chromatographed.

Detailed instructions are given for coating the glass capillary column (Figure 233). The PCBs are quantitated by using the percentage composition of the individual components in the PCB oils (see references 378–380 and Figure 234).

This method has good reproducibility and has a detection limit for the total amount of PCBs in the dried sample of at least 0.1 mg kg^{-1} and for DDT, DDD, and DDE limits of 0.01, 0.005, and 0.005 mg kg^{-1}, respectively.

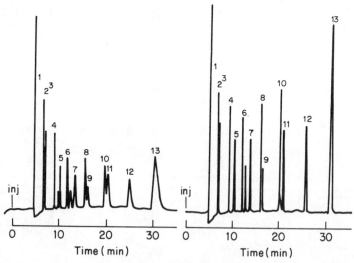

Figure 232 Illustration of the effect of changing injector glass tube. Left: standard solution run after about 100 injections of sewage sludge extracts. Right: the same solution run after changing the injector glass tubes. For identities of the peaks in the standard solution see Table 138. Reprinted with permission from Mattson and Nygren.[377] Copyright (1976) Elsevier Science Publishers B.V.

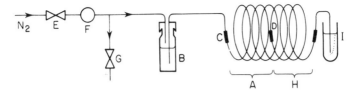

Figure 233 Set-up for coating of glass capillary columns. The column (A) is connected at C to the vial (B) with a PTFE tube 0.8 mm i.d. and at D to a 10 m capillary column (H). The nitrogen pressure in the column is regulated with the valve (E) and read on the manometer (F) on the gas bottle. The flow through the column is regulated with the splitter valve (G). The test tube (I) serves as a collector of waste, and also as a device for checking that gas passes through the column. Reprinted with permission from Mattson and Nygren.[377] Copyright (1976) Elsevier Science Publishers B.V.

Mattson and Nygren[377] devised a solvent extraction method for extracting PCBs from sewage sludge containing lipids. They point out that lipids and some other impurities in the crude extracts of sewage sludge can be destroyed by treatment with fuming sulphuric acid, either by shaking the acid[381] or by eluting on a fuming sulphuric acid–Celite column.[382,383] Dieldrin is decomposed by this treatment but DDT and its metabolites, DDD and DDE, are not (Table 138). Extracts of sewage sludges often contain large amounts of elemental sulphur, particularly after treatment with sulphuric acid. These interfere with early eluting compounds in the gas chromatographic step (Figure 235(a)).

Table 138 Effect of treatment of a solution of chlorinated hydrocarbons and the internal standard hexabromobenzene with fuming sulphuric acid (I), fuming sulphuric acid plus potassium cyanide (II), and potassium hydroxide (III) expressed as percentages of the compounds in an untreated solution

Compound name (no.)	Concentration (ng ml^{-1})	Method of treatment		
		I	II	III
α-BHC (1)	0.022	94	59	0
β-BHC (2)	0.072	94	70	6
Lindane (3)	0.026	92	55	11
Heptachlor (4)	0.026	94	90	104
Aldrin (5)	0.020	92	87	104
Heptachlor epoxide (6)	0.033	83	79	100
p,p'-DDMU (7)	0.090	100	91	190
Dieldrin (8)	0.048	0	0	103
p,p'-DDE (9)	0.022	104	100	437
p,p'-DDD (10)	0.085	100	95	0
o,p'-DTT (11)	0.096	98	95	0
p,p'-DDT (12)	0.103	100	95	0
Hexabromobenzene (13)	0.57	(100)	(100)	(100)

Reprinted with permission from Mattson and Nygren.[377] Copyright (1976) Elsevier Science Publishers B.V.

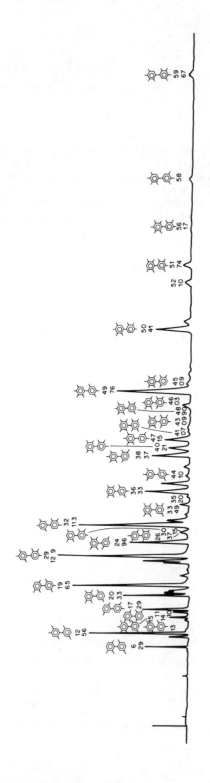

Figure 234 Chromatogram of Clophen A 60 run on a 60 m SF 96 glass capillary column at 185 °C. Reprinted with permission from Mattson and Nygren.[377] Copyright (1976) Elsevier Science Publishers B.V.

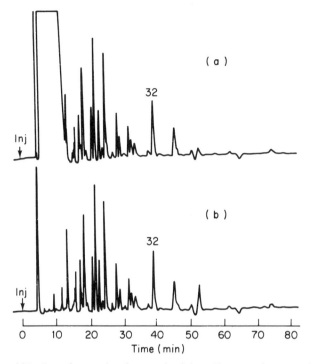

Figure 235 Interference by elemental sulphur. Extract of sewage sludge sample treated with 7% fuming sulphuric acid (a) and also treated with cyanide (b). Reprinted with permission from Mattson and Nygren.[377] Copyright (1976) Elsevier Science Publishers B.V.

Sulphur was removed by the Bartlett and Skoog[384] method in which the sulphur is reacted with cyanide in acetone solution to produce thiocyanate (Figure 235(b)). BHCs are decomposed to some extent, probably to pentachlorocyclohexane. An alternative procedure for the removal of sulphur utilizing barium hydroxide is also described. Alkali hydroxides should not be used as they cause dehydrochlorination of BHCs.[381] Lindane and its isomers are dehydrochlorinated to trichlorobenzenes[385] and are eluted together with the solvent. Cochrane and Maybury[386] have used the reaction with sodium hydroxide in methanol for the identification of BHCs. Dieldrin is not decomposed in the potassium hydroxide treatment and can thus be detected in the chromatogram of that aliquot. Some common chlorinated hydrocarbon pollutants and the internal standard hexabromobenzene were treated, according to the general procedure described above with sulphuric acid, potassium cyanide, and potassium hydroxide. The results of the recovery experiments are shown in Table 138. When using packed columns, a precolumn of sodium and potassium hydroxides will give the same effect as the potassium hydroxide treatment described above.[387] Mattsson and Nygren[377] have also tested a column with a packed alkaline post-column to remove the sulphur peak from the chromatogram. In the post-column DDT and DDD are dehydrochlorinated but this does not effect their retention times.

McIntyre et al.[388,389] described a method for the analysis of PCBs and chlorinated insecticides in sewage sludges in which homogenized samples are extracted with hexane, concentrated and cleaned up on an alumina/alumina plus silver nitrate column and eluted with hexane. After concentration of the eluent, PCB and organochlorine compounds were determined by a silica gel chromatographic procedure and gas chromatography.

Method

All solvents used in the analytical procedures were of superior quality (glass distilled). Glassware was cleaned by immersion in 'Decon 90' (Decon Laboratories, UK), rinsing with distilled water and oven drying. Samples of sewage sludge were collected in clean glass-stoppered bottles, homogenized thoroughly, and stored in a refrigerator at 4 °C. After concentration of the solvent extract, clean-up was carried out using an alumina/alumina plus silver nitrate[390] (to remove sulphur) column and hexane eluant (15 ml). After concentration of the eluant, separation of PCB and organochlorine insecticides was achieved using a modiication[391] of a silica gel column chromatographic procedure.[392] Aliquots of each of the two eluant fractions obtained were then analysed by electron capture gas liquid chromatography. The gas chromatography was fitted with a ^{63}Ni electron capture detector. Glass columns (2 cm × 3 mm i.d.) containing 1.5% OV-17 plus 1.95% QF-1 on 100–120 mesh Supelcoport were used for primary identification and 1% NPGS on 80–100 mesh Gas Chrom Q for confirmation. Carrier gas flows of 50 ml min^{-1}, argon–methane (95:5) were maintained through each column during analysis. The temperatures of the column oven, injection port, and detector block were 2000, 250, and 300 °C respectively. Concentrations of organochlorine insecticides in the extracts were calculated by measuring peak heights on the chromatograms and referring these data to previously constructed calibration curves. Concentrations of PCBs were calculated by measuring heights of the last four major peaks on the chromatogram, multiplying each value by the retention time (min) of that particular peak, summing the products, and referring to previously constructed calibration graphs. The method selected for clean-up produced suitably purified extracts for analysis by electron capture gas chromatography. Standard solutions of Aroclor 1260, γ-HCH, p,p'-DDE and dieldrin were carried through the clean-up procedure in order to assess recovery from the column, the results of which are indicated in Table 139. Examination of the table reveals the recoveries of the four determinands were quantitative, with 99.2% of the Aroclor 1260, 98.75% of the p,p'-DDE, 100.6% of the γ-HCH and 101% of the dieldrin being recovered. The degree of scatter (RSD) of the results was small, the highest being 4.4% for γ-HCH and the lowest 1.8% for Aroclor 1260.

McIntyre[388,389] found that the method by Holden and Marsden[392] resulted

Table 139 Recovery of determinands from clean-up column

Determinand	Weight applied (ng)	Mean weight recovered (ng)	Recovery (%)	Relative s.d. (%)	Range (ng)
Aroclor 1260	500	496.10	99.2	1.8	488.0–508.0
p,p'-DDE	100	98.75	98.75	4.1	94.0–103.0
γ-HCH	25	25.15	100.6	4.4	23.9–26.5
Dieldrin	120	121.2	101.0	3.0	118.4–126.5

Reprinted with permission from McIntyre et al.[388] Copyright (1980) *Science and Technology Letters*.

Table 140 Recovery of polychlorinated biphenyls and organochlorine insecticides from silica gel partition column

Compound	Weight applied (ng)	Mean* weight recovered (ng)	Mean recovery (%)	Relative s.d. (%)	Range (ng)
Aroclor 1260	500	499.18	99.8	1.6	488.2–506.5
p,p'-DDE	100	99.6	99.6	3.1	96.4–103.2
γ-HCH	25	25.6	102.5	2.9	24.9–26.4
Dieldrin	120	122.8	102.3	3.1	119.1–127.8

*Of four replicates.
Reprinted with permission from McIntyre et al.[388] Copyright (1980) *Science and Technology Letters*.

in good reproducible separation of PCB and organochlorine insecticides. PCB and DDE were found to emerge in the first 6 ml of hexane eluate, whilst the remaining organochlorine insecticides emerged in the diethyl ether–hexane eluate. A standard solution containing Aroclor 1260, p,p'-DDE, γ-HCH and dieldrin was carried through this separation procedure in order to assess recovery, the results of which are indicated in Table 140. The results indicate that recovery from the four compounds was quantitative, with recoveries of 99.8, 99.6, 102.5, and 102.3% for Aroclor 1260, p,p'-DDE, γ-HCH, and dieldrin, and displays little scatter, with relative standard deviations of 1.6, 3.1, 2.9, and 3.1% respectively for the four determinations.

McIntyre[388,389] examined the influence of solids concentration in the sewage sample on the extraction of PCBs and organochlorine insecticides and found that in the 0.2–10 g l^{-1} solids concentration range, determinations of Aroclor 1260, p,p'-DDE, γ-HCH, and dieldrin were not subject to a relative standard and deviation of greater than 7.6 and that the F-test was highly significant for all four determinations (Table 141). Examination of the percentage recoveries (Table 142) shows that Aroclor 1260 and DDE both exhibited significant increases in percentage recovery from 40.5 g l^{-1} total solids to 1.0 g l^{-1} total solids, that of Aroclor 1260 increasing from 71.4% to 96.3%, DDE from 52.1% to 61.8%. γ-HCH, again, displayed a different trend, with the increase in percentage recovery over the range of total solids concentrations not being

Table 141 Variation in determinand concentration with total solids concentration in unspiked sewage sludge

Determinand	Solids concentration (g l^{-1})	F-test result	Mean* determinand concentration (μg l^{-1})	Relative s.d. (%)	Range (μg l^{-1})
Aroclor 1260	40.5		3.46a[1]	4.4	3.36–3.64
	10.0	0.01	3.16a	2.5	3.04–3.20
	1.0		3.45a	6.2	3.28–3.72
	0.2		2.42b	4.9	2.32–2.56
p,p'-DDE	40.5		1.28a	0.45	1.27–1.28
	10.0		1.16b	2.8	1.12–1.20
	1.0	0.01	1.31a	2.9	1.28–1.36
	0.2		1.00c	3.3	0.96–1.04
γ-HCH	40.5		7.14a	2.8	6.54–7.74
	10.0	0.01	7.86ab	2.9	7.60–8.10
	1.0		8.50bc	2.9	8.14–8.66
	0.2		9.18c	7.6	8.14–9.60
Dieldrin	40.5		8.34a	2.3	8.14–8.52
	10.0	0.01	9.08ab	3.4	8.80–9.34
	1.0		9.78b	3.9	9.34–10.26
	0.2		9.86b	4.7	9.20–10.26

*Of four replicates.
[1]Means not followed by a common letter are significantly different at the 0.05 significance level.
Reprinted with permission from McIntyre et al.[388] Copyright (1980) *Science and Technology Letters*.

significant due to the degree of scatter of the results, as was the case with dieldrin. Overall, it can be concluded that the degree of scatter, reflected in the relative standard deviations, was less for both Aroclor 1260 and p,p'-DDE than for γ-HCH and dieldrin and that the solids concentration significantly influenced the percentage recovery of Aroclor 1260 and p,p'-DDE from spiked samples, but not those of γ-HCH and dieldrin.

McIntyre et al.[388,389] conclude that the extraction of PCBs and organochlorine insecticides is most efficient at a total solids concentration of 1 g l^{-1}, using the extraction procedure described above. The recovery of p,p'-DDE from subsamples was always found to be the lowest of the four determinands considered (61.8%), whilst recoveries of Aroclor 1260, γ-HCH, and dieldrin from the diluted sample averaged 96.3, 89.4, and 82.9% respectively at 1 g l^{-1} total solids.

Various other workers[393-399] have reported methods for the determination of PCBs and organochlorine insecticides in sewage and sewage sludges.

Chlorinated insecticides/PCBs in sediments and oysters

A method has been described[400] for the separation of PCBs from DDT and other chlorinated pesticides, thereby permitting the qualitative and

Table 142 Variation in percentage recovery of polychlorinated biphenyls and organochlorine insecticides with total solids concentrations

Determinand	Solids concentrations (g l^{-1})	F-test result	Mean* percentage recovery (%)	Relative (%)	Range (%)
Aroclor 1260	40.5		71.4a^1	9.7	65.0–80.9
	10.0	1.0	78.6ab	26.9	60.2–101.0
	1.0		96.3b	3.2	92.2–99.4
	0.2		61.3a	6.8	57.6–66.7
p,p'-DDE	40.5		49.4a	10.3	50.6–56.5
	10.0	0.1	52.1a	5.7	59.9–67.0
	1.0		62.7b	4.9	68.5–93.9
	0.2		61.8b	3.9	44.6–56.6
γ-HCH	40.5		74.8a	12.2	63.2–85.4
	10.0	NS	87.1a	7.9	87.6–95.0
	1.0		85.5a	10.4	78.7–97.6
	0.2		89.4a	10.6	78.7–99.1
Dieldrin	40.5		76.0a	5.4	70.9–80.2
	10.0	NS	78.9a	13.3	69.4–88.1
	1.0		82.9a	9.4	74.3–92.8
	0.2		69.8a	9.4	60.6–75.4

*Of four replicate determinations.
[1]Means not followed by a common letter are significantly different at the 0.05 significance level.
Reprinted with permission from McIntyre et al.[388] Copyright (1980) Science and Technology Letters.

semiqualitative analyses of each of these chemical groupings. The procedure involves adsorption chromatography on alumina and charcoal, elution with increasing fractional amounts of hexane on alumina columns, and with acetone–diethyl ether and benzene on charcoal columns. The PCB and pesticides are then determined by gas chromatography on the separate eluates without interference. This method is described further under gas chromatography–mass spectrometry.

Chlorinated insecticides/PCBs in whales and dolphins

Gaskin et al.[401] have described a gas chromatographic method for determining DDT, dieldrin, and PCBs in the organs of whales and dolphins. Total DDT in blubbers ranged from 1.25 to 7.4 ppm, dieldrin in blubber from 0.007 to 0.04 ppm, and PCB in blubber from 0.69 to 5.0 ppm.

Gas chromatography–mass spectrometry

This technique has been applied to water pollution[403] control and analysis of sewage sludges[404] and river sediment and oyster samples.[405]

Erikson and Pellizzari[404] analysed municipal sewage samples in the USA by a gas chromatography–mass spectrometry–computer technique for chlorinated insecticides and PCBs.

Method

The samples (~300 g) were extracted at pH 11 six times with a total of 350 ml chloroform to remove neutral and basic compounds. The extract was dried with sodium sulphate, vacuum filtered, and concentrated to 2 ml using a Kuderna Danish apparatus. In cases where the sample background interfered significantly, an aliquot of the sample was chromatographed on a 1.0 × 30 cm silica gel column (Snyder and Reinert[402]). PCBs and related compounds were eluted with 50 ml hexane; pesticides and other compounds were eluted with 50 ml toluene.

Sample methylation

Acidic components of the sludge samples were treated with diazomethane and dimethyl suphate (Keith[406-408]). Analysis of all samples for PCBs was accomplished using a Finnigan 3300 quadrupole g.c.–m.s. with a PDP/12 computer. The 180 cm × 2 mm i.d. glass column, packed with 2% OV-101 on Chromosorb W, was held at 120 °C for 3 minutes, programmed to 230 °C at 12° min^{-1} and held isothermally until all peaks had eluted. Helium flow was 30 ml min^{-1}. The ionization voltage was nominally 70 eV and multiplier voltages were between 1.8 and 2.2 kV. Full scan spectra were obtained from m/e 100–500. Samples were analysed under the following temperature conditions: up to 150 °C for 3 minutes, programmed to 230 °C at 80 min^{-1} isothermally until all peaks had eluted. PCBs were quantitated by g.c.–m.s.–computer using the selected ion monitoring mode to provide maximum sensitivity and precision. This technique has been used in similar work on polychlorinated naphthalenes (Erikson et al.[409,410]). Ten ions were selected for monitoring: one from the parent cluster for each of the chlorinated biphenyls ($C_{12}H_9Cl$ through $C_{12}Cl_{10}$). PCBs were quantitated using anthracene as external standard and a previously determined relative molar response (anthracene parent ion mass 178; 27 ng ml^{-1}). Anthracene does not interfere with PCB determination nor do PCBs or their fragment ions interfere with the determination of anthracene.

The retention time results for 35 chlorinated compounds found in sewage sludge are given in Table 143. Not all compounds could be identified. A large number of spectra contained what appeared to be chlorine isotope clusters which are not reported. This could be due to interferences, very low levels, or spurious peaks. Although no structure could be assigned, the mass spectra indicated possible structures for three compounds in Table 143. The compound containing two chlorines with mol. wt. = 187 (RT = 2.3–2.7 min) may have the molecular formula $C_8H_7Cl_2$ and could be a dichlorodihydroindole or related compounds. Two distinct compounds were observed with four chlorines and mol. wt. = 240. These compounds appear to be isomers of tetrachlorostyrene ($C_8H_4Cl_4$). The 4,4′-dichlorobenzophenone identification was confirmed by comparison of the retention time with an authentic sample. The two peaks identified as DDE isomers are probably the two common isomers, o,p′-DDE and p,p′-DDE which generally are separable by gas chromatography.

Table 143 Summary of chlorinated compounds found in sewage sludge

Compound[a]	Retention time (min)[b]
Dichlorobenzene	0.5
Mol. wt. = 194, Cl_1	0.8
Mol. wt. = 222, Cl_1	0.9
Trichlorobenzene	1.0
Chloroaniline (tent.)	1.2
Dichloroaniline	2.1–4.7 (1.1)
Tetrachlorobenzene	2.2 (0.7)
Mol. wt. = 187, Cl_2	2.3–2.7 (0.6)
Mol. wt. = 171, Cl_2	3.0
Mol. wt. = 240, Cl_4[e]	3.0, 4.3 (1.1)
Trichloroaniline[d]	3.2, 4.4
Dichloronaphthalene	3.7
Trichlorophenol	3.7
Mol. wt. = 302, Cl_1	4.6
Mol. wt. = 210, Cl_3	5.1
Chlorobiphenyl	6.2
Dichlorobiphenol[f]	6.3–8.2
Trichlorobiphenyl[f]	7.5–9.6
Mol. wt. = 192, Cl_1	7.6
Mol. wt. = 288, Cl_1	8.6–11.1 (5.8)
Tetrachloronaphthalene	8.6
Mol. wt. = 218, Cl_1 (tent.)	9.1
Mol. wt. = 256, Cl_1	9.1
Dichlorobenzophenone	9.2 (7.2)
Mol. wt. = 269, Cl_1	9.9
Mol. wt. = 256, Cl_2	10.1
Pentachlorobiphenyl[f]	10.2–11.1
Mol. wt. = 288, Cl_3	10.5
Mol. wt. = 280, Cl_1	10.7
Mol. wt. = 241, Cl_1	10.8
Mol. wt. = 285, Cl_1	12.6
DDE[e]	12.7, 13.2 (8.6)
Mol. wt. = 356, Cl_2	12.8
Mol. wt. = 397, Cl_1	15.0

[a]Unidentified compounds are listed with the apparent molecular weight and number of chlorines. If the identification of a compound is tentative, it is denoted by (tent.).
[b]Retention times are listed for the chromatographic temperature conditions, 12 °C for 3 min, then 12° min^{-1} to 230 °C, then hold. Values in parentheses are for chromatographic temperature conditions, 150 °C for 3 min, then 8° min^{-1} to 230 °C, then hold.
[d]Differences in retention times possibly indicate different isomers.
[e]Two separate isomers observed in some samples.
[f]Several isomers observed.
Reprinted with permission from Erikson and Pellizzari.[404] Copyright (1979) Springer-Verlag N.Y.

Table 144 Quantitation of PCBs in hexane eluate of neutral extract of sewage sludge

Compound	Amount found ($\mu g\ ml^{-1}$)	Sludge concentration ($\mu g\ l^{-1}$)
$C_{12}H_9Cl$	8.5[a]	57
$C_{12}H_8Cl_2$	220[a]	1500
$C_{12}H_7Cl_3$	760[a]	5100
$C_{12}H_6Cl_4$	470[a]	3100
$C_{12}H_5Cl_5$	57	380
$C_{12}H_4Cl_6$[b]	76	510
$C_{12}H_3Cl_7$[b]	24	160
$C_{12}H_2Cl_8$		
$C_{12}HCl_9$		
$C_{12}Cl_{10}$[b]	14	93
Total PCBs	1600	10,800

[a] Average of two determinations.
[b] Not identified in mass spectra summarized in Table 143. Identification confirmed by comparison of the intensities of two or more ions in the parent cluster.
Reproduced with permission from Erikson and Pellizzari.[404] Copyright (1979) Springer-Verlag N.Y.

Examination of neutral extracts of sewage sludge by gas chromatography–mass spectrometry revealed the presence of appreciable amounts of PCBs (Table 144).

Teichman et al.[405] separated PCBs from chlorinated insecticides in sediments and oyster and soil samples using gas chromatography coupled to mass spectrometry. PCBs were separated from DDT and its analogues and from the other common chlorinated insecticides by adsorption chromatography on columns of alumina and charcoal. Elution from alumina columns with increasing fractional amounts of hexane first isolated dieldrin and heptachlor from a mixture of chlorinated insecticides and PCBs. The remaining fraction, when added to a charcoal column, could be separated into two fractions, one containing the chlorinated insecticides, the other containing the PCBs, by eluting with acetone–diethyl ether (25:75) and benzene, respectively. The PCBs and the insecticides were then determined by gas chromatography on the separate column eluates without cross-interference.

Teichman et al.[405] used a gas chromatograph (Aerograph 1200) containing a glass column (6 ft × 0.125 in.) packed with 4% SE-30, 6% SP-4201 on Chromosorb W (100–120 mesh). They also used an Aerograph 204 gas chromatograph containing a glass column (6 ft × 0.125 in.) with 4% SE-30, 6% QF-1 on Chromosorb W (80–100 mesh). The operating conditions were:

Model:	1200	204
Column temperature	180 °C	185 °C
Injector temperature	215 °C	200 °C
Detector temperature	200 °C	200 °C
Nitrogen gas flow rate	25 ml min^{-1}	30 ml min^{-1}

Both instruments contained an electron capture detector with a tritium foil source. For gas chromatography–mass spectrometry, a Varian 1400 gas chromatograph coupled to a Finnegan 3000 mass spectrometer was used. The 1400 was equipped with a glass column (6 ft × 2 mm i.d.) packed with 4% SE-30, 6% SP-4201 on Supelcoport (100–120 mesh). The operating conditions were: column temperature, 210 °C; transfer-line temperature, 250 °C, gas jet separator temperature, 255 °C, flow rate of helium gas, 12 ml min^{-1}; sensitivity, 10^{-7} A/V; electron multiplier voltage, 2.25 kV; electron ionization current, 6.95 eV.

A summation of the elution of the chlorinated organic insecticides and the PCBs from the alumina column is given in Table 145. Heptachlor epoxide and dieldrin were removed from the column by extending the elution solvent beyond the 30 ml volume with an additional, but separate, elution volume of 30 ml. The PCBs remained an integral part of the mixture containing the insecticides in the first 30 ml of eluate. The elution pattern of alumina column fraction 1

Table 145 Percentage recovery of insecticides eluted from neutral alumina

Compound	First fraction				Second fraction
	0–15 ml	15–20 ml	20–25 ml	25–30 ml	30–60 ml
Lindane			10	90	
Heptachlor	100				
Aldrin	100				
Heptachlor epoxide					100
p,p'-DDE	100				
Dieldrin					100
p,p'-DDD			50	50	
p,p'-DDT	100				
PCBs	100				
γ-Chlordane	10	80	10		
α-Chlordane	80	20			

Reprinted with permission from Teichman et al.[405] Copyright (1978) Elsevier Science Publishers B.V.

Table 146 Percentage recovery of insecticides eluted from charcoal

Compound	First fraction: 90 ml acetone–diethyl ether (25:75)			Second fraction: 60 ml benzene	
	0–30 ml	30–60 ml	60–90 ml	0–30 ml	30–60 ml
Lindane	30	40	30		
Heptachlor	100				
Aldrin	100				
p,p'-DDE	50	50			
p,p'-DDT	80	20			
PCBs				80	20
Chlordane	100				

Reprinted with permission from Teichman et al.[405] Copyright (1978) Elsevier Science Publishers B.V.

on the charcoal column, Table 146, shows that the insecticides were separated from the PCBs by means of acetone–diethyl ether eluant. The PCBs were subsequently removed from the charcoal column with benzene. Known amounts of insecticides and PCBs (Aroclor 1254) were added to soils and oyster samples; the samples were analysed as described above to check the efficiency of the analytical procedure. Recoveries of the added chemicals to the soils and the

Table 147 Recovery of insecticides and PCBs from fortified soil samples

Insecticide	Fortified (ppm)	Recovered (ppm)	Recovered (%)
Lindane	0.0013	0.0011	84.6
Heptachlor	0.027	0.0022	81.5
Aldrin	0.0027	0.0023	85.2
Heptachlor epoxide	0.0027	0.0025	92.6
p,p'-DDE	0.0027	0.0025	92.6
Dieldrin	0.0027	0.0026	96.3
p,p'-DDD	0.0027	0.0027	100
p,p'-DDT	0.0027	0.0027	100
γ-Chlordane	0.0033	0.0035	106
α-Chlordane	0.0033	0.0037	112
PCB (Aroclor 1254)	0.066	0.066	100

Reprinted with permission from Teichman et al.[405] Copyright (1978) Elsevier Science Publishers B.V.

Table 148 Recovery of insecticides and PCBs from fortified oysters

Insecticide	Fortified (ppm)	Recovered (ppm)	Recovery (%)
Heptachlor	0.0027	0.0019	70.4
Aldrin	0.0027	0.0021	77.8
Heptachlor epoxide	0.0027	0.0028	102
α-Chlordane	0.0033	0.0033	100
PCB (Aroclor 1254)	0.066	0.045	68.2

Reprinted with permission from Teichman et al.[405] Copyright (1978) Elsevier Science Publishers B.V.

Table 149 Limits of detectability of insecticides and PCBs using the described procedure under conditions

Insecticide	Detectability (pp10^9)
Lindane	0.04
Heptachlor	0.05
Aldrin	0.06
Heptachlor epoxide	0.10
p,p'-DDE	0.14
Dieldrin	0.14
p,p'-DDD	0.25
p,p'-DDT	0.33
γ-Chlordane	0.10
α-Chlordane	0.11
Aroclors 1254, 1260 (PCBs)	6.5

Reprinted with permission from Teichman et al.[405] Copyright (1978) Elsevier Science Publishers B.V.

oysters were consistent and acceptable (Tables 147 and 148). The limits of detectability of the chemicals examined (Table 149) refer to those obtained from pure solutions and they are also applicable to samples extracts.

Kaiser[411] used gas chromatography–mass spectrometry to identify mirex (dechlorane, $C_{10}H_{12}$) in fishes in Lake Ontario, Canada. Under standard gas chromatographic conditions the peak due to this substance is superimposed on that of the PCBs, and, as a result, the presence of mirex may have been unrecognized and it may therefore have been misinterpreted, as a PCB isomer, by previous workers. The fish samples were digested with sulphuric acid. The purified extracts were analysed for their PCB contents by two parallel means:

(1) quantitative determination of PCBs by gas chromatography with electron capture detectors[412,413] and
(2) qualitative investigation of the gas chromatographic peaks by computerized gas chromatography–mass spectrometry.[414] The sample and analytical data obtained are summarized in Table 150.

Table 150 Sample data for PCB and mirex residues from two fishes from the Bay of Quinte, Lake Ontario, Canada

	Weight (g)	PCBs as Aroclor (ppm)			Mirex (ppm)
		1242	1254	1260	
Northern longnose gar (*Lepistosteus osseus* (L.)) 902 g					
Gonads	32	2.09	1.18	0.44	0.020
Viscera, fat	62	3.14	1.95	0.90	0.041
Liver	17	3.68	2.31	1.08	0.047
Northern pike (*Esox lucius* (L.)) 2930 g					
Pectoral to pelvic fin	950	ND	ND	ND	0.025
Post-anal fin	280	0.89	1.01	0.48	0.050

ND, value not determined.

One of the PCB peaks was found with a mass spectrometric fragmentation pattern different from that of known PCB isomers. The base peak of this compound had a mass-to-charge ratio (m/e) of 272 with an isotope cluster centred on this peak, unambiguously indicating a $(C_5Cl_6)+$ moiety. Mass spectrometric fragmentations showing this cluster are derived from compounds containing a perchlorocyclopentadiene unit in their molecule structure or, for a very few cases, from similar, highly chlorinated hydrocarbons. Compounds of this kind include such insecticides as aldrin, chlordane, dieldrin, endrin, endosulfan, heptachlor, kepone, mirex, pentac, and toxaphene. The identification of the unknown in the fish samples as mirex was established by a combination of gas chromatographic and mass spectrometric techniques. For several gas chromatographic conditions the retention volumes of the compounds aldrin, chlordane, endrin, dieldrin, heptachlor, and toxaphene were considerably smaller than that of mirex. The retention volume of mirex was identical with that of the observed compound, and endosulfan cannot be chromatographed under these conditions. The differentiation between mirex, kepone, and pentac was achieved by combined gas chromatography–mass spectrometry with a computer-controlled system. Thus a constant retention time for the unknown compound was assured. The mass spectrometer was set to observe seven small mass ranges, five of which are relevant to mirex or kepone fragments, or both. The m/e ranges were as follows: 220–225, 235–241, 253–258, 270–278, 353–361, 451–463, and 505–517, with integration times for each atomic mass unit of 30 ms. Table 151 lists the

Table 151 Mass spectra of mirex observed in fish samples and of mirex and kepone standards

		Relative intensities	
m/e	Fish sample	Mirex	Kepont
235	50	28	24
236	7	4	3
237	77	55	38
238	8	6	4
239	49	30	25
240	4	4	3
271	15	10	9
272	54	52	50
273	8	5	5
274	100	100	100
275	10	7	7
276	80	76	80
277	6	5	6
278	34	36	35
353	2	3	3
355	9	9	8
356	2	2	6
357	5	4	11
358	2	1	2
359	6	5	12
360	3	1	2
361	4	3	7
451	3	1	4
453	2	1	3
455	0	0	2
457	0	0	7
459	0	0	8
461	0	0	6
463	0	0	4
507	1	2	2
509	1	3	1
511	1	4	0
513	1	2	0

Only major peaks with $m/e \geq 235$ are presented.

observed intensities together with those of mirex and kepone standards. For $m/e < 360$ the mass spectra of mirex and kepone are quite similar. Both compounds have base peaks of m/e 272 due to the $(C_5Cl_6)^+$ ion. Mirex, however, has a different fragmentation pattern at the high mass end which is substantially demonstrated by the mass spectrum of the fish samples: peaks due to kepone (m/e 451–463) are absent, and those due to mirex (m/e 505–517) are observed. The molecular ions of mirex (m/e 540–544) are of very low relative intensities and were not investigated. All gas chromatographic and gas chromatographic–mass spectrometric data for the observed compound were in good agreement with those of authentic mirex.

Kaiser[411] points out that the analytical conditions employed in many laboratories do not allow for a separation of or differentiation between PCBs and mirex. In fact, in the ordinary analytical procedure, the mirex peak is exactly superimposed on one of the major Aroclor 1260 peaks. Since with the use of the electron capture detector it is impossible to differentiate between a PCB and coeluting mirex, there is a strong likelihood that the presence of mirex in many environmental samples has not been recognized.

Markin et al.[415] have also discussed the possible confusion between mirex and PCBs in the analyses of crabs, shrimp, fish, and fish products. In their method the samples were thoroughly scrubbed to remove mud, algae, and other residues; they were ground whole and mixed in a Waring blender to make a composite sample. Samples were prepared and analysed on a whole body basis as received. A 20 g subsample of the composite was removed and analysed as follows. The homogenized sample was extracted with a mixture of hexane and isopropanol, and the extract subjected to a concentrated sulphuric acid clean-up. The sulphuric acid destroys dieldrin, endrin, and organophosphorus insecticides, but the improvement in sensitivity by this clean-up was considered more than adequate compensation for the loss of these other insecticides. The final extract was cleaned up on a Florisil column and concentrated to the desired level for analysis. If PCBs were suspected in the first analysis, their presence usually being indicated by a series of characteristic peaks, the sample was reprocessed to separate the PCBs from the insecticides as described by Armour and Burke,[416] Gaul and Cruz-LaGrange,[417] and Markin et al.[418] After concentrating to the appropriate volume, the extract from both methods of clean-up were chromatographed on a Hewlett-Packard Model 402 dual column gas chromatograph equipped with dual electron capture detection. Each sample was analysed on two different columns: the first column was a mixture of 1.5% OV-17 and 1.95% QF-1 on Gas Chrom Q. The temperatures of the injector, oven, and detector were 250, 200, and 210 °C respectively. The second column was 2% DC-200 on Gas Chrom Q with injector, oven, and detector temperatures of 245, 175, and 205 °C respectively. Argon–methane at 80 ml min^{-1} was the carrier gas. Level of detection was 0.001 ppm for DDT and its metabolites, 0.005 ppm for mirex and 0.01 ppm for Aroclor 1260.

Markin et al.[415] comment that they found mirex in only a minority of the samples they analysed, contrary to results obtained by earlier workers. All samples containing mirex were from around Savannah, Georgia, an area with a history of concentrated mirex use among the most extensive in the United States. The recovery of mirex in only 12% of the samples, all from one area, could indicate that mirex is not so general nor widespread a contaminant of seafood as are PCBs and DDT. This does not correspond to earlier seafood studies[419-421] which reported that mirex occurred much more frequently and densely in many of these same collection sites. Probably the reason for the discrepancy between their study and earlier studies is the confusion of Aroclor 1260 with mirex. The retention time for the last peaks of Aroclor 1260 is almost identical to the retention time for mirex on most columns routinely used for analysing mirex.[418] Unless extensive additional clean-up procedures are employed, such as those used by Markin et al.,[415] it is almost impossible to separate these two peaks. In their study, if the problem of PCB confusion had not been recognized and the special clean-up procedure used, the PCB peaks probably would have been reported as mirex. The possibility of confusing PCBs with the organochlorine insecticides was not generally recognized until 1970 and methods for distinguishing PCBs from mirex were not developed for general use until 1971. Since the three previous studies which reported mirex in seafood were all conducted before 1971, they probably did not recognize the problem and may have inadvertently reported the PCB Aroclor 1260 as mirex.

Derivatization gas chromatography

Luckas et al.[422] have described a method for determining PCBs and chlorinated insecticides in environmental samples by the simultaneous use of electron capture gas chromatography and derivatization gas chromatography. The method is based on the different stabilities of chlorinated insecticides and PCBs towards magnesium oxide in a microreactor. Extracts of samples are injected twice, first into a regular gas chromatograph and then into a gas chromatograph equipped with a microreactor for derivatization. A 'basic' chromatogram and a 'derivatization' chromatogram are obtained and the combination of the two chromatograms provided a satisfactory solution.

Chemical derivatization of sample extracts is very convenient. The extracts containing insecticides and PCBs, after the first injection into the gas chromatograph, are treated with derivatization reagents, the insecticides being converted into derivatives while the PCBs remain unchanged. Table 152 demonstrates the stability of chlorinated insecticides and PCBs towards reagents for chemical derivatization.

Luckas et al.,[422] as a result of these considerations, developed their microreactor gas chromatographic technique in which derivatization is carried out *in situ*. Preheated magnesium oxide affects the rapid quantitative dehydrochlorination of saturated DDT metabolites to the corresponding DDT olefins.[423] The derivatization products immediately obtained in the gaseous phase by means of the microreactor (with nitrogen as the carrier gas and magnesium oxide as the catalyst) are comparable with the products of chemical

Table 152 Stability of chlorinated insecticides and PCBs

Substance	Treatment with conc. H_2SO_4*	Treatment with ethanolic KOH*
Aldrin	+	+
Dieldrin	−	+
Endrin	−	+
Endosulfan	−	−
HCH isomers	+	−
PCBs	+	+
p,p'-DDT	+	→p,p'-DDE
o,p-DDT	+	→o,p-DDE
p,p'-DDE	+	+
o,p-DDE	+	+
p,p'-DDD	+	→p,p'-DDMU
o,p-DDD	+	→o,p-DDMU
p,p'-DDMU	+	+
o,p'-DDMU	+	+

+, unchanged; −, decomposed (products of decomposition are not detected); →, dehydrochlorination to the olefin.
Reprinted with permission from Luckas et al.[422] Copyright (1978) Elsevier Science Publishers B.V.

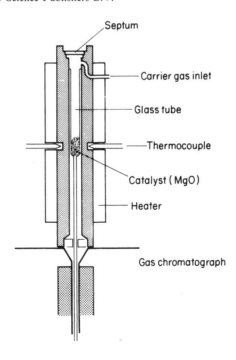

Figure 236 Microreactor for derivatization gas chromatography. Reprinted with permission from Luckas et al.[422] Copyright (1978) Elsevier Science Publishers B.V.

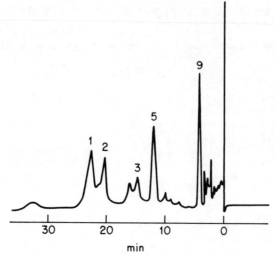

Figure 237 'Basic' chromatogram of an extract of a fish sample. G.c. conditions: 5% QF-1 on Gas Chrom Q, 100–120 mesh; glass column, 1.6 m × 3 mm i.d.; carrier gas, nitrogen at a flow rate of 60 ml min^{-1}; detector, ECD. Peaks: 1 = p,p'-DDT; 2 = p,p'-DDD; 3 = o,p-DDT; 5 = p,p'-DDE; 9 = γ-HCH. Reprinted with permission from Luckas et al.[422] Copyright (1978) Elsevier Science Publishers B.V.

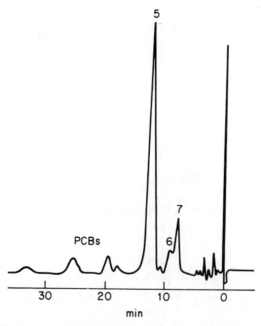

Figure 238 'Derivatization' gas chromatogram of an extract of a fish sample from Figure 237. G.c. conditions as in Figure 237. Conditions for derivatization: temperature of microreactor, 225 °C; catalyst, magnesium oxide. Peaks: 5 = p,p'-DDE; 6 = p,p'-DDMU; 7 = o,p-DDE. Reprinted with permission from Luckas et al.[422] Copyright (1978) Elsevier Science Publishers B.V.

derivatization with an alkali in the liquid phase, and substances that are stable to treatment with alkali are also not decomposed in the microreactor (Table 152). Two gas chromatographs with an all glass system and an electron capture detector were used. One chromatograph was equipped with a microreactor for the derivatization gas chromatography (Figure 236). Luckas et al.[422] used sample extracts obtained by digestion with perchloric acid—acetic acid extraction with n-hexane and clean-up with sulphuric acid.[425,426]

A 'basic' chromatogram of an extract of a fish sample is shown in Figure 237. The peaks of γ-HCH and DDT metabolites appear, but the background suffers from interference from peaks of PCBs. Figure 238 shows the 'derivatization' gas chromatogram of the same extract obtained by the procedure described above. After derivatization, the peaks of γ-HCH and the saturated DDT metabolites disappeared. The saturated DDT metabolites (p,p'-DDT, p,p'-DDD, and o,p-DDT) are converted quantitatively into the corresponding DDT olefins (p,p'-DDE, p,p'-DDMU, and o,p-DDE). The main peak in the 'derivatization' gas chromatogram represents the sum of p,p'-DDT and p,p'-DDE from the 'basic' chromatogram and is often sufficient for the determination of the total DDT content. The content of PCBs can be calculated in the 'derivatization' gas chromatogram without interference effects due to saturated DDT metabolites.

Rapid derivatization in the gaseous phase for the determination of DDT metabolites and PCBs was first carried out in 1973 by means of catalytic reduction (carbon skeleton chromatography with hydrogen as carrier gas). This method has the disadvantage that a flame ionization detector is used, which has insufficient sensitivity.[427,428]

Prescott and Cooke[429] have described the application of carbon skeleton gas chromatography to the analysis of environmental samples containing residues of organochlorine insecticides, PCBs and polychlorinated naphthalenes. Their results suggest that extraction by steam distillation followed by carbon skeleton gas chromatography with either a flame ionization or mass spectrometric detector is a practical method for the determination of organochlorine compounds in the environment. In this procedure, sulphur, nitrogen, oxygen, and halogens in the sample are replaced by hydrogen and the unsaturated bonds formed are saturated by hydrogenation over a palladium–platinum catalyst prior to passing the vapours into the gas chromatograph.

A gas chromatograph fitted with dual flame ionization detectors was employed. The catalyst was packed into the part of the column that passed through the injection port heater and was thus maintained at the required temperature. Hydrogen was used as the carrier gas. Products were identified by a mass spectrometer linked to the gas chromatograph. At low temperatures hydrogenation of the aromatic rings tended to occur. As the temperature of the catalyst was increased from 140 to 305 °C there was a progressive decrease in the formation of bicyclohexyl and phenylcyclohexyl and an increase in the yield of biphenyl. At 305 °C biphenyl was the only product from Aroclors. It is likely that a low temperatures loss of chlorine is followed by or coupled with hydrogenation of the aromatic rings. At higher temperatures a secondary reaction involving the dehydrogenation of the cyclic system is also present.

Dechloration of polychloronaphthalenes was much easier. At catalyst temperatures less than 280 °C hydrogenation of the rings was less pronounced although gas chromatographic–mass spectrometric studies indicated that some tetrahydronaphthalene was present. At 305 °C polychloronaphthalenes were quantitatively converted into naphthalene. Using this technique, polychlorinated terphenyls were converted into a mixture of o-, m-, and p-terphenyl at 305 °C. Using 5% platinum catalyst, conversion of polychloronaphthalenes was poor. At 205 °C naphthalene gave two compounds, which, from mass spectrometric studies, were suggested to be tetra- and decahydronaphthalene. As the temperature was increased the peak heights decreased and at 305 °C no peaks remained; presumably the naphthalene skeleton was completely destroyed at this temperature. The results for polychloronaphthalenes were similar to those for naphthalene. PCBs were also completely destroyed at 305 °C. At 280 °C small amounts of bicyclohexyl, phenylcyclohexyl, and biphenyl were eluted. As the temperature was decreased the amount of bicyclohexyl increased. A 180 °C conversion into bicyclohexyl was quantitative, and polychloronaphthalenes were converted only into decahydronaphthalene. Initially a 5% SE-30 column was used by Prescott and Cooke[429] to separate the biphenyl and naphthalene after catalysis. Later gas–solid chromatography with a rubidium chloride column was employed, as inorganic salts give excellent separations of hydrocarbons and also very good reproducibility over a long period of time. Prescott and Cooke[429] used this technique to determine PCBs and PCNs (polychlorinated terphenyls) in sediment samples taken in an estuary. The samples were extracted by blending with doubly distilled water and then steam distilling for 2 h. The contaminants were extracted into 20 ml of hexane. After steam distillation the sample was removed and dried to a constant mass. The unit was washed with hexane and the combined extract and washings were concentrated to 2 ml. For carbon skeleton gas chromatography a 3% palladium catalyst was used together with a 2% rubidium chloride column. Temperature programming was used to obtain good separation. Among the compounds identified were a range of polynuclear aromatic hydrocarbons including naphthalene, methylnaphthalene, biphenyl, dimethylnaphthalene, phenanthrene, anthracene, methylphenanthrene, fluoranthene, pyrene, and chrysene. Only trace amounts of diphenylethane, formed by dechlorination of DDT and related compounds, were found. This suggests that only trace amounts of these particular compounds were present. Biphenyl and naphthalene were found in the samples before catalysis and so biphenyl and naphthalene levels before and after catalysis were measured. Additional biphenyl or naphthalene was assumed to arise from dechlorination of PCBs or PCNs.

Gosink[430] and Aue[431] have also studied the determination of organochlorine insecticides and PCBs in water samples.

Column chromatography

The determination of chlorinated insecticides and polychlorinated biphenyl contaminants in the environment is usually carried out by application of gas chromatography. Gas chromatography has replaced thin-layer chromatography, which is not suitable for the detection of picogram amounts and high performance liquid chromatography has been used, although earlier work indicated that it had inadequate sensitivity and was rather time-consuming.[432-437]

Leoni[438] separated 50 organochlorine insecticides and PCBs into four groups by silica gel microcolumn chromatography. The separation was undertaken to simplify the chromatograms obtained by gas chromatography from contaminated samples. The sample (10–15 litres of surface water or 20–25 litres of potable water) was acidified with hydrochloric acid to pH 1–3, and subjected to continuous extraction, twice with light petroleum (boiling range 40–60 °C and once with benzene. The benzene extract was evaporated and the residue was dissolved in light petroleum and added to the other extracts. The combined extracts were concentrated to 15 ml and were partially purified by extraction with acetonitrile saturated with light petroleum (4×30 ml). The extract was diluted with 2% sodium chloride solution (700 ml) and the solution was again extracted with light petroleum (2×100 ml). This extract was evaporated and the residue was dissolved in 1 ml of hexane. This solution followed by 1 ml of hexane used for rinsing the container, was applied to a column (100 mm $\times$ 4.2 mm) of silica gel (Grace 950, 60–200 mesh, dried at 130 °C for 2 h then deactivated with 5% of water). The column was then percolated (at 1 ml min^{-1}) in turn with hexane (20 ml), benzene–hexane (3:2) (8 ml), benzene (8 ml) and ethyl acetate–benzene (1:1) (14 ml). Each eluate was evaporated and the residue was dissolved in 1 ml of hexane and subjected to gas chromatography on OV-17 as stationary phase with electron capture detection. Quantitative recovery was achieved for all insecticides except malathion, disulfoton, dimethoate, and phorate.

Sackmauer et al.[439] used columns filled with silicic acid–Celite to separate organochlorine insecticides from PCBs. The PCBs were eluted with petroleum ether. To elute insecticides from the column they used a mixture of acetonitrile and hexane and methylene chloride.

Preconcentration of chlorinated insecticides/ PCBs from water prior to analysis

Amberlite XAD-2 resin,[440-443] cellulose triacetate membrane filters,[444] and Tenax polymer[445] have all been used to obtain concentrated extracts from water samples.

Leoni et al.[445,446] observed that in the extraction of organochlorine insecticides and PCBs from surface and coastal waters in the presence of other pollutants such as oil, surface active substance etc., the results obtained with an absorption column of Tenax–Celite (a porous polymer; trade mark registered by Enka NV; developed by AKZO Research Labs, Arnhem, Netherlands) are equivalent to those obtained with the continuous liquid–liquid extraction technique. For natural waters that contain solids in suspension that absorb pesticides, it may be necessary to filter the water before extraction with Tenax and then to extract the suspended solids separately. Analyses of river and estuarine sea waters, filtered before extraction, showed the effectiveness of Tenax, and the extracts obtained for the pesticide analysis prove to be much less contaminated by interfering substances than the corresponding extracts obtained by the liquid–liquid technique. Leoni et al.[446] showed that for the extraction of organic micropollutants such as pesticides and aromatic polycyclic hydrocarbons from waters, the recoveries of these substances from unpolluted waters (mineral and drinking waters) when added at the level of 1 ppb averaged

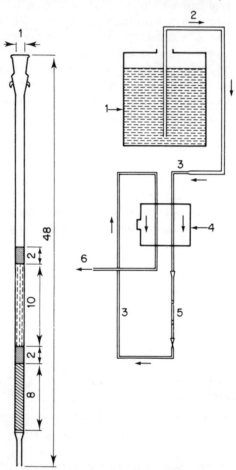

Figure 239 Equipment used for the extraction of micropollutants. 1, glass container; 2, glass tubes; 3, silicone rubber hose; 4, peristaltic pump; 5, glass absorption column; 6, extracted water discharge. Left: adsorption column (dimensions in cm). From bottom to top: Tenax, glass pellets, mixture of Celite and plugs of silanized glass wool, glass pellets. Reprinted with permission from Leoni et al.[446] Copyright (1975) Elsevier Science Publishers B.V.

90%. The equipment used for the extraction experiments is illustrated in Figure 239.

Water samples were passed through the peristaltic pump into the absorption column at a flow rate of about $3 \, l \, h^{-1}$. When the absorption was completed, the pesticides were eluted with three 10 ml volumes of diethyl ether, in such a way that the solvent also passed through the section of hose through which the water reached the column. Finally, the diethyl ether was dried over anhydrous sodium sulphate. The water container was washed with light petroleum to remove pesticide adsorbed on the glass and this solution, after concentration, was added to the column elutate. For the analysis of naturally polluted water, the mixed diethyl ether and light petroleum extract was evaporated, the residue dissolved on light petroleum and the solution purified by partitioning with acetonitrile saturated with light petroleum.[447,448] The resulting solution was evaporated just to dryness, the residue dissolved in 1 ml of n-hexane and insecticides and polychlorobiphenyls were separated into four fractions by deactivated silica gel microcolumn chromatography[447] (silica gel type Grace 950, 60–200 mesh).[449] The various eluates from the silica gel were then analysed by gas chromatography.[450] In order to evaluate the effectiveness of extraction from natural waters with the Tenax–Celite column, the samples were also extracted simultaneously by the liquid–liquid technique. Table 153 indicates the tests effected, adding pesticides and other pollutants to a mineral water (pH 6.8) that had been shown to be pesticide free prior to the analyses. Table 154

Table 153 Summary of trials on the Tenax extraction of pesticides from waters artificially polluted with pesticides (1 ppb) and other substances*

Trial no.	Volume of extracted water (l)	Pollutant	Adsorption column	Pesticides examined organo-chlorine	organo-phosphorus
1	10	0.1 ppm of Arkopal N-100	Tenax	+	+
2	10	0.1 ppm of alkylbenzene-sulphonate (Na salt)		+	+
3	10	3% NaCl (marine salt)		+	+
4	10	0.1 ppm of alkylbenzene-sulphonate + 1 ppm of mineral oil		+	+
5	10	1 ppm of mineral oil + 2 ppm of Tween 80		−	+
6	8			−	+
7a	14		Tenax–Celite	−	+
7b	10	1 ppm of mineral oil + 4 ppm of Tween 80		+	+
7c	10			+	+
7d	10			+	+
7e	10			−	+

*Standards mixtures of pesticides prepared in acetone were added to 8–15 l of water at a level of 100 μl of standard solution per litre of water, so that the concentration of the various pesticides was 1 ppb ($1 \, \mu g \, l^{-1}$).
Reprinted with permission from Leoni et al.[446] Copyright (1975) Elsevier Science Publishers B.V.

Table 154 Percentage recovery of insecticides added at the 1 ppb level from mineral or drinking waters artificially contaminated in the laboratory with oil, surfactants, etc.

Insecticides and trials	Adsorption with Tenax alone								Adsorption with Tenax–Celite							
Trial no.	1	2	3	4	5	6*	6†	7a	7b*	7b†	7c*	7c†	7d*	7d†	7e*	7e†
Hexachlorobenzene	93.5	82.5	94.7	102.9	‡				92.0	Abs	100.6	Abs	88.8–85.9	Abs	Abs	
Dieldrin	85.9	95.6	98.9	89.4					→	→	→	→	→	→		
Heptachlor	82.2	79.7	85.2	96.0												
o,p′-DDE	73.2	83.9	77.7	101.3					77.8	Abs	87.0	Abs	86.5	Abs		
o,p′-DDT	75.6	83.2	82.9	111.8					→	→	→	→	→	Abs		
p,p′-DDT	75.9	89.2	83.9	106.9					93.4	Abs	95.4	Abs	88.1	Abs		
γ-BHC													→	Abs		
β-BHC				87.7								85.8		Abs	Abs	
Ronnel	93.4	85.5	98.9	82.9									86.3–79.3	Abs	Abs	
Dursban					98.1	88.6	Abs	89.6							88.7	Abs
Malathion					12.0	20.1	64.9	78.0	50.0	33.3	32.8	38.1	32.6	30.7	82.2	20.0
Parathion	96.5	87.2	96.6	51.2	66.3	74.5	13.8	85.0							97.0	Abs
Parathion-methyl					27.3	39.9	43.0	81.0			59.1	27.2	71.4	23.2	91.9	Abs
Trithion	76.6	68.4	87.5												94.2	Abs
Ethion															90.5	Abs

*Percentage recovery of the column, hose, and container.
†Percentage recovery in the filtered water on the Tenax column.
‡Not determined.
Abs = absent.
Reprinted with permission from Leoni et al.[446] Copyright (1975) Elsevier Science Publishers B.V.

shows the results obtained. In the adsorption with Tenax alone, the first three trials gave satisfactory results, while in the presence of mineral oil (trials 4, 5, and 6) a considerable proportion of the organophosphorus pesticides (particularly malathion and parathion-methyl) was not adsorbed and was recovered in the filtered water. Test 7 (7a–7e) showed that this drawback can be overcome by placing, ahead of the layer of Tenax in the adsorption column, a layer of Celite 545 which, in order to prevent blocking of the column, is mixed with silanized glass wool plugs (Figure 239). A number of analyses of surface and estuarine sea waters were carried out by this modified Tenax column and simultaneously by the liquid–liquid extraction technique. To some of the samples taken, standard mixtures of pesticides were also added, each at the level of 1 ppb (i.e. in concentration from 13 to 500 times higher than that usually found in the waters analysed). One recovery trial also specifically concerned polychlorobiphenyls. The scheme of these tests is shown in Table 155 and the results obtained are given in Tables 156 and 157. Table 156 (trials 8, 9, and 11) shows that the two extraction methods, when applied to surface waters that were not filtered before extraction, yielded very similar results for many insecticides, with the exception of compounds of the DDT series, for which discordant results were frequently obtained (see the differences between trials 8a and b and 11a and b). Similarly, as shown in Table 157 (trial 11c), when the standard mixture of insecticides was added to a non-filtered water, i.e. containing suspended solids, the recoveries of active substances such as DDT and malathion were unsatisfactory. Some insecticides that are present in waters

Table 155 Summary of trials of Tenax and liquid–liquid extraction of insecticides from natural polluted water

Trial no.	sample	Trials effected
8	Tiber river water (30 l)	8a: 15 l, Tenax–Celite extraction
		8b: 15 l, liquid–liquid extraction
9	Sea-coast water (20 l)	9a: 10 l, Tenax–Celite extraction
		9b: 10 l, liquid–liquid extraction
10	Sea-coast water (30 l)	10a: 10 l, Tenax–Celite extraction
		10b: 10 l, liquid–liquid extraction
		10c: 10 l + 1.6 ppb standard of PCB* and Tenax–Celite extraction
11	Tiber river water (30 l)	11a: 10 l, Tenax–Celite extraction
		11b: 10 l, liquid–liquid extraction
		11c: 10 l + 1 ppb standard of pesticides and Tenax–Celite extraction
12	Tiber river water (30 l) filtered before extraction	12a: 10 l, Tenax–Celite extraction
		12b: 10 l, liquid–liquid extraction
		12c: 10 l + 1 ppb standard of pesticides and Tenax–Celite extraction

*PCB added as standard mixture of fenclor 42, 54, and 60 equivalent to 16 ppb of decachlorobiphenyl in total.
Reprinted with permission from Leoni et al.[446] Copyright (1975) Elsevier Science Publishers B.V.

Table 156 Detection of insecticides in surface water samples extracted by adsorption on Tenax–Celite (series a) and by the liquid–liquid technique (series b). Results expressed in ppt (ng l^{-1})

Insecticide identified	Trial no.							
	8a*	8b*	9a*	9b*	11a*	11b*	12b†	12b†
Hexachlorobenzene	5.4	5.8	3.3	3.8	2.6		8.1	6.6
Dieldrin	5.8	6.4			15.6	8.2	14.6	14.6
Heptachlor	1.5	1.7			0.9	0.9	7.1	2.7
p,p'-DDE	2.2	3.5			2.8	2.5	10.6	5.2
o,p'-DDT	5.0	12.2			13.0	12.3	24.8	19.8
p,p'-DDT	8.8	18.8	9.0	15.7	32.2	24.7	37.2	35.5
γ-BHC					5.2	4.0		
β-BHC	37.2	43.9	6.0	5.7	75.8	77.8	58.4	67.5
γ-BHC	9.7	13.6			19.4	17.3	10.5	15.5
Ronnel	18.9	21.7			2.1	2.1		
Dursban	27.5	29.4			41.2	43.2		
Diazinon	37.4	37.4					19.7	22.3
Malathion	26.5	31.7			27.0	31.0	27.3	27.3
Parathion					37.5	39.3		
Parathion-methyl	32.6	38.2			21.0	26.0		
Total (ppt)	218.6	263.5	18.3	25.2	296.3	289.1	218.3	217.0

*Waters not filtered before extraction.
†Waters filtered through paper before extraction.
Reprinted with permission from Leoni et al.[446] Copyright (1975) Elsevier Science Publishers B.V.

Table 157 Recovery test on extraction with Tenax–Celite of insecticides and polychlorobiphenyls added to surface waters at levels of 1.0 and 1.6 ppb respectively

Insecticide additives	Trial		
	10c*	11c*	12c†
PCB	100.2		
Hexachlorobenzene (HCB)		84.8	83.9
o,p'-DDT		38.0	81.8
p,p'-DDT		47.0	93.1
β-BHC		70.0	73.0
γ-BHC		84.9	92.0
Dieldrin		75.4	92.7
Parathion-methyl		82.3	101.0
Malathion		25.0	93.3

*Water not filtered before addition of standards and extraction.
†Water filtered through paper before addition of standards and extraction.
All results for corresponding non-treated samples are corrected. Results given are percentage recoveries.
Reprinted with permission from Leoni et al.[446] Copyright (1975) Elsevier Science Publishers B.V.

are partially adsorbed by suspended or sedimentable solids, therefore, such adsorbed amounts cannot be extracted by filtration procedures on polymers while, at least to a great extent, they can be extracted by procedures in which the water is extracted with solvents. Trials 12a and b (Table 156) show that when the two extraction techniques are applied to filtered water, almost identical results were obtained, and trial 12c (Table 157) shows that with Tenax extraction the recoveries of insecticides added to a filtered surface water before extraction were satisfactory. During the tests it was observed that amounts of up to 30% of some insecticides can be adsorbed by the walls of the hose and the container, so that at the end of the extraction it is essential to wash these surfaces with solvents.

Leoni et al.[445,446] conclude that the extraction of insecticides from waters by adsorption on Tenax, yields results equivalent to those by the liquid-liquid procedure when applied to mineral, drinking, and surface waters that completely or almost completely lack solid matter in suspension. For waters that contain suspended solids that can adsorb some insecticides in considerable amounts, the results of the two methods are equivalent only if the water has previously been filtered. In these instances, therefore, the analysis will involve filtered water as well as the residue of filtration. Compared with liquid-liquid extraction, the main advantages of Tenax are the considerable amount of time saved, the possible automation of the process and that gas chromatographic analysis shows the 'extracts' obtained with Tenax to be less contaminated by interfering substances. Another advantage of Tenax is that the product can be used 'as received', without preliminary treatment.

Voltammetric methods

Farwell et al.[451] used voltammetry to identify organochlorine insecticides, PCBs, polychlorinated naphthalenes, and polychlorinated benzenes. They list tables of reduction potentials for over 100 organochlorine compounds.

Polychlorinated terphenyls

The occurrence of these substances has been reported in river water, oysters, eels, and human fat.[452] The samples were extracted with hexane. The extracts were cleaned by column chromatography, either on alumina or Florisil. The cell extracts contained much oily residue and were cleaned over both Florisil and alumina. The analysis of polychlorinated terphenyls was carried out with a combination of a mass spectrometer and a gas chromatograph used in the mass fragmentography mode. Two m/e values were selected ($m/e = 436$ and $m/e = 470$) and from all samples mass fragmentograms were taken on these m/e values (see Figures 240 and 241). The gas chromatographic conditions were: column length 3 ft, ⅛ in. o.d. (Pyrex), filled 3% OV-17 on Varaport 30. Temperature programming was used: $6\,°C\,min^{-1}$, 200–285 °C for $m/e = 436$ and $8\,°C\,min^{-1}$, 200–285 °C for $m/e = 470$. The gas chromatographic conditions for the PCBs were: same column, temperature programming

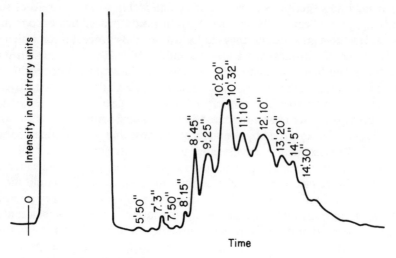

Figure 240 Mass fragmentogram on $m/e = 436$ of a standard of the polychlorinated terphenyl Clophen Harz (W). Reprinted with permission from Freudenthal and Greeve.[452] Copyright (1973) Springer Verlag, N.Y.

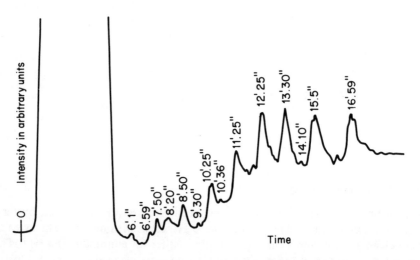

Figure 241 Mass fragmentogram on $m/e = 436$ of a sample of oyster. Reprinted with permission from Freudenthal and Greeve.[452] Copyright (1973) Springer Verlag, N.Y.

4 °C min^{-1}, 125–200 °C for $m/e = 256$; 6 °C min^{-1}, 125–200 °C for $m/e = 290$, and 6 °C min^{-1}, 150–250 °C for $m/e = 324$. Approximately 0.1 ppm of polychlorinated terphenyl was found in river water, 0.15 ppb in oyster tissue, 0.5 in eel tissue and up to 1 ppb in human fat, the corresponding figures for PCBs were 0.1, 0.2, 5, and 0.5–1 ppb.

Polybrominated biphenyls

Polybrominated biphenyls have been detected in environmental samples by a procedure involving ultraviolet irradiation followed by gas chromatography of the photodegradation products.[453-455]

REFERENCES

1. US Environmental Protection Agency, *Report No. EPA-600-4-74*, 1974T 108.
2. Wilson, A. J., Forester, J. and Knight, J. *US Fish Wildl. Serv. Circ.* 355 18–20 (1969). Centre for Estuaries and Research Gulf Breese, Florida. U.S.A.
3. US Environmental Protection Agency Analytical Quality Control Laboratory, US Government Printing Office, Washington DC, 63 pp. (24203) (1972).
4. Wilson, A. J. and Forester, J. *J. Ass. Off. Anal. Chem.*, **54**, 525 (1971).
5. Mills, P. A., Caley, J. F. and Grither, R. A. *J. Ass. Anal. Chem.*, **46**, 106 (1963).
6. Deubert, K. H. *Bull. Environ. Contam. Toxicol.*, **5**, 379(1970).
7. Croll, B. T. *Analyst (Lond.)*, **96**, 810 (1971).
8. Kahanovitch, Y. and Lahav, N. *Environ. Sci. Technol.*, **8**, 762 (1974).
9. Law, L. M. and Goerlitz, D. F. *J. Am. Oil Chem. Ass.*, **53**, 1276 (1969).
10. *A Guide to Marine Pollution*, Chs 1, 2, and 4 (E. D. Goldberg, editor). Gordon and Breach, New York, N.Y. (1972).
11. Braus, H., Middleton, F. M. and Watton, G. *Anal. Chem.*, **23**, 1160 (1951).
12. Middleton, F. M., Grant, W. and Rosen, A. A. *Ind. Engng Chem.*, **48**, 268 (1956).
13. Ludzok, F. J., Middleton, F. M. and Ettinger, E. B. *Sew. Ind. Wastes*, **30**, 662 (1958).
14. Polange, R. C. and Megregian, S. *J. AWWA*, **50**, 1214 (1958).
15. Middleton, F. M. and Lichtenberg, J. J. *Ind. Engng Chem.*, **52**, 99A (1960).
16. *Standard Methods for the Examination of Water and Waste Water*, APHA, AWWA WPCF, Washington DC (13th edn, 1971) p. 103.
17. Sproul, O. J. and Ryckman, D. W. *WPCF*, **33**, 1188 (1961).
18. Rosen, A. A. and Middleton, F. M. *Anal. Chem.*, **31**, 1729 (1959).
19. Booth, R. L., English, J. N. and McDermott, G. N. *J. Am. Wat. Wks Ass.*, **57**, 215 (1965).
20. Eichelberger, A. and Lichtenberg, J. J. *J. Am. Wat. Wks Ass.*, **25**, 63 (1971).
21. Eichelberger, J. W. and Lichtenberg, J. L. *J. Am. Wat. Wks Ass.*, **63**, 25 (1971).
22. Aue, W. A., Kapila, S. and Hastings, C. R. *J. Chromat.*, **73**, 99 (1972).
23. Burnham, A. K. *Anal. Chem.*, **44**, 139 (1972).
24. Ahling, B. and Jenson, S. *Anal. Chem.*, **42**, 1483 (1970).
25. Kahn, L. and Wayman, C. H. *Anal. Chem.*, **36**, 1340 (1964).
26. Goldberg, M. C., Delong, L. and Sinclair, M. *Anal. Chem.*, **45**, 89 (1973).
27. Kawahara, F. K. *et al.*, *WPCF*, **39**, 572 (1967).
28. *Methods for Organic Pesticides in Water and Wastewater*. 1972-759-31/2113, Region 5-II, EPA, US Govt Prntg Office, Washington, DC (1971).
29. Harvey, G. R. Absorption of chlorinated hydrocarbon from seawater by a crosslinked polymer. Woods Hole Oceanographic Institute, Woods Hole, Massachusetts. Unpublished manuscript (1972).
30. Brodtmann, N. V. *J. Am. Wat. Wks Ass.*, **67**, 558 (1975).
31. Breidenbach, A. W. *The Identification and Measurement of Chlorinated/Hydrocarbon Pesticides in Surface Water*. 1968-0-315-842, US Govt Prntg Office, Washington, DC.
32. Aspila, K. I., Carron, J. M. and Chau, A. S. Y. *J. Ass. Anal. Chem.*, **60**, 1097 (1977).

33. *Analytical Methods Manual* (1974). Inland Waters Directorate, Water Quality Branch, Ottawa, Ontario, Canada.
34. Miles, J. R. W., Tu, C. M. and Harris, O. R. *J. Econ. Entomol.*, **62**, 1334 (1969).
35. Chau, A. S. Y., Rosen, J. D. and Cochrane, W. P. *Bull. Environ. Contam. Toxicol.*, **6**, 225 (1971).
36. Chau, A. S. Y. *J. Am. Oil Colours Ass.*, **57**, 585 (1974).
37. Thompson, J. F. (Ed.). *Analysis of Pesticide Residues in Human and Environmental Samples*, US Environmental Protection Agency, Research Triangle Park, NC, sec. 10A 12/2/74, p. 9 (1974).
38. Eichelberger, J. W. and Lichtenberg, J. L. *Environ. Sci. Technol.*, **5**, 541 (1971).
39. McFarren, E. F., Lishka, R. J. and Parker, J. H. *Anal. Chem.*, **42**, 358 (1970).
40. Weil, L. *Gas-u- WassFach*, **113**, 64 (1972).
41. Weil, L. *Anal. Abstr.*, **24**, 1259 (1973).
42. Beyermann, K. and Eckrich, W. *Z. Anal. Chem.*, **265**, 1 (1974).
43. Mangani, F., Cresentini, G. and Bruner, F. *Anal. Chem.*, **53**, 1627 (1981).
44. Brodtmann, N. V. *Bull. Environ. Contam. Toxicol.*, **15**, 33 (1976).
45. Weil, L. and Ernst, K. E. *Gas-u- WassFach*, **112**, 184 (1971).
46. Engst, R. and Knoll, R. *Nahrung*, **17**, 837 (1973).
47. Thompson, J. F., Reid, S. J. and Kantor, E. J. *Arch. Environ. Contam. Toxicol.*, **6**, 143 (1977).
48. Sherma, J. and Shafik, T. M. *Arch. Environ. Contam. Toxicol.*, **3**, 55 (1975).
49. Pionke, H. B., Konrad, J. G., Chesters, J. G. and Armstrong, D. E. *Analyst (Lond.)*, **93**, 363 (1968).
50. Kadoum, A. M. *Bull. Environ. Contam. Toxicol.*, **3**, 65 (1968).
51. Kadoum, A. M. *Bull. Environ. Contam. Toxicol.*, **3**, 247 (1968).
52. Konrad, J. G., Pionke, H. B. and Chesters, G. *Analyst (Lond.)*, **90**, 490 (1969).
53. Askew, J., Ruzicka, J. H. and Wheals, B. B. *Analyst (Lond.)*, **94**, 275 (1969).
54. Johnson, L. C. *Bull. Environ. Contam. Toxicol.*, **5**, 542 (1970).
55. Law, L. M. and Goerlitz, D. F. *J. Ass. Off. Anal. Chem.*, **53**, 1276 (1970).
56. Herzel, F. *Arch. Hyg. Bakteriol*, **154**, 18 (1970).
57. Ahling, B. and Jenson, S. *Anal. Chem.*, **42**, 1483 (1970).
58. Ballinger, D. G. *Methods for Organic Pesticides in Water and Waste water*. US Govt Printing Office, Washington DC, 58 pp. (1971).
59. Ripley, B. D., Wilkinson, R. J. and Chau, A. S. Y. *J. Ass. Off. Anal. Chem.*, **57**, 1033 (1974).
60. ASTM. *Tentative Method of Test for Organochlorine Pesticides in Water*. Designation D3086-72T (1972).
61. Sackmauerevá, M., Pal'usova, O. and Szokolay, A. *Wat. Res.*, **11**, 551 (1977).
62. Sackmauerevá, M., Pal'usova, O. and Hluchan, E. *Vod. Hospod*, **10**, 267 (1972).
63. Szokolay, A., Uhnak, J. and Madaric, A. *Chem. zvesti*, **25**, 453 (1971).
64. Janak, J., Sackmauerevá, M., Szokolay, A. and Madaric, A. *Chem. zvesti*, **27**, 128 (1973).
65. Janak, J., Sackmauerevá, M., Szokolay, A. and Pal'usova, O. *J. Chromat.*, **91**, 545 (1974).
66. Mahel'ova, H., Sackmauerevá, M., Szokolay, A. and Kovac, J. *J. Chromat.*, **89**, 177 (1974).
67. Szokolay, A., Madaric, A., Uhnak, J. and Sackmauerevá, M. *Acta Hyg.*, 2415 (1970).
68. Szokolay, A. and Madaric, A. *J. Chromat.*, **42**, 509 (1969).
69. Mills, P. A. *J. Ass. Off. Agric. Chem.*, **42**, 734 (1959).
70. Mills, P. A. *J. Ass. Off. Agric. Chem.*, **51**, 29 (1968).
71. Burke, J. A. and Malone, B. *J. Ass. Off. Anal. Chem.*, **49**, 1003 (1960).
72. Suzuki, M., Yamato, Y. and Wanatabe, T. *Environ. Sci. Technol.*, **11**, 1109 (1977).

73. Suzuki, M., Yamato, Y. and Wanatabe, T. *Nippon Nogei Kagaku Kaishi*, **47**, 1 (1973).
74. *Chromopack News*, June 1979 No. 20.
75. American Public Health Association. *Standard Methods for the Examination of Water and Wastewater*, 15th Edition. Method 509A p. 493.
76. Method S73. Supplement to the 15th edition of *Standard Methods for the Examination of Water and Wastewater: Selected Analytical Methods* approved and cited by the United States Environmental Protection Agency. Published by American Public Health Association, American Water Works Association, Water Pollution Control Federation.
77. Lamar, W. L., Goerlitz, D. F. and Law, L. M. *Identification and Measurement of Chlorinated Organic Pesticides in Water by Electron-capture Gas Chromatography*, US Geol. Survey Water-Supply Paper 1817-B, 12 pp. (1965).
78. Lamar, W. L., Goerlitz, D. F. and Law, L. M. Determination of organic insecticides in water by electron-capture gas chromatography, in *Organic Pesticides in the Environment:* Am. Chem. Soc., *Advances in Chemistry*, ser. **60**, 187–199 (1966).
79. (US) Federal Water Pollution Control Administration. *FWPCA Method for Chlorinated Hydrocarbon Pesticides in Water and Wastewater* Cincinnati, Federal Water Pollution Control Adm., 29 pp. (1969).
80. Erney, D. R. *Anal. Lett.*, **12**, 501 (1979).
81. Chau, A. S. Y. *J. Ass. Off. Anal. Chem.*, **55**, 519 (1972).
82. Hesselberg, R. J. and Johnson, J. L. *Bull. Environ. Contam. Toxicol.*, **7**, 115 (1972).
83. LeRoy, M. L. and Goerlitz, D. F. *J. Ass. Off. Anal. Chem.*, **53**, 1276 (1970).
84. Weil, L. and Quentin, K. E. *Z. für Wasser and Abwasser Forschung*, **7**, 147 (1974).
85. Lauren, G. *Bull. Environ. Contam. Toxicol.*, **5**, 542 (1970).
86. Griffitt, K. R. and Craun, J. C. *J. Ass. Off. Anal. Chem.*, **57**, 168 (1974).
87. Södergren, A. *Bull. Environ. Contam. Toxicol.*, **10**, 116 (1973).
88. Woodham, R. L. and Collier, C. W. *J. Ass. Off. Anal. Chem.*, **54**, 1117 (1971).
89. Baird, R. B., Carmona, L. G. and Kuo, C. L. *Bull. Environ. Contam. Toxicol.*, **9**, 108 (1973).
90. Croll, B. T. *Chemy Ind.*, **40**, 1295 (1970).
91. Wood, G. W. *Anal. Abstr.*, **23**, 4219 (1972).
92. Levi, I. and Nowicki, T. W. *Bull. Environ. Contam. Toxicol.*, **9**, 20 (1973).
93. Dolan, J. W. and Hall, R. C. *Anal. Chem.*, **45**, 2198 (1973).
94. Novikova, K. F. *Zh. Vses. Khim. Obshch.*, **18**, 562 (1973).
95. Arias, C., Vidal, A., Vidal, C. and Maria, J. *An. Bromat.*, **22**, 273 (1970).
96. Markin, G. P., Hawthorne, J. C., Collins, H. L. and Ford, J. H. *Pesticides Monitoring Journal*, **7**, 139 (1974).
97. Kouyoumjian, H. H. and Uglow, R. F. *Environmental Pollution*, **7**, 103 (1974).
98. Williams, R. and Holden, A. V. National Institute of Oceanography, Wormley, Godalming, Surrey, UK. private communication.
99. Frank, R., Armstrong, A. F., Boeleus, R. G., Braun, H. E. and Douglas, C. W. *Pesticide Monitoring Journal*, **7**, 165 (1974).
100. Langlois, R. E., Stemp, A. R. and Liska, B. J. *J. Milk Food Technol.*, **27**, 202 (1954).
101. Hamence, J. H., Hall, P. S. and Caverly, D. J. *Analyst (Lond.)*, **90**, 649 (1965).
102. Mills, P. A. *J. Ass. Off. Agric. Chem.*, **42**, 734 (1959).
103. Willmott, F. W. and Dolphin, R. J. *J. Chromat. Sci.*, **12**, 695 (1974).
104. Maggs, R. J. *Column*, **2 (4)**, 5 (Pye Unicam Ltd) (1968).
105. Koen, J. G., Huber, J. F. K., Poppe, H. and den Boef, C. *J. Chromat. Sci.*, **8**, 192 (1970).
106. Frei, R., Lawrence, J. F., Hope, J. and Cassidy, R. H. *J. Chromat. Sci.*, **12**, 40 (1974).

107. Bombough, K. J., Levangie, R. F., King, R. N. and Abrahams, I. *J. Chromat. Sci.*, **8**, 657 (1970).
108. Eisenheiss, F. and Sieper, H. *J. Chromatography*, **83**, 430 (1973).
109. Maggs, R. J., Joynes, P. L. and Lovelock, J. E. *Anal. Chem.*, **43**, 1966 (1971).
110. Lubkowitz, J. A., Montoloy, D. and Parker, W. C. *J. Chromat.*, **76**, 21 (1973).
111. Devaux, P. and Guichon, G. *J. Chromat. Sci.*, **8**, 502 (1970).
112. Suzuki, T., Nagayoshi, H. and Kashiwa, T. *Agric. Biol. Chem.*, **38**, 279 (1974).
113. Suzuki, T. *Anal. Abstr.*, **26**, 2374 (1974).
114. Fehringer, N. V. and Westfall, J. E. *J. Chromat.*, **57**, 397 (1971).
115. Chen, Jo-Yun T. and Dority, R. W. *J. Am. Oil Colour Ass.*, **55**, 15 (1972).
116. Hutzinger, O., Jamieson, W. D. and Safe, S. *J. Am. Oil Colour Ass.*, **54**, 178 (1971).
117. Kawatski, J. A. and Frasch, D. L. *J. Ass. Off. Anal. Chem.*, **52**, 1108 (1969).
118. Armstrong, D. W. and Terrill, R. Q. *Anal. Chem.*, **51**, 2160 (1979).
119. Achari, R. G., Sandhu, S. S. and Warren, W. *J. Bull. Environ. Contam. Toxicol.*, **13**, 94 (1975).
120. Bevenue, A., Kelley, T. W. and Hylin, J. *J. Chromat.*, **54**, 71 (1971).
121. Keith, L. H. and Alford, A. L. *J. Ass. Off. Anal. Chem.*, **53**, 1018 (1970).
122. Ernst, W., Goerke, H., Eder, G. and Schaefer, R. C. *Bull. Environ. Contam. Toxicol.*, **15**, 55 (1976).
123. Ernst, W., Schaefer, H., Goerke, H. and Eder, G. *Z. Anal. Chem.*, **227**, 358 (1974).
124. Sadar, M. H. and Guilbauelt, G. G. *J. Agric. Food Chem.*, **19**, 357 (1971).
125. Taylor, R., Bogacka, T. and Krasnicki, K. *Chemia Analit.*, **19**, 73 (1974).
126. Taylor, R. and Bogacka, T. *Anal. Abstr.*, **17**, 1206 (1969).
127. Balayanis, P. G. *J. Chromat.*, **90**, 198 (1974).
128. Gooding, P. H., Philip, H. G. and Tawnk, H. S. *Bull. Environ. Contam. Toxicol.*, **7**, 288 (1972).
129. Gillespie, D. M., Eldredge, J. D. and Thompson, C. E. *Wat. Res.*, **9**, 817 (1975).
130. Wilson, A. J. *Bull. Environ. Contam. Toxicol.*, **15**, 515 (1976).
131. Neudorf, S. and Khan, M. A. Q. *Bull. Environ. Contam. Toxicol.*, **13**, 443 (1975).
132. Picer, N., Picer, M. and Strohal, P. *Bull. Environ. Contam. Toxicol.*, **14**, 565 (1975).
133. Woodham, D. W., Loftis, C. D. and Collier, C. W. *J. Agric. Food Chem.*, **20**, 163 (1972).
134. Lester, J. F. and Smiley, J. W. *Bull. Environ. Contam. Toxicol.*, **7**, 43 (1972).
135. Richards, J. F. and Fritz, J. S. *Talanta*, **21**, 91 (1974).
136. Simal, J., Crous Vidal, J., Maria-Charro Arias, A., Boado, M. A., Diaz, R. and Vilas, D. *An Bromat (Spain)*, **23**, 1 (1971).
137. Johnson, R. E. and Starr, R. I. *J. Agric. Food Chem.*, **20**, 48 (1972).
138. Criba, M. and Morley, H. V. *J. Agric. Food Chem.*, **16**, 916 (1968).
139. Konrad, J. G., Pionke, H. B. and Chesters, G. *Analyst (Lond.)*, **94**, 490 (1969).
140. Pionke, H. B. *Anal. Abstr.*, **17**, 2442 (1969).
141. Mahel'ova, H., Sackmauerevá, M., Szokolav, A. and Kovac, J. *J. Chromat.*, **89**, 177 (1974).
142. Yamato, Y., Suzuki, M. and Aktyama, T. *Bull. Environ. Contam. Toxicol.*, **14**, 380 (1975).
143. Strachel, B., Bactjen, K., Cetinkaya, M., Dueszein, J., Lahl, U., Lierse, K., Thiemann, W., Gahel, B., Kozicki, R. and Podbielski, A. *Anal. Chem.*, **53**, 1469 (1981).
144. Mosinska, K. *Pr. Inst. Przem. Org.*, **1971**, 253 (1972).
145. Devine, M. *J. Agric. Food Chem.*, **21**, 1095 (1973).
146. Gorbach, S., Haarring, R., Knauf, W. and Werner, H. *J. Bull. Environ. Contam. Toxicol.*, **6**, 40 (1971).
147. Chau, A. S. Y. and Terry, K. *J. Ass. Off. Anal. Chem.*, **55**, 1228 (1972).
148. Chau, A. S. Y. and Terry, K. *Anal. Abstr.*, **24**, 2513 (1973).

149. Chau, A. S. Y. *J. Ass. Off. Anal. Chem.*, **55**, 1232 (1972).
150. Chau, A. S. Y. and Terry, K. *J. Ass. Off. Anal. Chem.*, **57**, 394 (1974).
151. Bergmann, H. and Hellman, H. *Dtsch. Gewasserkundeliche Mitteilungen*, **24**, 31 (1981).
152. Porcaro, P. J. and Shubiak, P. *Anal. Chem.*, **44**, 1865 (1972).
153. Katz, S. E. and Strusz, R. F. *Bull. Environ. Contam. Toxicol.*, **3**, 258 (1968).
154. Andrade, P. S. L. and Wheeler, R. B. *Bull. Environ. Contam. Toxicol.*, **11**, 415 (1974).
155. Harless, R. L., Harris, D. E., Sovocool, W., Zehr, R. D., Wilson, N. K. and Oswald, E. O. *Biochemical Mass Spectrometry*, **5**, 232 (1978).
156. Laseter, J. L., DeLeon, I. R. and Remele, P. C. *Anal. Chem.*, **50**, 1169 (1978).
157. Kaiser, K. L. E. *Science*, **185**, 523 (1974).
158. Gether, J., Lunde, G. and Steinnes, E. *Anal. Chim. Acta*, **108**, 137 (1979).
159. Ahnoff, M. and Josefsson, B. *Anal. Chem.*, **46**, 658 (1974).
160. Lunde, G., Gether, J. and Steinnes, E. *Ambio*, **5**, 180 (1976).
161. Schmitt, A. and Zweig, G. *J. Agric. Food Chem.*, **10**, 481 (1962).
162. Jager, W. and Hagenmaier, H. *Z. für Wasser and Abwasser Forschung*, **13**, 66 (1980).
163. Van Steenderen, R. A. *Lab. Pract.*, **29**, 380 (1980).
164. McCahill, M. P., Conroy, L. E. and Maier, W. J. *Environ. Sci. Technol.*, **14**, 201 (1980).
165. Kühn, W., Fuchs, F. and Southeimer, H. *Z. für Wasser und Abwasser Forshung*, **10**, 192 (1977).
166. Crosby, D. G. and Bowers, O. *J. Agric. Food Chem.*, **16**, 839 (1968).
167. Frei, R. W., Lawrence, J. F. and Belliveau, P. E. *Z. Analyt. Chem.*, **254**, 271 (1971).
168. Handa, S. K. and Dikshit, A. K. *Analyst (Lond.)*, **104**, 1185 (1979).
169. Rolls, J. W. and Cortes, A. *J. Gas Chromat.*, **2**, 132 (1964).
170. Holden, E. R., Jones, W. M. and Beroze, M. *J. Agric. Food Chem.*, **17**, 56 (1969).
171. Cohen, I. C., Norcup, J., Ruzicka, J. H. A. and Wheals, B. B. *J. Chromat.*, **49**, 215 (1970).
172. Gutenmann, W. H. and Lisk, D. J. *J. Agric. Food Chem.*, **3**, 48 (1965).
173. Coburn, J. A., Riplay, B. D. and Chau, A. S. Y. *J. Ass. Off. Anal. Chem.*, **59**, 188 (1976).
174. Lewis, D. L. and Paris, D. F. *J. Agric. Food Chem.*, **22**, 148 (1974).
175. Ueji, M. and Kanazawa, J. *Japan Analyst*, **22**, 16 (1973).
176. Miskus, R., Gordon, H. T. and George, D. A. *J. Agric. Food Chem.*, **7**, 613 (1959).
177. Venesch, E. E. and Riveros, M. H. C. K. *J. Ass. Off. Anal. Chem.*, **54**, 128 (1971).
178. Rangaswamy, J. R. and Majumder, S. K. *J. Ass. Off. Anal. Chem.*, **57**, 592 (1974).
179. Reeves, R. G. and Woodham, D. W. *J. Agric. Food Chem.*, **22**, 76 (1974).
180. Hasler, C. F. *Bull. Environ. Contam. Toxicol.*, **12**, 599 (1974).
181. Lawrence, J. F. and Frei, R. W. *Anal. Chem.*, **44**, 2046 (1972).
182. Westlake, W. E., Monika, I. and Gunther, F. A. *Bull. Environ. Contam. Toxicol.*, **8**, 109 (1972).
183. Cohen, I. C., Norcup, J., Kuzicka, J. H. A. and Wheals, B. B. *J. Chromat.*, **49**, 215 (1970).
184. Thompson, J. F., Reid, S. T. and Kanton, E. T. *Arch. Environ. Contam. Toxicol.*, **6**, 143 (1977).
185. Sundaram, K. M. S., Szeto, S. Y. and Hindle, R. *J. Chromat.*, **177**, 29 (1979).
186. Coburn, J. A., Riplay, B. D. and Chau, A. S. Y. *J. Ass. Off. Anal. Chem.*, **59**, 188 (1976).
187. Thompson, J. F., Reid, S. J. and Kantor, E. T. *Arch. Environ. Contam. Toxicol.*, **6**, 143 (1977).
188. Anderson, J. L. and Chesney, D. J. *Anal. Chem.*, **52**, 2156 (1980).

189. Supplement to the 15th edition of *Standard Methods for the Examination of Water and Waste Water Selected Analytical Methods Approved and Cited by the US Environmental Protection Agency*. Prepared and Published by American Public Health Association, American Waterworks Association, Water Pollution Control Federation. Sept. 1978. Methods S60 and S63. Methods for Benzidine, Chlorinated Organic Compounds, Pentachlorophenol and Pesticides in Water and Waste Water. (INTERIM, Pending Issuance of Methods for Organic Analysis of Water and Wastes, Sept. 1978). Environmental Protection Agency, Environmental Monitoring and Support Laboratory (EMSL).
190. *ibid.*, Method S51.
191. De Loach, H. K. and Hemphill, D. D. *J. Ass. Off. Anal. Chem.*, **52**, 333 (1969).
192. De Loach, H. K. and Hemphill, D. D. *J. Ass. Off. Anal. Chem.*, **53**, 1129 (1970).
193. Brody, S. S. and Chaney, J. E. *J. Gas Chromat.*, **4**, 42 (1966).
194. Bache, C. A. and Lisk, D. J. *Anal. Chem.*, **37**, 1477 (1965).
195. Hartmann, C. H. *Aerograph Research Notes 1–6* (1966).
196. Kanazawa, J. and Kawahara, T. *Nippon, Nogei, Kagaku, Kaishi*, **40**, 178 (1966).
197. Guiffrida, L., Ives, N. F. and Bostwick, D. C. *J. Ass. Off. Anal. Chem.*, **49**, 8 (1966).
198. Kamen, A. *J. Gas Chromat.*, **3**, 336 (1965).
199. Claborn, H. V., Mann, H. D. and Vehler, D. D. *J. Ass. Off. Anal. Chem.*, **51**, 1243 (1968).
200. Burchfield, H. P., Rhoades, J. W. and Wheeler, R. J. *J. Agric. Food Chem.*, **13**, 511 (1965).
201. Burchfield, H. P., Johnson, D. E., Rhoades, J. W. and Wheeler, R. J. *J. Gas Chromat.*, **3**, 28 (1965).
202. Ives, N. F. and Guiffrida, L. *J. Ass. Off. Anal. Chem.*, **50**, 1 (1967).
203. Dressler, M. and Janak, J. *Colln. Czeck. Chem. Commun.*, **33**, 3970 (1968).
204. Dressler, M. and Janak, J. *Anal. Abstr.*, **18**, 2876 (1970).
205. Wooley, D. E. *Anal. Chem.*, **40**, 210 (1968).
206. Novak, A. V. and Malmstadt, H. V. *Anal. Chem.*, **40**, 1108 (1968).
207. Bowman, M. C. and Beroza, M. *Anal. Chem.*, **40**, 1448 (1968).
208. Bowman, M. C., Beroza, M. and Hill, K. R. *J. Ass. Off. Anal. Chem.*, **54**, 346 (1971).
209. Kamen, A. *J. Chromat. Sci.*, **7**, 541 (1969).
210. Skolnick, M. *J. Chromat. Sci.*, **8**, 462 (1970).
211. Grice, H. W., Yates, M. L. and David, D. J. *J. Chromat. Sci.*, **9**, 90 (1970).
212. Craven, D. A. *Anal. Chem.*, **42**, 1679 (1970).
213. Svojanousky, V. and Nebola, R. *Chemickchisty*, **67**, 295 (1973).
214. Greenhalgh, R. and Cochrane, W. P. *J. Chromat.*, **70**, 37 (1972).
215. Askew, J., Ruzicka, J. H. and Wheals, B. B. *Analyst (Lond.)*, **94**, 275 (1959).
216. Laws, E. P. and Webley, D. J. *Analyst (Lond.)*, **86**, 249 (1961).
217. Bates, J. A. R. *Analyst (Lond.)*, **90**, 453 (1965).
218. Suffet, I., Faust, D. S. and Carey, W. T. *Environ. Sci. Technol.*, **1**, 639 (1967).
219. Cook, C. E., Stanley, C. W. and Barney, J. E. *Anal. Chem.*, **36**, 2354 (1964).
220. Simmons, J. H. and Talton, J. O. G. *J. Chromat.*, **27**, 253 (1967).
221. Abbott, D. C., Burridge, A. S., Thomson, J. and Wells, K. S. *Analyst (Lond.)*, **92**, 170 (1967).
222. Suffet, I. H. and Faust, S. D. *J. Agric. Food Chem.*, **20**, 52 (1972).
223. Thompson, J. F., Reid, S. J. and Kantor, E. J. *Arch. Environ. Contam. Toxicol.*, **6**, 143 (1977).
224. Daughton, C. G., Crosby, D. G., Garnos, R. L. and Hseih, D. P. H. *J. Agric. Food Chem.*, **24**, 236 (1976).
225. Blanchet, P. F. *J. Chromat.*, **179**, 123 (1979).
226. Miles, J. R. W. and Harris, C. R. *J. Econ. Entomol.*, **71**, 125 (1978).

227. Verweij, A., Regenhardt, C. E. A. and Boter, H. L. *Chemosphere*, **8**, 115 (1979).
228. Daughton, C. G., Cook, A. M. and Alexander, M. *Anal. Chem.*, **51**, 1949 (1979).
229. Bargnoux, H., Pepin, D., Chahard, J. L., Vedrine, F., Petit, J. and Berger, J. A. *Analusis*, **5**, 170 (1977).
230. Thompson, J. F., Reid, S. J. and Kantor, E. J. *Arch. Environ. Contam. Toxicol.*, **6**, 144 (1977).
231. Bourne, S. *J. Environ. Sci. Hlth*, **B13**, 75 (1978).
232. Mallet, V. N., Berkane, K. and Caissie, G. E. *J. Chromat.*, **139**, 386 (1977).
233. Rodica, T. *Revta. Chem.*, **20**, 259 (1969).
234. Zycinski, D. *Roczn. panst. Zakl. Hig.*, **22**, 189 (1971).
235. Leoni, V. and Puccetti, G. *Farmaco Ed. prat.*, **26**, 383 (1971).
236. Schutzmann, O. and Barthel, R. *Anal. Abstr.*, **18**, 3474 (1966).
237. Kovacs, O. *Anal. Abstr.*, **13**, 2016 (1966).
238. Bidleman, T. F., Nowlan, B. and Frei, R. W. *Anal. Chim. Acta*, **60**, 13 (1972).
239. Jovanovic, D. A. and Prosic, Z. *Acta Pharm. jugosl*, **22**, 91 (1972).
240. Guven, K. C. and Aktugla, A. *Eczaclik Bult.*, **14**, 44 (1972).
241. Gauson, R. M., Robinson, D. W. and Goodman, A. *Environ. Sci. Technol.*, **7**, 1137 (1973).
242. Cranmer, M. F. and Peoples, A. *Anal. Biochem.*, **55**, 255 (1973).
243. Cranmer, M. F. and Peoples, A. *Anal. Abstr.*, **22**, 1779 (1972).
244. Karlhuber, B. A. and Eberle, D. *Anal. Chem.*, **47**, 1094 (1975).
245. Goodson, L. H., Jacobs, W. B. and Davis, A. *Analyt. Biochem.*, **51**, 362 (1973).
246. Bauman, O. *Anal. Abstr.*, **15**, 5550 (1968).
247. Report US Environ. Prot. Agency EPA-R2-72-010, Midwest Research Institute (1972).
248. Kawahara, F. K., Lichtenberg, J. J. and Eichelberger, J. W. *J. Wat. Pollut. Control Fedn*, **39**, 446 (1967).
249. Hindin, E., Horteen, M. J., May, D. S., Skrinde, R. T. and Dunstan, G. H. *J. Am. Water Wks Assoc.*, **54**, 88 (1962).
250. Skinde, R. T., Caskey, J. W. and Ellespie, C. K. *J. Am. Water Wks Assoc.*, **54**, 1407 (1962).
251. Teasley, J. T. and Cax, W. S. *J. Am. Water Wks Assoc.*, **55**, 8 (1963).
252. Moye, H. A. *J. Chromat. Sci.*, **13**, 268 (1975).
253. Paschal, D. C., Bicknell, R. and Dresbach, D. *Anal. Chem.*, **49**, 1551 (1977).
254. Junk, G. A. *J. Chromat.*, **99**, 745 (1974).
255. Venkataraman, S. and Sathyamurthy, V. *J. Ind. Water Wks Assoc.*, **11**, 351 (1979).
256. Weber, K. *Wat. Res.*, **10**, 237 (1976).
257. Bargnoux, H., Pepin, D., Chahard, J. L., Vedrine, F., Petit, J. and Berger, J. A. *Analusis*, **5**, 170 (1977).
258. Wolfe, N. L., Zepp, R. G., Gordon, G. A., Baughman, G. L. and Clive, D. M. *Environ. Sci. Technol.*, **11**, 88 (1977).
259. Rice, J. R. and Dishberger, H. J. *J. Agric. Food Chem.*, **16**, 867 (1968).
260. Deutsch, M. E., Westlake, W. E. and Gunther, F. A. *J. Agric. Food Chem.*, **18**, 178 (1970).
261. Dale, W. E. and Miles, J. W. *J. Agric. Food. Chem.*, **17**, 60 (1969).
262. Miller, C. W. and Funes, A. *J. Chromat.*, **59**, 161 (1971).
263. Shafik, M. T. *Bull. Environ. Contam. Toxicol.*, **3**, 309 (1968).
264. Howe, L. H. and Petty, C. F. *J. Agric. Food Chem.*, **17**, 401 (1969).
265. Otsuki, A. and Takaku, T. *Anal. Chem.*, **51**, 833 (1979).
266. Coburn, J. A., Valdamanis, J. A. and Chau, A. S. Y. *J. Ass. Off. Anal. Chem.*, **60**, 224 (1977).
267. Berkane, K., Caissie, G. E. and Mallet, V. N. *Proc. Symp. on Fenitrothion*, NRC (Canada) Assoc. Comm. Sci. Crit. Environ. Quality, Report 16073, NRCC, Ottawa, 95 (1977).

268. Daughton, C. G., Crosby, D. G., Garnos, R. L. and Hsieh, J. *J. Agric. Food Chem.*, **24**, 236 (1976).
269. Zitko, V. and Cunningham, T. D. *Bull. Environ. Contam. Toxicol.*, **14**, 19 (1975).
270. Mallet, V. N., Brun, G. L., MacDonald, R. N. and Berkane, K. *J. Chromat.*, **160**, 81 (1978).
271. Berkane, K., Caissie, G. E. and Mallet, V. N. *J. Chromat.*, **139**, 386 (1977).
272. Zakrevsky, J. G. and Mallet, V. N. *J. Chromat.*, **132**, 315 (1977).
273. Ripley, B. D., Hall, J. A. and Chau, A. S. Y. *Environ. Lett.*, **7**, 97 (1974).
274. Grift, N. and Lockhart, W. L. *J. Ass. Off. Anal. Chem.*, **57**, 1282 (1974).
275. Zakrevsky, J. G. and Mallet, V. N. *J. Chromat.*, **132**, 315 (1977).
276. NRC Associate Committee on Scientific Criteria for Environmental Quality, *Fenitrothion: The Effects of its Use on Environmental Quality and its Chemistry*, NRCC No. 14104, p. 106 (1975).
277. Mallet, V. and Brun, G. L. *Bull. Environ. Contam. Toxicol.*, **12**, 739 (1974).
278. Brun, G. L. and Mallet, V. *J. Chromat.*, **80**, 117 (1973).
279. Vogler, P. *Liminologica*, **7**, 309 (1970).
280. Goossen, J. T. H. and Kloosterboer, J. G. *Anal. Chem.*, **50**, 707 (1978).
281. Albro, P. W. and Fishbein, O. *J. Chromat.*, **69**, 273 (1972).
282. Webb, R. C. and McCall, A. C. *J. Chromat. Sci.*, **11**, 366 (1973).
283. Chriswell, C. D., Ericson, R. L., Junk, G. A., Lee, K. W., Fritz, J. S. and Svec, H. J. *J. Am. Wat. Wks Assoc.*, **69**, 669 (1977).
284. Musty, P. R. and Nickless, G. *J. Chromat.*, **100**, 83 (1974).
285. Musty, P. R. and Nickless, G. *J. Chromat.*, **120**, 369 (1976).
286. Lawrence, J. and Tosine, H. M. *Environ. Sci. Technol.*, **10**, 381 (1976).
287. Coburn, J. A., Valdmanis, I. A. and Chau, A. S. Y. *J. Ass. Off. Anal. Chem.*, **60**, 224 (1977).
288. Le'Bel, G. L. and Williams, D. T. *Bull. Environ. Contam. Toxicol.*, **24**, 397 (1980).
289. McNeil, E. E., Otson, R., Miles, W. F. and Rajabalee, F. J. M. *J. Chromat.*, **132**, 277 (1977).
290. Lebel, G. L., Williams, D. T., Griffith, G. and Benoit, F. M. *J. Ass. Off. Anal. Chem.*, **62**, 281 (1979).
291. Bauer, U. *Gas-u WassFach Wasser Abwass.*, **113**, 58 (1972).
292. US Environmental Protection Agency. US National Information Service, Springfield, Va. *Report No. PB 279547* (1977) (30381).
293. Inland Waters Directorate. *Analytical Methods Manual*. Water Quality Branch, Ottawa, Ontario, Part 2, *Organic Constituents* (1974).
294. Coburn, J. A., Valdmanis, I. A. and Chau, A. S. Y. *J. Ass. Off. Anal. Chem.*, **60**, 224 (1977).
295. Gesser, H. D., Chow, A., Davis, F. C., Uthe, J. F. and Reinke, J. *Anal. Lett.*, **4**, 883 (1971).
296. Bedford, J. W. *Bull. Environ. Contam. Toxicol.*, **12**, 662 (1974).
297. Ahnoff, M. and Josefsson, B. *Bull. Environ. Contam. Toxicol.*, **13**, 159 (1975).
298. Berg, O. W., Diosady, P. L. and Rees, G. A. V. *Bull. Environ. Contam. Toxicol.*, **7**, 338 (1972).
299. Sawyer, L. D. *J. Ass. Off. Anal. Chem.*, **61**, 272 (1978).
300. Webb, R. G. and McCall, A. C. *J. Chromat.*, **11**, 366 (1973).
301. Stalling, D. L. and Huckins, J. N. *J. Ass. Off. Anal. Chem.*, **54**, 801 (1971).
302. Webb, R. G. and McCall, A. C. *J. Ass. Off. Anal. Chem.*, **55**, 746 (1972).
303. Tas, A. C. and Kleipool, R. J. C. *Bull. Environ. Contam. Toxicol.*, **8**, 32 (1972).
304. Albro, P. W. and Fishbein, L. *J. Chromat.*, **69**, 273 (1972).
305. Webb, R. G. and McCall, A. C. *J. Chromat. Sci.*, **11**, 366 (1973).
306. Beezhold, F. L. and Stout, V. F. *Bull. Environ. Contam. Toxicol.*, **10**, 10 (1973).
307. Sissons, D. and Welti, D. *J. Chromat.*, **60**, 15 (1971).

308. Albro, P. W., Haseman, J. K., Clemmer, T. and Corbett, B. J. *J. Chromat.*, **136**, 147 (1977).
309. *Chromo-Pack News.* June 1979 No. 20. Chromopack Ltd. PO Box 3. 4330 AA Middelburg, The Netherlands.
310. Schulte, E. and Acker, L. *Z. Analyt. Chem.*, **268**, 260 (1974).
311. Sullivan, K. F., Altas, E. L. and Giain, C. S. *Anal. Chem.*, **53**, 1718 (1981).
312. Ahnoff, M. and Josefsson, B. *Anal. Lett.*, **6**, 1083 (1973).
313. Webb, R. G. and McCall, A. C. *J. Chromat. Sci.*, **11**, 366 (1973).
314. Skinner, R. F., Fins, W. F. and Banelli, E. J. *Finnigan Spectra*, **3**, No. 1 (1973).
315. Eichelberger, J. W., Harris, L. E. and Budde, W. J. *Anal. Chem.*, **46**, 227 (1974).
316. Karlruber, B. A., Hormann, W. D. and Ramsteinen, K. A. *Anal. Chem.*, **47**, 2453 (1975).
317. Ahnoff, M. and Josefsson, B. *Anal. Lett.*, **6**, 1036 (1973).
318. Elder, D. *Marine Pollution Bulletin*, **7**, 63 (1976).
319. Gaffney, P. E. *J. Wat. Pollut. Control*, **49**, 401 (1977).
320. Delfino, J. J. and Easty, B. *Anal. Chem.*, **51**, 2235 (1979).
321. Easty, D. B. and Wabers, B. A. *Tappi*, **61**, 71 (1978).
322. Easty, D. B. and Wabers, B. A. *Anal. Lett.*, **10**, 857 (1977).
323. Berg, O. W., Diosady, P. L. and Rees, G. A. V. *Bull. Environ. Contam. Toxicol.*, **7**, 338 (1972).
324. Hanai, T. and Walton, H. F. *Anal. Chem.*, **49**, 1954 (1977).
325. Brinkman, A. Th., Seetz, J. W. F. L. and Reymer, H. G. M. *J. Chromat.*, **116**, 353 (1976).
326. Brinkman, A. Th., De Kok, A., De Vries, G. and Reymer, H. G. M. *J. Chromat.*, **128**, 101 (1976).
327. Kaminsky, L. S. and Fasco, M. J. *J. Chromat.*, **155**, 363 (1978).
328. Gesser, H. D., Chow, A., Davis, F. C., Uthe, J. F. and Reinke, J. *Anal. Lett.*, **4**, 883 (1971).
329. Uthe, J. F., Reinke, J. and Gesser, H. *Environ. Lett.*, **3**, 117 (1972).
330. de Vos, R. H. and Peet, E. W. *Bull. Environ. Contam. Toxicol.*, **6**, 164 (1971).
331. Sackmauer, O. M., Pal'usova, O. and Szokolay, A. *Wat. Res.*, **11**, 551 (1977).
332. Farwell, S. O., Beland, F. A. and Geer, R. D. *Bull. Environ. Contam. Toxicol.*, **10**, 157 (1973).
333. Neely, W. B. *Science of the Total Environment*, **7**, 117 (1977).
334. Frederick, L. L. *J. Fish. Res. Bd Can.*, **32**, 1705 (1975).
335. Jan, J. and Malversic, S. *Bull. Environ. Contam. Toxicol.*, **6**, 772 (1978).
336. Olsson, M., Jenson, B. and Reutergard, L. *Ambio*, **7**, 66 (1978).
337. Szelewski, M. J., Hill, D. R., Spiegel, J. and Tifft, E. C. *Anal. Chem.*, **51**, 2405 (1979).
338. Kaiser, K. L. E. and Wong, P. T. S. *Bull. Environ. Contam. Toxicol.*, **11**, 291 (1974).
339. Herbst, E., Schennert, I., Kein, W. and Korte, F. *Chemosphere*, **6**, 725 (1977).
340. Furukawa, K., Tonomura, K. and Kamibayashi, A. *Appl. Environ. Microbiol.*, **35**, 223 (1978).
341. Zitko, V. and Choi, P. *PCB and Other Industrial Halogenated Hydrocarbons in the Environment.* Fish. Res. Board Can. Technical Report No. 272, Biological Station, St. Andrews, N.B. (1971).
342. Oller, W. L. and Cramer, M. F. *J. Chromat. Sci.*, **13**, 296 (1975).
343. Edwards, R. *Chem. Ind. (London)*, 1340 (1970).
344. Fishbein, L. *J. Chromat.*, **68**, 345 (1972).
345. Schulte, E., Thier, H. P. and Acker, L. *Deut. Lebensm-Rundsch.*, **72**, 229 (1976).
346. Göke, G. *Deut. Lebensm.-Rundsch.*, **71**, 309 (1975).
347. Ashling, B. and Jensen, S. *Anal. Chem.*, **42**, 1483 (1970).
348. Elder, G. *J. Chromat.*, **121**, 269 (1976).

349. Musty, P. R. and Nickless, G. *J. Chromat.*, **89**, 185 (1974).
350. Girenko, D. B., Klisenko, M. A. and Dishcholka, Y. K. *Hydrobiological Journal*, **11**, 60 (1975).
351. Department of the Environment. *Methods for the Examination of Waters and Associated Materials; Organochlorine Insecticides and Polychlorinated Biphenyls in Waters, 1978 Tentative Method (32313)*. HMSO, London, 28 pp. (1979).
352. Szelewski, M. J., Hill, D. R., Spiegel, S. J. and Tifft, E. C. *Anal. Chem.*, **51**, 2405 (1979).
353. Canagay, A. B. and Levine, P. L. US Environmental Protection Agency; Cincinnati, Ohio, *Report EPA. 600/2-79-166* (1979). 100 pp. (P22CAR).
354. Södergren, A. *J. Chromat.*, **160**, 271 (1978).
355. Södergren, A. *J. Chromat.*, **71**, 532 (1972).
356. Blomberg, L. *Chromatographia*, **8**, 324 (1975).
357. Larsson, K., Odham, G. and Södergren, A. *Marine Chem.*, **2**, 49 (1974).
358. Bacaloni, A., Goretti, G., Lagano, A. and Petronio, B. M. *J. Chromat.*, **175**, 169 (1979).
359. Nota, G., Goretti, G., Armenante, M. and Marino, G. *J. Chromat.*, **95**, 229 (1974).
360. Goretti, G., Liberti, A. and Nota, G. *Chromatogrpahia*, **8**, 486 (1975).
361. Vidal-Madiar, C., Bekamy, S., Gormand, M. F., Arpino, P. and Guiochon, G. *Anal. Chem.*, **49**, 768 (1977).
362. Goretti, G., Liberti, A. and Pili, G. *J. High Resolut. Chromatogr. Chromatogr. Commun.*, **1**, 143 (1978).
363. Goretti, G. and Liberti, A. *J. Chromat.*, **89**, 161 (1978).
364. Georlitz, D. F. and Law, L. M. *Method for Chlorinated Hydrocarbons in Water and Waste Water*. US Environmental Protection Agency. Interim method. p. 7 Sept (1978).
365. Supplement to the 15th edition of *Standard Methods for the Examination of Water and Waste Water*. Selected analytical methods approved and cited by the US Environmental Protection Agency. Prepared and Published by American Public Health Association, American Water Works Association, Water Pollution Control Federation. Method S78 Method for PCBs in water and waste water Sept (1978).
366. Miller, J. D., Thomas, R. E. and Schattenberg, H. J. *Anal. Chem.*, **53**, 214 (1981).
367. Goerlitz, D. F. and Law, L. M. *J. Ass. Off. Anal. Chem.*, **57**, 176 (1974).
368. Jensen, S., Renberg, L. and Reutergård, L. *Anal. Chem.*, **49**, 316 (1977).
369. Pearson, J. R., Aldrich, F. D. and Stone, A. W. *J. Agric. Food Chem.*, **15**, 938 (1967).
370. Ahling, B. and Jenson, S. *Anal. Chem.*, **42**, 1483 (1970).
371. Jenson, S., Johneis, A. G., Olsson, M. and Otterlind, G. *Ambio Spec. Rep.*, **1**, 71 (1972).
372. Ahnoff, M. and Josefsson, B. *Bull. Environ. Contam. Toxicol.*, **13**, 159 (1975).
373. Goerlitz, D. F. and Law, L. H. *Bull. Environ. Contam. Toxicol.*, **6**, 9 (1971).
374. Johnson, L. D., Waltz, R. H., Ussary, J. P. and Kaiser, F. E. *J. Ass. Off. Anal. Chem.*, **59**, 174 (1976).
375. Rudling, L. *Wat. Res.*, **4**, 533 (1970).
376. Renberg, L. *Anal. Chem.*, **46**, 459 (1974).
377. Mattson, P. E. and Nygren, S. *J. Chromat.*, **124**, 265 (1976).
378. Schulte, E. and Acker, L. *Z. Anal. Chem.*, **268**, 260 (1974).
379. Schulte, E. and Acker, L. *Naturwissenschaften*, **61**, 79 (1974).
380. Jenson, S. and Sundström, G. *Ambio*, **3**, 70 (1974).
381. Jenson, S., Johnels, A. G., Olsson, M. and Otterlind, G. *Ambio Spec. Rep.*, **1**, 71 (1972).
382. *Methods of Analysis of the Association of Official Analytical Chemists*, Association of Official Analytical Chemists, Washington DC 10th edn, p. 393 (1965).
383. Erne, K. *Acta Pharmacol. Toxicol.*, **14**, 158 (1958).

384. Bartlett, J. K. and Skoog, D. A. *Anal. Chem.*, **26**, 1008 (1954).
385. Zimmerli, B., Sulser, H. and Marek, B. *Mitt. Geb. Lebensmittelunters. Hyg.*, **62**, 60 (1971).
386. Cochrane, W. P. and Maybury, R. B. *J. Ass. Off. Anal. Chem.*, **56**, 1324 (1973).
387. Miller, G. A. and Wells, C. E. *J. Ass. Off. Anal. Chem.*, **52**, 548 (1969).
388. McIntyre, A. E., Perry, R. and Lester, J. N. *Environ. Technol. Lett.*, **1**, 157 (1980).
389. McIntyre, A. E., Lester, J. N. and Perry, R. *Analysis of Organic Substances of Concern in Sewage Sludge,* final report to the Department of the Environment for contracts DGR/480/66 and DGR/480/240, pp. 42–45, Imperial College, London, UK. Nov (1979).
390. Homes, D. C. and Wood, N. F. *J. Chromat.*, **67**, 173 (1972).
391. Wells, D. E. and Johnstone, S. H. *J. Chromat.*, **140**, 17 (1977).
392. Holden, A. V. and Marsden, K. *J. Chromat.*, **44**, 481 (1969).
393. Dube, D. J., Veith, G. D. and Lee, G. F. *J. Wat. Pollut. Contr. Fedn*, **46**, 966 (1974).
394. Lawrence, J. and Tosine, H. *Environ. Sci. Technol.*, **10**, 381 (1976).
395. Shannon, E. E., Ludwig, F. J. and Valdemanis, I. *Environment Canada Research Report No. 49,* Ottawa, 35 pp. (1976).
396. Bergh, A. K. and Peoples, R. S. *Sci. Total Environ.*, **8**, 197 (1977).
397. Harper, D. B., Smith, R. V. and Gotto, D. M. *Environ. Pollut.*, **12**, 223 (1977).
398. Lawrence, J. and Tosine, H. *Bull. Environ. Contam. Toxicol.*, **17**, 49 (1977).
399. Liur, D., Chawla, V. K. and Chau, A. S. Y. *Proc. 9th Ann. Conf. on Trace Substances in Environ. Health* Univ. Missouri, Columbia, Missouri, 189 (1975).
400. Teichnan, J., Bevenue, A. and Hylin, J. W. *J. Chromat.*, **151**, 155 (1978).
401. Gaskin, D. E., Smith, G. J. D., Arnold, P. W., Louisy, M. V., Frank, R., Moldrinet, M. and McWade, J. W. *J. Fish. Res. Can.*, **31**, 1235 (1974).
402. Snyder, D. and Reinert, R. *Bull. Environ. Contam. Toxicol.*, **6**, 385 (1971).
403. Bonelli, E. and Smith, R. D. *Effluent Water Treatment J.*, **12**, 87 (1972).
404. Erikson, M. D. and Pellizzari, E. D. *Bull. Environ. Contam. Toxicol.*, **22**, 688 (1979).
405. Teichman, J., Bevenue, A. and Hylin, J. W. *J. Chromat.*, **151**, 155 (1978).
406. Keith, L. *Analysis of Organic Compounds in Two Kraft Mill Waste Waters,* EPA-600/4-75-005 (1975).
407. Keith, L. *Environ. Sci. Technol.*, **10**, 555 (1976).
408. Keith, L. *Identification and Analysis of Organic Pollutants in Water,* ed. L. H. Keith. Ann Arbor Science, Ann Arbor, Ml, Chapt. 36 (1976).
409. Erickson, M. D., Zweidinger, R. A., Michael, L. C. and Pellizzari, E. D. *Environmental Monitoring Near Industrial Sites: Polychlorinated Naphthalenes.* EPA-560/6-77-019 (1977).
410. Erickson, M. D., Michael, L. C., Zweidinger, R. A. and Pellizzari, E. D. *Environ. Sci. Technol.*, **12**, 927 (1978).
411. Kaiser, K. L. E. *Science*, **185**, 523 (1974).
412. Reynolds, L. M. *Res. Rev.*, **34**, 27 (1971).
413. Chau, A. S. Y. and Wilkinson, W. J. personal communication; Pesticide Analytical Manual (Department of Health, Education and Welfare, Food and Drug Administration, Washington, DC, 1971). vols. 1 and 2; Sawyer, L. D. *J. Ass. Off. Anal. Chem.*, **56**, 1015 (1973).
414. Bonelli, E. J. *Anal. Chem.*, **44**, 603 (1972).
415. Markin, G. P., Hawthorne, J. C., Collins, H. L. and Ford, J. H. *Pesticides Monitoring Journal*, **7**, 139 (1974).
416. Armour, J. A. and Burke, J. A. *J. Ass. Off. Anal. Chem.*, **53**, 761 (1970).
417. Gaul, J. and Cruz-LaGrange, P. *Separation of Mirex and PCBs in Fish.* Laboratory Information Bulletin, Food and Drug Administration, New Orleans District (1971).

418. Markin, G. P., Ford, J. H., Hawthorne, J. C., Spence, J. H., Davies, J. and Loftis, C. D. *Environmental Monitoring for the Insecticide Mirex* USDA APHIS 81-83, 19 pp. Nov. (1972).
419. Butler, P. A. *Biol. Sci.*, **19**, 889 (1969).
420. McKenzie, M. D. *Fluctuations in Abundance of the Blue Crab and Factors affecting Mortalities.* South Carolina Wildlife Resources Division. Technical Rep. No. 1 45 pp. (1970).
421. Mahood, R. K., McKenzie, M. D., Middough, D. P., Bellar, S. J., Davis, J. R. and Spitsbergen, D. *A Report on the Cooperative Blue Crab Study in South Atlantic States.* US Department of the Interior. Bureau of Commercial Fisheries (Projects Nos 2-79-R-1, 2-81-R-1, 2-82-R-1) 32 pp. (1970).
422. Luckas, B., Pscheidl, H. and Haberland, P. *J. Chromat.*, **147**, 41 (1978).
423. Luckas, B., Pscheidl, H. and Haberland, D. *Nahrung*, **20**, K-K2 (1976).
424. Stanley, R. L. and le Favoure, H. T. *J. Ass. Off. Anal. Chem.*, **48**, 666 (1965).
425. Murphy, P. G. *J. Ass. Off. Anal. Chem.*, **55**, 1360 (1972).
426. Wenzel, H. and Luckas, B. *Nahrung*, **21**, 347 (1977).
427. Zimmerli, B., Marek, B. and Sulzer, H. *Mitt. Geb. Lebensmittelunters. Hyg.*, **64**, 70 (1973).
428. Zimmerli, B. *J. Chromat.*, **88**, 65 (1974).
429. Prescott, A. M. and Cooke, M. *Proc. Anal. Div. Chem. Soc.*, **16**, 10 (1979).
430. Gosink, T. A. *Environ. Sci. Technol.*, **9**, 630 (1975).
431. Aue, W. A. *J. Chromat. Sci.*, **13**, 329 (1975).
432. Aitzetmuller, K. *J. Chromat.*, **107**, 411 (1975).
433. Zimmerli, B. and Marek, B. *Mitt. Geb. Lebensmittelunters. Hyg.*, **66**, 362 (1975).
434. Rohleder, H., Staudacher, M. and Summermann, W., *Z. Anal. Chem.*, **279**, 152 (1976).
435. Hadorn, H. and Zurcher, K. *Mitt. Geb. Lebensmittelunters. Hyg.*, **61**, 141 (1970).
436. Armour, J. A. and Burke, J. A. *J. Ass. Off. Anal. Chem.*, **53**, 761 (1970).
437. Stijve, T. and Cardinale, E. *Mitt. Geb. Lebensmittelunters. Hyg.*, **65**, 131 (1974).
438. Leoni, V. *J. Chromat.*, **62**, 63 (1971).
439. Sackmauer, O. M., Pal'usova, O. and Szokolay, A. *Wat. Res.*, **11**, 551 (1977).
440. Harvey, G. R. *Report US Environment Protection Agency*, EPA-R2-73-177, 32 pp. (1973).
441. Niederschulte, U. and Ballschmiter, K., *Z. Anal. Chem.*, **269**, 360 (1974).
442. Richard, J. J. and Fritz, J. S. *Talanta*, **21**, 91 (1974).
443. Junk, G. A., Richard, J. J., Grieser, M. D., Witiak, D., Witiak, J. L., Arguello, M. D., Vick, R., Svec, H. J., Fritz, J. S. and Calder, G. V. *J. Chromat.*, **99**, 745 (1974).
444. Kurtz, D. A. *Bull. Environ. Contam. Toxicol.*, **17**, 391 (1977).
445. Leoni, V., Pucetti, G., Columbo, R. J. and Ovidio, O. *J. Chromat.*, **125**, 399 (1976).
446. Leoni, V., Pucetti, G. and Grella, A. *J. Chromat.*, **106**, 119 (1975).
447. Leoni, V. *J. Chromat.*, **62**, 63 (1971).
448. Johnston, L. V. *J. Ass. Off. Anal. Chem.*, **48**, 668 (1965).
449. Claeys, R. R. and Inman, R. D. *J. Ass. Off. Anal. Chem.*, **57**, 399 (1974).
450. Leoni, V. and Pucetti, G. *J. Chromat.*, **43**, 388 (1969).
451. Farwell, S. O., Beland, F. A. and Geer, R. D. *Bull. Environ. Contam. Toxicol.*, **10**, 157 (1973).
452. Freudenthal, J. and Greve, P. A. *Bull. Environ. Contam. Toxicol.*, **10**, 108 (1973).
453. Banks, K. A. and Bills, D. D. *J. Chromat.*, **33**, 450 (1968).
454. Potter, W. G. *LIB 1609*, FDA, Minneapolis District (1969).
455. Erney, D. R. *J. Am. Oil Colour Ass.*, **58**, 1202 (1975).

Chapter 4

Herbicides

INTRODUCTION

Toxicity is of prime importance when considering the effects of herbicides and, indeed, all organic chemicals on humans, and both chronic and acute toxicities must be taken into account. Compounds having low acute toxicities, but which accumulate in the body, may have much lower long-term acceptable intake levels than ones of much higher acute toxicity which do accumulate in the body. These materials may be taken into the body from air, food, or water by ingestion or adsorption through the skin and in many cases residues in water may not be the prime source of the compound. In assessing the effects of herbicides on water supplies it is important to consider not only the active ingredient but also its degradation products in soil, water, and plants, and the other materials contained in any particular formulation. This can be of extreme importance as some formulation solvents or degradation products may be far more noxious than the herbicide itself. Acceptable intake of toxic compounds is assessed from long-term feeding and dermal application experiments on those animals found to be most sensitive to the toxicant. The transference of these data to acceptable daily intakes to man is naturally a job for expert toxicologists and involves the setting of a suitable safety factor. This safety factor is necessary because man may be more sensitive to the toxicant than the most sensitive test animal, and because individuals may vary in their reactions. The size of the safety factor, which is determined by the nature of the compound, may be of the order of 1000. In setting acceptable levels of materials in potable water not only man must be considered but also any animals, plants, etc. which might use the supply.

Many organic chemicals can give rise to unpleasant tastes and odours in water when present at levels considerably less than 1 mg l^{-1}. Water contaminated in this way will be unacceptable to consumers even though it is perfectly safe from biological and toxicological aspects. Even pleasant tastes and odours will be rejected, as the water is obviously 'contaminated'. It is particularly important to consider complete herbicide formulations, rather than active ingredients with respect to taste and odour, as formulation solvents may have very low threshold odours, that is a minimum detectable level in water. For instance, diesel oil, which is used as a solvent particularly for 2, 4, 5-T ester formulations, has a threshold odour of 0.00005 mg l^{-2}. The herbicide or its formulation materials may react chemically with other materials present in, or added to the water thus producing or increasing odours. A particular example of this is the reaction of phenolic compounds with chlorine to produce much more odorous chlorinated phenols.

The use of certain herbicides in or near to water will give rise to rapid decomposition of the affected vegetation, which in turn can cause deoxygenation of the water. The principle significance for potable water supplies is the possibility of tastes and odours in the water. Deoxygenation may also cause dissolution of metallic compounds in the water, which under aerobic conditions would be insoluble. For instance iron and manganese might be taken up into solution and reprecipitation at a waterworks by aeration or chlorination. These metals might be removed at the works. It is more likely, however, that the reprecipitation will be a slow process resulting in their deposition in the water mains, reducing hydraulic efficiency, or that they will be carried in suspended form to the consumer's tap. Both metals in this form can give rise to stains on washing in continuous rinse spin driers or automatic washing machines at levels of about 0.1 mg l^{-1}.

The most obvious method of entry of herbicides into water is by their direct application to the water in order to control aquatic vegetation. In these cases the quantity of herbicide added will be known and the concentration likely to reach a waterworks can be measured with reasonably accuracy. Where emergent vegetation is being sprayed some of the material may be sprayed directly on to the water surface and some may run off the plants into the water. Any herbicide reaching the soil or the banks close to the water may or may not be available for leaching into the water course, depending on the nature of the material. If the herbicide remains in the plants after their death then it may enter the water when they decompose.

The above observations are also pertinent to field-applied herbicides, which may enter water by spray drift, by leaching from, or erosion of the soil or via rotting vegetation or silage. The quantities reaching the water by leaching will depend upon the herbicide, rainfall, and soil type. The terrain may also be important in that it will affect the pattern of leaching or run-off. The time that the herbicide persists in the soil is also important in that it will affect the length of time that pollution is likely to continue. All these factors will apply at the same time, making each herbicide application an individual event and generalizations must be treated with caution.

The question of accidental spillage of a concentrate or a diluted spray into water must also be considered, as must malpractices such as dumping of excess chemicals, washing out of empty containers in ponds and rivers, and improper disposal of containers. The herbicides may also be present in industrial or agricultural effluents.

With direct addition of herbicide into water the concentration downstream may be calculated in the case of a prolonged discharge, from a knowledge of the dilution to be expected. The case of a slug discharge is not so simple. A maximum figure may be calculated as for a continuous discharge, but spreading of the band of contamination will lower the figure. Only a detailed knowledge of the band spreading to be expected will enable an accurate figure to be calculated, and in the absence of this knowledge it would be wise to assume a maximum figure. Factors which will reduce the concentration of herbicide downstream and must be taken into account are:

1. the stability of the compound towards chemical and biological degradation, and its removal from the water by volatilization;
2. absorption into the bottom muds, suspended material, and living organisms.

Only a detailed knowledge of the mass balance of the compound between the various phases will enable accurate concentration levels to be calculated.

Where herbicides are entering the water indirectly, the problem is more complicated and the amount of herbicides reaching the water will be affected by adsorption on soil, degradation, volatilization, and dilution before entering the water body. In the absence of information to the contrary very adverse conditions would be assumed, i.e. that all the herbicide would be washed into the supply by one inch of rain falling in 24 hours. Thus, if the whole of a catchment area were to be sprayed with a herbicide at 2 lb per acre, then the run-off water would contain 2 lb per 22,600 gal of water (1 in of rain = 22,600 gal per acre) or 9 mg l^{-1}. If part of a catchment were to be sprayed then the dilution of the run-off on entering the main stream could be calculated. Similarly the diluting effects of reservoir storage could also be calculated. Simple dilution calculations on reservoirs can be misleading in that complete mixing is probably never achieved.

The Water Research Centre[1] has published information on the health effects of herbicides on potable water supplies and their possible acceptance levels.

The main types of herbicides can be classified into the following catagories:

(1) triazine type;
(2) substituted urea type;
(3) phenoxyacetic and types and halagenated derivatives;
(4) miscellaneous herbicides including:
  picloram,
  acarol,
  dichlorbenil,
  bipyridylium type,
  carbine (barban),
  benthiocarb,
  S-alkyl derivatives of N, N-dialkyl dithiocarbamates,
  diacamba,
  pyrazon,
  paraquat,
  diquat,

dalapon,
glyphosate;
(5) mixtures of herbicides and insecticides.

TRIAZINE TYPE HERBICIDES

Atrazine, 2-chloro-4-ethylamino-6-isopropylamino-1,3,5-triazine
Propazine
Simazine, 2-chloro-4,6-*bis*-ethylamino-1,3,5-triazine
Prometon
Prometryne
Atraton, 2-ethylamino-4-isopropylamino-6-methoxy-1,3,5-triazine
Ametryne, 2-ethylamino-4-isopropylamino-6-methylthio-1,3,5-triazine
Terbutryne
Terbuthylazine, 4-tert-butylamino-2-chloro-6-ethylamino,1,3,5-triazine
GS 26571, 2-amino-4-tert-butylamino-6-methoxy-1,3,5-triazine
GS 30033, 2 amino-4-chloro-6-ethylamino-1,3,5-triazine
Terbumeton
Secbumeton

Gas chromatography

Not unexpectedly, this is the method of choice for the analysis of herbicides. McKone et al.[2] compared gas chromatographic methods for the determination of atrazine (2-chloro-4-ethylamino-6-isopropylamino-1,3,5-triazine), ametryne (2-ethylamino-4-isopropylamino-6 methylthio-1,3,5-triazine, and terbutryne in water. The herbicides were extracted from water with dichloromethane and the dried extracts were evaporated to dryness at a temperature below 35 °C. By gas chromatography on a glass column (1 m × 4 mm) of 2% neopentyl glycol succinate on Chromosorb W (80–100 mesh) operated at 195 °C and with a RbBr-tipped flame ionization detector the three herbicides could be separated and 0.001 ppm of each detected. This method was found to be superior to spectrophotometric and polargraphic methods.

Purkayastha and Cochrane[3] compared electron capture and electrolytic conductivity detectors in the gas chromatographic determination of prometon, atratron (2-ethyl-4-isopropylamino-6-methoxy-1,3,5-triazine), propazine, atrazine (2-chloro-4-ethylamino-6-isopropylamino-1,3,5-triazine), prometryne, simazine (2-chloro-4,6-*bis*-ethylamino-1,3,5-triazine), and ametryne, (2-ethylamino-4-isopropylamino-6-methylthio-1,3,5-triazine) in inland water samples. They found that the electolytic conductivity detector seemed to have a wider application than a $_{63}$Ni electron capture detector; use of the latter detector necessitated a clean-up stage for all the samples studied. The conductivity detector could be used in analysis of soil and water without sample clean-up although maize samples required a clean-up. Good recoveries of atrazine added to water were obtained by extraction with dichloromethane, from soil samples with acetonitrile, methanol, or acetone and from maize samples with acetonitrile.

Ramsteiner et al.[4] compared alkali flame ionization, micro-coulometric, flame photometric, and electrolytic conductivity detectors for the determination of triazine herbicides in water, crops, and soil. Methanol extracts were cleaned up on an alumina column and 12 herbicides were determined by gas chromatography with use of conventional columns containing 3% of Carbowax 20 M on 80–100 mesh Chromosorb G.

Lawrence[5] compared various extraction procedures for the removal of triazine herbicides (atrazine, propazine, and simazine) from root crops prior to gas chromatagraphic determination using an electrolytic conductivity detector. The preferred method was extraction with 50% aqueous methanol followed by partitioning between water and chloroform; the chloroform extract was then used for direct gas chromatography. The hexane partition and column clean-up steps were unnecessary for the N-specific detection system, thus shortening the time of analysis. The recoveries of these pesticides from potato, carrot, turnip, beet and parsnip cross at levels down to 0·02 ppm were very good.

A gas chromatographic method has been issued by the US Environmental Protection Agency[6] for the determining at the $\mu g/l^{-1}$ level the following herbicides in water and waste water: ametryne, altraton, atrazine, prometon, prometryne propazine, secbumeton, simazine, and terbuthylazine. The method describes an efficient sample extraction procedure and provides, through use of column chromatography, a method for the elimination of non-pesticide interferences and the preseparation of pesticide mixtures. Identification is made by nitrogen-specific gas chromatographic separation, and measurement is accomplished by the use of an electrolytic conductivity detector or a nitrogen-specific thermionic detector.

It is pointed out that solvents, reagents, glassware, and other sample processing hardware may yield discrete artefacts and/or elevated baselines causing misinterpretation of gas chromatograms. All of these materials must be demonstrated to be free from interferences under the conditions of the analysis. Specific selection of reagents and purification of solvents by distillation in all-glass systems is required.

The interferences in industrial effluents are high and varied and often pose great difficulty in obtaining accurate and precise measurement of triazine pesticides. The use of specific detector supported by an optional column clean-up procedure will eliminate many of these interferences. Nitrogen-containing compounds other than the triazines may interfere.

The sample (1–2 l) is extracted with 60 ml methylene chloride and the extract concentrated in a Kuderna-Danish flask and the residue taken up in hexane. Florisil clean-up is applied if necessary and the extract gas chromatographed on a column comprising Carbowax 30 M (1%) supported on Gas Chrom Q (100–120 mesh). Figure 242 shows a typical gas chromatogram obtained by this procedure for a mixture of herbicides. Retention data is tabulated in Table 158.

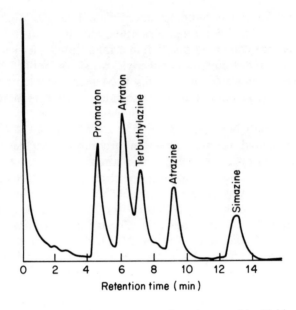

Figure 242 Gas chromatogram of a mixture of herbicides (see Table 158). Column packing, 1% Carbowax 20 M on Gas Chrom Q (100–120 mesh); Column temperature, 155 °C; carrier gas, helium at 80 ml min^{-1}; detector, electrolytic

Table 158 Retention ratios of various triazine pesticides relative to atrazine

Pesticide	Retention ratio
Prometon	0.52
Atraton	0.67
Propazine	0.71
Terbuthylazine	0.78
Secbumeton	0.88
Atrazine	1.00
Prometryne	1.10
Simazine	1.35
Ametryne	1.48

Absolute retention time of atrazine = 10.1 minutes.

Hormann et al.[7] monitored various European rivers for levels of atrazine, simazine, terbumeton, and dealkylated metabolites GS 26571 (2-amino-4-*tert*-butylamino-6-methoxy-1,3,5-triazine and G 30033 (2-amino-4-chloro-6-ethylamino-1,3,5-triazine). The compounds were extracted into dichloromethane and quantitated by gas chromatography with nitrogen-specific detection. Selected results were verified by gas chromatography with mass fragmentographic detection. The limits of detection were usually 0.4 mg m^{-3}.

Ten litre samples each were taken. Aliquots were transferred into aluminium bottles, deep frozen, and shipped to the laboratory. Until analysis samples were stored at −20 °C. Triazines stored at this temperature have been shown to be stable for many years in neutral aqueous solutions as well as in soil and crop materials.

One litre was thawed and the water was filtered to remove sediments. The filtrate was transferred to a 2 litre separatory funnel and extracted with three consecutive portions of dichloromethane which had been used previously to rinse the sample container. After passage over a plug of cotton to remove excess water, the extracts were combined and evaporated to dryness in a rotary evaporator, using a 30 °C water bath. The residue was dissolved in an appropriate volume, 0.5–2 ml, of a 1:1 mixture of hexane–ethanol and the solution was injected into a gas chromatograph equipped with a nitrogen-specific Hall or Coulson detector. Instrument parameters and operating conditions were as follows:

Column: glass 1 m long × 3 mm i.d. packed with either 3% Carbowax 20 M on 0.15–0.18 mm Gas-Chrom Q or 2% FFAP on 0.15–0.18 mm Gas Chrom G.

Temperatures: injector 250 °C; columns 190–210 °C isothermal; interface 250 °C; detector oven 800 °C.

Carrier gas: helium flowing at 60 ml min^{-1}.

Minimum detection levels were 0.1–0.4 mg m^{-3} except in a few cases where relatively high concentrations of interfering materials were present. Recoveries for all compounds were 80–120% at 5 mg l^{-1} and 10 mg l^{-1} fortification levels.

Several samples showing residues above the detection limit were confirmed by gas chromatography–mass spectometry. Instrument parameters and operating conditions were as follows:

Instrument: Finnigan Model 3000, equipped with programmable multiple ion monitor.

Masses selected: atrazine 215, simazine, 201, terbumeton 225.

Column: glass 1 m long × 2 mm i.d., packed with 2% SP 1000 on 0.15–0.18 mm Chromosorb G.

Temperatures: injector 240 °C; column 220 °C isothermal; separator 220 °C; transfer 180 °C; manifold 120 °C.

Electron energy: 70 eV.

Carrier gas: helium flowing at 30 ml/min^{-1}.

Wu et al.[8] carried out measurements of the enrichment of atrazine on the microsurface water of an estuary. These authors used a microsurface water sampling technique with a 16 mesh stainless steel screen collecting bulk samples from the top 100–150 μm pf the surface[9]. The enrichment of atrazine in the microsurface varied from none to 110 times, with the highest enrichment between mid-September and late October. Atrazine concentration in the actual microsurface was estimated to vary in the range 150–8850 μg l^{-1}.

Approximately 4 litres of microsurface or 15 litres of bulk surface water was taken for herbicide analysis. Five grams of calcium chloride were added to the water sample for each litre of water. The water sample was left overnight at about 5 °C and then filtered through a 142 mm Gelman type AE glass fibre filter to separate solid particles from the dissolved phase. The glass fibre filters, which had been previously cleaned with a water extraction procedure, were then extracted with a 1:3 mixture of toluene and dichloromethane. After filtration, solid particles were extracted with 200 ml of a 1:9:10 toluene–hexane–dichloromethane mixture in Soxhlet extractors. Herbicides in the dissolved phase were extracted in a separatory funnel first with 100 ml of dichloromethane. Grade V alumina (15 g H$_2$O and 85 g Al$_2$O$_3$) was used in the subsequent clean-up procedure. A gas chromatograph equipped with a Hall electrolytic detector in the nitrogen mode was used for quantitation. The chromatographic conditions were 1.8 m × 4 mm glass columns packed with 3% OC-17 on 80–100 mesh Gas Chrom Q, 190 °C column temperature, 65 ml min^{-1} helium flow rate.

Gas chromatography–mass fragmentography

Fenselau[10] has reviewed techniques based on the coupling of a gas chromatography to a mass spectrometer including mass fragmentography to the identification of organic compounds.

Karlhuber et al.[11] applied gas chromatography–mass fragmentography to some typical problems in pesticide residue analysis. Specific examples include two triazines (terbuthylazine and GS 26571), methoxyethanol, 3,5-dibromo-4-hydroxybenzoic acid (from bromofenoxim) and atrazine residues in sewage effluents. Using this technique 10 μg of terbuthylazine (4-tertbutylamino-2-chloro-6-ethylamino-1,3,5-triazine) and GS 26571 (2-amino-4-tertbutylamino-6-methoxy-1,3,5-triazine) could easily be determined even though the two compounds were not resolved on the gas chromatogram. The gas chromatography–mass fragmentography procedure was also used by Karlhuber et al.[11] to identify apparent atrazine residues in sewage water. The samples were cleaned up by the procedure of Ramsteiner.[12] An aliquot of sewage water was neutralized and extracted with dichloromethane. The extract was further cleaned up by passage through an alumina column. After concentration, atrazine was injected into the gas chromatograph equipped with a Coulson electrolytic conductivity detector.

Injection of a 12.5 μg sewage water aliquot into a gas chromatograph equipped with the nitrogen-specific Coulson electrolytic conductivity detector showed a peak with the same retention time as atrazine. From a standard injection an apparent atrazine concentration of 5.5 ppm in the sewage sample was calculated. The value found seemed extremely high and the sample was reinjected on the gas chromatograph with mass fragmentographic detection. No peak showed up in the chromatogram at the retention time of atrazine, indicating that there was less than 0.01 ppm of atrazine in the sewage water. The peak in the nitrogen-specific chromatogram, therefore, was a nitrogen-containing interference with the same retention time as atrazine. Using a 4 litre water sample containing 2 mg atrazine 82–98% recovery was obtained using this procedure.

Thin-layer chromatography

Zawadzka et al.[13] and Abbott[14] used thin-layer chromatography to determine simazine, atrazine, and prometryne herbicides in water and sewage. After extraction of a 250–1000 ml sample (water) or a 10–100 ml sample of sewage, with dichloromethane or ethyl ether at pH 9, the organic extract was condensed and applied to a column of basic aluminium oxide (activity III) and the herbicides were eluted with ether containing 0.5% of water. The eluate was condensed and applied to a layer of silica gel G impregnated with fluorescein. The chromatograms were developed with chloroform–acetone (9:1). The plates were dried, and the spots were located by spraying with 0.5% Brilliant green (C.I. Basic Green 1) in acetone and exposing to bromine vapour. The plates were evaluated planimetrically. For samples containing 5–100 μg of herbicide per litre the recoveries were between 83 and 97%.

Thin-layer chromatography has also been used[15] to determine ring-labelled [^{14}C] ametryne in water and soil. The herbicide was applied to silica gel plates as a methanel solution. After development the spots were located under ultraviolet radiation, removed, and treated with liquid scintillation solution for counting. The limit of detection was 5 ng.

Fishbein[16] has reviewed thin-layer, paper chromatographic, and gas and column chromatographic procedure for the determination of triazine herbicides.

SUBSTITUTED UREA TYPE HERBICIDES

Substituted urea herbicides with different substitutes X_1, X_2, Y_1, Y_2.

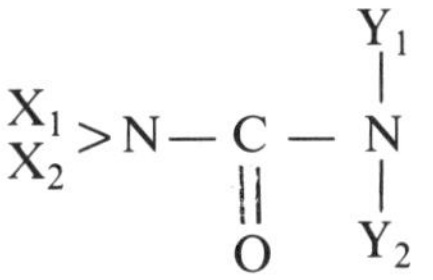

$$\begin{matrix} X_1 \\ X_2 \end{matrix} > N - \underset{\underset{O}{\|}}{C} - \underset{\underset{Y_2}{|}}{\overset{\overset{Y_1}{|}}{N}}$$

Substituents

Name	X_1	X_2	Y_1	Y_2
Buturon	H	4-Chlorophenyl	CH$_3$	CH(CH$_3$)C≡CH
Chlorbromuron	H	3-Chloro-4-bromophenyl	CH$_3$	-O-CH$_3$
Chlortoluron	H	3-Chlorotoluyl	CH$_3$	CH$_3$

continued

Continued

Name	Substituents			
	X_1	X_2	Y_1	Y_2
Diuron	H	3,4-dichlorlphenyl	CH_3	CH_3
Fenuron	H	Phenyl	CH_3	CH_3
Isoproturon	H	Cumenyl	CH_3	CH_3
Linuron	H	3,4-Dichlorophenyl	CH_3	$-O-CH_3$
Monuron	H	4-Chlorophenyl	CH_3	CH_3
Metabenzthia-zuron	CH_3	2-Benzothiazolyl	CH_3	H
Metoxuron	H	3-Chloro-6-metoxyphenyl	CH_3	CH_3
Neburon	H	3,4-Dichlorophenyl	CH_3	C_4H_9

Buturon	3-(4-Chlorophenyl)-1-methyl-1-(1-methylprop-2-ynyl)-urea
Chlorbromuron	3-(4-Bromo-3-chlorophenyl)-1 methoxy-1-unethylurea
Chlorotoluron	3-(3-Chlorotoluyl)-dimethyl urea
Chlorooxuron	3-[4-(4-Chlorophenoxy)phenyl]-1,1-dimethyl-urea
Diuron	3-(3,4-Dichlorophenyl-1)-1,1-dimethylurea
Linuron	3-[3,4-Dichlorophenyl]-1-methoxymethylurea
Monolinuron	3(4-Chlorophenyl)-1-methoxymethylurea
Monouron	3(4-Chlorophenyl)-1,1-dimethylurea
Siduron	1-(2-Methylcyclohexyl-3)-phenylurea
Fenuron	1,1-Dimethyl-3-phenylurea
Neburon	1-Butyl-3-(3,4-dichlorophenyl)-1-methylurea
Metabenzthiazuron	
Metoxuron	
Metobromuron	

Some of the urea herbicides have been separated on cellulose columns following hydrolysis to amines.[17] Gas chromatography using a nitrogen-specific detector has also been used.[18-21]. Gas chromatography of phenylurea herbicides is difficult because of their ease of thermal decomposition. Procedures have been reported in which careful control of conditions allows these compounds to be gas chromatographed intact.[20, 22] Alternatively the phenylurea herbicides can be hydrolized to the corresponding substituted anilines which are then determined by either gas chromatography directly[23] or as derivatives[24] or colorimetrically after coupling with a suitable chromophore.[25] More recently Deleu et al.[26] has used two-dimensional thin-layer chromatography and gas chromatography to separate and identify 11 urea herbicides in concentrations

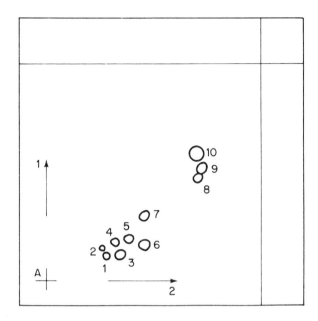

Figure 243 Two-dimensional chromatogram of: 1 = metoxuron; 2 = fenuron; 3 = monuton; 4 = isoproturon; 5 = chlorotoluron; 6 = diuron; 7 = metabenzthiazuron; 8 = neburon; 9 = linuron, chlorbromuron; 10 = buturon A. Reprinted with permission from Deleu et al.[26] Copyright (1977) Elsevier Science Publishers B.V.

Table 159 $R_f \times 100$ of urea herbicides

Herbicide	Solvent			
	(1)	(2)	(3)	(4)
Buturon	29	51	63	70
Chlorbromuron	31	50	56	69
Chlortoluron	7	19	19	37
Diuron	6	14	15	44
Fenuron	6	16	16	28
Isoproturon	6	18	16	31
Linuron	29	49	56	67
Monuron	4	12	12	33
Metabenzthiazuron	14	26	30	42
Metoxuron	4	11	10	30
Neburon	24	49	52	67

Reprinted with permission from Deleu et al.[26] Copyright (1977) Elsevier Science Publishers B.V.

down to 4 parts per 10^9 in natural waters. Four different adsorbents were compared. The eluting solvents were:

(1) diethyl ether–toluene (1:3);
(2) diethylether–toluene (2:1);
(3) equal volumes of (1) and (2); and
(4) chloroform–nitromethane (1:3).

Retention data are shown in Table 159 whilst Figure 243 shows a two-dimensional thin-layer chromatogram in which ten of the herbicides have been successfully resolved. Table 160 shows the detection systems employed.

Table 160 Detection of urea herbicides

Herbicide	Detection method			
	A	B	C	D
Buturon	+	+	+	+
Chlorbromuron	+	+	+	(+)
Chlortoluron	+	+	+	+
Diuron	+	+	+	(+)
Fenuron	+	−	+	+
Isoproturon	+	−	+	+
Linuron	+	+	+	(+)
Monuron	+	+	+	(+)
Metabenzthiazuron	+	−	−	−
Metoxuron	+	+	+	+
Neburon	+	+	+	−

A, Ultraviolet 254 lamp; B, Silver nitrate spray; C, Bratton–Marshall spray[27]; D, Diethylaminobenzaldehyde.
Reprinted with permission from Deleu et al.[26] Copyright (1977) Elsevier Science Publishers B.V.

Table 161 Gas chromatography of urea herbicides, retention times relative to neburon

Herbicide	T_R
Buturon	0.48
Chlororomuron	1.66
Chlorotoluron	0.62
Diuron	1.00
Fenuron	—
Isoproturon	—
Linuron	1.00
Metabenzthiazuron	—
Metoxuron	1.94
Monuron	0.44
Neburon	1.00

Reprinted with permission from Deleu et al.[26] Copyright (1977) Elsevier Science Publishers B.V.

To analyse river waters, the sample (0.5–1/l) was extracted with 50 ml chloroform and the chloroform gently evaporated off at 30–35 °C. The residue was dissolved in 250 µl chloroform prior to thin-layer chromatography. Identifications made by thin-layer chromatography were confirmed by gas chromatography of the extract using a ^{63}Ni electron capture detector. The column (1.8 m × 2 mm i.d.) consisted of 10% OV-17 and 10% OV-210 on Chromosorb W-HP (100–120 mesh) operated at 140 °C (injector temperature 250 °C, detector temperature 250 °C) using argon–methane 95.5 as carrier gas at 43 ml min^{-1}. Retention data for the eleven herbicides are tabulated on Table 161.

Guthrie et al.[28] also mentioned the use of thin-layer chromatography for the determination of diuron (3-(3,4-dichlorophenyl-1-methoxymethylurea) in river waters.

Farrington et al.[29] has described high performance liquid chromatography procedures for the determination in water, soil, and grain of residues of the following phenylurea herbicides:

chlorobromuron [3-(-4 bromo-3-chlorphenyl]-1-methoxy-1-methylurea)
chloroxuron, [3-4(-(4-chlorophenoxy)phenyl)]-1,1-dimethylurea
chlortoluron, 3-(3-chlorotoluyl)dimethylurea
diuron, [3-(3,4-dichlorophenyl-1)]-1,1-dimethylurea
linuron, [3-(3,4-dichlorophenyl)]-methoxymethylurea
metobromuron
monolinuron, [3-(4-chlorophenyl)-1]-methoxymethylurea
monouron [3-(4-chlorophenyl)]-1,1-dimethylurea

Farrington et al.[29] favour the high performance liquid chromatographic approach to the identification and determination of phenylurea herbicides because of the obvious limitations of alternate methods such as direct gas chromatography, gas chromatography of hydrolyses products or derivatives and spectrophotometry. They modified earlier work on liquid chromatographic methods.[30, 31]

Method

Farrington et al.[29] used a Waters Associates constant-volume solvent-delivery liquid chromatograph. A variable-wavelength ultraviolet monitor fitted with a 10 µl flow cell and set at 240 mm was used as a detector. A stainless steel column tube 300 × 4.6 mm i.d. was washed with chloroform and methanol and polished on the inner surface. One end was fitted with a ¼ × ¹⁄₁₆ in. column end fitting and the other was coupled to a 400 mm precolumn reservoir through a ¼ × ¼ in. Swagelock union. A slurry of 5 µm Spherisorb ODS was packed into the column by releasing 5000 psi of solvent (acetone) pressure. The column was prepared for stop-flow injection by removing the top few millimetres of packing and inserting a disc of stainless steel fine-mesh gauze, of 8 µm nominal porosity, a plug of silanized glass-wool and a top plug of porous PTFE.

Reagents

All reagents should be analytical reagent grade unless otherwise specified:
Dichloromethane, laboratory-reagent grade.
Methanol, spectograde.
Mobile phase: a solution of 60% methanol in water is prepared and 0.6% of ammonia solution (sp.gr.0.88) is added.
Sodium sulphate, anhydrous, granular.
Uron standard solutions: a standard solution in Spectograde methanol containing 0.25 mg l^{-1} of uron is prepared and diluted as necessary.

Procedure

Water

To 1 l of river water in a 2 l separating funnel is added 100 ml of dichloromethane and the funnel shaken for 30 s. The dichloromethane is run off and the extraction repeated twice, using 50 ml portions of dichloromethane. The extracts are dried by passing them through a column of anhydrous sodium sulphate (50 g) and the column is washed with 50 ml of dichloromethane. The extracts and washings are combined and the dichloromethane removed in a rotary evaporator with a waterbath at 55 °C. The flask is cooled and 5.0 ml of methanol added, swirling the flask to dissolve the residue; 5 µl of the solution are injected into the chromatograph. The uron content of the sample is determined using the procedure described below under grain.

Grain

A 50 g sample is ground and transferred into a 50 ml flat-bottomed flask. To this is added 100 ml of methanol and the flask is shaken on a wrist-action shaker for 1 h. The resulting slurry is filtered through a Whatman No. 1 filter paper using a reduced pressure. The flat-bottomed flask is washed with 50 ml of methanol, the washings are added to the filter funnel, left for 3–5 min, then the reduced pressure applied. The washing procedure is repeated with a further 50 ml of methanol. The extract and washings are combined and then the methanol removed using a rotary evaporator with a water bath at 55 °C. The residue is dissolved in dichloromethane using a total volume of 50 ml, and the dichloromethane extracts passed through a column of anhydrous sodium sulphate (50 g). The sodium sulphate is washed with 50 ml of dichloromethane, the extract and washings combined and evaporated to dryness at 55 °C in a rotary evaporator. The flask is cooled and 5.0 ml of methanol added, swirling the flask to dissolve the residue. The solution is filtered through a Whatman No. 42 filter paper. Using a flow rate of the mobile phase of 0.6 ml min^{-1}, 5 µl of the sample solution is injected into the liquid chromatograph. The uron content of the sample is calculated by comparing the peak height obtained with those obtained from 5 µl injections of standard solutions.

Soil

A sample of soil is air dried and 50 g transferred into a 500 ml flat-bottomed flask. The sample is extracted and filtered using the method described under Grain. The methanol is removed by using a rotary evaporator with a water bath at 55 °C. The flask is cooled and 5.0 ml of methanol are added, swirling to dissolve the residue. The solution is filtered through a Whatman No. 42 filter paper. Using a flow rate of 0.6 ml min^{-1}, 5 μl of extract are injected into the liquid chromatograph and the uron content determined using the procedure described under Grain.

The recoveries obtained for urons from fortified samples of wheat, soil and water are shown in Table 162. Samples of wheat and soil were fortified by adding known volumes of solutions containing (a) monuron, metobromuron, diuron, and chlorbromuron or (b) monolinuron, chlortoluron, linuron, and chlorooxuron. The lower recoveries from wheat may be attributable to residual oil, which remains after the evaporation of the dichloromethane, increasing the volume of the solvent as this oil is miscible with methanol.

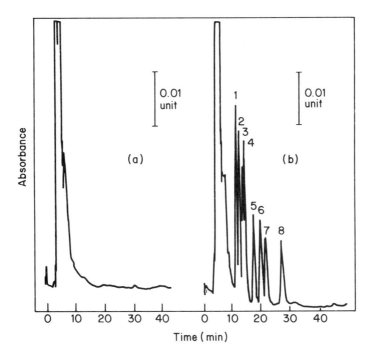

Figure 244 Typical chromatogram obtained from 5 μl injections of wheat extracts; (a) unfortified, and (b) fortified with uron herbicides at 2 mg kg^{-1}: 1, monuron; 2, monolinuron; 3, metobromuron; 4, chlortoluron; 5, diuron; 6, linuron; 7, chlorbromuron; and 8, chloroxuron. Reprinted with permission from Farrington *et al.*[29] Copyright (1977) Royal Society of Chemistry

Table 162 Recovery of urons from fortified samples

Uron	Wheat Fortified at 5 mg kg^{-1}		Fortified at 2 mg kg^{-1}		Fortified at 0.5 mg kg^{-1}		Soil fortified at 2 mg kg^{-1}		Water fortified at 0.1 mg kg^{-1}	
	Range	Mean	Range	Mean	Range	Mean	Range	Mean	Range	Mean
Chlorbromuron	91.0–95.5	92.5	87.5–93.0	91.0	84.5–92.5	88.0	98.0–101.0	99.5	98.0–100.0	99.0
Chlortoluron	89.5–94.0	92.0	87.0–92.5	89.5	85.5–94.0	89.0	100.0–103.0	101.5	97.0–100.5	99.0
Chloroxuron	87.5–91.5	89.5	87.0–91.0	88.5	86.5–94.0	89.5	97.0–103.0	99.0	94.5–100.5	97.0
Diuron	89.0–94.5	91.5	88.0–90.5	89.0	87.0–92.5	89.0	98.0–100.5	99.5	97.5–102.0	100.0
Linuron	90.5–94.5	92.5	86.5–93.5	89.0	85.5–96.0	91.5	94.0–100.5	97.5	96.0–100.5	98.5
Metobromuron	90.5–95.5	92.0	89.5–93.5	91.5	84.0–93.0	86.5	96.5–101.0	99.5	96.5–100.5	99.0
Monolinuron	90.0–95.5	92.5	89.0–94.0	92.5	87.0–95.5	90.0	100.5–103.0	102.0	97.5–100.0	98.5
Monuron	87.0–94.5	90.0	91.0–97.0	94.5	89.0–95.5	92.5	99.0–102.0	100.5	98.0–100.5	99.5

Five determinations were carried out on each sample. Results given are percentage recoveries.
Reprinted with permission from Farrington et al.[29]. Copyright (1977) Royal Society of Chemistry

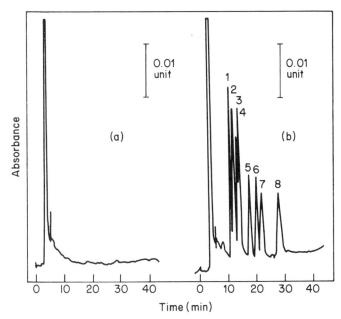

Figure 245 Typical chromatograms obtained from 5 μl injections of soil extracts; (a) unfortified; and (b) fortified with uron herbicides at 2 mg kg^{-1}. 1, Monuron; 2, monolinuron; 3, metobromuron; 4, chlortoluron; 5, diuron; 6, linuron; 7, chlorbromuron; and 8, chloroxuron. Reprinted with permission from Farrington et al.[29] Copyright (1977) Royal Society of Chemistry

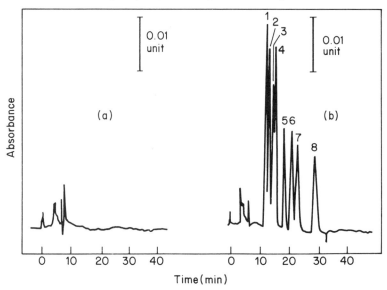

Figure 246 Typical chromatograms obtained from 5 μl injections of river water extracts; (a) unfortified and (b) fortified with uron herbicides at 0.1 mg kg^{-1}. 1, Monuron; 2, monolinuron; 3, metobromuron; 4, chlortoluron; 5, diuron; 6, linuron; 7, chlorbromuron, and 8, chloroxuron. Reprinted with permission from Farrington et al.[29] Copyright (1977) Royal Society of Chemistry

Typical chromatograms, obtained from extracts of wheat, soil, and river waters, are shown in Figures 244 and 246. The lower limits of detection were estimated to be 0.2 ppm for wheat, 0.2 ppm for soil, and 0.01 ppm for river water; below these levels both coextractives and signal noise interfere. Care was taken during the evaporation of solutions of urons in methanol to ensure that the temperature did not rise above 55 °C as at higher temperatures degradation occurs. A clean-up procedure was required to remove coextractives.[32]

Individual urea herbicides

Chlortoluron

Smith and Lord[33] have used liquid chromatography for the determination of chlortoluron [3-(3-chlortoluyl)-dimethylurea] residues in soil but report that diuron and monouron interfere in the chromatographic system used.

Diuron

In an electron capture method for estimating diuron [3-(3,4-dichlorophenyl-1)-1,-dimethylurea] in surface waters McKone and Hance[34, 35], extract the water sample (100 ml) with dichloromethane (2 × 25 ml) and the lower layers are combined and washed with water (5 ml). After filtering through cotton wool, the organic solvent is evaporated under reduced pressure at 35 °C. Saturated aqueous sodium chloride is added to the residue, the mixture is shaken, 2,2,4-trimethylpentane (5 ml) is added, and the mixture is shaken again. An aliquot of the organic layer is then subjected to gas chromatography in a stainless steel column packed with 5% of E301 (methyl silicone) on Gas Chrom Q (60–80 mesh). Recoveries from controls and from pond, canal, and river waters containing 0.001–1 ppm of diuron were about 94%, the coefficient of variation at the higher levels being about 7%.

Monolinuron (3-(4-chlorophenyl-1-methoxy) methylurea) and linuron (3-(3,4-dichlorophenyl) 1-1-methoxymethylurea).

Yuen[36] has described an absorptiometric method for determining these substances in herbicide formulations.

Siduron (1-(2-methylcyclohexyl-1-3-phenylurea))

This herbicide has been determined in fertilizers by flame ionization gas chromatography.[37] A chloroform extract of the sample was analysed on a glass column (4 ft × 3 mm) of 3% of JXR on 80–100 mesh Gas Chrom Q at 204 °C with helium as carrier gas (20 ml min^{-1}).

Metobromuron

This herbicide has been determined in tobacco leaves[38] by column chromatography on alumina and spectrophotometry of coupling products of their azo-derivatives.

1-Methoxy-1-methyl-3-phenylurea herbicide

In a method described by Rosales[39] the herbicide is hydrolysed in phosphoric acid to give aniline and *N*-methoxymethylamine, which together with certain impurities in the commercial product are titrated with sodium nitrite solution. Aniline and certain byproducts are also determined separately by gas chromatography on a glass column (2 m × 4 mm) containing 10% of silicone oil OV-17 on Chromosorb Q (80–100 mesh) temperature programmed from 100 to 200 °C at 4.5 °C min^{-1}, with thermal conductivity detection and helium carrier gas.

Mixed herbicide types

El-Deb[40] has used thin-layer chromatography to detect carbamate insecticides and urea herbicides in natural water. He tabulates the R_f values and spot colours of a number of these on silica gel with six different solvent systems. 4-Dimethylaminobenzaldehyde or diazotization and coupling with 1-naphthol were used to locate the spots.

Spengler and Jumar[41] used a spectophotometric method and thin-layer chromatography to determine carbamate and urea herbicide residues in sediments. The sample is extracted with acetone, the extract is evaporated *in vacuo* at 40 °C and the residue is hydrolysed with sulphuric acid. The solution is made alkaline with 15% aqueous sodium hydroxide and the liberated aniline (or substituted aniline) is steam distilled and collected in hydrochloric acid. The amine is diazotized and coupled with thymol, the solution is cleaned up on a column of MN 2100 Cellulose power and the azo-dye is determined spectrophotometrically at 440 nm (465 nm for the dye derived from 3-chloro- or 3,4-dichloroaniline) with correction for the extinction of a reagent blank. There was no significant loss in the extraction process. The acetone extract of the soil was then evaporated to dryness, and the residue dissolved in concentrated hydrochloric acid and purified by precipitation with sodium tungstate. The herbicide is extracted from the filtrate with chloroform and the extract is dried over sodium sulphate and evaporated *in vacuo*. The dried residue is dissolved in methanol and analysed by thin-layer chromatography on alumina with benzene–acetone (19:1) as solvent and 4-dimethylaminobenzaldehyde as spray reagent using visual comparison with standards. Recoveries were 100% for 5 ppm of added herbicide and ranged from 60 to 100%.

In a further thin-layer chromatographic method for determining carbamate and urea herbicides in water at the parts per 10^9 level Frei *et al.*[42] extracted a 500 ml sample with dichloromethane (2 × 50 ml) and evaporated the combined extract to 1 ml at room temperature in a rotary evaporator and then to dryness at 40 °C. The residue was dissolved in acetone (1 or 2 drops) and 0.5 ml of sodium hydroxide and heated to 80 °C for 30–40 min, cooled and shaken with 0.2 ml of hexane. 10 µl of the hexane layer is applied to a 0.25 mm layer of silica gel G-CaSO$_4$, 4 l of 0.2% dansyl chloride in acetone is applied to the sample spot, and the chromatogram developed by the ascending technique with benzene–triethylamine–acetone (75:24:1). The plate is sprayed with 20% triethanolamine in isopropyl alcohol or 20% liquid paraffin in toluene, then dried. The fluorescence of the spots of the dansyl derivatives of the aniline moieties is measured *in situ*. Results are reported for carbamate pesticides, e.g. propham, chlorpropham, and barban; and the urea pesticides, linuron, diuron, chlorbromuron, and fluometuron; detection limits are about 1 ng. Two-dimensional chromatography was used to eliminate interferences.

Cohen and Wheals[43] used a gas chromatograph equipped with an electron capture detector to determine ten substituted urea and carbamate herbicides in river water, soil and plant materials in amounts down to 0.001–0.05 ppm. The methods are applicable to those urea and carbamate herbicides that can be hydrolysed to yield an aromatic amine. A solution of the herbicide is first spotted on to a silica gel G plate together with herbicide standards (5–10 µg) and developed with chloroform or hexane–acetone (5:1). The plate containing the separated herbicide or the free amines is sprayed with 1-fluoro-1,4-dinitrobenzene (4% in acetone) and heated at 190 °C for 40 min to produce the 2,4-dinitrophenyl derivative of the herbicide amine moiety. Acetone extracts of the areas of interest are subjected to gas chromatography on a column of 1% of XE-60 and 0.1% of Epikote 1001 on Chromosorb G (AW-DCMS) (60–80 mesh) at 215 °C.

PHENOXYACETIC ACID TYPE HERBICIDES

2,4-Dester and sodium salts (2,4-dichlorophenoxyacetic acid)
2,4-DP(Dichloroprop); (2,4-dichlorophenoxypropionic acid)
2,4-DB(4-(2,4-dichlorophenoxy)-butyric acid)
MCPA(4 chloro-2-methylphenoxyacetic acid)
MCPB(4-(4-chloro-2-methylphenoxy)-butyric acid)
Silvex (2-(2,4,5-trichlorophenoxy)-propionic acid)
MCPP(mecoprop; mixture of mecoprop and 2-(2-chloro-4-methylphenoxy)-propionic acid)
2,4,5-T(2,4,5-trichlorophenoxyacetic acid)
Dicamba
Trifluralin(2-methoxy-3,6-dichlorobenzoic acid)
Methoxychlor
Fenoprop

Gas chromatography

Chemical derivatization of phenoxyalkanoic acid herbicides has been used as a means of forming less polar and more volatile compounds for the gas–liquid chromatographic analysis. Alkyl esters,[44-51] especially methyl esters,[46-51] have been used extensively for this purpose. Examples of earlier work utilizing diazomethane for the formation of methylesters of phenoxyalkanoic acid herbicides are discussed by Devine and Zweig[52] and Colas et al.[53]

Devine and Yip[52] adjust the water sample (1 litre) to pH 2 with hyrochloric acid and extract it with benzene (100, 50, and 50 ml). The extract is dried over sodium sulphate, concentrated to 0.1 ml and methylated by the addition of diazomethane in ethyl ether (1 ml). After 10 min, the volume is reduced to about 0.1 ml, acetone is added and an aliquot is analysed by gas chromatography on one of three columns: (1) 5% SE-30 on 60-80 mesh Chromosorb W at 175 °C or (2) 2% QF-1 on 90-100 mesh Anakrom ABS at 175 °C or (3) 20% Carbowax 20 M on 60-80 mesh Chromosorb W at 220 °C. In each instance nitrogen is the carrier gas and detection is by electron capture. The minimum detectable amount of pesticide in water was 2 parts per 10^9 for MCPA (4-chloro-2-methylphen-oxyacetic acid) and 0.01-0.05 part per 10^9 for 2,4-D(2,4-chlorophenoxyacetic acid) and its esters, 2,4,5-T(2,4,5-trichlorophenoxyacetic acid), dicamba, trifluralin (2-methoxy-3,6-dichlorobenzoic acid), and fenoprop. Recoveries were 50-60% for MCPA and dicamba and 80-95% or the other compounds.

Colas et al.[53] have described methods for the separation and determination of down to about 1 ppm of phenoxyalkanoic herbicides. Those present as salts or esters are hydrolysed by heating the water sample (1 litre) under reflux with sodium hydroxide for about 1 hour and the free acids are then extracted at pH 2 with chloroform or dichloromethane (recoveries usually about 70%). After evaporation of the solution to dryness, the residual free acids are dissolved in acetone and treated with diazomethane and the methyl esters are analysed on a temperature programmed column (1.5 m × 6 mm) containing 5% of silicone DOW 710 on Chromosorb W AW (45-60 mesh). Helium is used as carrier gas. For some separations other stationary phases are used. Typical results are presented for MCPA, MCPP and 2,4-D. Possible interference from phenols, chlorinated biphenyls, and surfactants is discussed.

Croll[54] has given details of the use of back-flushing with electron capture gas chromatography for the determination of phenoxyacetic acid type herbicides in water. Attention is paid to equalization of column resistance under operating and back-flushing conditions; baseline drift is thereby minimized. The system has been successfully used (with a variety of stationary phases, temperature ranging from 25 to 225 °C and nitrogen flow rates from 25 to 200 ml min^{-1}.

Larose and Chau[55] state that owing to the similar retention times of several common phenoxyacetic acid type of herbicides; the alkyl esters are subject to incorrect identification if several herbicides are present. Also, the sensitivity obtainable by means of electron capture detection of the alkyl esters by some herbicides, such as MCPA and MCPB is very poor and therefore the method is generally not suitable for the determination of these compounds in water. In addition, the methyl ester of MCPA has a very short retention time close to the solvent front and is prone to interference from sample coextractives, which usually appear in this region. In fact the MCPA methyl ester often cannot be detected even at higher levels because of overlapping with coextraction peaks when the same gas chromatographic parameters as for the determination of organochlorine pesticides are used. Hence other derivatives have been considered.

Among other derivatives such as those obtained from nitration,[56,57] bromination coupled with other reactions,[57,58] silylation using NO bis-trimethyl silylacetamide reagent,[59] and 2-chloroethylation,[60,61] derivatives obtained from reactions involving 1-bromomethyl-2,3,4,5,6-pentafluorobenzene[62,64] were found to have the most desirable characteristics, excellent detector response, and longer retention times. 1-Bromomethyl-2,3,4,5,6-pentafluorobenzene has been successfully applied to the determination of carbamates[66] and to the confirmation of organophosphorus pesticides[67,68] by forming the 2,3,4,5,6-pentafluorobenzyl derivatives of the phenolic and thiophenolic hydrolysis products. The conditions for the preparation of these derivatives of 2,4-D[69] and ten other herbicidal acids have also been studied.[65].

Chau and Terry[65,70] have discussed the disadvantages of the gas chromatography of methyl esters produced by reaction with diazomethane and have developed the reaction conditions for forming 2-chloroethyl (2-Cl) and pentafluorobenzyl (PFB) esters of phenoxyacetic acids. Agemian and Chau[71] have reported the method described below for determining low levels of 4-chloro-2-methylphenoxyacetic acid and 4-(4-chloro-2-methylphenoxy)-butyric acid in waters by derivatization with pentaflurobenzyl bromide. The increased sensitivity of the pentafluorobenzyl esters of these two herbicides over the 2-chloroethyl and methyl esters as well as their longer retention times make pentaflurorobenzyl bromide the preferred reagent.

These workers[71] used a gas-liquid chromatograph equipped with a nickel detector, a 6 ft × ¼ in. i.d. coiled glass column and an automatic sampler connected to a computing integrator for data processing. The column used was 3.5% m/m OV-101 and 5.5% m/m OV-210 on 80–100 mesh Chromosorb W, acid-washed and treated with dimethylchlorosilane and prepared as described by Chau and Wilkinson.[72] The operating conditions were as follows: injector temperature 220 °C, column temperature 220 °C, detector temperature 300 °C; carrier gas, argon–methane (9 + 1) at a flow rate of 60 ml min^{-1}.

Method

Reagents

The organic solvents used were of pesticide residue grade.

1-Bromomethyl-2,3,4,5,6-pentafluorobenzene: a 1% v/v solution in acetone is prepared in a 100 ml low actinic calibrated flask.

Potassium carbonate solution, 30% w/v.

Silica gel: silica gel (grade 950 for gas chromatography, 60–200 mesh. Fischer Scientific, Canada) is activated by heating at 130 °C overnight (14 h), then deactivated by adding distilled water 5% m/m.

Procedure

Extraction and derivative formation

A 1 l water sample in a 40 oz glass bottle is stirred on a magnetic stirrer using a PTFE stirring bar, so that the vortex formed at the surface almost reached the bottom of the bottle. Diluted sulphuric acid (1 + 1) is carefully added dropwise until the pH is 1 or less.

The 50 ml of dichloromethane is added and the bottle tightly covered with a PTFE or aluminium-lined cap. After stirring for 45 min the contents of the bottle are transferred to a 2 l separating funnel and shaken for 1 min. The organic layer is transferred to a clean 500 ml separating funnel. The aqueous layer is shaken with two additional 50 ml portions of dichloromethane and the organic layers transferred to the 500 ml separating funnel. The dichloromethane extract is washed by tumbling it a few times with 50 ml of organic-free distilled water (pH 6). The layers are allowed to separate and the lower, organic phase is transferred into a 500 ml round bottomed flask. The aqueous layer is washed in the same manner with two 10 ml portions of dichloromethane and the organic layers added to the rest of the organic extract in the round bottomed flask.

The dichloromethane is evaporated to 2–3 ml on a rotary evaporator and the evaporation finished on a stream bath under atmospheric pressure (water bath temperature kept below 40 °C). The residue is dissolved with several 2–3 ml portions of acetone, each time transferring the acetone quantitatively into a 15 ml graduated centrifuge tube, the total volume of acetone used not to exceed 10 ml. To the tube are added 50 μl of 1-bromomethyl-2,3,4,5,6-pentafluorobenzene solution and 30 μl of potassium carbonate solution. The tube is stoppered and shaken for a few seconds. The contents are left to react at room temperature for 5 h.

After reaction 2 ml of 2,2,4-trimethylpentane are immediately added and the solution evaporated to 1 ml with a gentle stream of nitrogen gas. Another 2 ml of 2,2,4-trimethylpentane are added and the evaporation repeated to a final volume of about 1 ml. This procedure removes traces of acetone, which can affect the column fractionation.

Column clean-up

The extract is cleaned up in order to remove excess reagent and contaminants introduced by the reagents. This clean-up also removes most of the organochlorine pesticides and polychlorinated biphenyls. Microcolumns are prepared by plugging clean disposable pipettes (5 cm × 5 mm i.d.) with a piece of glass wool that has previously been washed with acetone and dried. The columns are filled with 5 cm of deactivated silica gel and tapped with a rod to settle the solid.

The columns are prewetted with 5 ml of hexane and the hexane allowed to drain just to the top of the packing material. The concentrated sample extract plus the hexane rinsings are applied with a disposable pipette to the column.

Elution with 15 ml of a 25% v/v solution of benzene in hexane is carried out in order to remove excess of reagent and contaminants (fraction A). Elution with 8 ml of a 75% v/v solution of benzene in hexene is then carried out into a clean centrifuge tube (fraction B). The volume collected is recorded and this fraction analysed for the pentafluorobenzyl esters of MCPA and MCPB by means of gas chromatography. The detector response obtained for the pentafluorobenzyl ester of 4-chloro-2-methylphenoxyacetic acid (MCPA) is 25 times greater than that obtained for the corresponding 2-chloroethyl ester and 1000 times greater than that for the methyl ester, whereas the detector response obtained for the pentafluorobenzyl ester of 4-(4-chloro-2-methylphenoxy)-butyric acid (MCPB) is 500 times greater than that obtained for the corresponding 2-chloroethyl ester and 10,000 times greater than that for the methyl ester. Figure 247 depicts the relative sensitivity of the method for the three esters of these two herbicides.

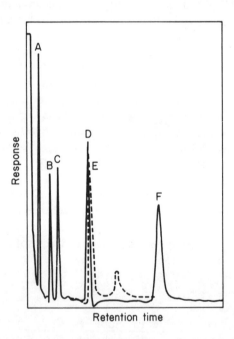

Figure 247 Chromatograms given by the methyl, 2-chloroethyl and pentaflurobenzyl esters of MCPA and MCPB: A, methyl ester of MCPA (25 mg μg^{-1}); B, methyl ester of MCPB (500 ng μl^{-1}). C, 2-chloroethyl ester of MCPA (500 pg μl^{-1}); D, PFB ester of MCPA (20 pg μg^{-1}; E, 2-chloroethyl ester of MCPB (25 ng μl^{-1}); F, PFB ester of MCPB (40 pg μl^{-1}.Sample size, 5 μl. Attenuation, 16. Reprinted with permission from Agemian and Chau.[71] Copyright (1976) Association of Official Analytical Chemists

Agemian and Chau[71] found that the few organochlorine pesticides that are eluted in the same fraction (fraction B, Table 163) as the pentafluorobenzyl derivatives of the phenoxyacetic acid herbicides do not interfere because they have distinct retention times (Table 164). Coburn and Chau[67, 68] have shown that organophosphorus pesticides do not interfere and Coburn et al.[66] have provided data to show that 24 of the most widely used phenols either are eluted

Table 163 Fraction of PCBs and organochlorine insecticides on a deactivated silica gel column

Compound	Amount eluted %	
	Fraction A*	Fraction B†
Aroclor 1254	100	—
Aroclor 1260	100	—
Lindane	98	—
Heptachlor	95	—
Aldrin	95	—
Heptachlor epoxide	102	—
γ-Chlordane	102	—
α-Chlordane	103	2
p,p-DDE	101	—
α-Endosulfan	101	2
Dieldrin	96	5
o,p'-DDT	100	—
Endrin	95	6
TDE	95	5
β-Endosulfan	—	104
p,p'-DDT	101	—
Methoxychlor	—	105

*Eluted in 25% v/v benzene in hexane
†Eluted in 75% v/v benzene in hexane
Reprinted with permission from Agemian and Chau.[71] Copyright (1976) Association of Official Analytical Chemists.

Table 164 Retention times relative to aldrin of compounds that occur in fraction B

Compound	Relative retention time
Aldrin	1.00*
PFB derivative of MCPA	1.77
Dieldrin	2.02
Endrin	2.30
TDE	2.30
β-Endosulfan	2.58
PFB derivative of MCPB	2.63
Methoxychlor	3.78

*Retention time = 5.2 min.
Reprinted with permission from Agemian and Chau.[71] Copyright (1976) Association of Official Analytical Chemists.

in fraction A with the PCBs and organochlorine pesticides or have distinct retention times from those of the pentafluorobenzyl esters of the two herbicides.

The whole of the above procedure, at a level of $0.5\,\mu g\,l^{-1}$ MCPA in 1 litre of distilled water, gave an average recovery of 75–80% with a coefficient of variation between 9 and 15%.

Agemian and Chau[73] using the above method compared the pentafluorobenzyl bromide reagent for forming esters and the boron trichloride-2-chloroethanol and dicyclohexylcarbodiimide-2-chloroethanol reagents for forming 2-chloroethyl esters of phenoxyacetic acid type herbicides, coupled with a complete solvent extraction system to obtain multiresidue methods for determining these compounds in natural waters at sub-ppb levels.

The eight herbicides studied by their workers were dicamba, MCPA, MCPB, 2,4-DB, picloram (4-amino-3,5,6-trichloropicolinic acid), 2,4-D, 2,4,5-T, silvex, and 2,4-DP.

Method

A gas chromatograph, equipped with ^{63}Ni detector 6 ft × ¼ in. o.d. coiled glass column and automatic sampler connected to a computing integrator for data processing was used. Operating conditions were as follows: injector temperature same as column temperature given below; detector 300 °C; argon–methane (9 + 1) carrier gas at 60 ml min^{-1}.

Columns: (1), 3.6% OV-101 and 5.5% OV-210 on 80–100 mesh Chromosorb W (acid-washed, treated with dimethylchlorosilane) (injector and column 210 °C); (2) 1.5% OV-17 and 1.95% QF-1 on 80–100 mesh Gas Chrom Q (injector and column 195 °C); (3) 3% OV-225 on 80–100 mesh Chromosorb Q.

Dicyclohexylcarbodiimide reagent: 2 g of the solid were dissolved in 100 ml pyridine in low actinic flask. This solution can be stored at room temperature for 3–4 months with no significant deterioration.

Boron trichloride reagent: research grade (or equivalent) boron trichloride gas was bubbled through triple-distilled 2-chloroethanol to about 10% concentration.

Potassium bicarbonate solution: 2 g potassium bicarbonate were dissolved in 100 ml deionized water, and extracted twice with benzene. The benzene extracts were discarded.

Hydrochloric acid: 20 ml concentrated hydrochloric acid were dissolved in water, and diluted to 100 ml.

Pentafluorobenzyl bromide reagent (1-bromoethyl-2,3,4,5,6-pentafluorobenzene): 1 ml reagent was transferred to the low actinic 100 ml volumetric flask and diluted to volume with acetone.

Potassium carbonate solution: 30 g anhydrous potassium ultrapure grade were dissolved in 100 ml water and extracted twice with benzene; the benzene extracts were discarded.

Silica gel: grade 950 for gas chromatography 60–200 mesh (Fisher Scientific, Don Mills, Ontario, Canada M3A 1A9), was activated by heating at 130 °C for 14 h, and deactivated by adding water 5% w/w.

Sodium sulphate: anhydrous reagent grade, was heated 12 h at 350 °C and stored in a clean glass bottle in a desiccator.

Extraction

A 1 l water sample is stirred in a 40 oz glass bottle on magnetic stirrer, using a Teflon stirring bar. Dilute sulphuric acid is carefully added (1 + 1) dropwise to pH 1. The contents of the bottle are transferred to a 2 l separatory funnel. The bottle is washed thoroughly with 200 ml ethyl acetate and the wash transferred to a 2 l separatory funnel and shaken vigorously for 1 min. The aqueous layer is returned to the 40 oz bottle and the organic layer transferred to a clean 500 ml separatory funnel. The aqueous contents of the 40 oz bottle are poured back into the original separatory funnel and the ethyl acetate extraction repeated twice with 50 ml portions. The organic layers are transferred to a 500 ml separatory funnel and the aqueous layer is discarded. 50 ml of 2% potassium bicarbonate solution are added to the organic layer and shaken vigorously for 2 min. The layers are separated and the aqueous layer is transferred to a clean 1 l separatory funnel. The organic layer is extracted twice with 30 ml potassium bicarbonate solution by shaking 1 min each time. The aqueous layers are transferred to a 1 l flask, discarding the organic layer.

The aqueous layer is acidified with dilute sulphuric acid (1 + 1) to pH 1; after evolution of carbon dioxide the aqueous layer is extracted with three portions of 50 ml methylene chloride by shaking 1 min each time. The organic layers are transferred to a clean 500 ml separatory funnel. The methylene chloride extract is washed by tumbling it a few times with 50 ml organic-free water (pH 6). The layers are allowed to separate and the lower organic phase is transferred into a 500 ml round bottom flask. The aqueous layer is washed in the same manner with two 10 ml portions of methylene chloride and the organic layers added to the rest of organic extract in the round bottom flask.

The methylene chloride extract is evaporated to 2–3 ml on a rotary evaporator, finishing the evaporation on a steam bath under atmospheric pressure (water bath temperature 35 °C). The residue is dissolved with several 2–3 ml portions of acetone, each time quantitatively transferring the acetone into a 15 ml graduated centrifuge tube. The total volume of acetone used should not exceed 10 ml.

Derivatization

Pentafluorobenzyl method To the acetone extract in a centrifuge tube, is added 200 µl pentafluorobenzyl bromide solution and 30 µl 30% potassium carbonate solution. The tube is stoppered and shaken for 2–3 s. The contents are allowed to react overnight at room temperature.

After reaction 2 ml iso-octane are added and the solution evaporated to 1 ml with a gentle stream of nitrogen. Another 2 ml iso-octane are added and the evaporation repeated to a final volume of about 1 ml. This procedure removes traces of acetone, which affect column fractionation.

Dicyclohexyl carbodiimide method   The acetone is evaporated in the 15 ml centrifuge tube with a gentle stream of nitrogen until just dry; 500 µl of reagent and 200 µl of 2-chloroethanol are added, the contents mixed well and the tube stoppered. The mixture is allowed to react overnight at 60 °C. After reaction, 5 ml benzene and 2–4 ml dilute hydrochloric acid (5 + 1) are added. The mixture is shaken well and the layers allowed to settle, discarding most of aqueous solution, using a disposable pipette. This process is repeated twice with water. As much aqueous washing as possible is withdrawn with a disposable pipette. Some benzene extract may be withdrawn in this process without affecting the accuracy of analysis. To the mixture is added 0.5–1 g anhydrous sodium sulphate. A known volume of benzene layer is transferred into a clean 15 ml centrifuge tube, 2 ml iso-octane are added and the extract evaporated to 1 ml under nitrogen in a 50–60 °C water bath. This is repeated with another 2 ml iso-octane to remove traces of benzene.

Boron trichloride method   The above method is followed, substituting 200 µl boron trichloride reagent for 500 µl dicyclohexylcarbodiimide reagent and 200 µl 2-chloroethanol. The mixture is reacted overnight at 60 °C. After reaction, 5 ml of benzene are added and the benzene extract washed with 4–5 ml potassium bicarbonate solution followed by 4–5 ml water. The benzene extract is dried over anhydrous sodium sulphate and the evaporation repeated with iso-octane as described in the dicyclohexyl carbodiimide method.

Column clean-up

A minicolumn is prepared by plugging a clean disposable pipette (10 cm × 5 mm i.d.) with a piece of glass wool that has previously been washed with acetone and dried. The column is filled with 5 cm silica gel (5% deactivated) and the column tapped in order to settle solid.

The column is prewetted with 5 ml hexane and hexane drained just to the top of the packing material. Using a disposable pipette, concentrated sample extract with hexane rinsings is applied to the column. The fractions are eluted as follows.

Pentafluorobenzyl esters

(1) Elution with 8 ml benzene–hexane (25 + 75) is carried out in order to remove excess reagent, and contaminants.

(2) Elution with 8 ml benzene–hexane (75 + 25) produces the fraction containing pentafluorobenzyl esters of dicamba, MCPA, 2,4-DP, 2,4-D, silvex, 2,4,5-T, MCPB, and 2,4-DB.

(3) Elution with 8 ml ethyl ether–benzene (10 + 90) produces the pentafluorobenzyl ester of picloram.

The volumes are adjusted to the nearest 1 ml. The volumes collected are recorded and the fractions analysed for pentafluorobenzyl esters of the above herbicides.

2-Chloroethyl esters

(1) Elution with 20 ml benzene–hexane (25 + 75) removes excess reagent and contaminants.

(2) Elution with 8 ml benzene produces the fraction containing 2-chloroethyl esters of dicamba, MCPA, 2,4-D, silvex, 2,4,5-T, MCPB, and 2,4-DB.

(3) Elution with 10 ml benzene removes some extraneous peaks which may interfere with picloram. This fraction is discarded.

(4) Elution with 8 ml ethyl ether–benzene (10 + 90) produces the fraction containing the 2-chloroethyl ester of picloram.

The extracts are concentrated under a gentle stream of nitrogen to the required volume and analysed for 2-chloroethyl esters of the above herbicides.

Agemian and Chau[73] found that methylene chloride had excellent properties for their requirements. It has a much higher dielectric constant than both ethyl acetate and ethyl ether and it is much less soluble in water than these solvents. Furthermore, it has a higher density than water, so it can be easily separated in the last step of the procedure. Ethyl acetate was chosen over ethyl ether for the first step in their extraction scheme because of its higher dielectric constant.

Gas chromatographic examination of the extracts showed that although the pentafluorobenzyl esters have somewhat longer retention times than the 2-chloroethyl esters, the separations obtained for the ten herbicides with both esters were very similar and much superior to the separations of the methyl esters. Figure 248C shows the good separation obtained on the OV-17/QF-1 column. It is also evident that the retention times are twice as long for the pentafluorobenzyl esters as for the 2-chloroethyl esters under the same conditions on this column. The pentafluorobenzyl reaction is less specific than the 2-chloroethyl reaction. The former reagent reacts under basic conditions with phenols and carboxylic acids.[74-76] Fortunately pentafluorobenzyl esters of most environmental phenols elute in the clean-up fraction.

The boron trichloride and dicyclohexylcarbodiimide reactions are more specific to these herbicidal compounds, but boron trichloride does not esterify dicamba and gives a low yield for picloram.

Agemian and Chau[73] tried silica gel[77] and Florisil[78] to clean up the pentafluorobenzyl and methyl esters of phenoxyacid herbicides, respectively. Different activities of silica gel, Florisil, alumina, and charcoal were tested for suitability. The esters studied were successfully recovered from both silica gel and Florisil, but not from alumina and charcoal, probably due to decomposition on the column. Since Florisil was the stronger adsorbent and required stronger solvents for elution, silica gel was chosen for use in the method described above.

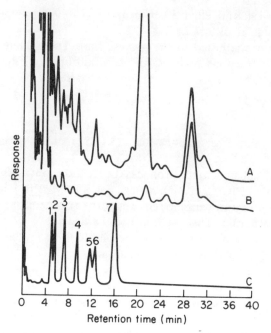

Figure 248 Comparison of gas chromatograms of 2-esters of 7 herbicides. Chromatogram from OV-17/QF-1 column; A, dicyclohexylcarbodiimide reaction blank concentrated to 1 ml and cleaned up as given in text (fraction 2); B, BCl_3 reaction blank concentrated to 1 ml and cleaned up as given in text (fraction 2); C, composite standard for 2-chloroethyl esters of 1, 250 pg MCPA; 2, 20 pg 2,4-DP; 3, 30 pg 2,4-D; 4, 10 pg silvex; 5, 2.5 ng MCPB: 6, 20 pg 2,4,5-T; 7, 100 pg 2,4-DB, all per microlitre. Reprinted with permission from Agemian and Chau.[73] Copyright (1977) Association of Official Analytical Chemists

Fractionation studies revealed that both esters of picloram were much more polar than the esters of the seven other herbicides studies, so it was possible to fractionate the herbicides into two groups, one containing seven herbicides and one containing picloram.

Table 165 shows the detection limits achieved for the herbicides by the three methods. The boron trichloride method is the most sensitive for most of the important herbicides and would be the choice for environmental water analysis. Solutions from the dicyclohexylcarbodiimide and pentafluoruobenzyl reactions should be concentrated only to 10 ml rather than 1 ml because of the poor blanks. This is reflected in the less sensitive detection.

The data in Table 166 show that the boron trichloride reaction gives good recovery and satisfactory precision for the standard detection limits. These are practical limits which are defined to be ten times higher than any detectable signal. The recovery for MCPB was somewhat low, and also has the highest detection limit, $2.5 \mu g\, l^{-1}$. The boron trichloride reaction is not very suitable

Table 165 Practical detection limits ($\mu g \, l^{-1}$) for herbicides in 1 l water sample: three methods of derivative formation*

Herbicide	Pentafluorobenzyl	Dicyclohexylcarbodiimide	BCl_3
Dicamba	0.1	0.2	NR†
MCPA	0.1	2.5	0.25
2,4-DP	0.1	0.2	0.02
2,3,6-TBA	0.1	0.2	NR†
2,4-D	0.2	0.3	0.03
Silvex	0.2	0.1	0.01
MCPB	0.25	25	2.5
2,4,5-T	0.25	0.2	0.02
2,4-DB	0.5	0.1	0.1
Picloram	0.5	0.5	0.05

*Pentafluorobenzyl and dicyclohexylcarbodiimide methods are based on concentration to 10 ml, BCl_3 method is based on concentration to 1 ml.
†NR, no reaction.
Reprinted with permission from Agemian and Chau.[73] Copyright (1977) Association of Official Analytical Chemists.

Table 166 Recoveries of eight herbicides from fortified water samples, extraction method and BCl_3/2-chloroethanol derivatization technique

Herbicide	Level in 1 l H_2O (μg)	No. of trials	Mean rec. (%)	Rel. s.d. (%)
MCPA	0.25	8	93	8
2,4-DP	0.02	8	100	7
2,4-D	0.03	8	102	10
Silvex	0.01	7	109	7
MCPB	2.5	8	64	10
2,4,5-T	0.02	7	87	9
2,4-DB	0.1	6	79	17
Picloram	0.05	7	78	15

Reprinted with permission from Agemian and Chau.[73] Copyright (1977) Association of Official Analytical Chemists.

for this herbicide. If a multiresidue analysis is not required, then the pentafluorobenzyl reaction reported by Agemian and Chau[73] should be used.

Carnac[79] used a modified technique of dynamic distribution in liquid-liquid systems used for concentration of traces of organic substances in water. An organic solvent is placed on granules of copolymer of styrene and divinylbenzene. The technique has been used for determination of phenoxyalkancarboxylic acids by gas-liquid chromatography with a flame ionization detector using chloroform as the solvent; the limit of detection is 5–10 $\mu g \, l^{-1}$.

Fredeen et al.[80] have described a gas chromatographic method for determining methoxychlor residues at the sub ppm level in water (sand and fauna) following injections of this substance into a river. Two litre aliquots of water samples were filtered under suction to remove suspended soils. The filtrate was extracted with 200 ml hexane three times. The combined hexane extract was dried over anhydrous sodium sulphate, concentrated to a suitable volume and analysed by electron capture–gas-liquid chromatography. Operating conditions for the gas chromatograph were as follows:

Column: aluminium 5 ft × ¼ in i.d. packed with 4% SE-30 and 6% QF-1 on 80–100 mesh Chromosorb W.
Detectors: electron capture with tritium ionizing source.
Temperatures: column 185 °C, injector 200 °C, detector 200 °C.
Carrier gas: oxygen-free nitrogen.
Flow rate: 40 ml min^{-1}.
Electrometer: range 1, sensitivity 4.

Under these conditions retention time for aldrin was 3.0 minutes. Retention time of other organochlorine insecticides relative to aldrin were: heptachlor 0.94; heptachlor epoxide 1.81; *p,p'*-DDE 2.47; dieldrin 2.81; *p,p'*-DDD 3.25; *p,p'*-DDT 3.78; and *p,p'*-methoxychlor 7.25.

A combination of gas chromatography with mass spectmometry was used to confirm the identity of the methoxychlor peak on the gas chromatogram.

Electrophoresis and thin-layer chromatography

Purkayastha[81] examined the applicability of paper electrophoresis to nine ionizable chlorinated phenoxyacetic acid type herbicides including 2,4-D, 2,4,5-T, MCPA, fenoprop, dicamba (2-methoxy-3,6-dichlorobenzoic acid), and picloram (4-amino-3,5,6-trichloropicolinic acid). Solutions were applied to paper moistened with pyridine–acetic acid buffer solution of pH 3.7, 4.4, or 6.5 and a voltage of 2–4 kV was applied. After 30 min the paper was air-dried, sprayed with ammoniacal silver nitrate solution, and exposed to u.v. radiation. Experimental variations that increased the mobility of the spots included decreasing the ionic concentration of the buffer by dilution, increasing the applied voltage from 2 to 4 kV (potential gradients of 50–100 V cm^{-1}), and adding a foreign electrolyte (e.g. potassium nitrate) to the buffer. Addition of methanol to the buffer resulted in decreased mobility as well as a variation in the relative mobilities of the compounds.

Bogacka and Taylor[82] determined 2,4-D and MCPA herbicides in water using thin-layer chromatography. In this method a one litre sample of filtered water is treated with 50 g of sodium chloride and 5 ml of hydrochloric acid and the herbicides are extracted into ethyl ether (200, 100, and 100 ml). The extract is dried with anhydrous sodium sulphate, concentrated to a few ml and passed through a column (180 mm × 15 mm) of silicic acid with 90% methanol–acetic acid (9:1) as stationary phase and the herbicides are eluted with 150 ml of light petroleum saturated with the methanol–acetic acid mixture. The first 30 ml of eluate is rejected. The remaining eluate is evaporated to dryness, and the residue is dissolved in ether and concentrated to about 0.1 ml before thin-layer chromatography on silica gel G–Keiselguhr G (2:3) (activated for 30 min at 120 °C) with light petroleum–acetic acid–liquid paraffin (10:1:2) as solvent. The developed plates are air-dried, sprayed with 0.5% silver nitrate solution, and dried then sprayed with 2 M potassium hydroxide–formaldehyde (1:1), dried at 130–135 °C for 30 min, sprayed with nitric acid, and observed in u.v. illumination. For the determination the spots are compared with standards. Evaluation by the method of standard addition gave recoveries of 95.1% and 88.8% with standard deviations of 14.2% and 14.3% for 2,4-D and MCPA respectively.

These workers[83] also examined the thin-layer chromatography of 2,4-DP (dichlorprop) and MCPP (mixture of mecoprop and 2-(2-chloro-4-methylphenoxy) propionic acid). In this method the ethyl ether extract of the sample is purified on a column of silicic acid and the herbicides are separated by thin-layer chromatography on silica gel-Kieselguhr (2:3) with light petroleum–acetic acid–kerosene (10:1:2) as solvent. The sensitivity is 3 µg of either compound per litre, the average recoveries of dichlorprop and MCPP are 85.7% and 87.4%, respectively, and the corresponding standard deviations were 13.9% and 15.5%.

Meinard[84] described a new chromogenic reagent for the detection of phenoxyacetic acid herbicides on thin-layer plates. The separated phenoxyacetic acids are detected (as violet spots on a white background) by spraying the plate with a solution of chromotropic acid ((4 g) in water (40 g) and sulphuric acid (56 g)) then heating at 160 °C. The limits of detection for 2,4-D, 2,4,5-T, and MCPA range from 0.05 to 0.2 µg per spot, compared with 1–5 µg by spraying with silver nitrate reagent followed by exposure to u.v. radiation.

Liquid-liquid extraction

Suffet[85] has evaluated liquid-liquid extraction techniques for separating phenoxyacetic acid herbicides from river water. He used the p-value (defined as the fraction of the total solute that distributes itself in the non-polar phase of an equivolume solvent pair) concept in the development of equations, based on liquid-liquid extraction theory, relating the number of extractions and the water-to-solvent ratios for the maximum recovery of the herbicide. Calculations show that a pesticide with a p-value of 0.90 or greater in an aqueous system can be 95% extracted from the aqueous phase by up to five extractions with a total volume of solvent up to 500 ml. He confirmed his equations by measurements with 2,4-D. By the application of this concept to the simultaneous quantitative extraction of phenoxyacetic acid herbicides from water Suffet[86] showed that the best solvents for 2,4-D and 2,4,5-T and their butyl and isopropyl esters are ethyl ether or ethyl acetate (2,4-D and esters) and benzene (2,4,5-T and esters). Thus a 90% recovery of 2,4-D from 1 litre of an aqueous solution is obtained by a two-stage serial extraction with 200 and 50 ml of ethyl acetate under p-value conditions. Turbid samples should be filtered before extraction.

Extraction

Phenoxyacetic acid herbicides can be selectively extracted by extracting first into a suitable organic solvent and then back-extracting the strongly acidic herbicides into a weakly basic aqueous solution. This type of extraction scheme separates the compounds of interest from neutral compounds and is commonly used for separating acidic from non-acidic compounds. Although aqueous basic solutions of 2% w/v potassium hydroxide and 2% w/v potassium carbonate successfully back-extracted these herbicides, 2% w/v sodium bicarbonate is preferred because this weak base does not extract weakly acidic compounds, including many environmental phenols.

The most commonly used solvents for extracting phenoxyalkanoic acids have been ethyl ether[87-90] and chloroform, although benzene has also been used.[91,92] Suffet[85,86] showed the superiority of ethyl acetate and ethyl ether over benzene, chloroform, carbon tetrachloride, and hexane for these compounds. Methylene chloride has an excellent solubility for these compounds.

Benzene is highly toxic, and has a high boiling point and relatively low dielectric constant compared with the above solvents. This solvent gave consistently low extraction efficiencies for 2,4-D and dicamba. This supports Suffet's[86] p-value for 2,4-D of 0.195 for benzene compared with 0.996 and 0.990 for ethyl acetate and ethyl ether, respectively. Chloroform is also unsuitable because it is more toxic and has a lower dielectric constant than the other three solvents and its vapours cause anomalous responses when it is used near a gas chromatograph with an electron capture detector.

Miscellaneous

Marshall[92] used two methods for the infrared analysis of dicamba–MCPA and dicamba–2,4-D formulations. The 'indirect' method involved precipitation of the herbicides with hydrochloric acid and extraction with chloroform. The chloroform extract was evaporated to dryness, the residue was dissolved in acetone and the herbicides were determined by measuring infrared extinctions at the relevant wavelengths. The 'direct' method involved dissolving the sample in acetone and measuring infrared extinctions. Although both methods gave good precision, the 'indirect' method was the more accurate.

Bogacka[93] used 4-aminophenazone as a reagent for the spectrophotometric determination of phenoxyacetic acid herbicides (2,4-D, dichlorprop, MCPA) in water. The herbicides are extracted from an acidified 1 litre sample of water with ethyl ether. The extract is evaporated and the residue is heated for 1 hour with 10 g of pyridine hydrochloride at 207–210 °C for 2,4-D, or at 225–230 °C for dichlorprop or MCPA. The resulting phenol derivative is steam distilled into aqueous ammonia (1 M) extracted with light petroleum (after acidification of the distillate) and re-extracted into 0.05 M aqueous ammonia for coupling with 4-aminophenazone in the presence of potassium ferricyanide. The extinction of this solution is measured at 515, 505, or 515 nm for 2,4-D, dichlorprop or MCPA respectively. The respective sensitivities are 20, 20, and 80 μg l^{-1} of water and the corresponding standard deviations of the recovery are 4.0%, 8.5%, and 3.5%. The method is not suitable for the determination of mixed herbicides.

MISCELLANEOUS HERBICIDES

Picloram (4-amino-3,5,6-trichloropicolinic acid)

Abbott et al.[94] described a pyrolysis unit for the determination of picloram and other herbicides in water, soil and forage. The determination is effected by electron capture–gas chromatography following thermal decarboxylation of the herbicide. Hall et al.[95] reported further on this method. The decarboxylation products are analysed on a column (5 mm i.d.) the first 6 inches of which is packed with Vycor chips (2–4 mm), the next 3.5 ft with 3% of SE-30 on Chromosorb W (60–80 mesh) and then 2 ft with 10% of DC-200 on Gas Chrom Q (60–80 mesh). The pyrolysis tube, which is packed with Vycor chips, is maintained at 385 °C. The column is operated at 165 °C with nitrogen as carrier gas (110 ml min^{-1}). The method when applied to ethyl ether extracts of water and soil gives recoveries of 93 ± 4 and $90 \pm 5\%$ respectively. Dennis et al.[96] have reported on the accumulation and persistence of picloram in surface waters and bottom deposits.

Acarol (isopropyl-4,4′-dibromobenzilate)

Cannizzara et al.[97] have carried out gas-liquid radio chromatography of this ^{14}C herbicide present as residues in weathered soil.

Dichlorbenil (2,6-dichlorobenzonitrile)

The persistence of this herbicide in a farm pond has been studied.[98]

Carbamate herbicides

Carbine (barban) (4-chlorobut-2-ynyl-3-chlorocarbanilate)

To determine traces of this herbicide in water, Bosyakova et al.[101] distilled a 1 litre sample of water with 30 ml of concentrated sulphuric acid and collected about 950 ml of distillate. The residue was hydrolysed by adding 100 ml of 5% sodium hydroxide solution and boiling under reflux for 1 hr. The herbicide was quantitatively hydrolysed to m-chloroaniline. The solution was steam distilled into a receiver containing 10 ml of 10% hydrochloric acid. From 500 ml of distillate a 10 ml aliquot was withdrawn. 1 ml of 0.4 M hydrochloric acid and 1.1 ml of 0.1 M sodium nitrite were added, the solution was cooled to 5 °C, 0.5 ml of 0.02% solution of 2-naphthol in 10% sodium hydroxide solution was added, and the colour intensity of the solution was referred to a calibration graph.

Benthiocarb (s-(4-chlorobenzyl-N,N-dielthylthiolcarbamate)

Suzuki et al.[102] have discussed the determination of this herbicide in rivers and agricultural drainage.

S-*alkyl derivatives of* N,N-*dialkyl dithiocarbamates*

Onuska and Boos[103] have determined these substances in waste water samples by a gas chromatographic–mass spectrometric method. Separation of the derivatives was carried out on a column (355 cm × 2.1 mm) of 10% of Apiezon L on Varaport 80 (80–100 mesh) operated at 250 °C with helium as carrier gas and flame ionization detection. Separation was also achieved on a similar column as above, but with the temperature maintained at 190 °C for 4.5 min and then programmed at 4 °C min^{-1} to 270 °C, the helium flow rate being 28 ml min^{-1}. The effluent from the second column was examined by mass spectrometry–gas chromatography.

Dicamba (2-methoxy-3,6-dichlorobenzoic acid)

This herbicide is discussed in methods for its codetermination with phenoxyacetic acid herbicides. Norris and Montgomery[104] have described a procedure for the determination of traces of dicamba and 2,4-D in streams after forest spraying. Dicamba and its metabolites (3,6-dichlorosalicylic acid and 5-hydroxydicamba) were determined gas chromatographically. For analysis a 500 ml aliquot of stream water was acidified to pH 1 with hydrochloric acid and extracted with three 150 ml portions of diethyl ether. Ether extracts were concentrated to 20 ml and methylated with diazomethane in ether. The ether extracts were then concentrated to 1 ml and injected into a gas chromatograph equipped with a microcoulometric detector. The 1.8 m × 6.25 mm glass column was packed with 60–80 mesh Gas-Chrom Q coated with 6% OV-1. The column was operated at 165 °C with a nitrogen flow of 68 ml min^{-1}. The retention time of dicamba was 2.6 minutes.

Methylation converts 3,6-dichlorosalicylic acid metabolite to dicamba and the 5-hydroxydicamba metabolite has a retention time of 6.6 minutes.

Pyrazon (5-amino-4-chloro-2-phenyl-3-pyridizone)

This pre- and post-emergent herbicide has been determined in water by spectrophotometric, thin-layer chromatographic methods[105–108] and by high performance liquid chromatography.[109] The high performance liquid chromatographic method is described below as it illustrates very well the applicability of this technique to trace organics analysis in water.

Method

Pyrazon was isolated from water samples (500 ml) by rotary evaporation to dryness *in vacuo*, extraction of the solid residue with methanol (2 × 25 ml) and further evaporation of the methanol extract (to approx. 2 ml). Final concentration (to 0.5 ml) was achieved by removal of methanol under a stream of nitrogen.

The equipment used consisted of two Model 6000A solvent delivery systems and a Model 660 gradient former (Waters Ass.) and a Model CE 212 variable wavelength u.v. monitor (Cecil Instruments) operated at 270 nm. Syringe injections were made through a stop-flow septumless injection port. The column (15 cm × 7 mm i.d.) was packed in an upward manner with Spherisorb-ODS by a slurry procedure using acetone as slurry medium. A linear gradient was established from two solvent mixtures consisting of (a) 10% methanol in 0.1% acetic acid in water and (b) 80% methanol in 0.1% acetic acid in water. The initial concentration was 35% b in a and the final concentration was 100% b with the gradient terminated after 20 min. The flow rate was maintained at 2.0 ml min^{-1} throughout the analysis.

Crathorne and Watts[109] determined the recovery efficiencies of pyrazon from water by analysing samples of river water spiked at levels of 10, 50, 100, and 200 μg l^{-1}. The recovery efficiency varied from 17% at the 10 μg l^{-1} level to over 90% recovered at 200 μg l^{-1}. A calibration curve was constructed for quantitation of water samples (Figure 249).

The gas chromatogram for a standard solution of pyrazon in methanol (Figure 251) shows a major peak due to pyrazon and a minor peak at a longer retention

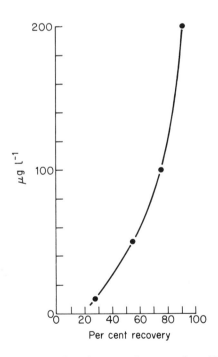

Figure 249 Calibration curve for extraction efficiency of pyrazon from surface water samples. Reprinted with permission from Crathorne and Watts.[109] Copyright (1977) Elsevier Science Publishers B.V.

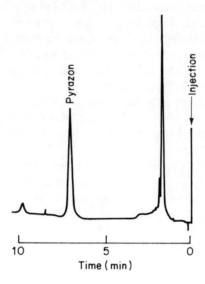

Figure 250 Gas chromatogram of a river water extract spiked at 100 μg l^{-1} with pyrazon. Reprinted with permission from Crathorne and Watts.[109] Copyright (1979) Elsevier Science Publishers B.V.

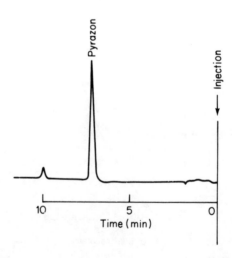

Figure 251 Gas chromatogram of a standard solution of pyrazon. Reprinted with permission from Crathorne and Watts.[109] Copyright (1979) Elsevier Science Publishers B.V.

time. The minor peak was present in all of the standard solutions analysed, the spiked water extracts and in the polluted surface water extracts.

The gas chromatogram from the analysis of a spiked water extract (Figure 250) shows the peak for pyrazon clearly separated from a large, essentially unretained peak. This large peak was usually observed in the analysis of water extracts and appears to arise from excluded components.

The gas chromatogram of a polluted river water extract (Figure 252) exhibits the peak due to pyrazon and the minor peak.

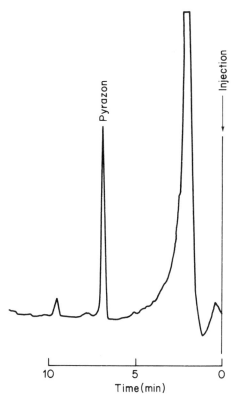

Figure 252 Gas chromatogram of a polluted river water extract. Reprinted with permission from Crathorne and Watts.[109] Copyright (1979) Elsevier Science Publishers B.V.

Paraquat (1,1′-dimethyl-4,4-bipyridinium chloride and diquat (1,1′-ethylene-2,2-bipyridylium bromide)

Calderbank and Yuens[110] and Pope and Benner[111] have described a spectrophotometric method for the determination of paraquat in water and soil in amounts down to 0.1 ppm. Paraquat, trifluralin, and diphenamid have also been determined gas chromatographically in water[112, 113] and in soil and agricultural run-off water.[114]

Söderquist and Crosby[112] added to the water sample (100 ml), sulphuric acid (3 ml) and platinum dioxide (25 mg) and hydrogen was bubbled through for 1 h, whereby paraquat is converted into 1,1'-dimethyl-4,4'-bipiperidyl. This is extracted with dichloromethane (3 × 50 ml) in the presence of 11 ml of 50% sodium hydroxide solution and the combined extract is treated with 0.01 N hydrochloric acid (4 ml) and evaporated in a rotary evaporator at 50–55 °C. The aqueous residue is transferred with 1 ml of 0.01 N hydrochloric acid to a 15 ml screw-cap tube and shaken with 50% sodium hydroxide solution (0.5 ml) and carbon disulphide (1 ml). Aliquots of the carbon disulphide phase (1–10 μl) are injected on to a glass column 66 ft × 0.125 in) packed with 10% of Triton X-100 and 1% potassium hydroxide on AW-DCMS Chromosorb G (70–80 mesh), and operated at 150 °C with nitrogen as carrier gas (30–40 ml min^{-1}) and flame ionization detection. The calibration graph is rectilinear for up to 1 ppm of paraquat. The limit of detection is 0.1 ppm but recovery is only 36–43% although reproducible.

To determine paraquat in agricultural run-off water and soil (Payne et al.[114]) separate the sediment from the sample (2 litres) by adding calcium chloride to aid flocculation, leaving the mixture overnight in a refrigerator for the sediment to settle, then decanting and filtering through a Whatman No. 42 paper under suction on a Buchner funnel. The wet sediment and soil core samples, are mixed for 4 h with dichloromethane in a Soxhlet extractor to remove trifluralin and diphenamid. A 1 litre aliquot of the filtrate is extracted with dichloromethane (100, 50, 50, and 50 ml). The dichloromethane extracts are concentrated by evaporation and the trifluralin and diphenamid are determined by direct injection, without further purification, on to a glass column (6 ft × 0.25 in. o.d.) packed with 10% DC-200 on Gas Chrom Q and operated at 220 °C with helium as carrier gas (100 ml min^{-1}) and a Coulson electrolytic-conductivity detector (N mode). Paraquat is determined in the dried sediment and core samples, and in a 500 ml aliquot of the filtrate, by a modification of a conventional colorimetric method. Recoveries of the three substances were between 86 and 96% from soil and 82 and 95% from water.

Khan[115] has described a method for determining paraquat and diquat in soils involving catalytic dehydrogenation of the herbicide followed by gas chromatography and also a pyrolytic method.[116]

Cannard and Criddle[113] have described a rapid pyrolysis–gas chromatography method for the simultaneous determination of paraquat and diquat in pond and river waters in amounts down to 0.001 ppm. These workers emphasize the precautions necessary to avoid errors due to adsorption of the herbicides on to glassware.

Method

Apparatus

A Perkin-Elmer F30 gas chromatograph having a standard injection-port modification for a Chemical Data Systems (CDS) Pyroprobe 190 with silica pyrolysis tubes.

Best results were obtained under the following conditions: temperature 1000 °C; pyrolysis time 5 s; ramp 2.0 °C ms^{-1}, and probe insertion distance, maximum. Gas chromatographic conditions were as follows: column 10% Carbowax 20 M and 2% potassium hydroxide on Celite (80–100 mesh) in a 600 × 3.5 mm i.d. glass column; column temperature 190 °C; injection port temperature 110 °C; carrier gas, nitrogen at a flow rate of 40 ml min^{-1}. A flame ionization detector was used. A standard herbicide solution (100 µl 1 ppm) was run in order to obtain a response factor. Standard volumes of herbicide solution (up to 100 µl) were injected continuously into the centre of a silica tube and the tube was heated continuously at 100–110 °C in an air stream from a hot air blower.

The tube was then inserted into the coil probe of the Pyroprobe and the sample pyrolysed under optimum conditions. Gas chromatographic analysis of the pyrolysate was then carried out.

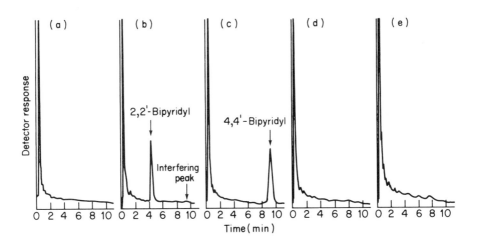

Figure 253 Pyrolysis patterns obtained from 100 µl samples of various aqueous systems: (a) deionized water; (b) diquat solution, 0.3 ppm; (c) paraquat solution, 0.5 ppm; (d) River Cwynnedd; and (e) River Ebbw. Reprinted with permission from Cannard and Criddle.[113] Copyright (1975) Royal Society of Chemistry

The method for the determination of paraquat and diquat is based on the following reactions:

Paraquat $\xrightarrow{\Delta}$ N⟨○⟩-⟨○⟩N + 2CH$_3$Cl

Diquat $\xrightarrow{\Delta}$ ⟨○⟩-⟨○⟩ + CH$_2$BrCH$_2$Br
         N   N

Although other reactions occur that give smaller fragments, it will be apparent (Figure 253 (b) and (c)) that the pyrolysis of both compounds produces few products with relative molecular masses comparable to those of the free bases, a feature which renders the method particularly suitable for both quantitative and qualitative analysis. However, for best results the procedure as described above, must be strictly adhered to.

The detection limits for the method as applied to river waters are governed by two main factors: the size of sample that can conveniently be introduced into the pyrolysis tube and the ability of the column to resolve the bipyridyl peaks from those due to other pyrolysis products. Of a total of nine samples of river waters most gave a simple pyrolysis pattern (Figure 253). The most complex pattern obtained shows that no interference with paraquat will occur, and that only slight interference with diquat is likely. However a small diquat pyrolysis peak (Figure 253) can interfere to a slight extent with the 4,4'-bipyridyl peak derived from paraquat but the value for paraquat may be simply corrected when appropriate, as the size of the interfering peak is proportional to the size of the 2,2'-bipyridyl peak derived from diquat.

Coha[99] used the ring oven technique to estimate traces of paraquat and diquat in water. Morfamquat and diquat have been determined by reduction at the dropping mercury electrode.[100]

Dalapon (2,2-dichloropropionic acid)

Earlier gas chromatographic methods for determining this growth regulator include those of Getzendaner[117] and Frank and Demint.[118]

The method of Getzendaner is applicable to plant tissues and body fluids and doubtlessly to water samples. The sample was extracted with ethyl ether and the residue was analysed by gas chromatography on a glass column (4 ft × 2 mm) of 4% of LAC-2R-446 plus 0.5% of phosphoric acid on Gas Chrom S (60–80 mesh) at 100 °C with a nitrogen as carrier gas (85 ml min^{-1}) and electron capture detection. Recoveries of about 90% were obtained for 10 ppm of the herbicide.

The Frank and Demint[118] method is directly applicable to water samples. After addition of solid sodium chloride (340 g l^{-1}) and aqueous hydrochloric acid (1:1) to bring the pH to 1, the sample was extracted with ethyl ether and the organic layer was then extracted with 0.1 M sodium bicarbonate (saturated with sodium chloride and adjusted with sodium hydroxide to pH 8). The aqueous solution adjusted to pH 1 with hydrochloric acid was extracted with ether and after evaporation of the ether to a small volume dalapon was esterified at room temperature by addition of diazomethane (0.5% solution in ether) and then applied to a stainless steel column (5 ft × 0.125 in) packed with Chromosorb P (60–80 mesh) pretreated with hexamethyldisilazane and then coated with 10% FFAP. The column was operated at 140 °C, with nitrogen carrier gas (30 ml min^{-1}) and electron capture detection. The recovery of dalapon ranged from 91 to 100%; the limit of detection was 0.1 ng. Herbicides of the phenoxyacetic aid type did not interfere; trichloroacetic acid could be determined simultaneously with dalapon.

In a more recently published method (Van der Poll and de Vos[119]), for the determination of dalapon in natural water and plant tissues the herbicide is first esterified with 3-phenolpropanol-1 then determined by electron capture–gas chromatography. As little as 0.001 mg dalapon per litre of water can be determined by this method. These workers used as gas chromatograph with two ^{63}Ni electron capture detectors. The detectors were operated in the pulse mode at 50 V. Two columns, both glass, were used to determine the ester. Column A was packed with 3% OV-210 on Gas Chrom Q (80–100 mesh), column B with 1.8% OV-1 and 2.7% OV-210 on Gas Chrom Q (80–100 mesh). Carrier gas: nitrogen, flow rate, 70 ml min^{-1} on both columns. The temperatures of the column oven, injectors and detectors were 160, 205, and 275 °C respectively.

Method

Extraction of natural water

A 500 ml volume of water was acidified with 30 ml 5% orthophosphoric acid. After standing for 15 min the mixture was filtered on a Buchner funnel by suction. The filter was washed twice with 20 ml water. The aqueous extract was saturated with 185 g sodium chloride and then extracted with 100, 50, and 50 ml diethyl ether respectively. The combined diethyl ether extracts were dried over anhydrous sodium sulphate.

After adding 1.0 ml 3-phenylpropanol-1 to the ether extract, the latter was evaporated to *ca.* 5 ml. The concentrate was transferred to a reagent tube and further evaporated until no diethyl ether was left. The residue was saturated with dry hydrogen chloride gas (5 min). A micro-Snyder condenser was put on top of the tube which was then placed in a steam bath for 5 min. After cooling, 25 ml hexane were pipetted through the condenser into the tube which was stoppered and shaken vigorously for 1 min. Then the mixture was transferred to a separating funnel. Subsequently, 20 ml water were added and the funnel shaken. The water layer was discarded. The hexane layer was extracted twice more with 20 ml water and then filtered over anhydrous sodium sulphate. A small volume of the hexane extract was transferred to a silica gel column and allowed to sink in. The column was rinsed twice with 1 ml hexane, eluted with 50 ml hexane and the eluate discarded. Finally, the column was eluted with 70 ml hexane. This eluate was concentrated to about 5 ml and transferred to a graduated test tube. The liquid was further concentrated to 1 ml, using a gentle stream of dry air, then an aliquot was injected into the gas chromatograph. A method is also described for determining dalapon in homogenized plant tissues.

Recoveries of between 94 and 103% were obtained. The response of the detector was linear up to nanogram amounts of the 3-phenylpropyl ester of dalapon injected. As little as 0.02 mg dalapon per kg plant tissue and 0.001 mg dalapon per litre natural water can be determined by this method. In Figures 254 and 255 typical chromatograms are shown of untreated fortified samples of potatoes and natural water, analysed by the method described.

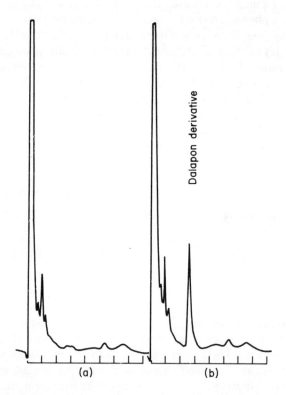

Figure 254 Chromatogram on a column packed with 3% OV-210 on Gas Chrom Q (80–100 mesh) of 100 µg potatoes (a) and 100 µg potatoes fortified with 0.001 mg l^{-1} dalapon (b). Reprinted with permission from Van der Poll and de Vas.[119] Copyright (1980) Elsevier Science Publishers B.V.

Glyphosate (*N*-phosphonomethylglycine)

This herbicide is manufactured by Monsanto, and marketed under the name 'Round-up'. Bronstad and Freistad[120] have described a polarographic method for

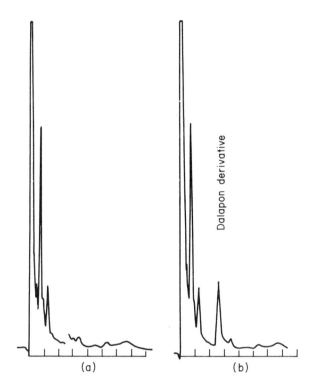

Figure 255 Chromatogram on a column packed with 3% OV-210 on Gas Chrom Q (80–100 mesh) of 100 μg natural water (a) and 100 μg natural water fortified with 0.001 mg l^{-1} dalapon (b). Reprinted with permission from Van der Poll and de Vas.[119] Copyright (1980) Elsevier Science Publishers B.V.

determining glyphosate residues in natural waters based on the polarography of the *N*-nitroso derivative.

Method

Apparatus

A Princeton Applied Research, Model 174A, polarographic analyser was connected to an M 174/70 drop timer head, utilizing a three electrode system. The working electrode was a dropping-mercury electrode with a drop time of 1 s. The auxiliary electrode was a platinum wire and the reference electrode a saturated calomel electrode isolated from the bulk of the solution by a salt-bridge tube. The electrolysis cell had a capacity of 50 ml.

A tube 300 mm long × 20 mm i.d. with a PTFE stopcock, sintered-glass disk and an 80 mm × 30 mm i.d. reservoir on top served as the chromatographic column.

Reagents

All reagents should be of analytical reagent grade.
Dowex 1-X8 anion-exchange resin, 50-100 mesh, chloride form.
Potassium bromide solution, 25% m/v.
Sodium nitrite solution 0.2 M, a fresh solution prepared daily.
Ammonium sulphamate solution. 1 M.

Glyphosate standard solution, dissolved by heating on a water bath, 0.7 g of glyphosate analytical standard 99.8% purity in 100 ml of water containing 10 drops of 1 + 1 sulphuric acid. The solution is stable for at least 3 months at room temperature. From this stock solution appropriate dilutions are made with water.

Procedure

An anion-exchange column is prepared by introducing 15 ml of prewashed Dowex 1-X8 resin contained as a slurry in approximately 10 ml of deionized

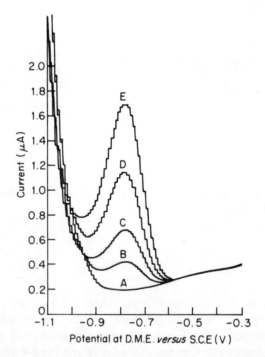

Figure 256 Current-potential graphs for glyphosate nitrosamine corresponding to: A, 0; B, 35; C, 70; D, 140; and E, 210 μg of glyphosate in total electrolyte volume of 45 ml. Reprinted with permission from Bronstad and Freistad.[120] Copyright (1978) Royal Society of Chemistry

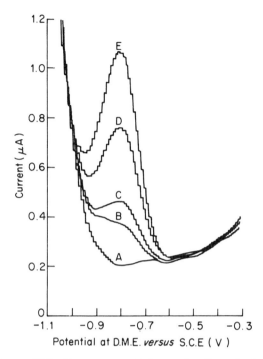

Figure 257 Current-potential graphs for glyphosate nitrosamine after anion-exchange treatment and nitrosation of various amounts of glyphosate added to 1 l of tap water. A, 0 (reagent blank); B, 35; C, 70; D, 140; and E, 210 µg added. Reprinted with permission from Bronstad and Freisad.[120] Copyright (1978) Royal Society of Chemisty

water, into a chromatographic column. The resin is washed, first with 100 ml of 0.5 M hydrochloric acid and then with 300 ml of deionized water and the washings discarded.

One litre water sample is allowed to percolate through the column at a rate of 500–600 ml h^{-1} followed by a 200 ml of deionized water and the percolates discarded. The column is eluted with 50 ml of 0.1 M hydrochloric acid at the same rate as above. The first 10 ml are discarded and the next 40 ml of eluate are collected in a measuring cylinder. The resin should not be re-used.

To the 40 ml eluate fraction is added 2 ml of sulphuric acid (1 + 1), 1 ml of potassium bromide, and 1 ml of sodium nitrite solution. The solution is swirled and left for 15 min, then 1 ml of ammonium sulphamate solution is added to destroy excess of nitrite and the contents of the measuring cylinder are poured into the electrolysis cell. The solution is deaerated by bubbling high-purity nitrogen gas through the cell for 3 min. Nitrogen is passed over the surface of the solution throughout the polarography. Measurements in the differential pulse mode are taken using pulse heights of 50 mV and a scan rate of 10 mV s^{-1}. As shown in Figure 256 glyphosate nitrosamine has a single well defined, differential pulse wave with a peak potential of -0.78 V. Figure 257 shows typical polargrams for single samples of tap water carried through the entire procedure after fortification at various levels. Compared with the polargrams in Figure 256 there is a slight change in the half-wave potential, which can be caused by dilution of the 0.1 M hydrochloric acid eluant with residual water from the column. This result also indicate that the method is little influenced by compounds naturally present in waters. A glyphosate concentration of 70 μg l^{-1} gives a distinct glyphosate nitrosamine wave and the response from a solution of half of this concentration is significantly different from that of the unfortified sample.

MIXTURES OF HERBICIDES AND ORGANOCHLORINE INSECTICIDES

Mixtures of herbicides

Abbott and Wagstaff[121] of the Laboratory of the Government Chemist UK have described a thin-layer chromatographic method for the identification of 12 acidic herbicides and 19 nitrogenous herbicides (carbamates, substituted ureas, and triazines).

Smith and Fitzpatrick[122] have also described a thin-layer method for the detection in water and soil of herbicide residues, including atrazine, barban, diuron, linuron, monouron, simazine, trifluralin, bromoxynil, dalapon, dicamba, MCPB, mecoprop, dicloram, 2,4-D, 2,4-DB, dichlorprop, 2,4,5-T, and 2,3,6-trichlorobenzoic acid.

Neutral and basic herbicides were extracted from water made alkaline with sodium hydroxide or from soil, with chloroform; extracts of soil were cleaned up on basic alumina containing 15% of water. Acidic herbicides were extracted with ethyl ether from water acidified with hydrochloric acid or from an aqueous extract of soil prepared by treatment with 10% aqueous potassium chloride that was 0.05 M in sodium hydroxide and filtration into 4 M hydrochloric acid. The concentrated chloroform solution of neutral and basic herbicides was applied to a pre-coated silica gel plate containing a fluorescent indicator and a chromatogram was developed two-dimensionally with hexane–acetone (10:3) followed after drying by chloroform–nitromethane (1:1). The spots were detected in u.v. radiation. Atrazine, barban, diuron, linuron, monouron, simazine, and trifluralin were successfully separated and were located as purple spots on a green fluorescent background. The ether extracts were dried over sodium sulphate, concentrated, and applied to a similar plate, which was developed two-dimensionally with chloroform–anhydrous acetic acid (19:1) followed after drying by benzene–hexane–anhydrous acetic acid (5:10:2). The spots were detected by spraying with bromocresol green. Bromoxynil and (as the acids) dalapon, dicamba, MCPA, MCPB, mecoprop, dicloram, 2,4-D, 2,4-DB, dichlorprop, 2,4,5-T, and 2,3,6-trichlorobenzoic acid were seen as yellow spots on a blue background. The limits of detection were 1 ppm in soil and 0.1 ppm in natural water.

Schulten et al.[123-125] have identified many pesticides and their metabolites using field desorption mass spectrometry. However, these workers did not apply the method to environmental samples. More recently Yamato et al.[126] have described a combination of gas chromatography with field desorption mass spectrometric analysis applied to benzthiocarb (S-(4-chlorobenzyl-N,N-diethylthiolcarbamate), oxidiazon (2-t-butyl-4-(2,4-dichloro-5-isopropoxyphenyl) Δ^2-1,3,4-oxadiazolin-5-one), and CNP (2,4,6-trichlorophenyl-4'-nitrophenyl-ether), herbicides in water, and other environmental samples. They demonstrated the usefulness of this technique for screening unknown compounds in the natural environment.

The gas chromatography operating conditions are shown in Table 167. FD mass spectra was obtained using a JEOL JMS 01SG-2 mass spectrometer with a combined EI/FI/FD source. Mass resolution was 1200.

Table 167 Gas chromatographic conditions for the determination of herbicides

Compounds	Benzthiocarb	CNP	Oxidiazon
Column (3 mm × 2 m Gas Chrom Q 80–100 mesh)	2% Apolar 10 C	2% DEGS 0.5% PA	5% ALG
N_2 carrier flow rate (ml min^{-1})	40	100	120
Range (V)	0.01	0.04	0.04
Sensitivity (m)	10	10	10
Column temp. (°C)	180	215	210
Detector temp. (°C)	250	215	210
Injector temp. (°C)	200	230	20
Detector	FPD[a]	ECD[b]	ECD[b]

[a]Flame photometric detector.
[b]Electron capture detector.

Surface water samples were taken in 3 litre Erlenmeyer flasks which were capped with aluminium foil. The water samples (100 l) were brought to the laboratory immediately and filtered through G-4 (150–200 mesh) and/or G-3 (100–150 mesh) Buchner glass filters to remove sediment which influenced the flow rate of the water sample through the XAD-2 column. The filtrate was passed through an XAD-2 column (1 × 10 cm) in which the resin was supported with cleaned and silanized glass wool plugs below and above the resin. The flow rate was adjusted at about 40–50 ml min^{-1}. Herbicides which were adsorbed on the resin were eluted with 70 ml of diethyl ether according to the method of Yamato et al.[127] The solvent layer was separated from the aqueous layer, because the small volume of water residing in the column was coeluted with the solvent. Then the solvent layer was dried by passing through a column of anydrous sodium sulphate (2 × 5 cm) and several rinsings to the column wall and the drying medium were added to the dried solvent layer. The dried solvent layer was subsequently concentrated to 2–3 ml with a Kuderna-Danish evaporative concentrator. No further clean-up was carried out. The concentrator was injected into a gas chromatograph for the measurement of residue concentrations in water sample. Then the concentrate was evaporated off and the residue was again dissolved in 0.5 ml of acetone. Field description mass spectra of standard benzthiocarb, CNP, and oxadiazon are shown in Figure 258. These spectra were recorded at an emitter heating current of 0 mA and show no fragment ions but do show isotope ions which are characteristic of organochlorine compounds.

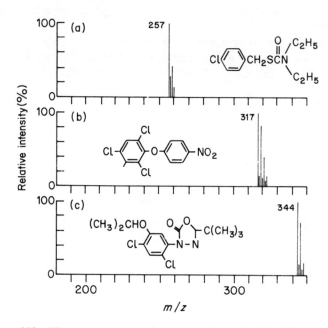

Figure 258 FD mass spectra of (a) benzthiocarb, (b) CNP, and (c) oxadiazon

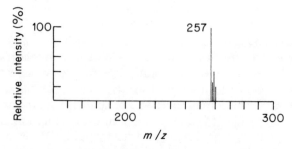

Figure 259 FD mass spectrum of herbicide residue extracted from river water

Figure 259 shows the field desorbption mass spectrum of the herbicide residue extracted from a river water sample. The spectrum shows an ion at m/z 257 corresponding to the molecular ion of benzthiocarb. The isotope ions observed possess almost the same relative intensities to the molecular ions as for authentic benzthiocarb. Residues of CNP and oxidiazon were also extracted; the levels in the final solution (0.5 ml) were 459 ppm for benzthiocarb, 15 ppm for CNP, and 33 ppm for oxadiazon, respectively. However, molecular ions of CNP and oxadiazon were not observed in the spectrum. This may be due to the low residue levels of these compounds.

An extract from an agricultural drainage water was also examined by field desorbption mass spectrometry. The spectrum is presented in Figure 260. A clear signal appears at m/z 257 and 344 corresponding to molecular ions of benzthiocarb and oxadiazon, respectively. In addition, the isotope ion patterns were almost identical with those of standards. Therefore, the possibility of the presence of these compounds in the sample was indicated.

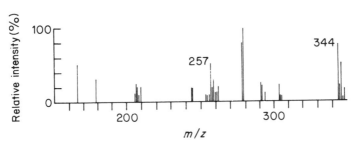

Figure 260 FD mass spectrum of herbicide residues extracted from agricultural drainage water

However, an ion at m/z 317 corresponding to the molecular ion of CNP was not present. Although the extract obtained from 100 l of water was not cleaned up, those contaminants did not interfere with confirmation of the presence of oxidiazon and benzthiocarb.

Erney[128] had described a photochemistry technique using ultraviolet irradiation followed by gas chromatography to confirm the identity of organochlorine insecticides and some herbicides. This is dealt with in the appropriate section.

REFERENCES

1. Croll, B. T. Water Research Centre Technical Memorandum TM 119. *Herbicides and Potable Water Supplies*, January (1976).
2. McKone, C. E., Byast, T. H. and Hance, R. J. *Analyst (Lond.)*, **97**, 653 (1972).
3. Purkayastha, R. and Cochrane, W. P. *J. Agric. Food Chem.*, **21**, 93 (1973).
4. Ramsteiner, K., Hoermann, W. D. and Eberle, D. O. *J. Ass. off. Anal. Chem.*, **57**, 192 (1974).

5. Lawrence, J. F. *J. Agric. Food Chem.*, **22**, 137 (1974).
6. Method S 69. *Method for Triazine Pesticides in Water and Wastewater. Methods for Benzidine, Chlorinated Organic Compounds, Pentachlorophenol and Pesticides in Water and Wastewater* (INTERIM Pending Issuance of Methods for Organic Analysis of Water and Wastes, Environmental Protection Agency, Environmental Monitoring and Support Laboratory (EMSL)). September (1978).
7. Hormann, W. D., Tournayre, J. C. and Egli, H. *Pesticide Monitoring Journal*, **13**, 128 (1979).
8. Wu, T. L., Lambert, L., Hastings, D. and Banning, D. *Bull. Environ. Contam. Toxicol.* **24**, 411 (1980).
9. Garrett, W. D. *Limnol Oceanography*, **10**, 602 (1965).
10. Fenselau, C. *Appl. Spectrosc.*, **28**, 305 (1974).
11. Karlhuber, B., Hormann, W. and Ramsteiner, K., *Anal. Chem.*, **47**, 2450 (1975).
12. Ramsteiner, K., Hermann, W. D. and Eberle, D. O. *J. Ass. Oil Anal. Chem.*, **57**, 192 (1974).
13. Zawadzka, H., Adamezewska, M. and Elbanowska, H. *Chemia. Analit.*, **18**, 327 (1973).
14. Abbott, O. *Anal. Abstr.*, **13**, 5917 (1966).
15. Frei, R. W. and Duffy, J. R. *Mikrochim. Acta*, **3**, 480 (1969).
16. Fishbein, L. *Chromat Rev.*, **12**, 167 (1970).
17. Geissbuhler, H. and Gross, D. *J. Chromat.*, **27**, 296 (1967).
18. Jarczyk, H. *J. Pflanz. Nachr. Bayer*, **25**, 21 (1972).
19. Jarczyk, H. *J. Pflanz. Nachr. Bayer*, **28**, 334 (1975).
20. McKone, C. E. and Hance, R. J. *J. Chromat.*, **36**, 234 (1968).
21. Lowen, W. K., Bleidner, W. E., Kirkland, J. J., Pease, H. L. and Zweig, G. (Eds) *Analytical Methods for Pesticides, Plant Growth Regulators and Food Additives* Volume IV *Herbicides*, Academic Press, New York, p. 157, (1964).
22. Khan, S. U., Greenhalgh, R. and Cochrane, W. P. *Bull. Environ. Contam. Toxicol.*, **13**, 602 (1975).
23. Kirkland, J. J. *Anal. Chem.*, **34**, 428 (1962).
24. Lolke, H. *Pesticide Science*, **5**, 749 (1974).
25. Freistad, Ho. *J. Ass. Off. Anal. Chem.*, **57**, 221 (1974).
26. Deleu, R., Barthelemy, J. P., and Copin, A. *J. Chromat.*, **134**, 483 (1977).
27. Onley, J. H. and Yip, G. *J. Ass. Off. Agric. Chem.*, **52**, 545 (1969).
28. Guthrie, R. K., Cherry, D. S. and Ferebee, R. N. *Wat. Res. Bull.*, **10**, 304 (1974).
29. Farrington, D. S., Hopkins, R. G. and Ruzicka, J. H. A. *Analyst (Lond.)*, **102**, 377 (1977).
30. Kirkland, J. J. *Chromat. Sci.*, **7**, 7 (1969).
31. Sidwell, J. A. and Ruzicka, J. H. A. *Analyst*, **101**, 111 (1976).
32. Onley, J. H. and Yip, G. *J. Ass. Off. Anal. Chem.*, **52**, 526 (1969).
33. Smith, A. E. and Lord, K. A. *J. Chromat.*, **107**, 407 (1975).
34. McKone, C. E. and Hance, R. J. *Bull. Environ. Contam. Toxicol.* **4**, 31 (1969).
35. McKone, C. E. and Hance, R. J. *Anal. Abstr.*, **17**, 3849 (1969).
36. Yuen, S. H. *Analyst (Lond.)*, **95**, 408 (1970).
37. Ealy, D. B., German, R. R., Kaiser, F. R. and Scroggs, R. E. *J. Ass. Off. Anal. Chem.*, **57**, 60 (1974).
38. Corbaz, R., Artho, P., Ceschini, M., Hausermann, M. and Plantefere, J. C. *Bectr. Tabakforsch.*, **5**, 80 (1969).
39. Rosales, J. Z. *Anal. Chem.*, **256**, 194 (1971).
40. El-Deb, M. A. *J. Ass. Off. Anal. Chem.*, **53**, 756 (1970).
41. Spengler, D. and Jumar, A. *Arch. Pflanzenschutz*, **7**, 151 (1971).
42. Frei, R. W., Lawrence, J. F. and LeGay, D. S. *Analyst (Lond.)*, **98**, 9 (1973).
43. Cohen, I. C. and Wheals, B. B. *J. Chromat.*, **43**, 233 (1969).
44. Meagher, W. R. *J. Agric. Food Chem.*, **14**, 374 (1966).

45. McCone, C. E. and Hance, R. J. *J. Chromat.*, **69**, 204 (1972).
46. Yip, G. *J. Am. Oil Colour Chemists Ass.*, **45**, 367 (1962).
47. Yip, G. *J. Am. Oil Colour Chemists Ass.*, **47**, 1116 (1964).
48. Suzuki, S. H. and Malina, M. *J. Am. Oil Colour Chemists Ass.*, **48**, 1164 (1965).
49. Goerlitz, D. F. and Lamar, W. L. *Determination of Phenoxy Acid Herbicides in Water by Electron-Capture and Microcoulometric Gas Chromatography*, Geological Survey Water-Supply, Paper 1817-C U.S. Government Printing Office, Washington, DC (1967).
50. Yip, G. *J. Am. Oil Colour Chemists Ass.*, **54**, 966 (1971).
51. Howard, S. F. and Yip, G. *J. Am Oil Colour Chemists Ass.*, **54**, 970 (1971).
52. Devine, J. N. and Zweig, G. *J. Ass. Off. Anal. Chem.*, **52**, 187 (1969).
53. Colas, A., Lerenard, A. and Royer, J. *Chim. Anal.*, **54**, 7 (1972).
54. Croll, B. T. *Analyst (Lond.)*, **96**, 810 (1971).
55. Larose, R. H. and Chau, A. S. Y. *J. Ass. Off. Anal. Chem.*, **56**, 1183 (1973).
56. Bache, C. A., Lisk, D. J. and Loos, M. *J. Ass. Off. Agric. Chem.*, **47**, 348 (1964).
57. Takahaski, M., Numata, T. and Takano, J. *Noyaku Kagaku*, **2**, 51 (1974).
58. Gutenmann, W. H. and Lisk, D. J. *J. Ass. Off. Agric. Chem.*, **46**, 859 (1963).
59. Baur, J. R. and Baker, R. D. *J. Ass. Off. Anal. Chem.*, **54**, 713 (1970).
60. Gutenman, W. H. and Lisk, D. J. *J. Ass. Off. Agric. Chem.*, **47**, 353 (1964).
61. Woodham, D. W., Mitchell, W. G., Loftis, C. D. and Collier, C. W. *J. Agric. Food Chem.*, **19**, 186 (1971).
62. Kawahara, F. K. *Anal. Chem.*, **40**, 1009 (1968).
63. Kawahara, F. K. *Anal. Chem.*, **40**, 2073 (1968).
64. Kawahara, F. K. *Environ, Sci. Technol.*, **5**, 235 (1971).
65. Chau, A. S. Y. and Terry, K. *J. Ass. Off. Anal. Chem.*, **58**, 1294 (1975).
66. Coburn, J. A., Ripley, B. D. and Chau, A. S. Y. *J. Ass. Off. Anal. Chem.*, **59**, 188 (1976).
67. Coburn, J. A. and Chau, A. S. Y. *J. Ass. Off. Anal. Chem.*, **57**, 1272 (1974).
68. Coburn, J. A. and Chau, A. S. Y. *Environ. Lett.*, **10**, 225 (1976).
69. Johnson, L. G. *J. Ass. Off. Anal. Chem.*, **56**, 1503 (1973).
70. Chau, A. S. Y. and Terry, K. *J. Ass. Off. Anal. Chem.*, **59**, 633 (1976).
71. Agemian, H. and Chau, A. S. Y. *J. Ass. Off. Anal. Chem.*, **59**, 732 (1976).
72. Chau, A. S. Y. and Wilkinson R. *Bull. Environ, Contam. Toxicol.*, **7**, 93 (1972).
73. Agemian, H. and Chau, A. S. Y. *J. Ass. Off. Anal. Chem.*, **60**, 1070 (1977).
74. Kawahara, F. K. *Anal, Chem.*, **40**, 1009 (1968).
75. Kawahara, F. K. *Anal. Chem.*, **40**, 2073 (1968).
76. Kawahara, F. K. *Environ. Sci. Technol.*, **5**, 235 (1971).
77. Johnson, L. G. *J. Ass. Off. Anal. Chem.*, **56**, 1503 (1973).
78. *Analytical Methods Manual* Inland Waters Directorate Water Quality Branch Ontario, Canada, (1974).
79. Carnac, V. D. *Zhurnal Analiticheskoi Khimmi*, **30**, 2444 (1975).
80. Fredeen, F. J. H., Saha, J. G. and Balba, M. H. *Pesticides Monitoring Journal*, **8**, 241 (1975).
81. Purkayastha, R. *Bull. Environ. Contam. Toxicol.*, **4**, 246 (1969).
82. Bogacka, T. and Taylor, R. *Chemia. Analit.*, **15**, 143 (1970).
83. Bogacka, T. and Taylor, R. *Chemia. Analit.*, **16**, 215 (1971).
84. Meinard, C. *J. Chromat.*, **61**, 173 (1971).
85. Suffet, I. H. *J. Agric. Food Chem.*, **21**, 288 (1973).
86. Suffet, I. H. *J. Agric. Food Chem.*, **21**, 591 (1973).
87. Woodham, D. W., Mitchell, W. G., Loftis, C. D. and Collier, C. W. *J. Agric. Food Chem.*, **19**, 186 (1971).
88. Bache, C. A., Lisk, D. J. and Loos, M. A. *J. Am. Oil Colour Chem. Ass.*, **47**, 348 (1964).
89. Mattson, P. E. and Kirsten, W. J. *J. Agric. Food Chem.*, **16**, 908 (1968).

90. Smith, A. E. and Fitzpatrick, A. *J. Chromat.*, **57**, 303 (1971).
91. Devine, J. M. and Zweig, G. *J. Am. Oil Colour Chem. Ass.*, **52**, 187 (1969).
92. Marshall, M. *J. Ass. Off. Anal. Chem.*, **54**, 706 (1971).
93. Bogacka, T., *Chemia. Analit.*, **16**, 59 (1971).
94. Abbott, S. D., Hall, R. C. and Giam, G. S. *J. Chromat.*, **45**, 317 (1969).
95. Hall, R. C., Giam, G. S. and Merkle, M. G. *Anal. Chem.*, **42**, 423 (1970).
96. Dennis, D. S., Gillespie, W. H., Moxey, R. A. and Shaw, R. *Arch. Environ. Contam. Toxicol.*, **6**, 421 (1977).
97. Cannizzara, R. D., Cullen, T. E. and Murphey, R. T. *J. Agric. Food Chem.*, **18**, 728 (1970).
98. Rice, C. P., Sikka, H. C. and Lynch, R. S. *J. Agric. Food Chem.*, **22**, 533 (1974).
99. Coha, O. *Anal. Lett.*, **2**, 623 (1969).
100. Volke, J., and Volkova, V. *Collin Czech. Chem. Commun.*, **34**, 2037 (1969).
101. Bosyakova, E. N., Bukharbaeva, A. S., Utebekova, N. R. and Shabanov, I. M. *Trudy Inst. kraev. Patol. M-VO Z dravooklr kaz. SSR* **16**, 28 (1969). Ref *Zh. Khim.* 19GD (1969) 16 Abstract No. 16G210.
102. Suzuki, M., Yamoto, Y. and Akiyama, T. *Wat. Res.*, **11**, 275 (1977).
103. Onuska, F. I. and Boos, W. R. *Anal. Chem.*, **45**, 967 (1973).
104. Norris, L. A. and Montgomery, M. L., *Bull. Environ. Contam. Toxicol.*, **13**, 1 (1975).
105. Drescher, N. *Methodensammiung zur Rückstandsanalytik von Pflanzenschutzmitteln*, Verlag Chimie, Weinheim (1972).
106. Drescher, N. *Bestimmung der Ruckstande von Pyramin in Pflanze und Boden*, 8–9 January, 1964, BASF, Ludwigshafen am Rhein, 78. (1964).
107. Zaborowska, W., Witkowska, I. and Kozak, H. *Rocz. Panstu. Zakl. Hig.*, **24**, 735 (1973).
108. Palusova, O., Sackmauerova, M. and Madarix, A. *J. Chromat.*, **106**, 405 (1975).
109. Crathorne, B. and Watts, C. D. *J. Chromat.*, **169**, 436 (1979).
110. Calderbank, A. and Yuens, O. *Analyst (Lond.)*, **90**, 99 (1965).
111. Pope, J. D. and Benner, J. E. *J. Ass. Off. Anal. Chem.*, **57**, 202 (1974).
112. Söderquist, C. J. and Crosby, *Bull. Environ. Contam. Toxicol.* **8**, 363 (1972).
113. Cannard, A. J. and Criddle, W. J. *Analyst (Lond.)*, **100**, 848 (1975).
114. Payne, W. R., Pope, J. D. and Benner, J. E. *J. Agric. Food Chem.*, **22**, 79 (1974).
115. Khan, S. L. *J. Agric. Food Chem.*, **22**, 863 (1974).
116. Martens, M. A. and Heyndricks A. *J. Pharm. Belg.*, **29**, 449 (1974).
117. Getzendaner, M. E. *J. Ass. Off. Anal. Chem.*, **52**, 824 (1969).
118. Frank, P. A. and Demint, R. J. *Environ. Sci. Technol.*, **3**, 69 (1969).
119. Van der Poll, J. M. and de Vos, R. H. *J. Chromat.*, **187**, 244 (1980).
120. Bronstad, J. O., and Freistad, H. O. *Analyst (Lond.)*, **101**, 820 (1978).
121. Abbott, D. C. and Wagstaff, P. J. *J. Chromat.*, **43**, 361 (1969).
122. Smith, A. E. and Fitzpatrick, A. *J. Chromat.*, **57**, 303 (1971).
123. Schulten, H. R. and Beckey, D. H. *J. Agric. Food Chem.*, **21**, 272 (1973).
124. Schulten, H. R., Prince, H., Beckey, H. D., Tomberg, W. and Korte, F. *Chemosphere*, **2**, 22 (1979).
125. Schulten, H. R. *J. Agric. Food Chem.*, **24**, 743 (1976).
126. Yamato, Y., Suzuki, M. and Wanatabe, T. *Biomedical Mass Spectrometry*, **6**, 205 (1979).
127. Yamato, Y., Suzuki, M. and Wanatabe, T. *J. Ass. Off. Anal. Chem.*, **61**, 1135 (1978).
128. Erney, D. R. *Anal. Lett.*, **12**, 501 (1979).

Index

Abate, determination of, 391–393, 410
Acarol, determination of, 497, 529
Acenaphthalene, determination of, 172
Acyclic isoprenoids in hydrocarbons, 26–31
Air, polyaromatic hydrocarbons in, 153
Algae, hydrocarbons in, 8
Aldrin, determination of, 280, 285, 287, 306, 310, 312, 315, 317, 337, 342, 345–346, 445
  thin-layer chromatography of, 346
Ametryne, determination of, 498, 499, 503
Amidithion, determination of, 391–393
Aminocarb, determination of, 299, 365, 369
  thin-layer chromatography of, 309
Aminofenitrothion, determination of, 411–413
Anionic detergents
  biodegradation of, 231, 235
  determination of, 218–247
  g.l.c. of, 240–242
  h.p.l.c. of, 242–247
  i.r. spectroscopy of, 231–236
  polarography of, 237–240
Anthracene, determination of, 131, 143, 153–155, 157, 172, 174, 185
Arenes, determination of, 127
Arochlor, determination of, 341, 416–482
Aromatics
  in hydrocarbons, 7–8, 11, 40–41, 48, 53–74
  in water, 100, 102, 112, 116–129
  i.r. spectroscopy of, 53–72
  u.v. spectroscopy of, 116–123, 128
Arsenic in hydrocarbons, 75
Atomic absorption spectroscopy of
  anionic detergents, 218–222
  hydrocarbons, 75
  non-ionic detergents, 249–253
Atraton, determination of, 499
Atrazine, determination of, 498–503
Aviation fuels, identification of, 7–8

Azine herbicides, thin-layer chromatography of, 503
Azinphos ethyl, determination of, 382–391
Azinphos methyl, determination of, 382–391, 395–398, 400

Barban, determination of, 369, 497, 514, 529
  thin-layer chromatography of, 369
Bayrusil, determination of, 413–415
Beaches, hydrocarbons in, 20–21
Benthiocarb, determination of, 497, 529, 543–545
Benzo(b)fluoranthracene, determination of, 181, 198
7H-Benz(de)anthracene-7-one, determination of, 160
Benzanthrene, determination of, 166
3,4 benzfluoranthrene, determination of, 181
11,12 benzfluoranthrene, determination of, 181
Benzo(a)anthracene, determination of, 13, 184–186
Benzo(d)anthracene, determination of, 158
Benzo(j)anthracene, determination of, 132, 137, 158, 160, 181, 184–186, 194, 198
Benzo(k)anthracene, determination of, 132, 194
Benzo(k)fluorene, determination of, 181
Benzo(ghi)perylene, determination of, 132, 134, 158, 161, 166, 178, 181, 184, 194, 198
1,12-Benzperylene, determination of, 181
3,4-Benz(a)pyrene, determination of, 181
Benzo(a)pyrene, determination of, 130–132, 153, 158–161, 166, 169–172, 178–180, 184, 186, 194, 195–196, 198
Benzo(e)pyrene, determination of, 131, 153–155, 157, 177, 181, 184–185

BHC, determination of, 30, 284, 301, 304–307, 310, 312, 342, 346–347, 434, 445
α-BHC, determination of, 30, 284, 304, 307, 310, 312
β-BHC, determination of, 342
γ-BHC (Lindane), determination of, 284, 289, 301, 304–307, 310, 312, 342, 444
δ-BHC, determination of, 304–307, 310, 312
Bicyclohexyl, determination of, 426
Biodegradation of
 anionic detergents, 231, 235
 cationic detergents, 249
 hydrocarbons, 6, 10–11, 23, 121
 non-ionic detergents, 275
 PCBs, 432–433
Bitumen, identification of, 7
Bromophos, determination of, 382–391, 400
Bromoxynil, determination of, 542
m-S-Butylphenyl methyl(phenylthio) carbamate, determination of, 364
Buturon, determination of, 503–504

Captan, determination of, 299, 334
Carbamate insecticides
 determination of, 292–301, 361–371, 513
 thin-layer chromatography of, 513–514
Carbaryl
 determination of, 299, 361–363, 365, 369, 400
 thin-layer chromatography of, 361
Carbine, determination of, 497, 529
Carbofuran, determination of, 299, 365, 367
Carbophenothion, determination of, 315, 382–391, 400
Carbophenoxon, determination of, 299
Cationic detergents
 biodegradation of, 249
 determination of, 247–249
α-Chlordane, determination of, 289, 315
γ-Chlordane, determination of, 289, 315
Chlordecone, determination of, 349
Chlordene, determination of, 315, 319
Chlorinated insecticides, 474–475
 h.p.l.c. of, 340–341
 photodecomposition of, 315, 334
Chlorine, determination of, 349–361
Cholestane, -4-methyl sterols in, 24–31
Cholinesterase, determination of organophosphorus compounds, 400–401
Chlorfenvinphos, determination of, 342, 381–391

Chlorobenside, determination of, 333
Chlorobromuron, determination of, 348, 503–504, 509–512, 514
4-Chloro-2-methyl(phenoxy)acetic acid, determination of, 516
4-Chloro-2-methyl(phenoxy)butyric acid, determination of, 516
Chlorooxuron, determination of, 504, 509–512
Chloropropham
 determination of, 369, 514
 thin-layer chromatography of, 369
Chlorotoluron, determination of, 503–504, 509–512
Chrysene, determination of, 131, 143, 181, 184, 186
cis-chlordane, determination of, 318
cis-nonachlor, determination of, 324
Coal tar oils, identification of, 44
Column chromatography of
 hydrocarbons, 41
 polyaromatic hydrocarbons, 168–169, 426–432, 474–475
COP phosphate, determination of, 413–416
Coral, determination of, 413–415
Coumaphos, determination of, 382–391, 413–415
Crops, triazine herbicides in, 499
Crude oil
 identification of, 10–20
 i.r. spectroscopy of, 56–64
Cruformate, determination of, 294, 382–391
Crustacae, Dursban in, 410
Cutting oils, identification of, 6
Cycloparaffins in hydrocarbons, 40

Dactal, determination of, 326
Dalapon, determination of, 498, 536–538, 542
2,4-D, determination of, 337, 514, 515, 520, 526–528, 542
2,4-DB, determination of, 514, 520, 542
DDD, determination of, 284, 304, 315, 329–330, 336–337, 341–345, 435, 444–445
DDE, determination of, 304–306, 309–310, 312, 315, 331, 341, 342–345, 426, 434–435, 445, 458
DDT, determination of, 309–310, 336–337, 341–347, 426, 435, 444–445, 449–451
Decachlorobiphenyl, determination of, 426

Demefox, determination of, 382–391
Demeton-*O*, determination of, 395–398
Demeton-*S*, determination of, 382–391, 395–398
Demeton-*S*-lethyl, determination of, 382–391
Demeton-*S*-methyl, determination of, 381–391
-3-Desoxytriterpanes in hydrocarbons, 24–31
Dialkylthiocarbonate *S*-alkyl derivatives, determination of, 530
Diazinon, determination of, 284, 294, 315, 382–393, 395, 398, 400
Dibenz(ah)anthracene, determination of, 181, 184
Dibrom, determination of, 382–391
Dicamba, determination of, 497, 514–515, 520, 526, 528, 530, 542
Dichlorobenil, determination of, 326
p,p'-Dichlorobenzophenone, determination of, 332
Dichlorobiphenyls, determination of, 424
1,1-Dichloro-2, 2-*bis*-(4-chlorophenyl)ethylene, determination of, 347
Dichloram, determination of, 542
Dichlorfenthion, determination of, 382–391
Dichlorprop, determination of, 542
Dichlorvos, determination of, 382–393, 399
Dicotel, determination of, 342
Dieldrin, determination of, 280, 284–285, 289, 292, 306, 309–310, 312, 315, 319, 337, 342, 345–346, 458–460
Diesel oil, hydrocarbons in, 7, 11, 132
Diethyl phosphate, determination of, 389–391
Diethylthiophosphate, determination of, 389–391
Dimethoate, determination of, 294, 381–391, 399–400
N,N'-dimethylcarbamate, determination of, 364
Dimethyl phosphate, determination of, 389–391
Dimethyl thiophosphate, determination of, 389–391
Dioxathion, determination of, 315
Diphenamid, determination of, 534
Diquat, determination of, 497
Disulfoton, determination of, 299, 382–391, 395–398
Diuron, determination of, 369, 504, 509–512, 514, 542

Dolphins, PCBs in, 461
2,4-DP, determination of, 514, 520, 527
Dursban, determination of, 391–393, 409–410

Emulsifiers in oil spills, 11
Emission spectrography of hydrocarbons, 76
Endrin, determination of, 280, 284, 289, 306–307, 315, 321, 342, 345
Endosulphan, determination of, 320, 321, 348, 400, 444
Ethion, determination of, 315, 382–393, 399
Ethyl benzene, determination of, 127
24-Ethyl cholestane in hydrocarbons, 24–31

Farnesane, identification of, 8
Fenchlorphos, determination of, 382–395
Fenitron, determination of, 411–413
Fenitrooxon, determination of, 412–413
Fenitrothion, determination of, 383–393, 400, 412–413
Fenoprop, determination of, 515, 526
Fenthion, determination of, 389, 400
Fenuron
  determination of, 369, 504
  thin-layer chromatography of, 369
Fish
  chlorinated insecticides in, 301–305, 331–337
  dieldrin in, 346
  dursban in, 410
  hydrocarbons in, 41
  mirex in, 349, 467–470
  organophosphorus compounds in, 331
  PCBs in, 432
Flow calorimetry, hydrocarbons in water, 115
Fluometuron, determination of, 514
Fluoranthrene, determination of, 130–132, 153–154, 158, 161–169, 166, 178, 181, 184–186, 194, 196, 198
Fluorescence emission spectroscopy of
  hydrocarbons, 4, 72–74, 112–114, 123–127
  polyaromatic hydrocarbons, 153–172
Fluorine, determination of, 177
Folpet, determination of, 299, 333
Food
  chlorinated insecticides in, 336–337
  organophosphorus insecticides in, 331
Foschlor, determination of, 339

Fuel oil
  fluorescence of, 73
  hydrocarbons in, 10–11, 41–52
  identification of, 2, 13–15
  i.r. spectroscopy of, 56, 62
  polyaromatic hydrocarbons in, 132
Fuel oil in
  marine organisms, 79
  water, 98–102

Gas chromatography–mass spectrometry of
  azine, herbicides, 501–503
  chlorinated insecticides, 461–470
  DDD, 342–344
  DDE, 342–344
  DDT, 342–344
  hydrocarbons, 21, 24–40
  Kepone, 349
  mirex, 349
  PCBs, 423, 425–426, 461–470
  polyaromatic hydrocarbons, 146–153
Gas chromatography of
  abate, 391–393, 410
  acarol, 529
  aldrin, 280, 285, 287, 306, 310, 312, 315, 317, 445
  ametryne, 498, 503
  amidothion, 391–393
  aminocarb, 299, 365
  aminofenitrothion, 411–413
  anionic detergents, 204–242
  arochlor, 416–482
  arochlor 1016, 416, 420, 421, 428, 430, 445
  arochlor 1221, 428, 430, 445, 446
  arochlor 1232, 420–421, 445, 446
  arochlor 1242, 417, 420, 421, 423, 444, 445
  arochlor 1243, 417
  arochlor 1248, 423, 445
  arochlor 1254, 417, 420, 421, 423, 424, 426, 428, 430, 445
  arochlor 1256, 444
  arochlor 1260, 424, 437, 444, 445, 458–459
  atraton, 499
  atrazine, 4, 98–503
  azinphosethyl, 382–391
  azinphosmethyl, 382–391, 395–398
  benthiocarb, 543–545
  BHC, 284, 301, 304–307, 310, 312, 347, 434, 444–445, 457
  α-BHC, 284, 301, 304–307, 310, 312, 347
  β-BHC, 304, 307, 310, 312
  γ-BHC, 284, 289, 301, 304–307, 310–312, 344–347, 444
  δ-BHC, 304, 305–307, 310, 312
  bicyclohexyl, 426
  bromophos, 382–391
  m-S-butylphenylmethyl)phenylthio)carbamate, 364
  captan, 299, 344
  carbaryl, 299, 361–363, 365–367
  carbofuran, 299, 365–368
  carbophenoxon, 299
  carbophenthion, 315, 382–391
  chlordane, 445
  α-chlordane, 289, 315
  γ-chlordane, 280, 287, 315
  chlordene, 315, 319
  chlorbenside, 333
  chlorfenvinphos, 381–391
  4-chloro-2-methylphenoxyacetic acid, 516
  4-(4-chloro-2-methylphenoxy)-butyric acid, 516
  cis-chlordane, 318
  cis-nonachlor, 324
  coumaphos, 382–391
  crufomate, 294, 382–391
  2,4-D, 515, 520
  dacthal, 326
  dalapon, 536–538
  2,4-DB, 520
  2,4-DD, 520
  DDD, 284, 304, 315, 329, 435, 444–445, 454
  DDE, 301, 304–306, 309–310, 312, 315, 331, 426, 434–451, 445, 454, 459
  DDT, 280, 292, 301, 304, 307, 309, 312, 331, 336–337, 347, 426, 435, 444–445, 449, 451, 454
  decachlorobiphenyl, 4, 26
  demeton-O, 395–398
  demeton-S, 382–391, 395–398
  demeton-S-ethyl, 382–391
  demeton-S-methyl, 381
  demetox, 382–392
  diazinon, 284, 294, 315, 382–393, 395–398
  dibrom, 382–391
  dicamba, 515
  dichlorfenthion, 382–391
  dichlorobenil, 326
  p,p'-dichlorobenzophenone, 332
  dichlorobiphenyls, 424

Gas chromatography of *(continued)*
1,1-dichloro-2, 2-*bis*-(4-chlorophenyl) ethylene, 347
dichlorvos, 382–393
dieldrin, 280, 284–285, 289, 292, 306, 309–310, 312, 319, 337, 346, 458–460
diethyl phosphate, 389–391
diethyl thiophosphate, 389–391
dimethoate, 294, 381–391
dimethyl phosphate, 389–391
dimethyl thiophosphate, 389–391
dioxathion, 315
diquat, 534–536
disulfoton, 229, 382–391, 395–398
diuron, 512
dursban, 391–393, 409–410
endosulphan, 320–321, 348, 444
endosulphan sulphate, 321
endrin, 280, 284, 289, 306–307, 315, 321
ethion, 315, 382–393
farnesane, 8
fenchlorphos, 391–395
fenitron, 411–413
fenitrooxon, 412–413
fenitrothion, 382–393, 412–413
fenoprop, 515
fenthion, 389
folpet, 299, 333
glyphosate, 538–542
guthion, 391–393
heptachlor, 348
hexachlorophene, 348
hydrocarbons, 77–97, 114–115, 127
3-ketocarbofuran, 366–367
lindane, 284–285, 287, 292, 299, 315, 347, 433–434, 445
haloxon, 382–391
heptachlor, 284–285, 287, 306–307, 310, 313, 322, 445
heptachlor epoxide, 284–285, 292, 306–307, 310, 312, 315, 445
2,2′,3,3′,4,4′,6′-heptachlorobiphenyl, 424
hexachlorobenzene, 304, 327, 347
hexachlorocycloheptadiene, 315
2,2′,3,4′,5,6′-hexachlorobiphenyl, 424
2,2′,3,3′,6,6′-hexachlorobiphenyl, 424
hexachlorocyclopentadiene, 315
1-hydroxychlordene, 323
isodofenphos, 391–393
isobenzan, 324
isodrin, 315, 323

malathion, 284, 294, 315, 382–391, 395–398, 409
MCPA, 515, 520
MCPB, 515, 520
MCPP, 515
mecarbam, 382–391
mecarphon, 395
menazon, 382–391
metalkamate, 299, 368
methidathion, 391–393
methiocarb, 299, 368
methomyl, 363–365
methoxychlor, 285, 289, 292, 306, 315, 346, 347, 445, 525–526
L-methoxy-1-methyl-3-phenyl urea, 513
methyl trithion, 315
metmercapturon, 366–367
mevinphos, 294, 382–391
mexacarbate, 365
mirex, 289, 309–310, 445, 467–470
mobam, 366–367
molinate, 361
monocrotophos, 299
monochlorobiphenyls, 424
morphothion, 382–391
1-naphthol, 361
nitrocresol, 412–413
octachloroepoxide, 325
organochlorine insecticides, 280–337, 377–378
organophosphorus insecticides, 346–347, 378–380
organosulphur compounds, 377–378
oxidiazon, 543–545
oxydemeton-methyl, 299, 382–391
parathion, 315, 382–391, 401–402
parathion ethyl, 395–398
parathion methyl, 315, 395–398
paraquat, 534–536
PCB-chlorinated insecticides, 433–461
PCBs, 327, 416–482
pentachloro-nitrobenzene, 445
pH3 containing insecticides, 393–395
phenoxyacetic acid herbicides, 514–526
phenkaptan, 382–393
phorate, 382–391
phosalone, 382–391
phosphamidon, 294, 382–393
photodieldrin, 309
phytane, 8
picloram, 520, 529
polyaromatic hydrocarbons, 130–146
polybrominated biphenyls, 481–482
polychlorinated terphenyls, 481–482

Gas chromatography *(continued)*
  pristane, 8
  prometon, 498-499
  prometryne, 498-499, 503
  propazine, 498, 499
  propoxur, 299, 363, 365-366, 368
  pyrimithate, 381-391
  pyrazon, 531-533
  ronnel, 391-393
  S-alkyl derivatives of N,N'-dialkyl-thiocarbamates, 530
  schradan, 382-391
  sevin, 361-363
  secbutmeton, 499
  siduron, 512
  silvex, 520
  simazine, 498-500, 503
  strobane, 445
  sulfotep, 382-391
  2,4,5-T, 520
  TDE, 306, 309, 336-337, 426
  terbumeton, 500
  terbuthylazine, 499, 502
  terbutryne, 498
  tetrachlorobiphenyls, 424
  tetradifon, 332
  thioazin, 382-391
  thiodan, 284
  toxaphene, 315, 445
  trans-chlordane, 318
  trans-nonachlor, 325
  treflan, 284
  trichlorfon, 382-391
  trichlorobiphenyls, 424
  trifluralin, 445
  trithion, 391-393
  vamidothion, 382-391
  vapona, 391-393
Gas oil, hydrocarbons in, 7, 44
Gas stripping of hydrocarbons, 88-97
Gel permeation chromatography of polynuclear hydrocarbons, 131, 194
Glyphosate, determination of, 498, 538-542
Greases
  determination of, 200-205
  hydrocarbons in, 43, 44
  i.r. spectroscopy of, 61-66, 205
Guthion, determination of, 391-393

Haloxon, determination of, 383-391
Head space analysis, hydrocarbons, 12, 31-40, 79-88

Heptachlor, determination of, 284, 285, 287, 306-307, 310, 315, 322, 342, 348, 445
Heptachlor epoxide, determination of, 284-285, 292, 306-307, 310, 312, 315, 445
2,2',3,3',4,5,6'-Heptachlorobiphenyl, determination of, 424
Hexachlorobenzene, determination of, 304, 327
Hexachlorobicycloheptadiene, determination of, 315
2,2',3,3',6,6'-Hexachlorobiphenyl, determination of, 424
2,2'3,4'5,6'-Hexachlorobiphenyl, determination of, 424
Hexachlorocyclopentadiene, determination of, 315
Hexachlorophane, determination of, 348
Herbicides
  phenoxyacetic acid type, 514-528
  substituted urea type, 503-514
  triazine type, 498-502
H.p.l.c. of
  anionic detergents, 242-247
  chlorinated insecticides, 340-341
  chlorobromuron, 509-512
  chlorotoluron, 509-512
  chloroxuron, 509-512
  diuron, 509-512
  linuron, 509-512
  monouron, 509-512
  parathion ethyl, 402
  parathion methyl, 402
  polyaromatic hydrocarbons, 172-194
  PCBs, 426-431
  urea herbicides, 507-512
  zectran, 364
Horiba Oil Analyser, 201-205
Hopanes in hydrocarbons, 24-31
Hydrocarbons
  atomic absorption spectrometry of, 75
  acyclic isoprenoid hydrocarbons in, 26-31
  aromatics in, 40-41, 48, 53-74
  arsenic in, 75
  asphaltenes in, 4
  biodegradation of, 6, 10-11, 23, 121
  cholestane in, 24-31
  column chromatography of, 41
  cycloparaffins in, 40
  3-desoxytriterpenes in, 24-31
  determination of, in
    beaches, 20-21

Hydrocarbons *(continued)*
  fish, 41
  marine oil, 10–20, 51–52
  rain water, 91–96
  sea water, 51–52, 91–97, 113, 127
  sediments, 37–38, 40, 117–126
  sewage, 41, 79
  tissue, 37–38
  water, 76–116
  emission spectrography of, 76
  fluorescence spectroscopy of, 4, 72–74, 112–114, 123–127
  gas chromatography of, 3–4, 6–40, 77–97, 114–115
  head space analysis of, 12, 31–40, 79–88
  hopanes in, 24–31
  infrared spectroscopy of, 4, 8–10, 13, 20, 53–72, 97–108, 114–115, 128–129
  indium in, 75
  iron in, 75
  isoprenoids in, 21–31
  liquid chromatography of, 31–40
  manganese in, 75
  mass spectrometry of, 11, 13, 76, 115
  mercury in, 75
  metals in, 74–76
  24-methyl cholestane in, 24–31
  4-methyl steranes in, 24–31
  molybdenum in, 75
  naphthenes in, 41
  neutron activation analysis of, 74
  nickel in, 16–20, 26, 74–76
  nitrogen in, 20, 76
  olefins in, 53–72
  oscillographic polarography of, 75
  paper chromatography of, 41–52, 112
  pentacyclic triterpenes in, 26–31
  polyaromatic hydrocarbons in, 40, 48
  phytane in, 20, 22–31
  polyhydrydroxyhopanes in, 24, 31
  pristane in, 20, 22–31
  Raman spectroscopy of, 9, 56
  γ-ray spectrometry of, 75
  rhenium in, 75
  selenium in, 75
  solvent extraction of, 76–82
  stearanes in, 21–31
  sulphur in, 20–31, 74–76
  thermal fragmentation of, 8
  thin-layer chromatography of, 4, 41–52, 108–112, 114
  trace metals in, 4
  tricyclic diterpanes in, 26–31
  trinorhopanes in, 28–31
  triterpenes in, 21–31
  ultraviolet spectroscopy of, 11, 72
  vanadium in, 16–20, 26, 74–76
  wax in, 20
  weathering of, 4–5, 7, 10–12, 20–22, 30–40, 54, 73, 76
  X-ray fluorescence spectroscopy of, 16–20, 75
  zinc in, 75
1-Hydroxychlordene, determination of, 323

Iodofenphos, determination of, 391–393
Iron in hydrocarbons, 75
Indium in hydrocarbons, 75
Indeno (1,2,3-cd) pyrene, determination of, 132, 158, 160, 177, 181, 194, 198
Infrared spectroscopy of
  anionic detergents, 231–236
  aromatics, 53–72
  crude oils, 56–64
  hydrocarbons, 13, 20, 53–72, 97–108, 114–115, 128–129
  fuel oils, 56, 62
  greases, 61–66, 205
  lubricating oils, 62
  naphthas, 53–55
  non-ionic detergents, 272
  olefins, 53–72
Isobenzan, determination of, 324, 342
Isodrin, determination of, 323
Isoprenoids, determination of, 9, 21–31
Isoproturon, determination of, 504

Kepone, determination of, 349
Kerosene, determination of, 7, 8, 115
3-Ketocarbofuran, determination of, 366–367

Leaves, trichlorphon in, 347–348
Lindane, (γ-BHC)
  determination of, 284–285, 287, 292, 299, 315, 337, 369, 443–445
  thin-layer chromatography of, 369
Linuron, determination of, 369, 504, 509–512, 514, 542
Lipids, PAHs in, 172
Lubricating oils
  identification of, 7, 8
  i.r. spectroscopy of, 62
  polyaromatic hydrocarbons in, 132

Malathion, determination of, 294, 315, 383–393, 395–398, 400, 409

Manganese in hydrocarbons, 75
Marine sediments, oil in, 10–20, 41–52, 79, 108
Mass spectrometry of
  herbicides, 543–545
  hydrocarbons, 11-13, 76, 115
  non-ionic detergents, 264–268
  polyaromatic hydrocarbons, 194
  PCBs, 432–433
Mevinphos, determination of, 383–391
Mecarbam, determination of, 383–391
Mecarphon, determination of, 395
Mecoprop, determination of, 542
Menazon, determination of, 383–391
Mercury in hydrocarbons, 75
Metabenzthiauron, determination of, 504
Metalkamate, determination of, 299, 368
Metals in hydrocarbons, 74–76
Methamidophos, determination of, 299
Methidathion, determination of, 391–393, 400
Methiocarb
  determination of, 299, 368–369
  thin-layer chromatography of, 369
Methomyl, determination of, 363–364
Methoxychlor, determination of, 285, 289, 292, 306, 315, 346–347, 445, 525–526
1-Methoxy-1-methyl-3-phenylurea, determination of, 513
N-Methyl carbamates, determination of, 364
S-Methyl fenitrothion, determination of, 411
Metmercapturon, determination of, 366–367
Metobromuron, determination of, 504, 509–513
N-Methylcarbamate, determination of, 364
24-Methylcholestane in hydrocarbons, 24–31
1-Methyl phenanthrene, determination of, 175, 177–178
4-Methyl stearanes, determination of, 26–31
Methyltrithion, determination of, 315
Metoxuron, determination of, 504
Mevinphos, determination of, 294
Mexacarbate, determination of, 365, 369
  thin-layer chromatography of, 369
Mineral oil in water, 101–102
Mirex
  biodegradation of, 349
  determination of, 289, 309, 310, 445, 467–470

gas chromatography–mass spectrometry of, 349
Mobam, determination of, 366–367
Molinate, determination of, 361
Molluscs
  dieldrin in, 346
  organochlorine insecticides in, 283–284
Molybdenum in hydrocarbons, 75
Monochlorobiphenyls, determination of, 424
Monocrotophos, determination of, 299
Monolinuron, determination of, 509–512
Monouron
  determination of, 369, 504, 509–512, 542
  thin-layer chromatography of, 369
Monouron TCA, determination of, 369
Morpthothion, determination of, 383–391
Motor oils
  hydrocarbons in, 43
  fluorescence of, 73
Mud, dursban in, 410
Mussels, polyaromatic hydrocarbons in, 131

Naphthalene, determination of, 153–154
Naphthas, i.r. spectroscopy of, 53–55
Naphthenes in hydrocarbons, 41
1-Naphthol, determination of, 361
Neburon, determination of, 369–504
Neutron activation analysis, 74, 349–361
Nickel in hydrocarbons, 20, 26, 74–76
Nitrocresol, determination of, 412–413
Nitrogen in hydrocarbons, 20, 76, 342, 412–413
N.m.r. spectroscopy of
  chlorinated insecticides, 342
  non-ionic detergents, 272
Non-ionic detergents, atomic absorption spectroscopy of, 249–253
  biodegradation of, 275
  column chromatography of, 263–274
  determination of, 249–276
  i.r. spectroscopy of, 272
  n.m.r. spectroscopy of, 272
  polarography of, 277
  thin-layer chromatography of, 272
  ultraviolet spectroscopy of, 272, 275

Octachloroepoxide, determination of, 325
Oil in
  marine sediments, 108
  trade effluents, 111–113
Olefins
  in hydrocarbons, 53–72

Olefins *(continued)*
　i.r. spectroscopy of, 53–72
Olive oil, composition of, 43
Optical luminescence spectroscopy of polyaromatic hydrocarbons, 131
Organochlorine insecticides
　column chromatography of, 337–341
　determination of, 377–378
　gas chromatography of, 280–377
　n.m.r. of, 342
　thin-layer chromatography of, 304, 341–342
Organophosphorus insecticides
　determination of, 292–301, 346–347, 377–378
　thin-layer chromatography of, 382–391, 399–400
Organosulphur compounds, determination of, 377–378
Oscillographic polarography of hydrocarbons, 75
Oxidiazon, determination of, 543–545
Oxy-demeton methyl, determination of, 299, 383–391
Oysters
　PCBs in, 460–461
　polyaromatic hydrocarbons in, 171
　polychlorinated terphenyls in, 481–482

Paraffin
　fluorescence of, 73
　hydrocarbons in, 44
　identification of, 7
Paraquat, determination of, 497, 534–536
Parathion, determination of, 315, 337, 383–391, 400–409
Parathion ethyl, determination of, 395–398, 402
Parathion methyl, determination of, 315, 395–400, 402
PCBs
　biodegradation of, 432–437
　column chromatography of, 426–432, 474–475
　determination of, 341, 416–482
　gas chromatography–mass spectrometry of, 423, 425–426, 461–470
　gel permeation chromatography of, 194
　h.p.l.c. of, 426–431
　perchlorination of, 423
　polarography of, 432
　preparative gas chromatography of, 423
　voltammetry of, 481
PCBs determination in oysters, 464–467

Pentachloronitrobenzene, determination of, 445
Pentacyclic triterpenes, determination of, 26–31
Perchlorination of DCBs, 423
Perylene, determination of, 161, 166, 172, 178–181, 185–186, 194
Petroleum
　fluorescence of, 73
　hydrocarbons in, 44
　identification of, 7
Phenanthracene, determination of, 131, 174, 178
Phenkaptan, determination of, 383–393, 400
Phenoxyacetic acid herbicides
　gas chromatography of, 514–526
　thin-layer chromatography of, 526–527
Phorate, determination of, 383–391
pH 3 containing insecticides, determination of, 393–395
Phosalone, determination of, 383–391
Phosphamidon, determination of, 294, 383–393, 400
Photodecomposition of organochlorine insecticides, 315–334
Photodieldrin, determination of, 309
Phytane, determination of, 8, 20, 22–31
Picloram, determination of, 497, 520, 526, 529
Plants, *m-S*-butylphenyl methyl(phenylthio) carbamates in, 304
Polarography of
　anionic detergents, 237–240
　non-ionic detergents, 277
　PCBs, 432
Polyaromatic hydrocarbons
　column chromatography of, 168–169
　fluorescence spectroscopy of, 153–172
　gas chromatography of, 130–146
　gas chromatography–mass spectrometry of, 146–153
　gel permeation chromatography of, 130
　h.p.l.c. of, 172–194
　isotope dilution gas chromatography of, 130
　mass spectrometry of, 194
　optical luminescence spectroscopy of, 130
　thin-layer chromatography of, 131, 140–141, 166–169, 194–200
Polyaromatic hydrocarbons determination in
　air, 153

Polyaromatic hydrocarbons determination in *(continued)*
  fish, 171–172
  hydrocarbons, 40
  lipids, 172
  mussels, 131, 146, 171
  oysters, 171
  sea water, 172
  sediments, 146, 152, 166, 171–172
  water, 129–200
  wood, 131, 171
Polybrominated biphenyls, determination of, 483
Polychlorinated benzenes, determination of, 432
Polychlorinated naphthalenes, determination of, 423, 473
Polychlorinated terphenyls, determination of, 481–482
Polyhydroxyhopanes, determination of, 24–31
Polyurethane foam organics absorbent, 102–104, 141, 194–197, 423, 431
POP phosphate, determination of, 415–416
Preparative gas chromatography of PCBs, 423
Pristane in hydrocarbons, 8, 20, 22–31
Prometon, determination of, 498–499
Prometryne, determination of, 498–499, 503
Propham, determination of, 514, 569
Propazine, determination of, 498–499
Propoxur, determination of, 299, 369
  thin-layer chromatography of, 369
Pyrazon, determination of, 497, 530–533
Pyrene, determination of, 130–131, 134, 153–154, 157, 177–178
Pyrimithate, determination of, 381–391

Quintozene, determination of, 342

Rain water
  calorinated insecticides in, 292
  hydrocarbons in, 91–96
Raman spectroscopy of hydrocarbons, 9, 56
γ-Ray spectrometry of hydrocarbons, 75
Rhenium in hydrocarbons, 75
Ronnel, determination of, 391–393

Sarin, determination of, 400–401
Sand, hydrocarbons in, 10–20

Sea water
  anionic detergents in, 218–221, 223, 229–231, 246
  cationic detergents in, 249–255
  endosulphan in, 348
  hydrocarbons in, 10–20, 95–97, 113, 115–116, 127, 348
  organochlorine insecticides in, 280–283, 287–292
  parathion in, 407–408
  PCBs in, 425, 435
  polyaromatic hydrocarbons in, 172
Secbumeten, determination of, 499
Sediments
  anionic detergents in, 236
  carbamate herbicides in, 513
  DDT in, 345
  dursban in, 409–410
  hydrocarbons in, 10–20, 37–38, 40, 51–52, 117–126
  methomyl in, 363–364
  organochlorine insecticides in, 301–304, 306, 345
  polyaromatic hydrocarbons in, 146
  PCBs in, 426, 448, 464–467
  urea herbicides in, 513
Selenium in hydrocarbons, 75
Sevin, determination of, 361–363
Sewage
  abate in, 410
  anionic detergents in, 222, 235, 239–240
  azine herbicides in, 502–503
  hydrocarbons in, 41, 79
  organochlorine insecticides in, 461–464
  PCBs in, 158, 426, 432, 449–464
  polyaromatic hydrocarbons in, 158
  sulphur in, 41
Soil
  aldrin in, 346
  ametryne in, 503
  m-S-butylphenyl, methyl(phenylthio)carbamate in, 364
  DDT in, 345
  dieldrin in, 346
  herbicides in, 534
  N-methyl carbamate in, 364
  organochlorine insecticides in, 284, 306, 312, 336–337, 345
  organophosphorus insecticides in, 331
  PCBs in, 464–467
  trichlorphor in, 347–348
  urea herbicides in, 509
Solvent extraction of hydrocarbons, 76–82

Siduron
  determination of, 369, 504, 512
  thin-layer chromatography of, 369
Silver, determination of, 514, 520
Simazine, determination of, 498–500, 503, 542
Squalane, determination of, 152
Steranes, determination of, 21–31
Strobane, determination of, 445
Styrene, determination of, 127
Sulfotep, determination of, 383–391
Sulphur in hydrocarbons, 20–21, 26, 41, 74–76
Swep
  determination of, 369
  thin-layer chromatography of, 369

2,4,5-T, determination of, 514–515, 520, 526–527, 542
Tar, determination of, 11
TDE, determination of, 306–309, 336–337, 426
Terbumeton, determination of, 500
Terbutryne, determination of, 498
Terbutylazine, determination of, 499, 502
Tetrachlorobiphemyls, determination of, 424
Tetradifon, determination of, 332
Tetrasul, determination of, 342
Thermal fractionation of hydrocarbons, 8
Thin-layer chromatography of
  abate, 410
  aldrin, 346
  aminocarb, 369
  azine herbicides, 503
  barban, 369
  carbamate insecticides, 513–514
  carbaryl, 361, 369
  chlorobromuron, 514
  chloropropham, 369, 515
  diuron, 369, 514
  fluometuron, 514
  fenitron, 411–413
  fenuron, 369
  herbicides, 542–543
  hydrocarbons, 4, 41–52, 108, 112, 114
  linuron, 369, 514
  malathion, 409
  methiocarb, 369
  N-methyl carbamates, 364
  mexacarbate, 369
  monouron, 369
  neburon, 369
  non-ionic detergents, 272, 275

organochlorine insecticides, 304, 341–342, 348
organophosphorus insecticides, 382–391, 399–400
PCBs, 432
phenoxyacetic acid herbicides, 526–527
polyaromatic hydrocarbons, 131, 140–141, 166–169, 194–200
propham, 369, 514
propoxur, 369
pyrazon, 530–531
siduron, 369
swep, 369
trichlorfon, 307, 347
urea herbicides, 504–507, 513–514
Thiodan, determination of, 284
Thionazin, determination of, 383–391
Tissue, hydrocarbons in, 37–38
Toluene, determination of, 128
m-Tolyl methylcarbamate, determination of, 363
Total organic chlorine (TOCl), determination of, 360–361
Toxaphene, determination of, 315, 445
Trans-chlordane, determination of, 318
Transformer oils, identification of, 6
Trans-nonachlor, determination of, 325
Trichlorfon, determination of, 347–348, 383, 399
Trichlorobiphenyls, determination of, 424
2,3,6-Trichlorobenzoic acid, determination of, 542
Tricyclic diterpanes, determination of, 26–31
Trifluralin, determination of, 342, 445, 514–515, 534, 542
Triphenylamine, determination of, 131
Trisnorhopanes, determination of, 24–31
Triterpenes, determination of, 21–31
Trithion, determination of, 391–393
Turbojet fuel, identification of, 7
TVO oil, identification of, 7

Ultraviolet (UV) spectroscopy of
  anionic detergents, 247
  aromatic hydrocarbons, 11, 116–123, 128
  hydrocarbons, 72
  non-ionic detergents, 272
Urea herbicides
  gas chromatography of, 504–507
  h.p.l.c. of, 507–512
  thin-layer chromatography of, 504–507, 513–514

Urea complex formation, hydrocarbons, 10

Vamidothion, determination of, 383–391
Vanadium in hydrocarbons, 16–20, 26, 74–76
Vapona, determination of, 391–393
Vegetable oils, determination of, 112
Vegetation, dursban in, 410
Voltammetry of PCB, 481

Water, hydrocarbons in, 76–116
Water plants, organochlorine insecticides in, 301–304

Wax in hydrocarbons, 10, 20
Weathering of hydrocarbons, 4–5, 7, 10–12, 20–22, 36–40, 54, 73
Whales, PCBs in, 461
White spirit, identification of, 7
Wood, polyaromatic hydrocarbons in, 131

X-ray fluorescence spectrometry of hydrocarbons, 16–20, 75

Zectran, determination of, 364
Zinc in hydrocarbons, 75